Methods in Enzymology

Volume 364
NUCLEAR RECEPTORS

METHODS IN ENZYMOLOGY

EDITORS-IN-CHIEF

John N. Abelson   Melvin I. Simon

DIVISION OF BIOLOGY
CALIFORNIA INSTITUTE OF TECHNOLOGY
PASADENA, CALIFORNIA

FOUNDING EDITORS

Sidney P. Colowick and Nathan O. Kaplan

Methods in Enzymology

Volume 364

Nuclear Receptors

EDITED BY

David W. Russell

DEPARTMENT OF MOLECULAR GENETICS
UNIVERSITY OF TEXAS SOUTHWESTERN MEDICAL CENTER
5323 HARRY HINES BOULEVARD
DALLAS, TEXAS

David J. Mangelsdorf

DEPARTMENT OF PHARMACOLOGY
HOWARD HUGHES MEDICAL INSTITUTE
UNIVERSITY OF TEXAS SOUTHWESTERN MEDICAL CENTER
5323 HARRY HINES BOULEVARD
DALLAS, TEXAS

ELSEVIER
ACADEMIC
PRESS

Amsterdam   Boston   Heidelberg   London   New York   Oxford
Paris   San Diego   San Francisco   Singapore   Sydney   Tokyo

Permissions may be sought directly from Elsevier's Science & Technology Rights
Department in Oxford, UK: phone: (+44) 1865 843830, fax: (+44) 1865 853333,
e-mail: permissions@elsevier.com.uk. You may also complete your request on-line
via the Elsevier Science homepage (http://elsevier.com), by selecting "Customer
Support" and then "Obtaining Permissions."

Academic Press
An Elsevier imprint.
525 B Street, Suite 1900, San Diego, California 92101-4495, USA
http://www.academicpress.com

Academic Press
84 Theobalds Road, London WC1X 8RR, UK
http://www.academicpress.com

International Standard Book Number: 0-12-182267-2

PRINTED IN THE UNITED STATES OF AMERICA
03 04 05 06 07 08   9 8 7 6 5 4 3 2 1

Table of Contents

Section I. Analysis of Nuclear Receptor Ligands

Section II. Structure/Function Analysis of Nuclear Receptors

Section V. Use of Animal Models to Study Nuclear Receptor Function

Contributors to Volume 364

Article numbers are in parentheses following the names of contributors.
Affiliations listed are current.

MARIA ACENA-NAGEL (20), *Department of Neurology, University of Colorado Health Sciences Center, Denver, Colorado 80262*

LORI AMMA (24), *Department of Human Genetics, Mount Sinai School of Medicine, New York, New York 10029*

JOHAN AUWERX (17), *Institut de Génétique et de Biologie Moléculaire et Cellulaire (IGBMC), CNRS/INSERM/Université Louis Pasteur, B.P. 163, F-67404 Illkirch, France*

BRUCE BLUMBERG (1), *Department of Developmental and Cell Biology, University of California, Irvine, California 92697-2300*

CÉCILE CALLEJA (22), *Institut de Génétique et de Biologie Moléculaire et Cellulaire, CNRS/INSERM/ULP, Collège de France, BP 10142, 67404 Illkirch Cedex, France*

PIERRE CHAMBON (22), *Institut de Génétique et de Biologie Moléculaire et Cellulaire, CNRS/INSERM/ULP, Collège de France, BP 10142, 67404 Illkirch Cedex, France*

CHING-YI CHANG (7), *Department of Pharmacology and Cancer Biology, Duke University Medical Center, Durham, North Carolina 27710*

BENOIT CHAPELLIER (22), *Institut de Génétique et de Biologie Moléculaire et Cellulaire, CNRS/INSERM/ULP, Collège de France, BP 10142, 67404 Illkirch Cedex, France*

GUOJUN CHENG (25), *Med. Nutr., Karolinska Institute, Huddinge University Hospital, Novum, Huddinge, S-14186, Sweden*

HELEN CHO (12), *Howard Hughes Medical Institute, Gene Expression Laboratory, The Salk Institute for Biological Studies, La Jolla, California 92037*

ANN L. CRADDOCK (18), *Departments of Internal Medicine and Pathology, Wake Forest University School of Medicine, Medical Center Boulevard, Winston-Salem, North Carolina 27157*

PAUL A. DAWSON (18), *Departments of Internal Medicine and Pathology, Wake Forest University School of Medicine, Medical Center Boulevard, Winston-Salem, North Carolina 27157*

GENEVIÈVE DEBLOIS (19), *Molecular Oncology Group, McGill University Health Centre, Montréal, Québec, Canada H3A 1A1*

ULF DICZFALUSY (2), *Karolinska Institutet, Department of Medical Laboratory Sciences and Technology, Division of Clinical Chemistry, Huddinge University Hospital C1.74, SE-141 86 Huddinge, Sweden*

ROBIN E. DODSON (20), *Department of Biochemistry, University of Illinois, Urbana, Illinois 61801*

RONALD M. EVANS (12), *Howard Hughes Medical Institute, Gene Expression Laboratory, The Salk Institute for Biological Studies, La Jolla, California 92037*

STEVEN A. FARBER (23), *Department of Microbiology and Immunology, Kimmel Cancer Center, Thomas Jefferson University, Philadelphia, Pennsylvania 19107*

ELISABETH FAYARD (17), *Institut de Génétique et de Biologie Moléculaire et Cellulaire (IGBMC), CNRS/INSERM/Université Louis Pasteur, B.P. 163, F-67404 Illkirch, France*

DOUGLAS FORREST (24), *Department of Human Genetics, Mount Sinai School of Medicine, New York, New York 10029*

TIMOTHY R. GEISTLINGER (13), *Departments of Pharmaceutical Chemistry and Cellular and Molecular Pharmacology, University of California San Francisco, California 94143-0446*

NORBERT B. GHYSELINCK (22), *Institut de Génétique et de Biologie Moléculaire et Cellulaire, CNRS/INSERM/ULP, Collège de France, BP 10142, 67404 Illkirch Cedex, France*

VINCENT GIGUÈRE (19), *Molecular Oncology Group, McGill University Health Centre, and Departments of Biochemistry, Medicine and Oncology, McGill University, Montréal, Québec, Canada H3A 1A1*

KATHRYN M. GOOLSBY (20), *Department of Biochemistry, University of Illinois, Urbana, Illinois 61801*

FELIX GRÜN (1), *Department of Developmental and Cell Biology, University of California, Irvine, California 92697-2300*

MATTHEW G. GUENTHER (14), *Division of Endocrinology, Diabetes, and Metabolism, Departments of Medicine and Genetics, and The Penn Diabetes Center, University of Pennsylvania Medical Center, Philadelphia, Pennsylvania 19104*

JAN-ÅKE GUSTAFSSON (25), *Med. Nutr., Karolinska Institute, Huddinge University Hospital, Novum, Huddinge, S-14186, Sweden*

R. KIPLIN GUY (13), *Departments of Pharmaceutical Chemistry and Cellular and Molecular Pharmacology, University of California San Francisco, California 94143-0446*

BIN HE (8), *Laboratories for Reproductive Biology and Department of Pediatrics, and the Department of Biochemistry and Biophysics, University of North Carolina, Chapel Hill, North Carolina 27599-7500*

SHIU-YING HO (23), *Department of Microbiology and Immunology, Kimmel Cancer Center, Thomas Jefferson University, Philadelphia, Pennsylvania 19107*

HUEY-JING HUANG (7), *Department of Pharmacology and Cancer Biology, Duke University Medical Center, Durham, North Carolina 27710*

ARUP KUMAR INDRA (22), *Institut de Génétique et de Biologie Moléculaire et Cellulaire, CNRS/INSERM/ULP, Collège de France, BP 10142, 67404 Illkirch Cedex, France*

MICHELLE JANSEN (7), *Department of Pharmacology and Cancer Biology, Duke University Medical Center, Durham, North Carolina 27710*

STACEY A. JONES (4), *Nuclear Receptor Functional Analysis, High Throughput Biology, GlaxoSmithKline, Research Triangle Park, North Carolina 27709-3398*

IWAN JONES (24), *Department of Human Genetics, Mount Sinai School of Medicine, New York, New York 10029*

KIMON C. KANELAKIS (10), *Department of Pharmacology, University of Michigan Medical School, Ann Arbor, Michigan 48109-0632*

SHIGEAKI KATO (21), *The Institute of Molecular and Cellular Biosciences, The University of Tokyo, Yayoi 1-1-1, Bunkyo-ku, Tokyo 113-0032, Japan*

JOHN A. KATZENELLENBOGEN (3), *Department of Chemistry, University of Illinois, Urbana, Illinois 61801-3792*

MATTHEW W. KELLEY (24), *Section on Developmental Neuroscience, National Institute on Deafness and Other Communication Disorders, National Institutes of Health, 5 Research Court, Rockville, Maryland 20850*

THOMAS KIESSELBACH (25), *Med. Nutr., Karolinska Institute, Huddinge University Hospital, Novum, Huddinge, S-14186, Sweden*

STEVEN A. KLIEWER (4), *Department of Molecular Biology, University of Texas Southwestern Medical Center, Dallas, Texas, 75390*

HANA KOUTNIKOVA (17), *CareX S.A., F-67000 Strasbourg, France*

TATIANA KOZLOVA (27), *Howard Hughes Medical Institute, Department of Human Genetics, University of Utah, Salt Lake City, Utah 84112-5331*

JOSÉE LAGANIÈRE (19), *Molecular Oncology Group, McGill University Health Centre, and Department of Biochemistry, McGill University, Montréal, Québec, Canada H3A 1A1*

PAMELA J. LANFORD (24), *Section on Developmental Neuroscience, National Institute on Deafness and Other Communication Disorders, National Institutes of Health, 5 Research Court, Rockville, Maryland 20850*

ELIZABETH LANGLEY (8), *Institute for Biomedical Research, National University of Mexico, Mexico City, Mexico*

VINCENT LAUDET (6), *Laboratoire de Biologie Moléculaire et Cellulaire, UMR CNRS 5665, Ecole Normale Supérieure de Lyon, 46 allée d'Italie, 69364 Lyon Cedex 07, France*

MITCHELL A. LAZAR (14), *Division of Endocrinology, Diabetes, and Metabolism, Departments of Medicine and Genetics, and The Penn Diabetes Center, University of Pennsylvania Medical Center, Philadelphia, Pennsylvania 19104*

YOON-KWANG LEE (9), *Department of Molecular and Cellular Biology, Baylor College of Medicine, One Baylor Plaza, Houston, Texas 77030*

DAVID Y. LEE (16), *Department of Biochemistry and Molecular Biology, University of Southern California, Los Angeles, California 90089*

JÜRGEN LEHMANN (17), *CareX S.A., F-67000 Strasbourg, France*

MEI LI (22), *Institut de Génétique et de Biologie Moléculaire et Cellulaire, CNRS/INSERM/ULP, Collège de France, BP 10142, 67404 Illkirch Cedex, France*

ERIK G. LUND (2), *Merck Research Laboratories, RY80W-250, Rahway, New Jersey 07065*

HAN MA (16), *Inflammatory and Viral Diseases Unit, Roche Bioscience, Palo Alto, California 94304*

SOHAIL MALIK (15), *Laboratory of Biochemistry and Molecular Biology, Rockefeller University, New York, New York 10021*

CHENGJIAN MAO (20), *Department of Biochemistry, University of Illinois, Urbana, Illinois 61801*

ALEXANDER MATA DE URQUIZA (26), *Ludwig Institute for Cancer Research, Stockholm Branch, S-171 77 Stockholm, Sweden*

DONALD P. MCDONNELL (7), *Department of Pharmacology and Cancer Biology, Duke University Medical Center, Durham, North Carolina 27710*

DANIEL METZGER (22), *Institut de Génétique et de Biologie Moléculaire et Cellulaire, CNRS/INSERM/ULP, Collège de France, BP 10142, 67404 Illkirch Cedex, France*

DAVID D. MOORE (9), *Department of Molecular and Cellular Biology, Baylor College of Medicine, One Baylor Plaza, Houston, Texas 77030*

RAMESH NARAYANAN (11), *Department of Molecular and Cellular Biology, Baylor College of Medicine, Houston, Texas 77030*

LILY NG (24), *Department of Human Genetics, Mount Sinai School of Medicine, New York, New York 10029*

NGOC-HA NGUYEN (5), *Departments of Pharmaceutical Chemistry and Cellular and Molecular Pharmacology, University of California, San Francisco, California 94143-2280*

STEFAN NILSSON (25), *Med. Nutr., Karolinska Institute, Huddinge University Hospital, Novum, Huddinge, S-14186, Sweden*

JOHN D. NORRIS (7), *Department of Pharmacology and Cancer Biology, Duke University Medical Center, Durham, North Carolina 27710*

MICHAEL PACK (23), *Department of Medicine, University of Pennsylvania, Philadelphia, Pennsylvania 19104*

DEREK J. PARKS (4), *Systems Research, GlaxoSmithKline, Research Triangle Park, North Carolina, 27709-3398*

THOMAS PERLMANN (26), *Ludwig Institute for Cancer Research, Stockholm Branch, S-171 77 Stockholm, Sweden*

WILLIAM B. PRATT (10), *Department of Pharmacology, University of Michigan Medical School, Ann Arbor, Michigan 48109-0632*

MARC ROBINSON-RECHAVI (6), *Laboratoire de Biologie Moléculaire et Cellulaire, UMR CNRS 5665, Ecole Normale Supérieure de Lyon, 46 allée d'Italie, 69364 Lyon Cedex 07, France*

ROBERT G. ROEDER (15), *Laboratory of Biochemistry and Molecular Biology, Rockefeller University, New York, New York 10021*

BRIAN G. ROWAN (11), *Department of Biochemistry and Molecular Biology, Medical College of Ohio, Toledo, Ohio 43614*

SHIGEHIRA SAJI (25), *Med. Nutr., Karolinska Institute, Huddinge University Hospital, Novum, Huddinge, S-14186, Sweden*

HIDEKI SAKAGUCHI (25), *Med. Nutr., Karolinska Institute, Huddinge University Hospital, Novum, Huddinge, S-14186, Sweden*

THOMAS S. SCANLAN (5), *Departments of Pharmaceutical Chemistry and Cellular and Molecular Pharmacology, University of California, San Francisco, California 94143-2280*

YONGFENG SHANG (16), *Department of Biochemistry and Molecular Biology, Peking University Health Science Center, Beijing 100083, People's Republic of China*

DAVID J. SHAPIRO (20), *Department of Biochemistry, University of Illinois, Urbana, Illinois 61801*

MICHAEL R. STALLCUP (16), *Department of Pathology, University of Southern California, Los Angeles, California 90089-9092*

KEN-ICHI TAKEYAMA (21), *The Institute of Molecular and Cellular Biosciences, The University of Tokyo, Yayoi 1-1-1, Bunkyo-ku, Tokyo 113-0032, Japan*

ANOBEL TAMRAZI (3), *Department of Chemistry, University of Illinois, Urbana, Illinois 61801-3792*

CARL S. THUMMEL (27), *Howard Hughes Medical Institute, Department of Human Genetics, University of Utah, Salt Lake City, Utah 84112-5331*

LING WANG (25), *Med. Nutr., Karolinska Institute, Huddinge University Hospital, Novum, Huddinge, S-14186, Sweden*

MARGARET WARNER (25), *Med. Nutr., Karolinska Institute, Huddinge University Hospital, Novum, Huddinge, S-14186, Sweden*

NANCY L. WEIGEL (11), *Department of Molecular and Cellular Biology, Baylor College of Medicine, Houston, Texas 77030*

ZHANG WEIHUA (25), *Med. Nutr., Karolinska Institute, Huddinge University Hospital, Novum, Huddinge, S-14186, Sweden*

ELIZABETH M. WILSON (8), *Laboratories for Reproductive Biology and Department of Pediatrics, and the Department of Biochemistry and Biophysics, University of North Carolina, Chapel Hill, North Carolina 27599-7500*

WEI XU (12), *Howard Hughes Medical Institute, Gene Expression Laboratory, The Salk Institute for Biological Studies, La Jolla, California 92037*

HIKARI A. I. YOSHIHARA (5), *Departments of Pharmaceutical Chemistry and Cellular and Molecular Pharmacology, University of California, San Francisco, California 94143-2280*

Preface

In 1975 Bert O'Malley and Joel Hardman edited five volumes in the *Methods in Enzymology* series on the topic of hormone action. Included in this now classic collection was a volume entitled "Steroid Hormones," which contained close to fifty papers describing the methods of the day used to study nuclear hormone receptors. Assays for steroid hormone binding, receptor purification, biological response systems, and steroid metabolism constituted the arsenal then available to delve into the myriad functions of nuclear receptors. These methods served the receptor community admirably and their application led to the next revolution in the field, which occurred in the mid-1980s when the laboratories of Pierre Chambon, Ronald Evans and Keith Yamamoto reported the first isolation of cDNAs encoding the mammalian estrogen and glucocorticoid receptors. These remarkable achievements ushered in the current era of nuclear receptor research, and with it came the development of new methods to study receptor structure and function.

This volume represents a compilation of these newer methods. They span the gamut of current receptor research, from chemical methods for ligand purification, synthesis and quantitation, to the characterization of mice that lack one or more receptor encoding genes. How to utilize receptor-relevant information in the proliferating genome sequences of higher eukaryotic organisms is detailed together with molecular methods to identify networks of receptor-responsive target genes. Newer versions of biochemical methods for receptor purification and assay are found, as are those that report in exquisite detail how receptor interacting proteins such as chaperones and coregulators are obtained. The study of receptor biology is not confined to a single organism, and to this end several articles address the use of species as disparate as flies and fish to gain insight into receptor function.

The present group of authors also represents a Who's Who of the new generation of nuclear receptor researchers and each of the represented laboratories has contributed significantly to our current knowledge base. Although none of the old lions from the earlier *Methods in Enzymology* volume on receptors were able to contribute to the present collection, a substantial debt is owed them for laying the foundations on which all of us stand.

Tomes of this nature are not assembled in a vacuum, and we offer heartfelt thanks to the many authors for their hard work and bonhomie; our editors Shirley Light and Noelle Gracy at Academic Press, who, under duress from Keith Yamamoto, John Abelson, and Mel Simon, conceived this volume; and our secretaries, Lidia Galvan and Betsy Layton, who are standing in line for canonization.

DAVID W. RUSSELL
DAVID J. MANGELSDORF

METHODS IN ENZYMOLOGY

VOLUME XLI. Carbohydrate Metabolism (Part B)
Edited by W. A. WOOD

VOLUME XLII. Carbohydrate Metabolism (Part C)
Edited by W. A. WOOD

VOLUME XLIII. Antibiotics
Edited by JOHN H. HASH

VOLUME XLIV. Immobilized Enzymes
Edited by KLAUS MOSBACH

VOLUME XLV Proteolytic Enzymes (Part B)
Edited by LASZLO LORAND

VOLUME XLVI. Affinity Labeling
Edited by WILLIAM B. JAKOBY AND MEIR WILCHEK

VOLUME XLVII. Enzyme Structure (Part E)
Edited by C. H. W. HIRS AND SERGE N. TIMASHEFF

VOLUME XLVIII. Enzyme Structure (Part F)
Edited by C. H. W. HIRS AND SERGE N. TIMASHEFF

VOLUME XLIX. Enzyme Structure (Part G)
Edited by C. H. W. HIRS AND SERGE N. TIMASHEFF

VOLUME L. Complex Carbohydrates (Part C)
Edited by VICTOR GINSBURG

VOLUME LI. Purine and Pyrimidine. Nucleotide Metabolism
Edited by PATRICIA A. HOFFEE AND MARY ELLEN JONES

VOLUME LII. Biomembranes (Part C: Biological Oxidations)
Edited by SIDNEY FLEISCHER AND LESTER PACKER

VOLUME LIII. Biomembranes (Part D: Biological Oxidations)
Edited by SIDNEY FLEISCHER AND LESTER PACKER

VOLUME LIV. Biomembranes (Part E: Biological Oxidations)
Edited by SIDNEY FLEISCHER AND LESTER PACKER

Section I

Analysis of Nuclear Receptor Ligands

[1] Identification of Novel Nuclear Hormone Receptor Ligands by Activity-Guided Purification

By FELIX GRÜN and BRUCE BLUMBERG

Introduction

Members of the nuclear hormone receptor superfamily share a common architecture typically including a highly conserved DNA-binding domain (DBD) and a ligand-binding domain (LBD). Many of these ligand-modulated transcription factors have specific endogenous ligands and act as ligand sensors that regulate gene expression during development, cellular differentiation, reproduction, and lipid homeostasis. Those receptors lacking known endogenous ligands are termed as "orphan receptors." The identification of natural ligands for this class of nuclear receptors is an important goal in understanding their biology. Much progress has been made in recent years identifying ligands for orphan receptors (reviewed in Refs. 1–4). It is thought that nearly all nuclear receptors evolved from an ancestral estrogen receptor.[5,6] Based on this observation and the existence of conserved LBD sequences, it is not unreasonable to hypothesize that many orphan receptors are ligand-dependent. If many or most orphan receptors do indeed have endogenous ligands, the immediate question that arises is why these ligands have not yet been identified.

Several potential contributing factors for the slow pace of identification of natural ligands can be considered. One is that many previous screens of natural or synthetic ligands are inherently biased toward known bioactive

[1] T. T. Lu, J. J. Repa, and D. J. Mangelsdorf, Orphan nuclear receptors as eLiXiRs and FiXeRs of sterol metabolism, *J. Biol. Chem.* **17**, 17 (2001).

[2] J. J. Repa and D. J. Mangelsdorf, The role of orphan nuclear receptors in the regulation of cholesterol homeostasis, *Annu. Rev. Cell Dev. Biol.* **16**, 459–481 (2000).

[3] T. M. Willson, S. A. Jones, J. T. Moore, and S. A. Kliewer, Chemical genomics: functional analysis of orphan nuclear receptors in the regulation of bile acid metabolism, *Med. Res. Rev.* **21**(6), 513–522 (2001).

[4] W. Xie and R. M. Evans, Orphan nuclear receptors: the exotics of xenobiotics, *J. Biol. Chem.* **276**(41), 37739–37742 (2001).

[5] J. W. Thornton, Evolution of vertebrate steroid receptors from an ancestral estrogen receptor by ligand exploitation and serial genome expansions, *Proc. Natl. Acad. Sci. USA* **98**(10), 5671–5676 (2001).

[6] J. W. Thornton and R. DeSalle, A new method to localize and test the significance of incongruence: detecting domain shuffling in the nuclear receptor superfamily, *Syst. Biol.* **49**(2), 183–201 (2000).

compounds or derivatives thereof and are thus limited in their scope. Ligands or their immediate metabolic precursors may be constitutively present within a cell and have fast turnover rates within narrow concentration ranges. Perturbation through exogenous application of test ligands may therefore be difficult without rigorous control over experimental conditions. Ligand synthesis may also be regulated in a very restricted manner in time and space during development. Use of an inappropriate source material may therefore preclude successful purification.

The strategy that we have applied successfully in identifying novel nuclear receptor ligands is based on the key principle of activity-guided purification followed by structure determination. Prior knowledge of an activity's relation to known nuclear receptor ligands is therefore not pertinent. Following sample preparation, extracts are fractionated by suitable HPLC methods and tested for their ability to modulate transcriptional activity of specific nuclear receptors. Central to the success of this approach is a receptor bioassay that has the appropriate sensitivity to detect rapidly ligand-dependent transactivation in small-scale formats. A particular benefit of the modular structure of nuclear receptors is the ability to construct chimeras that retain the specific ligand-dependent nature of a receptor, but facilitate high-throughput screening by utilizing a common reporter construct. Chimeras between the yeast GAL4 DBD and nuclear receptor LBDs have proven invaluable for developing such assays. Reporter constructs exhibit exquisite sensitivity to added ligand in the presence of the chimeric receptors and are independent of endogenous receptor expression making them excellent tools for transactivation screens. Depending on the specific receptor used, quantitative determination of ligand concentrations between 10^{-4} and 10^{-11} M can be measured in tissue culture assays. Therefore, the activity profile of a specific chimeric receptor in response to ligand mixtures (e.g., HPLC fractions) allows the candidate ligand to be purified to homogeneity and its structure determined in the absence of prior information about its chemical nature.

Although the information presented below is drawn primarily from our experiences of novel retinoid receptor ligand purification, the chapter is organized with the goal of presenting general strategies and protocols that are expected to be widely applicable for lipophilic ligands isolated from both tissue sources or dilute xenobiotic ligands present as environmental endocrine disruptors. Special focus is given to sample preparation and extraction techniques, HPLC fractionation methods, and the receptor activation transcription assay. A brief outline of mass spectrometric approaches to structure determination is also included as an introduction on how to proceed when an activity has been purified and to give a sense of the scale in material required.

Sample Preparation and Extraction Methods

The methods described below will use retinoids as example compounds. It is important to note, however, that the approach is rather general and can readily be adapted for other types of compounds. Extracting and identifying compounds that activate retinoid receptors is straightforward, although not trivial.[7] The primary considerations are that one must identify an appropriate source of material and then extract undegraded and unmodified candidate compounds that are then fractionated and tested for retinoid activity. In the case of embryos, one typically pools embryos from stages wherein the receptor is expressed. Tissues expressing the receptors of interest are also good candidates. Lastly, environmental samples such as water or sediments may be used as sources of potentially unknown retinoids.

A good starting point is to utilize a method that recovers the unknown compound quantitatively from the source material, while maintaining its stability, chemical form, and biological activity. This extract should be tested directly for biological activity and also crudely fractionated using semi-preparative scale HPLC and the fractions tested for activity. It is not always the case that activity can be detected in the crude extract; however, if no activity is detected in the fractions, then the extraction method must be changed or optimized. This step ensures that (1) there actually is an activity to purify and (2) a reference method exists that can be used to validate subsequent preparative methods such as solid-phase extraction or solvent partition. It is important to emphasize that the activity must be successfully identified at this stage before proceeding to large-scale purification.

General Considerations

For ligand extracts prepared from tissue samples, or that require special consideration in terms of sensitivity to light, temperature, or biochemical reactivity, we routinely use a homogenous liquid–liquid technique.[7,8] The method is both rapid and gentle requiring no prolonged extraction steps, elevated temperatures, or pH extremes. It is particularly well suited to protein-rich samples since deproteination occurs during extraction and the extracts are compatible with direct HPLC injection. Extraction volumes are kept relatively small and this method can be easily scaled to accommodate

[7] B. Blumberg, J. Bolado, Jr., F. Derguini, A. G. Craig, T. A. Moreno, D. Charkravarti, R. A. Heyman, J. Buck, and R. M. Evans, Novel RAR ligands in Xenopus embryos, *Proc. Natl. Acad. Sci. (USA)* **93**, 4873–4878 (1996).

[8] S. W. McClean, M. E. Ruddel, E. G. Gross, J. J. DeGiovanna, and G. L. Peck, Liquid-chromatographic assay for retinol (vitamin A) and retinol analogs in therapeutic trials, *Clin. Chem.* **28**, 693–696 (1982).

larger volumes up to several liters. The sample is first homogenized with 0.4 volumes of acetonitrile:*n*-butanol (1:1) using a polytron, Dounce, or other appropriate homogenizer. Homogenizing for 1 min using the polytron at maximum speed is adequate. Aqueous samples, such as serum, can be vortexed vigorously for several minutes. Phase separation is accomplished by addition of 0.3 volumes of saturated dibasic potassium phosphate solution. The samples are homogenized or vortexed for an additional minute and then centrifuged at $10,000 \times g$ for 1–10 min at 4°C to separate the layers. After centrifugation, the organic phase is removed for HPLC separation. Proteins form a gelatinous phase-lock between the aqueous (bottom) and the organic (top) layers and are thereby effectively removed. Addition of the antioxidant *tert*-butylated hydroxytoluene (BHT) at 1 μM during the extraction is recommended to reduce nonspecific oxidation of the sample.

Solutions
Acetonitrile:1-butanol (1:1 v/v)
Saturated K_2HPO_4 solution, pH 7.5
1 mM BHT in methanol
Phosphate-buffered saline (PBS)

Example for small sample volumes (*Eppendorf tube scale*)
0.8 ml Aqueous samples
0.32 ml Acetonitrile:1-butanol (1:1)
0.28 ml K_2HPO_4
1.4 ml Total volume

Centrifuge at 14,000 rpm in a benchtop microcentrifuge for 1 min. Approximately, 200 μl of organic extract is obtained that can be injected directly for HPLC analysis without further sample cleanup.

Example for embryos, tissues or solid material
20 ml Homogenate
8 ml Acetonitrile:1-butanol
6 ml Saturated K_2HPO_4
34 ml Total

For embryos, tissues, and freeze-dried samples, the material should be homogenized for at least 1 min in a 10-fold excess of PBS before addition of acetonitrile–butanol. After addition of the saturated potassium phosphate, the material is vortexed, transferred to polypropylene tubes and centrifuged at $10,000 \times g$ for 10 min at 4°C in a Sorval SA-600 rotor or similar rotor. The top organic layer is transferred to a new tube and centrifugation is repeated to remove any carry over of particulate matter. The recovered material can be concentrated by rotary evaporation and reconstituted in acetonitrile–butanol or simply injected onto the HPLC column in appropriately sized aliquots (see below).

Alternative Organic Extraction Methods

Acetonitrile–butanol extraction is convenient and usually preferred for compounds of unknown chemical properties or volatility. It is not the best method for extracting acids or if the material is known to be stable and not volatile. Eichele and co-workers have published extensively on the extraction and characterization of retinoic acids (reviewed in Wedden et al.[9]). In addition, there are a variety of other types of liquid–liquid organic extractions, e.g., the Folch method (methanol–chloroform 2:1).[10] It should be noted that the use of chloroform is best avoided due to its toxicity and frequent acidity. Dichloromethane gives similar extractions and is not acidic.

Solid-Phase Extraction

Materials
500 g Amberlite XAD-2 resin (Rohm & Haas)
Large Soxhlet extractor
Rotary evaporator
Water filtration canisters (paper or glass fiber prefilter, connectors)
Pond pump
Large Buchner funnel
Acid-washed detergent-free glassware
4 liter Methanol
4 liter Acetone
4 liter Hexane
4 liter Dichloromethane
30 liter Double-distilled deionized water
20 cm Glass fiber filters
Glass wool

[9] S. Wedden, C. Thaller, and G. Eichele, Targeted slow-release of retinoids into chick embryos, Methods Enzymol. **190**, 201–209 (1990).
[10] J. Folch, M. Lees, and G. H. S. Stanley, A simple method for the isolation and purification of total lipids from animal tissues, J. Biol. Chem. **226**, 497–509 (1957).

For preparative extractions of large aqueous samples, e.g., serum or environmental water samples, liquid–liquid extraction becomes impractical due to the large volumes of organic solvents required. A suitable solid-phase extraction method with broad affinity for hydrophobic organic compounds (HOC) is therefore desirable. We have successfully employed XAD-2 resin beads (Rohm & Haas) to capture retinoid receptor activators from environmental water samples. Alternatively, other high-flow matrices (e.g., HP-20, XAD-7, C18, silica) with affinity for the receptor activity in question may be substituted. It may be necessary to try several matrices before finding one that is suitable. XAD-2 resin as supplied by the manufacturer requires extensive cleanup prior to use in order to remove residual synthetic reaction products. The Environmental Protection Agency (EPA) has developed and standardized methods for the preparation and use of solid-phase resins such as XAD-2,[11] and we have adopted these without change. Briefly, resin is extracted sequentially with methanol, acetone, hexane, and methylene chloride for 24 hr each in a Soxhlet extractor, followed by the reverse order of solvents for sequential 4 hr extractions to bring the resin back to a polar solvent miscible with water. The methanol is replaced by washing the beads six times with 10 volumes of double-distilled water. Beads are stored under water in EPA-certified glass sample bottles until use. Storage should be for 3 months or less and care should be taken to avoid mechanical damage during column packing. The final 4-hr hexane extract is used as an XAD-2 blank.

XAD-2 may be used for small-scale as well as large-scale extractions. Embryos or tissues to be used for solid-phase extraction are homogenized in deionized water or phosphate buffer at approximately 50 mg/ml and the extracts are clarified by centrifugation or filtration. The supernatant is extracted by stirring with preconditioned XAD-2 at a ratio of 1 volume of resin slurry to 5 volumes of supernatant. The mixture is stirred at room temperature for 4 hr with a propeller stirrer, and the beads recovered by decanting the supernatant and removing as much liquid as possible by vacuum filtration. The resin is rinsed with 10 volumes of deionized water and the liquid is again removed by filtration. Adsorbed material is recovered by stirring with 10 volumes of methanol for 2 hr, and the methanol recovered as above. The process is repeated with 10 volumes of acetone and the acetone recovered. Both methanol and acetone soluble materials are combined, clarified by filtration, rotary evaporated, and reconstituted in dimethyl sulfoxide or chloroform–methanol (1:2 ratio) under argon.

[11] E. Crecelius and L. Lefkovitz, HOC Sampling Media Preparation and Handling; XAD-2 Resin and GF/F Filters, Standard Operating Procedure MSL-M-090-00 ed. US EPA, 1994.

For very large-scale preparative purposes (e.g., environmental samples), XAD-2 resin is packed into polycarbonate water filtration canisters modified to contain resin and is fitted with 3/4-inch connectors. Glass wool is packed into the central perforated drainage tube to prevent resin from being flushed out. The inverted canister is partially filled with water and a slurry of resin poured around the central drain tube. Excess water is aspirated from the center taking care to avoid introduction of air bubbles that can lead to channeling and a reduction in the exposure of beads during pumping. Glass wool is packed on top of the resin bed and the canister base with O-ring is screwed tight. Upright packed canisters should be gently flushed with clean deionized water to verify that the canister does not leak water from the seals or resin from the drain outlet. Canisters premodified for chemical media may be obtained from aquarium suppliers (e.g., Filtronics, Oxnard, CA), although these have somewhat lower capacity. Three canisters each containing approximately 500 ml of packed resin can be connected in series to retain a suitable flow rate of 2–3 liter/min. An additional canister containing a paper or a glass fiber prefilter is connected between the pump and the first canister to remove particulates when collecting environmental water samples. The whole assembly is connected by 3/4-inch polyethylene tubing and secured using metal screw collar retainers. A typical pond or aquarium circulating pump (e.g., Eheim, Little Giant, Iwaki, Hydrothruster) with a magnetic drive is utilized to provide the flow. The canister valves are set to provide a steady flow of 2–3 liter/min, which is suitable to prevent undue compression of the resin bed. After several hours, the flow rate should be checked and the prefilter exchanged if fouling is a serious problem. The intake hose is placed within a weighted bucket on the lake bed to provide protection from sediment disturbances, weeds, and other larger debris during setup. The columns are run for 24–48 hr to allow the XAD-2 resin to scavenge HOCs from the water source. Columns are then sealed and shipped to the lab for resin processing.

Resin slurry (500 ml per extraction) is transferred to a 4-liter glass beaker for extraction. Excess water is removed by aspiration. The resin is then sequentially extracted three times with 1 liter each of methanol, acetone, and methylene chloride for 1 hr at room temperature with gentle stirring using a propeller stirrer (Fisher), filtered through a glass fiber filter on a large Buchner funnel and concentrated on a rotary evaporator. The concentrated extract is redissolved in a minimum of organic solvent, e.g., methanol and adjusted to approximately 50–75% solvent/25–50% aqueous buffer depending on the point at which precipitation of solutes is noticeable. The extract is cleared by centrifugation. Any precipitate is back-extracted with 75% methanol/25% buffer. Remaining insoluble material should be discarded.

The pooled organic phases are loaded in aliquots onto a reversed-phase C18 column equilibrated in buffer (typically 50 mM ammonium acetate, pH 6.5) until the entire extract is on the column. Subsequently, a linear gradient of buffer–methanol–chloroform is employed at a flow rate appropriate for the size of column used and fractions are collected. Under these conditions, the materials do not begin to elute until the organic component of the mobile phase reaches a critical level for each type of compound. In this way, handling time is minimized and the requirement for drying down large volumes of organic extracts is eliminated.

Solvent Partition as a Prepurification Step

Once a sample is known to contain activity, it is often valuable to determine whether it can be partially purified by solvent partition prior to chromatography. This enables one to reduce the complexity of a mixture, which can reduce greatly the time spent in optimizing later HPLC separations. Solvent partition employs sequential extractions between immiscible solvents of different chemical properties. One begins with a known amount of retinoid activity then adds the sample to a large excess of methanol (> 20 volumes) in a separatory funnel. An equal volume of iso-octane or *n*-heptane (*n*-heptane is easier to work with) is then added and the mixture is shaken vigorously. After phase separation, the lower methanol phase is removed and reextracted with another aliquot of *n*-heptane, shaken as above, and the phases separated. After three extractions, the *n*-heptane and methanol phases are rotary evaporated to dryness, reconstituted, and tested for activity. If the activity partitions into the nonpolar phase it is used directly for HPLC purification. If it partitions into the polar methanol phase, the next step is to partition the activity between ethyl acetate and water using the same procedure. If the activity partitions into the ethyl acetate phase it is then further purified by HPLC. If it partitions into the aqueous phase, further partitioning between water and *n*-butanol is conducted and the resulting extracts tested for activity. In testing each of these partition steps, it is important to quantitate carefully the fraction of the active component recovered. If it is not recovered quantitatively then solvent partition is unsuitable as a prepurification step.

Reversed-Phase HPLC

We utilize a Waters 600 Delta HPLC system running Millennium 32 chromatography manager software. The key components of the system are dual Rheodyne 7725i injection valves for semi-prep and analytical injection switching, a Waters 600E pump and 996 photodiode array detector, and a Pharmacia SuperFrac fraction collector. We favor Vydac 218TP (fully endcapped) or 201TP (non-endcapped) C18 columns for reversed-phase chromatography although the quality of modern columns makes the exact choice of manufacturer and chemistry a personal preference. Guard columns should be used throughout to protect and prolong the lifetime of the columns. A good set of columns for extract purification comprises a 25 × 250 mm preparative, 10 × 250 mm semi-preparative and 4.6 × 250 mm analytical columns. A set of columns and injection syringes dedicated for ligand purification is well worth the investment to avoid contamination from standards. In all cases, PDA monitoring of a wide spectrum of wavelengths at one spectrum per second is employed. When methanol is used as the solvent, we typically scan a window of 220–600 nm. Acetonitrile is optically clear at low wavelengths, therefore a window of 190–600 nm is used.

Crude extracts are first fractionated on the preparative column. The column is preequilibrated by executing mock runs of the gradient elution method 2–3 times until the baseline is stable and clear of contaminant peaks. It is important to scrupulously clean the injector ports between runs, especially if the HPLC system has been used previously for analysis of standards. We always perform a blank run prior to sample loading to enable collection of solvent only controls for each later fraction. The column is loaded under initial flow conditions, typically 95–100% of an aqueous buffer, e.g., 50 mM ammonium acetate, pH 6.5. Multiple aliquots (2 ml or less) are injected every 3 min with baseline monitoring. If breakthrough of injected compounds is observed, the percentage of methanol in the sample or volume of the injected aliquot is reduced. After the final injection, the column is allowed to return to baseline absorbance value before starting the gradient run. An extract from 500 ml packed XAD-2 resin is normally split into 2–3 preparative runs to avoid column overloading and the collected fractions pooled.

Gradient elution, as outlined in Table I, cycles the column to 100% dichloromethane and back to the initial conditions within 80 min. All of the retained compounds will elute within 60 min. Fractions are collected at 1-min intervals (flow rate 8 ml/min). We typically add BHT to the empty tubes to achieve a final concentration of 1 μM. This addition ensures that BHT is continuously present. After collection, the samples are flushed with argon, stored at $-80°C$, and protected from light until analyzed in the reporter assay.

Active fractions from the preparative runs are pooled, diluted 3:1 with aqueous buffer, reinjected, and subjected to further fractionation by HPLC. As a second column, we use a Vydac 218TP510 semi-preparative C18 column. The eluting solvent is an acetonitrile gradient from 0 to 100%. One-minute fractions (flow rate 2 ml/min) are collected and 100 μl

TABLE I

GRADIENT CONDITIONS

Time (min)	Flow (ml/min)	% A	% B	% C	% D	Change
(A) Preparative C-18 column with buffer–methanol–dichloromethane gradient						
0	8	100	0	0	0	—
5	8	30	70	0	0	Linear
35	8	0	100	0	0	Linear
40	8	0	100	0	0	Linear
50	8	0	40	60	0	Linear
60	8	0	100	0	0	Jump
65	8	100	0	0	0	Jump
80	0	100	0	0	0	Jump
(B) Semi-preparative C-18 column with buffer–acetonitrile–dichloromethane gradient						
0	3	100	0	0	0	—
5	3	80	0	0	20	Linear
40	3	0	0	0	100	Linear
42.5	3	0	0	0	100	Linear
50	3	0	0	100	0	Linear
55	3	0	0	0	100	Jump
65	3	100	0	0	0	Jump
80	0	100	0	0	0	Jump
(C) Analytical C-18 column with buffer–mixed methanol/acetonitrile (1:1) gradient						
0	1	100	0	0	0	—
2.5	1	40	30	0	30	Linear
30	1	30	35	0	35	Linear
35	1	0	50	0	50	Linear
40	1	100	0	0	0	Jump
60	0	100	0	0	0	Jump

Eluent	Composition
A	50 mM Ammonium acetate, pH 6.5
B	100% Methanol
C	100% Dichloromethane
D	100% Acetonitrile

aliquots are removed for assays. Bona fide activity peaks should show consistent patterns of activation between column runs and some indication of dose-dependency when diluted or compared with adjacent fractions.

Additional steps can be performed on high-quality analytical columns from this point onwards as the amount of material will no longer saturate and impede column performance. Choices of column chemistry and mobile phase will be dependent on the specific ligand and need to be addressed on a case-by-case basis. C18 columns are ideally suited for fractionating nonpolar retinoids. Our preferred analytical column is a Vydac 201TP54 developed with a shallow methanol gradient. The slight difference in solvent polarity may result in a change in order of elution of specific coeluting components, which together with careful optimization of gradient conditions is often sufficient to give baseline separation of compounds observed in the PDA UV spectrum. Additional changes in parameters such as buffer pH or counterion and column temperature may be useful. Organic acids, such as retinoic acid, demonstrate a significant shift in retention time (several minutes) when changing from neutral to acidic conditions due to the change in ionization states. Acetic acid (1%) is useful as a relatively gentle modifier and can be removed easily during subsequent sample processing; however, care must be taken to ensure that the fractions retain activity under these conditions. Neutralization with buffer following elution is recommended to prevent losses from acid catalysis.

Caveats

The initial stages of any purification should be undertaken with the utmost care to avoid the degradation and modification of potentially unstable compounds. Such steps include eliminating rotary evaporation where feasible to avoid oxidation and potential loss of volatile compounds, ensuring that contact with oxygen is minimized, the continuous presence of suitable antioxidants (e.g., BHT), and rapid testing of fractions for biological activity. It is often advisable to purchase screw-capped test tubes and blanket the samples with argon prior to freezing them. Neutral buffers without primary amines are favored to minimize chemical modification and degradation. Whenever feasible, all extractions, fractionation, transfections, and other analyses should be performed using subdued lighting. Once some idea of the lability of the compounds under study is obtained, these precautions may be modified accordingly.

How Much Material is Required for Chemical Characterization?

One difficulty often encountered when identifying new receptor ligands is that the activation assays can identify nM or pM levels of compounds, whereas tens of μM or more may be required for chemical analysis. Generally speaking, one requires milligrams of pure compound to characterize an unprecedented carbon skeleton by mass spectrometry, ^{1}H- and ^{13}C-NMR. This requirement may be reduced if an NMR with micro- or nanoinverse detection probes is available. For example, we required less than 20 μg of pure material to characterize 3-hydroxyl ethyl benzoate glucosamine by ^{1}H-NMR and tandem mass spectrometry.[12] Typically, tens of micrograms are required to identify a known compound using mass spectrometry.

Activity-Guided Fractionation

One of the most important components of our ligand identification method is the activity-guided fractionation. The purification is inherently unbiased with respect to any *a priori* knowledge of the compound's relation to known ligands since it is guided solely by following the compound's potential to transactivate the receptor LBD. Each fraction is tested for activity in the cotransfection assay and active fractions selected for further study. The identity of the compound is determined after it is purified to homogeneity through chemical characterization. This approach enables the identification of compounds with unusual fractionation or absorption properties. To date, we have identified and characterized novel ligands for the retinoic acid receptor,[7] and identified the orphan receptor BXR as a specific receptor for benzoates[12] using activity-guided fractionation.

For high-throughput screening of fractions, we use a calcium phosphate/DNA precipitate protocol adapted for a 96-well format. This low-cost transfection method saves on reagents, and when combined with multichannel dispensers and plate readers, allows a single operator to assay easily as many as several thousand data points per experiment.

Success in ligand purification is determined in large part by the sensitivity and robustness of the reporter assay. A key requirement is that the sensitivity of the assay must be comparable to the concentration of the active component in the material being studied. For receptors that have high-affinity ($\sim$nM K_d) ligands such as RAR, assay sensitivities in the subnanomolar range are optimal and readily attained (Fig. 1). This sensitivity allows one to detect even weak activation and to minimize the use of precious material. To achieve such sensitivity, considerable attention to the optimization of assay conditions is essential. Particularly important is to optimize the transfection efficiency, which is strongly influenced by the pH of the BES phosphate buffer, incubator CO_2 levels, and buffering capacity of the growth medium. For this reason, empirical determination of optimal conditions using the conditions below as a starting point of reference should be conducted. Consistent results can be attained but require rigorous

[12] B. Blumberg, H. Kang, J. Bolado, Jr., H. Chen, A. G. Craig, T. A. Moreno, K. Umesono, T. Perlmann, E. M. De Robertis, and R. M. Evans, BXR, an embryonic orphan nuclear receptor activated by a novel class of endogenous benzoate metabolites, *Genes Dev.* **12**(9), 1269–1277 (1998).

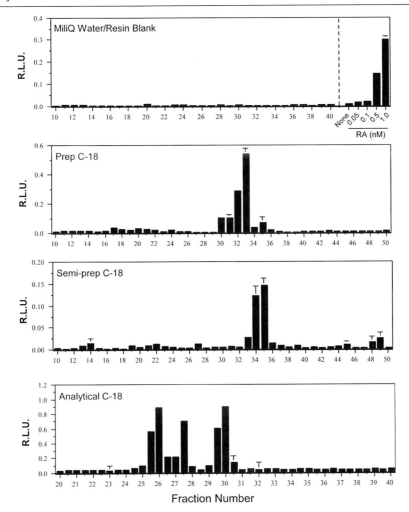

FIG. 1. Activity-guided fractionation of Minnesota lake water. HOCs from a Minnesota lake with a high incidence of malformed amphibians or control Milli-Q water were captured by solid-phase XAD-2 resin extraction, fractionated by C18 reverse-phase HPLC, and tested in receptor activation assays. Aliquots from 1 min fractions were tested for GAL4-RARα-mediated transactivation of luciferase reporter constructs in transient transfection assays and compared with all-trans retinoic acid standards. An environmental retinoid activity is detected in fractions 30–33 of the preparative C18 column. Fractions 32 and 33 were pooled and rechromatographed on the semi-preparative C18 column. Fractions 34 and 35 were pooled and rechromatographed on the analytical C18 column, where the activity resolved into three distinct peaks. No activity is seen in the corresponding water and resin controls (top); retinoic acid standards could be detected in the subnanomolar range (top). Bars represent the means ± S.E.M. of triplicates normalized to β-galactosidase controls and expressed as relative luciferase units (R.L.U.).

adherence to a laboratory standard experimental procedure. Individual operators will frequently obtain suboptimal performance even with the same reagents.

Calcium chloride solution and BES buffer are prepared as 500 ml 2× stocks. The optimal pH of the BES phosphate buffer stock should be determined for the specific conditions, i.e., cell line, medium, and incubation times, used for the ligand screen. This optimization is done most easily by test transfections with CMX-β-galactosidase expression plasmid after incremental addition of 5 μl aliquots of 0.5 N NaOH or HCl to 1 ml samples of the 2× BES buffer. The 500 ml stock buffer is subsequently adjusted volumetrically based on the sample that gives the best results. The buffer is stable at room temperature for 6 months or can be frozen as 10 ml aliquots at −20°C.

2× BES pH 6.95 buffer

NaCl	8.18 g
BES	5.33 g
150 mM Na$_2$HPO$_4$	5.0 ml
(pH the solution carefully to 6.95)	
Distilled H$_2$O to	500 ml
Filter sterilize	

2× CaCl$_2$ solution

1 M Tris, pH 7.5	0.5 ml
0.5 M EDTA, pH 8.0	0.1 ml
CaCl$_2$·2H$_2$O	18.37 g
Distilled H$_2$O to	500 ml
Filter sterilize	

Plasmid DNAs are purified by alkaline lysis followed by double banding in CsCl density gradients and stored at 4°C as concentrated (>1 mg/ml) stocks in TE/10 buffer (10 mM Tris, pH 7.5, 0.1 mM EDTA). Sufficient plasmid DNA (2–5 mg) should be prepared to allow regular large-scale transfections over several months. Column-based purification methods may be used but these typically cost at least 10-fold more per mg of plasmid obtained than CsCl density-gradient centrifugation.

Transfection Protocol

1. COS-7 cells are grown to 80% confluency in 10-cm tissue culture petri dishes. Cells are removed with 0.01% trypsin/0.3 mM EDTA in PBS and pelleted by centrifugation at 1500 × g for 5 min.

2. Count and seed 96-well tissue culture plates at 5000 cells/well dispensed in 100 μl 10% FBS/DMEM.

3. Incubate cells for > 5 hr prior to transfection to allow cells to reattach. Cells should be at a density of 30–50% confluence.

4. Prepare calcium phosphate/DNA precipitate within the range of 10–20 μg DNA/ml calcium phosphate solution. The plasmid mix will include a GAL4-DBD/receptor-LBD fusion effector plasmid, a luciferase reporter plasmid containing GAL UAS promoter sequences [e.g., tk(MH100)$_4$-luc], a constitutively expressed β-galactosidase reporter control (e.g., CMX-β-gal), and carrier DNA. The optimal ratio of receptor effector plasmid to luciferase reporter should be determined by preliminary titration experiments. For receptors with known ligands, a complete dose–response curve will indicate conditions of maximum sensitivity and fold induction. Frequently, the optimum effector plasmid amount will lie in the range of 50–1000 ng/plate at a ratio between 1:5 and 1:10 of effector–reporter. We use typically a ratio of 1:5:5:4 of receptor–luciferase reporter–β-galactosidase control–carrier DNA, i.e., 1 μg pCMX-GAL4-RARα, 5 μg tk-(MH100)$_4$-luc, 5 μg pCMX-β-gal: 4 μg pBluescript for a total of 15 μg/96-well plate. The plasmid DNAs are mixed into 0.6 ml 2 $\times$ CaCl$_2$ solution and then added dropwise with vigorous vortexing to 0.6 ml 2 $\times$ BES buffer. A fine precipitate should form. Allow the precipitate to mature for 5 min, but not longer than 15 min, with occasional vortexing.

5. Mix with 10 ml prewarmed 10% FBS/DMEM medium and dispense 100 μl per well. A fine, even precipitate coating the cells should be visible after several minutes. A large clumpy precipitate or a sparse precipitate is an indication that the transfection conditions are not optimal.

6. Incubate plates for 5–24 hr at 37°C in a humidified incubator at 5% CO$_2$ in air.

7. After transfection, the medium is aspirated from each well and cells washed gently twice with 200 μl prewarmed PBS to remove precipitate and serum. Medium is replaced with 100 μl/well serum-free ITLB/DMEM or 10% charcoal-stripped FBS/DMEM.

8. Test fractions from HPLC separations are prepared as follows. Aliquots of 100–500 μl from each fraction are transferred to siliconized Eppendorf tubes and rotor evaporated unheated in a SpeedVac. Samples are resuspended in 8 μl of 100% ethanol or DMSO, mixed with 400 μl ITLB/DMEM medium (or other suitable medium), and 100 μl dispensed into each of triplicate wells of the transfected 96-well cell culture plate. When available,

a standard curve of a positive ligand control should be included on every 96-well plate.

9. Incubate plates for a further 24–48 hr at 37°C in a humidified 5% CO_2 incubator to allow for maximum ligand-dependent transcriptional responses. An initial test from 12 to 48 hr is suggested to verify the best incubation times as the final luciferase activity will reflect the integration of multiple factors. These include effects of ligand metabolism (including potential metabolic activation or breakdown), receptor protein levels, decay of basal luciferase levels after switching to minimal or defined medium.

10. Media is removed by careful aspiration and plates washed once with PBS.

11. 150 μl/well complete cell lysis buffer is added and plates shaken on a Titer Plate Shaker for 10–30 min.

Luciferase and β-Galactosidase Assays

Cell lysis buffer

1 M Tris–PO_4, pH 7.8	25 ml
Glycerol	150 ml
CHAPS	20 g
Phosphotidyl choline (lecithin from egg-yolk)	10 g
BSA	10 g
Distilled H_2O to	1000 ml

Stir the lecithin on a hot plate to 70°C in 100 ml deionized water until completely dissolved. Add the remaining components except BSA. BSA is dissolved separately in 50 ml distilled water and added when the temperature is below 50°C. Filter when warm through a 0.45 μm filter. Store at 4°C. The lysis solution should be a pale straw yellow color. Add the remaining reagents shortly before use.

Add fresh per 15 ml lysis solution

0.1 M EGTA, pH 8.0	600 μl
1 M $MgCl_2$	120 μl
1 M DTT	15 μl
0.2 M PMSF in methanol	30 μl

Luciferase assay reagent (per 10 ml)

200 mM Tricine–NaOH, pH 7.8	1 ml
Mg^{2+} stock solution	0.1 ml

0.5 M EDTA	2 μl
1 M DTT	0.3 ml
2.5 mM Coenzyme A	1 ml
20 mM ATP	0.5 ml
1 mM Luciferin	1 ml
Distilled H_2O to	10 ml

Mg^{2+} stock solution

$(MgCO_3)_4 \cdot Mg(OH)_2 \cdot 5H_2O$	10.394 g
$MgSO_4 \cdot 7H_2O$	13.162 g
Distilled H_2O to	160 ml
Stir until dissolved	

25 mM Coenzyme A stock solution
0.01 g in 5 ml H_2O
Dispense into aliquots and store at $-20°C$

100 mM ATP stock solution
0.11 g in 5 ml H_2O
Dispense into aliquots and store at $-20°C$

10 mM D-Luciferin in methanol
Protect from light, store at $-20°C$

β-Galactosidase assay buffer

β-Gal base solution	10 ml
β-Mercaptoethanol	30 μl
O-Nitrophenyl-β-galactopyranoside	10 mg

β-Gal base solution

$Na_2HPO_2 \cdot 12H_2O$	21.49 g
$NaH_2PO_4 \cdot 2H_2O$	6.24 g
KCl	0.75 g
1 M $MgCl_2$	1 ml
Distilled H_2O to	1000 ml

Autoclave and store at room temperature

1. For luciferase and β-galactosidase assays, 50 μl lysate aliquots are transferred to solid white luminometer and clear, flat-bottom 96-well plates.
2. Dispense 100 μl/well luciferase assay reagent to the samples in the luminometer plates, mix briefly on a microtiter plate shaker, then

read on a 96-well plate luminometer (e.g., Torcon R-7 or Dynex MLX) set to measure in cycle mode. The stabilized luciferase reaction has a half-life of approximately 5 min permitting quantitation of the entire plate.

3. For β-galactosidase activity determination, dispense 100 μl β-galactosidase assay buffer to each well, mix briefly on a microtiter plate shaker and incubate at 37°C (e.g., in a bacterial incubator). Follow the color development by periodic measurement on a 96-well plate spectrophotometer set at 405 nm. When the OD reaches 0.5–1.0 AU (15–60 min), stop the reaction by addition of 100 μl 1 M sodium bicarbonate, mix briefly and remeasure the OD. Note the time.

Luciferase and β-galactosidase data are exported to Excel spreadsheet templates for data analysis and graphing. Background values for luciferase and β-galactosidase activity are determined from mock-transfected wells and can be subtracted from the experimental values. Individual wells are expressed as luciferase values normalized to β-galactosidase activity and plotted as Relative Luciferase Units per O.D. 405 nm per minute. Triplicate assays are averaged and plotted as the mean $\pm$ S.E.M. Fold activation is expressed as the activation observed relative to the average for solvent only controls.

Structure Determination by Mass Spectrometry

After sequential rounds of HPLC purification and receptor activity-guided assays, the active fraction(s) should be sufficiently enriched for preliminary structural studies by mass spectrometry. Consultation with an experienced mass spectrometrist is strongly recommended in planning an analytical strategy and in the interpretation of data when attempting to elucidate the chemical structure of an unknown compound. While every situation presents unique challenges, most structure characterizations follow a general scheme as outlined below:

1. Determine complexity of active fraction and identify molecular ion of active component.
2. Obtain exact mass measurement and derive molecular formula.
3. Obtain electron-impact (EI) spectra for structural analysis and library search.
4. Compliment structural analysis with ^{1}H- and ^{13}C-NMR studies where appropriate. Mass spectrometry is unlikely to provide identification of structural isomers unless these can be chromatographically separated and compared with authentic standards.

The two biggest hurdles facing the investigator concern purity and quantity. Some of the most common mass spectrometry techniques outlined below utilize additional chromatographic separation with gradients of either solvent (LC) or temperature (GC). These methods can often provide sufficient resolution to obtain spectra on individual components in mixtures. Purification to homogeneity, as judged for instance by peak symmetry and peak purity plots (UV absorbance ratios) of chromatographic peaks, is therefore not strictly necessary but remains desirable if possible. The main benefit of a homogenous preparation is increased sensitivity. While mass spectrometry can be an exquisitely sensitive methodology for detection, structure determination of unknown compounds requires sufficient material for method development and collection of high-quality spectra. Quantities of 1 nmole per compound are usually adequate for gathering basic information, but complete structure determination may well require more, especially if compounds of interest require derivatization prior to analysis.

Exact Mass Measurement by ES-MS

Our strategy initially is to analyze unknown samples by electrospray ionization TOF mass spectrometry (ES-MS) in positive ion mode by flow-injection analysis on a Micromass LCT instrument. The method requires no special sample preparation, is compatible with most HPLC fractionation solvents, can analyze broad classes of organic compounds, and is sensitive in the submicromolar range. Compounds with one or more readily ionizable functional groups (e.g., carboxyl-, amino-, or hydroxyl) should give satisfactory spectra. Samples are prepared in a minimal volume of methanol, typically 100 μl in conical borosilicate glass autosampler tubes, and 50 μl injected directly from a Gilson 231XL autosampler into the methanol solvent stream (200 μl/min). Mass spectra between 100 and 1000 m/z are collected, as this range will likely contain the putative small lipophilic ligands of interest. Inclusion of a small amount of water and/or acid can help in ionization, but may be omitted. Solvent only and resin blanks are run prior to any samples to aid in identification and elimination of common solvent impurities and resin contaminants.

If the activity is sufficiently pure to give a clear indication of a candidate ligand, the exact mass of the compound is determined (measurements made within 5 ppm) and a molecular formula is derived. Unless indicated otherwise by the distribution and intensities of isotopic peaks, we constrain molecular formula searches to the common organic elements C, H, O, N, S, and P. In the absence of acid, sodiated ions (M + 23) are frequently the predominant species, so Na should also be included.

For more complex spectra containing multiple substances, or peaks of weak ion intensities with limited quantities of sample, we have successfully obtained simultaneous exact mass measurements of multiple peaks by calibrating the spectra with several internal standards. Caution should be applied in assignment of mass peaks to HPLC chromatographic data simply on the basis of comparing ion intensity with UV absorbance since in electrospray the ionization of a compound is determined by its functional groups and chemical nature. Therefore, the detection limits of different compounds vary substantially in a manner different from their optical properties. In addition, ionization of impurities may competitively inhibit the detection of sample components when injected in the direct ES-MS mode, thus making interpretation of absolute ion intensities between fractions only semi-quantitative. Nevertheless, a comparison of mass spectra from adjacent HPLC fractions can be useful in eliminating those peaks that do not follow the activity profile and in prioritizing candidate compounds. If available, an in-line narrow bore HPLC system coupled to ES-MS is ideal for maintaining sensitivity, subtraction of solvent, and contaminant background spectra and when attempting to match mass peaks to specific UV chromatographic peaks. We use a Micromass Q-TOF2 instrument with an Agilent 1100 series HPLC system for these purposes. This mass spectrometer has the added benefit of providing structural information when operated in the MS-MS mode that can be compared with putative synthetic standards of the candidate ligands.

Electron-Impact Mass Spectrometry

Electron-impact (EI) fragmentation spectra are highly reproducible and predictive with specific functional groups and substructures yielding distinctive fragmentation patterns. Therefore, the ability to generate such spectra on the target compounds can be highly beneficial. EI spectra can be screened against mass spectral databases, e.g., Wiley, or used to reconstruct a compound's substructures by looking for the sequential loss of probable fragmentation ions from the molecular ion. In addition, analysis by gas chromatography-mass spectrometry (GC-MS) allows collection of high-quality spectra of individual components from mixtures of compounds. Our primary instrument for these purposes is a Thermo-Finnegan Trace MS system. Samples are initially captured onto a DB-5 fused silica capillary column and components are sequentially volatized with a temperature gradient (linear from 50 to 290°C at 10°C/min, hold at 290°C for 20 min). Complex mixtures of closely related compounds can be resolved by retention time, and the method provides for an additional principle of separation distinct from solvent elution times from HPLC columns. Prior to analysis, HPLC fractions must undergo solvent exchange into a nonpolar solvent, e.g., hexane or methylene chloride, to avoid undue damage to the GC column. Samples are evaporated under a stream of dry nitrogen or rotor evaporated, resuspended in 100–500 μl of dry solvent, and any residual water removed by drying the sample over anhydrous sodium sulfate. For very dilute samples, the solvent volume can be reduced further. Samples dissolved in as little as 10 μl can be used with conical autosampler vials. GC analysis is particularly well suited for the analysis of small volatile lipophilic molecules, but does have important limitations with respect to certain types of natural product characterizations. Thermal instability and excessive fragmentation may prevent collection of useful spectra. In addition, natural products containing multiple polar groups, e.g., those with more than one hydroxyl group, may be difficult to volatize. Carboxylic acids are refractory generally to analysis unless derivatized. Labile hydroxyl groups should be derivatized to form, for example, trimethylsilyl (TMS) ethers or other suitable groups to improve volatility, thermal stability, and fragmentation characteristics. Since derivatization is likely to destroy biological activity, it should be the last step in the analytical chain of purification. Numerous derivatization protocols can be found in any good analytical text.[13]

Concluding Remarks

The recent identification of orphan receptor ligands has increased greatly our understanding of such important processes as cholesterol and bile acid metabolism and the xenobiotic response. Most of the remaining orphan receptors have apparent ligand-binding pockets, hence it is likely that there are several ligand-binding pockets with endogenous ligands. If past trends continue, they will lead to surprising new insights into the homeostatic regulation of cellular biochemistry and novel signaling pathways. Perhaps, a perceived lack of expertise underlies the reluctance of molecular, cellular, and developmental biologists to venture into unfamiliar scientific disciplines, such as natural product isolation and structure determination. However, as access to easy-to-use advanced analytical instrumentation (e.g., mass spectrometry) becomes more commonplace, so too should the desire to incorporate these methods into the standard repertoire of techniques available within any modern biology laboratory. We hope that the strategies outlined above serve to illustrate that ligand identification using activity-guided purification is within the reach of most laboratories interested in fully characterizing nuclear hormone receptor biology.

[13] D. R. Knapp, "Handbook of Analytical Derivatization Reactions," 1st Ed. Wiley-Interscience, John Wiley & Sons, Inc., New York, 1979.

Acknowledgment

We thank Drs. Jochen Buck, Fadila Derguini, William Fenical, Heonjoong Kang, and John Greaves for past advice on natural product extraction, purification, and characterization. Research on receptor ligand identification in the author's laboratory was supported by grants from the US EPA (STAR G9D1 0090) and the NIH (GM60572).

[2] Quantitation of Receptor Ligands by Mass Spectrometry

By Erik G. Lund and Ulf Diczfalusy

Biological Significance and Available Methods

The human genome contains approximately 50 members of the nuclear receptor family.[1] It is now clear that members of this class of proteins serve as important signal transducers, responding to the appearance of signaling molecules with an increase or decrease in the cellular expression of target genes. This behavior is typified by the classic steroid hormone receptors, which were among the first nuclear receptors to be identified and characterized.

Nuclear receptors can be subdivided into several classes, and according to one subdivision they may be classified as either being endocrine receptors, adopted orphan receptors, or orphan receptors.[2] The first group consists of well-characterized endocrine hormone receptors such as the vitamin A and D receptors, and the steroid and thyroid hormone receptors. These proteins typically have very high-affinity specific ligands. The second group consists of receptors for dietary or endogenous ligands present at higher concentration, which often bind several members of a class of molecules as ligands. This group is exemplified by the PPARα/δ/γ receptors, which bind fatty acids and their derivatives, the LXRα/β receptors, which bind oxysterols, and FXR, which interacts with bile acids. The third group consists of receptors without identified ligands, and it is possible that many receptors of this class actually do not have endogenous ligands. With the identification of ligands for many different nuclear receptors, a pattern can be recognized in that they are all generally relatively small, lipophilic compounds, reflecting the necessity for extracellular ligands to cross the plasma membrane for access to the receptor. This characteristic distinguishes nuclear receptors from cell surface receptors, which often have proteins or peptides as natural ligands.

[1] J. M. Maglich, A. Sluder, X. Guan, Y. Shi, D. D. McKee, K. Carrick, K. Kamdar, T. M. Willson, and J. T. Moore, *Genome Biol.* **2**, 1 (2001).
[2] A. Chawla, J. J. Repa, R. M. Evans, and D. J. Mangelsdorf, *Science* **294**, 1866 (2001).

Given the relative homogeneity of the characteristics of nuclear receptor ligands, similar methodologies can be applied to the analysis of many molecules in this group. We describe methods for the determination of oxysterols, LXR ligands, by gas chromatography-mass spectrometry (GC-MS). Two of the methods include an optional saponification step and purification on a disposable column prior to GC-MS analysis. GC-MS is the method of choice for detection because of its specificity and sensitivity, which enables the determination of compounds in complex mixtures over a large concentration range. In addition, relatively inexpensive and easy-to-maintain benchtop instrumentation is available. Analysis by GC requires the analyte to be vaporized without decomposing and this is generally the case for nuclear receptor ligands. HPLC allows for greater versatility but in our experience even HPLC-MS-MS does not have sufficient sensitivity for the determination of oxysterols in biological samples.

Content of Oxysterols in Tissues

The concentration of oxysterols in tissues generally increases with increasing cholesterol concentration. As treatments with diets or drugs may influence the cholesterol content of tissues, the concentration of an oxysterol in a tissue is preferably expressed as a ratio of oxysterol to cholesterol concentration. The ratio of a specific oxysterol to cholesterol may vary within wide limits between different tissues. For example, the ratio of 27-hydroxycholesterol to cholesterol is reported to be between 100 and 400 ng/mg in tissues like liver, heart, spleen, and kidney, while the corresponding ratio in lung tissue is 600[3] and may reach 18,000 in atherosclerotic plaques (Table I). Likewise, the ratio of 24-hydroxycholesterol to cholesterol in liver, spleen, and kidney is less than 100 ng/mg, while the ratio in brain is between 500 and 2000 (Table I).[4] Typical concentrations of selected oxysterols in human plasma from normolipidemic volunteers are also shown in Table I. The plasma concentrations of 24- and 27-hydroxycholesterol increase with plasma cholesterol concentration.[5–7] A pronounced age-dependency of plasma 24-hydroxycholesterol

[3] A. Babiker, O. Andersson, D. Lindblom, J. van der Linden, B. Wiklund, D. Lütjohann, U. Diczfalusy, and I. Björkhem, *J. Lipid Res.* **40**, 1417 (1999).

[4] D. Lütjohann, O. Breuer, G. Ahlborg, I. Nennesmo, Å. Sidén, U. Diczfalusy, and I. Björkhem, *Proc. Natl. Acad. Sci. USA* **93**, 9799 (1996).

[5] I. Björkhem, D. Lütjohann, U. Diczfalusy, L. Ståhle, G. Ahlborg, and J. Wahren, *J. Lipid Res.* **39**, 1594 (1998).

[6] R. Harik-Khan and R. P. Holmes, *J. Steroid Biochem.* **36**, 351 (1990).

[7] L. Bretillon, D. Lütjohann, L. Ståhle, T. Widhe, L. Bindl, G. Eggertsen, U. Diczfalusy, and I. Björkhem, *J. Lipid Res.* **41**, 840 (2000).

TABLE I

CONCENTRATIONS OF SELECTED OXYSTEROLS IN HUMAN TISSUES

Oxysterol	Plasma (nmol/liter)	Liver[a]	Brain[a]	Atherosclerotic plaques[a]
4β-Hydroxycholesterol	72[b]	21[b]	165[b]	150[i]
7α-Hydroxycholesterol	107[c]	806[d]	n.d.	1440[j]
24-Hydroxycholesterol	191[e]	90[f]	2190[f]	410[j]
25-Hydroxycholesterol	5[c]	n.d.	n.d.	760[j]
27-Hydroxycholesterol	382[c]	390[g]	130[f]	18,800[j]
3β-Hydroxy-5-cholestenoic acid	283[h]	n.d.	n.d.	n.d.

n.d., not determined.

[a] ng oxysterol/mg cholesterol.

[b] K. Bodin, L. Bretillon, Y. Aden, L. Bertilsson, U. Broomé, C. Einarsson, and U. Diczfalusy, *J. Biol. Chem.* **276**, 38685 (2001).

[c] S. Dzeletovic, O. Breuer, E. Lund, and U. Diczfalusy, *Anal. Biochem.* **225**, 73 (1995).

[d] A. Honda, T. Yoshida, N. Tanaka, Y. Matsuzaki, B. He, J. Shoda, and T. Osuga, *J. Gastroenterol.* **30**, 651 (1995).

[e] L. Bretillon, D. Lütjohann, L. Ståhle, T. Widhe, L. Bindl, G. Eggertsen, U. Diczfalusy, and I. Björkhem, *J. Lipid Res.* **41**, 840 (2000).

[f] D. Lütjohann, O. Breuer, G. Ahlborg, I. Nennesmo, Å. Sidén, U. Diczfalusy, and I. Björkhem, *Proc. Natl. Acad. Sci. USA* **93**, 9799 (1996).

[g] A. Babiker, O. Andersson, D. Lindblom, J. Van der Linden, B. Wiklund, D. Lütjohann, U. Diczfalusy, and I. Björkhem, *J. Lipid Res.* **40**, 1417 (1999).

[h] A. Babiker and U. Diczfalusy, *Biochim. Biophys. Acta* **1392**, 333 (1998).

[i] Calculated from O. Breuer, S. Dzeletovic, E. Lund, and U. Diczfalusy, *Biochim. Biophys. Acta* **1302**, 145 (1996).

[j] Calculated from M. Crisby, J. Nilsson, V. Kostulas, I. Björkhem, and U. Diczfalusy, *Biochim. Biophys. Acta* **1344**, 278 (1997).

concentrations is observed, with higher concentrations during the first few years of life and lower concentrations after puberty.[4] This age-dependency reflects the balance between cerebral production and metabolism of 24-hydroxycholesterol in the liver.[7] While most oxysterols do not show gender differences, the plasma concentration of 27-hydroxycholesterol has been reported to be higher in males (444 nmol/liter) than in females (325 nmol/liter).[8] Several plasma oxysterols are present mainly as esters of long-chain fatty acids[9]; and 4β-hydroxycholesterol, 7α-hydroxycholesterol, 24(S)-hydroxycholesterol, and 27-hydroxycholesterol are all esterified to more than 70%,[8,10] while 3β-hydroxy-5-cholestenoic acid is not present

[8] S. Dzeletovic, O. Breuer, E. Lund, and U. Diczfalusy, *Anal. Biochem.* **225**, 73 (1995).

[9] L. L. Smith, J. I. Teng, Y. Y. Lin, P. K. Seitz, and M. F. McGehee, *J. Steroid Biochem.* **14**, 889 (1981).

[10] K. Bodin, L. Bretillon, Y. Aden, L. Bertilsson, U. Broomé, C. Einarsson, and U. Diczfalusy, *J. Biol. Chem.* **276**, 38685 (2001).

TABLE II

CONCENTRATIONS OF SELECTED OXYSTEROLS IN MOUSE TISSUES

	Plasma (nmol/liter)	Liver[a]	Brain[a]
4β-Hydroxycholesterol	200[b]	n.d.	40[b]
24-Hydroxycholesterol	55[b]	70[b]	2800[b]
25-Hydroxycholesterol	~10[c]	<100[c]	n.d.
27-Hydroxycholesterol	200[b]	17[b]	19[b]

n.d., not determined.
[a]ng oxysterol/mg cholesterol.
[b]Unpublished data [U. Diczfalusy (2001)].
[c]Adapted from J. Li-Hawkins, E. G. Lund, S. D. Turley, and D. W. Russell, *J. Biol. Chem.* **275**, 16536 (2000).

in an esterified form.[11] Pronounced species differences in tissue oxysterol concentrations exist. The level of 27-hydroxycholesterol in rabbit plasma is typically 7.5 nmol/liter[12] compared to 382 nmol/liter in human plasma.[8] Concentrations of selected oxysterols in mouse tissues are shown in Table II.

Methodology: Internal Standards

A distinct advantage of using mass spectrometry for quantitative measurements of oxysterols is the feasibility of using deuterium labeled oxysterols as internal standards; ideally, one internal standard for each oxysterol to be analyzed. The deuterium labeled internal standard, e.g., [^{2}H$_6$]4β-hydroxycholesterol for the determination of 4β-hydroxycholesterol is added to the biological sample prior to sample preparation and analysis. Due to its near identical physicochemical properties to the natural oxysterol, it can be assumed that the internal deuterated standard will be degraded or lost to the same extent as the analyte. This behavior is particularly important for accurate quantitation of oxysterols in biological samples, since a chromatographic step for the separation of oxysterols from the bulk of cholesterol is required. In the final step of the procedure, the purified sample is analyzed by GC-MS. Specific ions corresponding to the labeled and unlabeled 4β-hydroxycholesterol, respectively, are monitored and displayed as separate gas chromatograms. The ratio of the heights of the peaks for 4β-hydroxycholesterol and [^{2}H$_6$]4β-hydroxycholesterol is then used for calculation of the amount of 4β-hydroxycholesterol present in the original sample by interpolation against a standard curve.

[11] A. Babiker and U. Diczfalusy, *Biochim. Biophys. Acta* **1392**, 333 (1998).
[12] M. Crisby, J. Nilsson, V. Kostulas, I. Björkhem, and U. Diczfalusy, *Biochim. Biophys. Acta* **1344**, 278 (1997).

Ideally, a deuterium labeled analogue of each analyte is used as internal standard; however, such compounds are generally not commercially available and must be synthesized. In the method described below, a mixture of several deuterated oxysterols is added as internal standard. The synthesis of these compounds is described in Dzeletovic *et al.*[8] and references therein. In the absence of individual deuterated standards, one nonendogenous oxysterol may be used as an internal standard for all analytes. Such a simplification will, however, make the assay sensitive to random variation in treatment between samples. A better compromise is to use a mixture of all or some of deuterated 4β-hydroxycholesterol, 7β-hydroxycholesterol, 7-oxocholesterol, 5,6-epoxycholesterol, and 25-hydroxycholesterol standards for measurement of 4β-, 7α- and 7β-hydroxycholesterol, 7-oxocholesterol, cholesterol-5(α or β),6-epoxide, and side-chain oxygenated oxysterols, respectively. Synthesis of these five internal standards is relatively straightforward and is described below. [^{2}H$_6$]7α-Hydroxycholesterol and [^{2}H$_6$]7β-hydroxycholesterol are also available commercially (C/D/N Isotopes, Pointe Claire, Quebec, Canada; http://www.cdniso.com). If side-chain oxygenated oxysterols are to be determined, only deuterated 25-hydroxycholesterol is required as a standard. As an alternative of last resort, 6α-hydroxycholestanol is a nonendogenous oxysterol that can be obtained from Steraloids (Cat. No. C3750-000), and could be used as an internal standard for all analytes of interest.

Synthesis of [^{2}H$_6$]4β-Hydroxycholesterol

Dissolve 25 mg selenium dioxide in 20 μl of water. Heat until completely dissolved. Add 500 μl glacial acetic acid prewarmed to 80°C. Quickly add 50 mg [^{2}H$_6$]cholesterol[13] in 250 μl toluene (prewarmed to 80°C). Incubate for 1 hr at 80°C. Add 100 mg sodium acetate and heat for a few minutes. Let the mixture incubate for 10 min. Transfer the mixture to a separatory funnel. Rinse the tube with toluene and diethyl ether. Add 30–40 ml diethyl ether and 10 ml 0.9% NaCl. Shake well. Discard the water phase. Check the pH, which should be neutral. Wash the ether phase twice with 10 ml 0.9% NaCl. Evaporate the ether extract and dissolve the residue

---

[13] Deuterated cholesterol can be obtained from Aldrich ([2,2,3,4,4,6-^{2}H$_6$]cholesterol, Cat. No. 48,857); C/D/N Isotopes ([2,2,3,4,4,6-^{2}H$_6$]cholesterol, [26,27-^{2}H$_6$]cholesterol, and [25,26, 27-^{2}H$_7$]cholesterol; http://www.cdniso.com); or Medical Isotopes ([2,2,3,4,4,6-^{2}H$_6$]cholesterol and [26,27-^{2}H$_6$]cholesterol; http://www.medicalisotopes.com). While all of these compounds could be used for synthesis of deuterated oxysterols [2,2,3,4,4,6-^{2}H$_6$]cholesterol is likely to lose one or two deuterium atoms during synthetic conversion to oxysterols and mass spectrometric fragmentation. We have used [26,27-^{2}H$_6$]cholesterol for our synthesis but [25,26,27-^{2}H$_7$]cholesterol is equally feasible.

in ether–ethanol (9:1, v/v). Purify the product by TLC (toluene–ethyl acetate, 3:7, v/v). Visualize the $[^2H_6]4\beta$-hydroxycholesterol with iodine vapor and circle the band with a pencil. Let the iodine evaporate and scrape off the band using a razor blade. Extract the silica three times by vortexing vigorously with 0.5 ml methanol and pellet the silica by low-speed centrifugation. Combine the extracts and remove any residual silica by passing the solution through filter paper. Rinse the filter paper with more methanol, combine with the original extract and evaporate under a gentle stream of an inert gas (nitrogen or argon).

Synthesis of $[^2H_6]$Cholesterol-5α,6α-epoxide

Dissolve 10 mg of $[^2H_6]$cholesterol[13] in 5 ml chloroform at room temperature under stirring. Add 15 mg of 3-chloroperoxybenzoic acid (*meta*-chloroperbenzoic acid, MCPBA; Aldrich Cat. No. 27,303-1, $\sim 75\%$ pure) and continue stirring for 2 hr. Wash three times with 5-ml aliquots of saturated sodium bicarbonate (*Caution*: carbon dioxide formation may cause elevated pressure), and then with water until neutral. Separate the organic and aqueous phases carefully and evaporate the organic phase under a gentle stream of an inert gas (nitrogen or argon). Dissolve the residue in a small amount of toluene and purify the compound by preparative TLC, using toluene–ethyl acetate, 3:7 (v/v) as the mobile phase. The TLC purification is particularly important in this case since residual MCPBA in the internal standard will oxidize any cholesterol present in the biological sample to be analyzed.

Synthesis of $[^2H_6]7\beta$-Hydroxycholesterol and $[^2H_6]7$-Oxocholesterol

Dissolve 10 mg $[^2H_6]$cholesterol in 0.5 ml acetic anhydride. Boil under reflux for 1 hr. Add 400 μl acetic acid and 220 mg sodium acetate. Adjust temperature to 58°C. Slowly add 11 mg CrO_3 and let the mixture react at 58°C for 4.5 hr. Cool the mixture to 25°C and transfer to a separatory funnel. Add 5% $NaHCO_3$ (in water) until a basic pH is reached. Extract the solution with diethyl ether. Wash the ether extract with water until a neutral pH is obtained. Purify the product (7-oxocholesterol acetate) by TLC (mobile phase: toluene–ethyl acetate 9:1, v/v) as above. For synthesis of $[^2H_6]7$-oxocholesterol, saponify the purified acetate by stirring in 0.5 M sodium hydroxide in water–ethanol (2:8, v/v) overnight. Neutralize the mixture with hydrochloric acid, extract the $[^2H_6]7$-oxocholesterol with diethyl ether, and purify by TLC (toluene–ethyl acetate 3:7, v/v). For synthesis of $[^2H_6]7\beta$-hydroxycholesterol, dissolve 10 mg of the purified acetate in 5 ml anhydrous ether with continuous stirring. Add 10 mg lithium aluminum hydride (*Caution*: reacts violently with water) and stir the suspension for 1 hr. Destroy excess lithium aluminum hydride by slowly adding 5 ml water-saturated diethyl ether, and then 1 ml ethanol. Acidify the solution with hydrochloric acid, extract with diethyl ether, wash with water and purify by preparative TLC using silica plates and toluene–ethyl acetate (3:7, v/v). This procedure will generate a 1:9 mixture of 7α- and 7β-hydroxycholesterol.

Synthesis of [^{2}H$_3$]25-Hydroxycholesterol

Prepare a solution of [^{2}H$_3$]methyl magnesium iodide in diethyl ether at room temperature by mixing 100 μl of [^{2}H$_3$]iodomethane (Sigma 17,603-6) with approximately 50 mg of magnesium powder and one very small crystal of iodine in 4 ml of anhydrous diethyl ether with continuous stirring and with a cooling condenser attached. The reaction will commence spontaneously after a few minutes and is likely to generate enough heat to bring the ether to boil. Stir the solution under refluxing condition until the boiling stops. Then, under continued stirring, add dropwise a solution of 50 mg 3β-acetoxy-27-nor-cholest-5-en-25-one (Cat. No. 1125-3, Research Plus, Bayonne, NJ) dissolved in 2 ml of anhydrous diethyl ether (*Caution*: this reaction mixture will react violently with water). After 15 min of stirring, terminate the reaction with the dropwise addition of 2 ml water-saturated diethyl ether followed by 1 ml ethanol. After extraction with diethyl ether and extensive washing with water, purify the [^{2}H$_3$]25-hydroxycholesterol by preparative silica gel TLC using toluene–ethyl acetate (3:7, v/v) as the mobile phase.

All synthesized internal standards are best quantitated by GC using a flame ionization detector. The response factor is determined by GC analysis of the corresponding unlabeled oxysterols. These can be obtained from several sources, including Sigma, Steraloids (http://www.steraloids.com), and Research Plus (http://www.researchplus.com).

Sample Preparation

Oxysterols are present together with cholesterol in tissues and the molar ratio of cholesterol to oxysterol is often 10^4–10^6. This large excess of cholesterol may cause problems during sample preparation since cholesterol is easily autoxidized to numerous different oxysterols. It is especially important to pay attention to this problem when analyzing oxysterols that may be formed both by nonenzymatic and enzymatic mechanisms, e.g., 4β-hydroxycholesterol, 7α-hydroxycholesterol, and 25-hydroxycholesterol. One way to minimize the spontaneous oxidation problem is to separate oxysterols from cholesterol early during the sample preparation procedure. Antioxidants and metal chelators also may be added to the sample to minimize nonenzymatic cholesterol oxidation during work-up. Since oxysterols are present as fatty acyl esters in many tissues, a saponification step is usually included in the sample preparation procedure. The conditions of this hydrolytic step should be carefully chosen to obtain a high yield and to avoid artifacts such as decomposition of oxysterols.[14] Side-chain oxidized oxysterols, including 22-, 24-, and 27-hydroxycholesterol, have been identified as sulfate esters in feces from infants,[15] and 24-hydroxycholesterol was identified as a 3-sulfate-24-glucuronide in serum and urine from patients with cholestatic liver disease.[16] Analysis of such conjugates requires an acid solvolysis step and treatment with glucuronidase to obtain the free steroid.[17] Before final GC-MS analysis of oxysterols, carboxylic acid groups are converted into methyl esters and hydroxyl groups are derivatized to trimethylsilyl ethers or *tert*-butyldimethylsilyl ethers.

Procedures

Below are described three methods that we have developed and used for determination of oxysterols in plasma and tissues. Two of the methods are based on isotope-dilution mass spectrometry, utilizing deuterium labeled internal standards. The third method utilizes a homolog structure lacking one methylene unit as an internal standard.

Determination of Oxysterols in Human Plasma[8]

Oxysterols are present in plasma both in free form and esterified to long-chain fatty acids. This method determines total plasma oxysterols (free + esterified), but may be used to determine free oxysterols simply by omitting the alkaline hydrolysis (saponification) step. For determination of tissue oxysterols, we recommend making a homogenate of the tissue and extracting aliquots of this homogenate with chloroform–methanol (2:1, v/v; 10 ml/400 mg tissue). Add the internal standard at the time of extraction and not at the homogenization step to avoid nonspecific adsorption.

The following nine oxysterols are determined: 7α- and 7β-hydroxycholesterol, 7-oxocholesterol, cholesterol-5α,6α-epoxide, cholesterol-5β,6β-epoxide, cholestane-3β,5α,6β-triol, 24-, 25-, and 27-hydroxycholesterol. To 1 ml of plasma (prepared in EDTA, 4.7 mM, potassium salt), a mixture of deuterium labeled internal standards is added containing 200 ng each of

[14] P. W. Park, F. Guardiola, S. H. Park, and P. B. Addis, *J. Am. Oil Chem. Soc.* **73**, 623 (1996).
[15] J.-Å. Gustafsson and J. Sjövall, *Eur. J. Biochem.* **8**, 467 (1969).
[16] L. J. Meng, W. J. Griffiths, H. Nazer, Y. Yang, and J. Sjövall, *J. Lipid Res.* **38**, 926 (1997).
[17] I. Björkhem, U. Andersson, E. Ellis, G. Alvelius, L. Ellegård, U. Diczfalusy, J. Sjövall, and C. Einarsson, *J. Biol. Chem.* **276**, 37004 (2001).

[^{2}H$_6$]7α-hydroxycholesterol, [^{2}H$_6$]7β-hydroxycholesterol, [^{2}H$_6$]7-oxocholesterol, [^{2}H$_6$]cholesterol-5α,6α-epoxide, [^{2}H$_6$]cholestane-triol, [^{2}H$_3$]24-hydroxycholesterol, [^{2}H$_3$]25-hydroxycholesterol, and [^{2}H$_4$]27-hydroxycholesterol. Add 50 μg of the antioxidant butylated hydroxytoluene (BHT) to each sample. Hydrolyze oxysterol esters by adding 10 ml 0.35 M ethanolic potassium hydroxide and incubating with stirring under an argon atmosphere for 2 hr at room temperature. Adjust the pH to 7 by adding 30 μl of phosphoric acid, and extract the mixture with 18 ml chloroform + 6 ml 0.9% (w/v) NaCl solution. Evaporate the organic phase and dissolve the residue in 1 ml toluene. Cholesterol is subsequently separated from oxysterols by solid-phase extraction. Apply the sample in 1 ml toluene to a 100 mg Isolute silica cartridge (International Sorbent Technology, Mid Glamorgan, UK), which has been conditioned with 2 ml hexane. Wash the cartridge with 1 ml hexane. Elute cholesterol with 8 ml 0.5% 2-propanol in hexane and discard. Elute oxysterols with 5 ml 30% 2-propanol in hexane. Evaporate the solvent and convert oxysterols to trimethylsilyl ethers by adding 350 μl pyridine–hexamethyldisilazane–trimethylchlorosilane (3:2:1, v/v/v), and incubating at 60°C for 30 min. Evaporate the solvent and dissolve the residue in 100 μl hexane. Analyze by GC-MS using an HP-5MS capillary column (30 m × 0.25 mm, 0.25 μm phase thickness). The temperature program for the gas chromatograph is as follows: 180°C for 1 min, 20°/min to 250°C, 5°/min to 300°C, 300°C for 8 min. Flow rate: 0.8 ml helium/min. Injector (splitless mode) and transfer line are kept at 270°C and 280°C, respectively. The mass selective detector is run in the selective ion monitoring (SIM) mode using ions m/z 456 (7α-, 7β-, and 27-hydroxycholesterol), 474 (cholesterol-5α,6α-epoxide and cholesterol-5β,6β-epoxide), 546 (cholestane-3β,5α,6β-triol), 413 (24-hydroxycholesterol), 131 (25-hydroxycholesterol), 472 (7-oxocholesterol) for the analytes, and m/z 462, 460 ([^{2}H$_4$]27-hydroxycholesterol), 480, 552, 416, 134, 478 for the corresponding deuterium labeled internal standards.

Determination of 3β-Hydroxy-5-cholestenoic Acid in Plasma[11,18]

The sterol 3β-hydroxy-5-cholestenoic acid is present entirely as the free acid in plasma, and thus no saponification step is necessary. Add 200 ng norcholestenoic acid (internal standard; the synthesis of this compound and of cholestenoic acid is described in Lund *et al.*;[19] cholenic acid (Sigma C2650) is an acceptable substitute for norcholestenoic acid) to 1 ml of plasma and

[18] I. Björkhem, O. Andersson, U. Diczfalusy, B. Sevastik, R. Xiu, C. Duan, and E. Lund, *Proc. Natl. Acad. Sci. USA* **91**, 8592 (1994).

[19] E. Lund, O. Andersson, J. Zhang, A. Babiker, G. Ahlborg, U. Diczfalusy, K. Einarsson, J. Sjövall, and I. Björkhem, *Arterioscler. Thromb. Vasc. Biol.* **16**, 208 (1996).

extract with methanol–chloroform (2:1, v/v). Evaporate the solvent and dissolve the residue in 0.5 ml chloroform. Apply the sample to a Bond Elut NH2 cartridge (Varian Sample Preparation Products, Harbour City, CA). Elute neutral lipids with 4 ml chloroform–isopropanol (2:1, v/v) and discard the eluate. Elute 3β-hydroxy-5-cholestenoic acid with 4 ml 2% acetic acid in diethyl ether, evaporate the solvent, and convert into methyl ester by treatment with 2,2-dimethoxypropane (0.7 ml) in acid methanol (1 ml, 11 μl HCl) for 20 min at 55°C. Convert hydroxyl groups to trimethylsilyl ethers (see above) and dissolve in 100 μl hexane. Analyze by GC-MS using a HP-5MS capillary column (30 m × 0.25 mm, 0.25 μm phase thickness). Temperature program: 180°C for 1 min, 35°/min to 270°C, 20°/min to 300°C, 300°C for 20 min. The sum of the 488, 398, and 359 ions (m/z) is recorded for the internal standard norcholestenoic acid, and the sum of the 502, 412, and 373 ions (m/z) is calculated for 3β-hydroxy-5-cholestenoic acid.

Determination of 4β-Hydroxycholesterol in Plasma[10]

Add 100 ng [^{2}H$_6$]4β-hydroxycholesterol (internal standard) and 10 μg BHT to 1 ml plasma. Saponification and extraction are done exactly as described above (see section entitled "Determination of oxysterols in human plasma"). The oxysterol fraction is derivatized to *tert*-butyldimethylsilyl ether by treatment with 100 μl *tert*-butyldimethylsilyl-dimethylformamide (Supelco Inc., Bellafonte, PA) at 22°C overnight. Add 1 ml water and extract twice with 1 ml ethyl acetate. Evaporate the solvent and dissolve the residue in 100 μl hexane. (A higher sensitivity is obtained when 4β-hydroxycholesterol is derivatized to the *tert*-butyldimethylsilyl ether derivative compared to the trimethylsilyl ether derivative.) Analyze the sample by GC-MS using a HP-5MS capillary column (30 m × 0.25 mm, 0.25 μm phase thickness. Temperature program: 180°C for 1 min, 35°C/min to 270°C, 20°C/min to 310°C, 310°C for 17 min. Flow rate: 0.8 ml helium/min. Injector (splitless mode) and transfer line temperature: 270°C. Use the mass spectrometer in the SIM mode, and record ions at (m/z) 573 (4β-hydroxycholesterol) and 579 ([^{2}H$_6$]4β-hydroxycholesterol).

Common Procedures for the Quantitative Determination of Different Oxysterols

The ratio of the peak heights of analyte to deuterated standard in combination with a standard curve is used to determine the amount of oxysterol in the original sample. Here we use 4β-hydroxycholesterol as an example but the procedure is representative for the different oxysterols. In Fig. 1, typical ion chromatograms for [^{2}H$_6$]4β-hydroxycholesterol (m/z 579) and 4β-hydroxycholesterol (m/z 573) obtained from analysis of a plasma

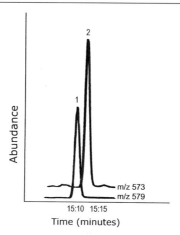

FIG. 1. Ion chromatograms from analysis of a plasma sample. Shown are 1, peak for the internal standard, [^{2}H$_6$]4β-hydroxycholesterol (m/z 579); 2, peak for the analyte, 4β-hydroxycholesterol (m/z 573).

sample are shown. The ratio of the peak heights of the unlabeled to deuterated compound is calculated, and with the help of a standard curve (Fig. 2) the amount of oxysterol in the original sample is determined. The standard curve is obtained by analyzing mixtures of varying amounts of unlabeled 4β-hydroxycholesterol with a constant amount of [^{2}H$_6$]4β-hydroxycholesterol, preferably the same amount that was added to the biological sample. It is not necessary to run the standards through the chromatography step before analysis; however, it is advisable to make a new standard curve for each batch of analyses. Typically the standards are run prior to the samples, and both the plotting of the curve and the subsequent analysis of samples is done automatically by the chromatography software.

A compound of particular interest with respect to LXR activation is 24(S),25-epoxycholesterol, which is postulated to be formed by a shunt in the cholesterol biosynthetic pathway,[20] and which is a potent activator of LXR.[21] When the GC injector is used in a splitless mode, this compound partly decomposes, giving rise to several distinct peaks. This decomposition does not seem to happen when a split injector is used; however, the sensitivity is reduced several or more fold. We have attempted to analyze 24(S),25-epoxycholesterol with the use of [^{2}H$_6$]24(S),25-epoxycholesterol as an internal standard. Despite obtaining a signal for the deuterated internal standard, we have been unable to detect unlabeled

[20] J. A. Nelson, S. R. Steckbeck, and T. A. Spencer, J. Am. Chem. Soc. **103**, 6974 (1981).
[21] B. A. Janowski, P. J. Willy, T. R. Devi, J. R. Falck, and D. J. Mangelsdorf, Nature **383**, 728 (1996).

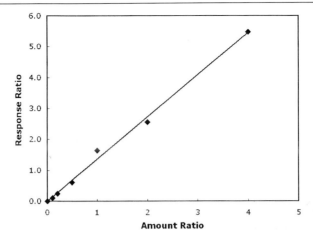

FIG. 2. Standard curve for 4β-hydroxycholesterol constructed from analysis of seven different standard points. Each standard point was created by mixing increasing amounts of the analyte, $[^2H_0]4\beta$-hydroxycholesterol, with a fixed amount of the internal standard, $[^2H_6]4\beta$-hydroxycholesterol in ratios from 0 to 4 (analyte/internal standard). The standard points were analyzed by GC-MS as described in section "Procedures". The ratio between peak height for the analyte and peak height for the internal standard (the response ratio) is plotted against the ratio between the amounts of analyte and internal standard (the amount ratio).

$24(S),25$-epoxycholesterol in multiple biological samples, including those from mouse and human liver, human plasma, and human monocyte-derived macrophages. At present we cannot state with certainty whether this failure was due to the absence of detectable quantities of $24(S),25$-epoxycholesterol in the analyzed samples, or due to methodological problems such as thermal decomposition. Another explanation for this discrepancy may be that the importance of the shunt pathway for the synthesis of $24(S),25$-epoxycholesterol varies substantially between tissues, strains, and/or species.

Pitfalls

Oxysterols in biological samples are typically present at concentrations several orders of magnitude lower than cholesterol. The major concern when performing a comprehensive determination of oxysterols is, therefore, to avoid artefactual formation of oxysterols by autoxidation of cholesterol in the sample. This avoidance requires that antioxidants be added, and that the sample be manipulated in an oxygen-free environment as much as possible before separation of cholesterol and oxysterols. Kudo et al.[22] and

[22] K. Kudo, G. T. Emmons, E. W. Casserly, D. P. Via, L. C. Smith, J. St. Pyrek, and G. J. Schroepfer, Jr., *J. Lipid Res.* **30**, 1097 (1989).

Breuer and Björkhem[23] have performed careful experiments to determine which oxysterols are formed *in vivo* in humans and rats, respectively. Whereas Kudo *et al.* analyzed oxysterols under a controlled atmosphere where air was excluded, Breuer and Björkhem metabolically labeled oxysterols *in vivo* by keeping rats in an [^{18}O]-containing atmosphere, permitting the identification of oxysterols formed *in vivo* by their incorporation of ^{18}O. The results cast doubt about the physiologic relevance of some of the oxysterols most readily formed by autoxidation, most notably 5,6-oxygenated oxysterols. Fortunately, however, the oxysterols most implicated in LXR activation, including 4β-hydroxycholesterol, 20(S)-hydroxycholesterol, 22(R)-hydroxycholesterol, 24(S)-hydroxycholesterol, 24(S),25-epoxycholesterol, 25-hydroxycholesterol, and 27-hydroxycholesterol, are not readily formed by autoxidation and air exclusion is not necessary for the reliable quantification of these compounds.[24] Removal of cholesterol is still crucial, since the presence of excess cholesterol will create a chromatographic background that significantly reduces sensitivity.

Generally, the cartridge chromatography step described here is robust and uncomplicated; however, the side-chain hydroxylated oxysterols elute immediately after cholesterol and a compromise must be made between removing all residual cholesterol and recovering all oxysterols. Striking this balance is less of a problem when using an individual deuterated internal standard for each oxysterol, but it can be a significant source of error otherwise. If side-chain hydroxylated oxysterols are to be analyzed without the corresponding internal standards, elute cholesterol with 6 ml rather than 8 ml of 0.5% isopropanol in hexane. This modification ensures complete recovery of side-chain oxysterols, and the residual cholesterol present will not impair subsequent chromatographic behavior.

Conclusions

The identification of oxysterols as ligands for the nuclear receptors, LXRα and LXRβ, has created a need for accurate methods for determination of these compounds in complex mixtures. The analysis of oxysterols is demanding, since they are present at very low levels in tissues where the concentration of cholesterol may be several orders of magnitude higher than those of the oxysterols. Nevertheless, by careful sample handling to avoid cholesterol autoxidation, and by using GC-MS with appropriate internal standards, reliable determinations of oxysterols are

---

[23] O. Breuer and I. Björkhem, *J. Biol. Chem.* **270**, 20278 (1995).

[24] Whereas 4β-hydroxycholesterol, 20(S)-hydroxycholesterol, and 25-hydroxycholesterol are known autoxidation products of cholesterol, they are formed to negligible extent during sample preparation in air.

possible. Isotope-dilution techniques are applicable to many other small organic compounds of interest in biology and medicine.

Acknowledgment

This study was supported by grants from the Swedish Heart Lung Foundation.

[3] Site-Specific Fluorescent Labeling of Estrogen Receptors and Structure–Activity Relationships of Ligands in Terms of Receptor Dimer Stability

By ANOBEL TAMRAZI and JOHN A. KATZENELLENBOGEN

Introduction

The estrogen receptor (ER) is a ligand-regulated transcription factor that belongs to the nuclear receptor (NR) superfamily and acts as a dimeric species. There are two subtypes of ER, ERα and ERβ, which are both mainly regulated by the endogenous estrogen, estradiol (E$_2$). ER modulation is involved in the development and regulation of reproductive, cardiovascular, and bone health, in addition to controlling various aspects of cognitive function.[1] In addition to maintaining homeostasis in many tissues, an excessive activity of ER has been correlated with the development and proliferation of certain breast and uterine carcinomas.[2] In clinical settings, ER activity is modulated with exogenous estrogen and antiestrogen ligands as hormone replacement therapy (HRT) and as anticancer agents.[3]

Ligands that bind to the ER can be categorized into three pharmacological classes: agonists, mixed agonist–antagonists, and pure antagonists. The mixed agonist–antagonists are also referred to as selective estrogen receptor modulators (SERMs), because of their tissue-selective agonist or antagonist activities. Both *in vitro* and *in vivo* model systems suggest that ligand-induced ER conformations and coregulator (coactivator or corepressor) recruitment profiles to ER dimers play a crucial role in the complex tissue-selective modulation of ER activity.[4,5] Here, we report the development of a novel *in vitro* ER model system for evaluating

---

[1] D. P. McDonnell and J. D. Norris, *Science* **296**, 1642 (2002).
[2] I. Persson, *J. Steroid Biochem. Mol. Biol.* **74**, 357 (2000).
[3] F. Cosman and R. Lindsay, *Endocr. Rev.* **20**, 418 (1999).
[4] D. M. Kraichely, J. Sun, J. A. Katzenellenbogen, and B. S. Katzenellenbogen, *Endocrinology* **141**, 3534 (2000).
[5] Y. Shang and M. Brown, *Science* **295**, 2465 (2002).

the pharmacological character of ER ligands based on their stabilization of ER dimers, assessed using site-specific fluorescent-labeled receptors.

Principle of Site-Specific Fluorescent-Labeled NRs

Tsien and coworkers have used fluorescent NRs (as fusion constructs with green fluorescent protein (GFP) and its derivatives) to monitor coregulator recruitment inside live cells.[6] Our site-specific fluorescent labeled NR lacks a large GFP fluorophore (about 25,000 Da[7]) and is instead covalently labeled with a small fluorophore (about 400–800 Da) at a single cysteine residue. Site-specific fluorescent NRs have at least three distinct advantages. First, the relatively small fluorophore can be site-specifically attached to a reactive-cysteine residue at a conformationally sensitive region of the receptor (while retaining the native functionality of the NR) as a sensor of its environment, to monitor ligand- and coregulator-induced conformational changes in the NR. Second, a large variety of cysteine-specific fluorophores can be used so that different fluorescent techniques can be employed to study the various modes of NR modulation by ligands and coregulators. Third, direct labeling of proteins with different fluorophores allows for kinetic and thermodynamic measurements of protein–protein interactions to be conducted without the interference of large fusion proteins, fluorescent antibodies, or fluorescent streptavidin complexes.[8]

Engineering of Single Reactive-Cysteine ER Constructs

Proteins can be covalently labeled in a site-specific manner through a reactive (solvent-exposed) cysteine residue.[9] The ligand-binding domain (LBD) of ERα contains four cysteines: 381, 417, 447, and 530. One of these is deeply buried (C447) and is unreactive toward cysteine-specific modifying agents.[10,11] Cysteine 381 has intermediate reactivity, whereas both cysteines 417 and 530 readily react with cysteine-specific modifying agents.[10,11] Studies indicate that these three residues are either partially (C381) or fully solvent exposed (C417 and C530), and are therefore candidates for site-specific labeling.

[6] J. Llopis, S. Westin, M. Ricote, J. Wang, C. Y. Cho, R. Kurokawa, T. Mullen, D. W. Rose, M. G. Rosenfeld, R. Y. Tsien, C. K. Glass, *Proc. Natl. Acad. Sci. U.S.A.* **97**, 4363 (2000).
[7] R. Y. Tsien, *Annu. Rev. Biochem.* **67**, 509 (1998).
[8] G. Zhou, R. Cummings, J. Hermes, and D. E. Moller, *Methods* **25**, 54 (2001).
[9] A. Waggoner, *Methods Enzymol.* **246**, 362 (1995).
[10] G. B. Hegy, C. H. L. Shackleton, M. Carlquist, T. Bonn, O. Engstrom, P. Sjoholm, and H. E. Witkowska, *Steroids* **61**, 367 (1996).
[11] S. W. Goldstein, J. Bordner, L. R. Hoth, and K. F. Geoghegan, *Bioconjug. Chem.* **12**, 406 (2001).

Mutational studies suggest that either a cysteine to alanine or a cysteine to serine mutation at positions 381, 417, or 530 has minimal effect on ER activity.[12–17] We chose serine as a conservative replacement for solvent-exposed cysteines in our constructs. To enable site-specific fluorophore labeling of the ER alpha ligand-binding domain (ERα-LBD), we have, in all cases, mutated cysteine 381 to serine and then separately mutated either cysteine 417 or cysteine 530 to serine, leaving a single reactive cysteine, either at 530 or 417, respectively. For convenience, we have designated these ER constructs as *C530* and *C417* (with the bold italics type to indicate that they are mutant ERs with a single reactive cysteine at that particular residue). Cysteine to serine mutations at positions 381, 417, and 530 were introduced in a pET15b human ERα-LBD construct (304–554) using Quick Change Site-Directed Mutagenesis Kit with Pfu Turbo™ DNA polymerase (Stratagene Inc., La Jolla, CA) and the appropriate oligonucleotides. The corresponding cysteine to serine ERα-LBD constructs were subcloned in NdeI/BamHI sites of a newly double digested pET15b construct and sequenced.[18]

Preparation of Site-Specific Fluorescent ERs

Expression of Single Reactive-Cysteine ER Constructs

Our ER-LBD constructs in pET15b are expressed and purified as described previously[19] in accordance with the Novagen pET System Manual (Madison, WI) with the following adaptations. Overnight culture of BL21(DE3)pLysS *Escherichia coli* transformed with pET15b–ERα-LBD (304–554) construct is diluted 100-fold in Lurea Broth (LB) with 100 μg/ml ampicillin and shaken at 37°C. When the OD$_{600}$ reading reaches 0.4–0.6, isopropyl-β-D-thiogalactopyranoside (IPTG) is added to a final concentration of 1 mM, and the culture is shaken at 28°C for 3 hr to induce expression of *C530* and *C417* ERα-LBD. After induction, the bacteria are collected and can be stored at −20°C for 1–2 months.

[12] A. Stoica, E. Pentecost, and M. B. Martin, *J. Cell. Biochem.* **79**, 282 (2000).

[13] S. Neff, C. Sadowski, and R. Miksicek, *Mol. Endocrinol.* **8**, 1215 (1994).

[14] J. C. Reese and B. S. Katzenellenbogen, *J. Biol. Chem.* **266**, 10880 (1991).

[15] M. Gangloff, M. Ruff, S. Eiler, S. Duclaud, J. M. Wurtz, and D. Moras, *J. Biol. Chem.* **276**, 15059 (2001).

[16] J. C. Reese, C. H. Wooge, and B. S. Katzenellenbogen, *Mol. Endocrinol.* **6**, 2160 (1992).

[17] R. L. Rich, L. R. Hoth, K. F. Geoghegan, T. A. Brown, P. K. LeMotte, S. P. Simons, P. Hensley, and D. G. Myszka, *Proc. Natl. Acad. Sci. U.S.A.* **99**, 8562 (2002).

[18] A. Tamrazi, K. E. Carlson, J. R. Daniels, K. M. Hurth, and J. A. Katzenellenbogen, *Mol. Endocrinol.* **16**, 2706 (2002).

[19] K. E. Carlson, I. Choi, A. Gee, B. S. Katzenellenbogen, and J. A. Katzenellenbogen, *Biochemistry* **36**, 14897 (1997).

Affinity Purification and Site-Specific Labeling of ER with
a Single Fluorophore

Our ER **C530** and **C417** constructs have an N-terminal His_6-tag through which they can be purified over nickel–nitrilotriacetic acid resin (Ni–NTA agarose, Qiagen Inc., Valencia, CA). We find it convenient to label these ER preparations while they are attached to the Ni–NTA resin. In this manner, the receptor can be purified to near homogeneity before labeling, and excess cysteine-specific fluorophore can be removed from the receptor simply by washing the resin, prior to elution of the labeled receptor. We conduct all the purification, labeling, wash, and elute steps using batch mode with 15-ml centrifuge tubes (Corning Incorporated, Corning, NY). To separate ER-resin mixtures from supernatant, we spin the tubes at $1000 \times g$ for 2 min using a Sorvall RC5B Plus centrifuge with a SH-300 swinging bucket rotor (Kendro Laboratory Products, Newtown, CT) and carefully decant the supernatant. We find that the inclusion of the nonnucleophilic thiol reductant, tris(carboxyethyl)phosphine (TCEP), to maintain the receptor in a fully reduced state, is important for efficient labeling. The protocols for receptor purification and labeling are listed below:

1. Freeze-thaw the bacterial pellet three times before resuspending in binding buffer (0.1 mM TCEP, 5 mM imidazole, 0.5 M NaCl, 20 mM Tris–HCl, pH 7.9 adjusted at room temperature) at volumes 10-fold the weight of the pellet and sonicate 10–20 sec to shear the DNA.
2. Centrifuge the mixture at $30,000 \times g$ for 30 min, and save the supernatant.
3. Conduct a $[^3H]E_2$-binding study[19] on the supernatant to estimate the concentration of active receptor loaded onto the Ni–NTA resin (about 0.3 mg of receptor per 0.1 ml of packed resin).
4. Allow the His_6-tagged receptor to adsorb to Ni^{2+} resin in batch mode with 1 hr incubation at 4°C and occasional mixing.
5. Wash the resin two times (1 ml buffer per 0.1 ml of packed resin) using wash buffer (0.1 mM TCEP, 30 mM imidazole, 0.25 M NaCl, 10 mM Tris–HCl, pH 7.9) with a 5 and a 30 min incubation at 4°C.
6. Wash the resin two times (1 ml buffer per 0.1 ml of packed resin) using labeling buffer (0.1 mM TCEP, 50 mM Tris–HCl, 10% glycerol, pH 7.0) with a 5 min incubation at 4°C. (*Note:* Labeling is conducted at pH 7.0 to minimize nonspecific labeling of primary amines at the N-terminal and lysine residues.[9,20])

[20] R. R. Haugland, "Molecular Probes, Inc., Catalog" (K. D. Larison, ed.), 5th Ed. Molecular Probes, Eugene, Oregon, 1992–1994.

7. Add 3 ml of labeling buffer per 0.1 ml packed resin (this volume ensures μM concentrations of fluorophore and receptor during labeling) before adding a 30:1 stoichiometric ratio of cysteine-specific fluorophore (from a mM stock solution in DMF) to active ER (obtained from step 3), and incubate overnight at 4°C with occasional mixing.

8. Remove excess cysteine-specific fluorophore by washing the ER-resin mixture six times with fluorophore wash buffer (50 mM Tris–HCl, 10% glycerol, 10 mM β-mercaptoethanol, pH 8.0).

9. Elute the labeled receptor (1 ml buffer per 0.1 ml packed resin) with elute buffer (1 M imidazole, 0.5 M NaCl, 20 mM Tris–HCl, 10 mM β-mercaptoethanol, pH 7.9) or alternatively with strip buffer (100 mM EDTA, 0.5 M NaCl, 20 mM Tris–HCl, 10 mM β-mercaptoethanol, pH 7.9) with 1 hr incubation at 4°C with occasional mixing. Dialyze the labeled receptor two times against storage buffer (50 mM Tris–HCl, 10% glycerol, 10 mM β-mercaptoethanol, 0.02% NaN$_3$, pH 8.0).

To preserve the natural state of the fluorophore, we try to minimize exposure to light by routinely covering all reactions containing fluorophores or fluorescent receptor with aluminum foil during the labeling procedure. Our fluorescent receptors can be stored at 0°C for up to two months with minimal loss of [^{3}H]E$_2$-binding capacity. The level of labeling, determined by matrix-assisted laser-desorption ionization mass spectroscopy (MALDI MS), is 90–95% monolabeling, with no multiple labeling evident.

Characterization of the Native Activity of Fluorescent ER

It is critical to assess for the native functional state of the receptor after it has been labeled with a fluorophore at a particular cysteine residue. In the case of ERα, cysteines 417 and 530 are routinely labeled with cysteine-specific compounds before obtaining X-ray crystallographic data on wild-type receptor, with no detrimental effects observed in the dual labeled receptor.[11] In addition, reports of full-length ERs with multiple cysteine to alanine or cysteine to serine mutations at residues 381, 417, and 530 suggest near wild-type binding affinity and transcriptional activity.[14,15,17] We have used ligand-binding affinity and coactivator recruitment profiles to demonstrate the preservation of native functionality in our *C417* and *C530* site-specific fluorescent ERα-LBDs.

The estradiol-binding affinities of unlabeled, fluorescein-, and tetra-methylrhodamine-labeled ER double mutants are listed in Table I. We have found that our unlabeled and fluorophore-labeled cysteine to serine double

TABLE I

CHARACTERIZATION OF FLUORESCENT-LABELED ERα-LBDS FOR
ESTRADIOL-BINDING AFFINITY

ERα-LBD (304–554)	$[^3H]E_2K_d$ (nM)
wt-unlabeled	0.16
C417-Unlabeled	0.58
C417-Acceptor	0.54
C417-Donor	0.39
C530-Unlabeled	0.52
C530-Acceptor	0.46
C530-Donor	0.24

Donor = fluorescein; Acceptor = tetramethylrhodamine. *C417* denotes
C381S/C530S double mutant. *C530* denotes C381S/C417S double mutant.

mutant ERα-LBDs exhibit near wild-type affinities for estradiol.
Interestingly, the attachment of a fluorophore to cysteine 417 or 530 even
appears to increase the affinity of our labeled receptors for E_2 relative to that
of the unlabeled receptors (Table I).

To verify that the labeled ERα-LBD mutants adopt characteristic
agonist and antagonist conformations when complexed with various ligands,
we performed fluorescence-based coactivator recruitment experiments.
A glutathione *S*-transferase (GST) steroid receptor coactivator-1 (SRC-1
residues 629–831, containing NR box regions 1–3) protein construct was
immobilized on glutathione Sepharose (GSH) resin and used to "pull-
down" fluorophore-labeled ERs that had been equilibrated with various ER
ligands. Both *C530*-fluorescein and *C417*-tetramethylrhodamine (TMR)
receptors retain wild-type ligand-induced functional interactions with this
SRC-1 coactivator peptide, with interaction being induced by agonist
ligands and inhibited by antagonist ligands (Fig. 1). Thus, both the
mutational changes and fluorescent labeling appear to have no detrimental
effects on the characteristic activities of ER as measured by ligand binding
and coactivator recruitment assay.

Choice of Fluorophores and Fluorescent Techniques

To site-specifically label a cysteine residue (with a free sulfhydryl group)
in proteins, one needs a cysteine-specific fluorophore containing an
iodoacetamide or maleimide functional group. Fortunately, there are a
large number of cysteine-specific fluorophores to choose from, many of
which are listed in the Molecular Probes (Eugene, OR) or Amersham

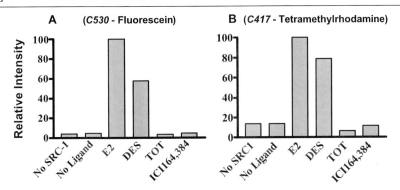

FIG. 1. Fluorescent coactivator recruitment assay. GST-SRC-1 (residues 629–831) was immobilized on GSH resin and used to pull-down ligand-bound fluorescein (donor) labeled *C530* (panel A), or tetramethylrhodamine (acceptor) labeled *C417* (panel B) ERs. E_2, estradiol; DES, diethylstilbestrol; TOT, *trans*-4-hydroxytamoxifen.

Biosciences (Piscataway, NJ) catalogs. There is also an equally large number of fluorescent techniques that can be employed with various fluorophores selected. For this review, we will focus on fluorescence resonance energy transfer (FRET) and various FRET-pair fluorophores that we have used to generate site-specific fluorescent NRs.

There are several excellent reviews on the theory behind FRET;[21,22] this review will focus on the use of this technique for monitoring the modulation of NRs by ligands and coactivators. FRET develops only when an excited donor fluorophore is in close proximity to an acceptor fluorophore (typically in the range of 10–75 Å), so that it can transfer its energy.[21] This technique enables the monitoring of a change in distance between proteins labeled with two different fluorophores. Therefore, one can monitor protein–protein interactions and conformational changes in proteins. Using site-specific fluorescent-labeled receptors, we have used FRET to monitor ligand- and coactivator-modulation in the thermodynamic and kinetic stability of ER dimers (for details see "Using Site-Specific Fluorescent Receptors to Monitor ERα Homodimer Stability"). FRET can also be used to monitor NR conformational changes measured through changes in the intermonomer distances between particular labeled-cysteine residues in the receptor.

The fluorescence spectra of ER labeled with fluorescein (a FRET donor) and TMR (a FRET acceptor) are shown in Fig. 2. Fluorescein- and

[21] P. R. Selvin, *Methods Enzymol.* **246**, 300 (1995).
[22] P. R. Selvin, *Nat. Struct. Biol.* **7**, 730 (2000).

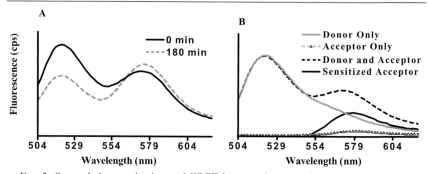

FIG. 2. Spectral characterization and FRET between donor- and acceptor-ER monomers. (A) Spectra show the development of a FRET signal through a decrease in donor intensity (521 nm) and an increase in acceptor intensity (580 nm) as donor–acceptor ERα-LBD dimers are formed over time. (B) Donor-only and acceptor-only spectra, excited at the donor excitation (488 nm), are shown. The FRET signal through the "sensitized" acceptor emission is shown after a subtraction of donor-only spectrum from the equilibrated donor–acceptor reaction.

TMR-labeled ERα-LBD emission bands are broad, and as a result, fluorescein emission overlaps considerably with the TMR maximum emission (580 nm). By contrast, there is minimal overlap of TMR emission at the fluorescein maximum emission (521 nm). The FRET signal can be followed by the enhanced emission from acceptor (sensitized acceptor emission) or by the decreased emission from donor, depending on the pair of fluorophores chosen.[21] For this FRET pair, we found that it is cleaner to monitor FRET as a decrease in donor intensity, rather than an increase in acceptor intensity (because the latter would require correction for the overlap of the donor emission).

When monitoring FRET through a decrease in donor intensity, it is critical to conduct controls showing that the generated signal is due to donor- and acceptor-labeled protein–protein interaction and not due to nonspecific quenching. One control would be to monitor both donor and acceptor emission spectra and show that the generated FRET signal leads to both a decrease in donor intensity and an increase in acceptor intensity (Fig. 2). Another control would be to show that the FRET signal is reversible by disrupting the protein–protein interaction and recovering the decrease in donor intensity. We have also used fluorescein and Cy5 as a FRET pair, which allow for direct monitoring of both donor and acceptor emission intensities without the need for any spectral overlap corrections. Due to lower spectral overlap, however, this FRET pair is limited to detecting changes over smaller distances compared with the fluorescein and TMR FRET pair.

Using Site-Specific Fluorescent Receptors to Monitor ERα Homodimer Stability

Nuclear receptors (NRs) form strong dimers that are essential for their function as transcription factors. NRs that bind steroid ligands (i.e., ER) typically function as dimers, either homodimers between identical receptor monomers or between closely related subtypes. NRs that bind nonsteroidal ligands, however, typically function as heterodimers with the retinoid X receptor (RXR).[23,24] ER has been reported to exist as a dimer even in the absence of ligand, and the dimer interaction of liganded ERα-LBD is strong and resistant to high levels of denaturants.[25] ER dimers have been estimated, by indirect methods, to have an equilibrium dissociation constant (K_d) of about 2–3 nM,[26] although dimer affinity varies from NR to NR.[26–28] There is evidence that the strength of the dimer interaction is regulated by ligand binding, although this issue has not been studied in a systematic fashion.[26–28] Using site-specific fluorescent-labeled receptors, we have developed convenient FRET-based methods for measuring the thermodynamic and kinetic stability of ERα-LBD dimers.

Ligand Effects on the Thermodynamic Stability of ERα-LBD Dimers

With fluorescein- (donor) and TMR-labeled (acceptor) receptors, we can use FRET to directly measure the thermodynamic stability (affinity) of ERα-LBD dimers. To measure the K_d of ERα-LBD dimer affinity, a fixed low concentration of donor-labeled receptor (0.1 nM, the detection limit of our fluorometer for fluorescein) is titrated with increasing concentrations of the acceptor-labeled receptor. After equilibrium is reached, FRET between donor–acceptor dimers is measured based on donor intensity,[21] with the percent FRET reaching a maximum level as all the donor-labeled monomers form dimers with acceptor-labeled monomers. To measure FRET we used a Spex Fluorolog II (model IIIc) cuvette-based fluorometer with Data

[23] D. J. Mangelsdorf, C. Thummel, M. Beato, P. Herrlich, G. Schutz, K. Umesono, B. Blumberg, P. Kastner, M. Mark, and P. Chambon, *Cell* **83**, 835 (1995).

[24] J. M. Olefsky, *J. Biol. Chem.* **276**, 36863 (2001).

[25] M. Salomonsson, J. Häggblad, B. W. O'Malley, and G. M. Sitbon, *J. Steroid Biochem. Mol. Biol.* **48**, 447 (1994).

[26] M. E. Brandt and L. E. Vickery, *J. Biol. Chem.* **272**, 4843 (1997).

[27] D. M. Tanenbaum, Y. Wang, S. P. Williams, and P. B. Sigler, *Proc. Natl. Acad. Sci. U.S.A.* **95**, 5998 (1998).

[28] H. Wang, G. A. Peters, X. Zeng, M. Tang, W. Ip, and S. A. Khan, *J. Biol. Chem.* **270**, 23322 (1995).

Max 2.2 software (Spex Industries, Inc., Edison, NJ). The protocols for measuring ER dimer affinity are listed below:

1. Prepare a fresh stock solution of 0.1 nM donor-labeled ER, 0.3 mg/ml chicken ovalbumin (as a carrier protein), 1 μM ligand (or vehicle) in Tris–glycerol pH 8.0 buffer (50 mM Tris–HCl, 10% glycerol), and place 700 μl of this stock into separate tubes (keep all reactions at 4°C and protect from light).

2. Remove 10 μl from a serial dilution of acceptor-labeled ER solution with 0.3 mg/ml chicken ovalbumin in Tris–glycerol pH 8.0 buffer, and add to the 700 μl solution of 0.1 nM donor-labeled ER (1.5% dilution).

3. Because ERα-LBD dimer dissociation rates are rather slow (see "Ligand Effects on the Kinetic Stability of ERα-LBD Dimers"), allow 5–8 hr at room temperature (in the dark) for reactions to reach equilibrium.

4. Place 600 μl of each sample into a 5.0 × 5.0 mm quartz fluorescence cuvette and measure donor emission intensity at 521 nm while exciting at 488 nm, with the sample chamber held constant at 25°C.

We also conduct the following signal corrections for each sample: dark counts correction (to correct for any fluctuations in the PMT), signal/reference correction (the sample signal divided by a reference signal to correct for fluctuations in light source intensity), and blank subtraction. The percent FRET was then fitted to a simple single-term binding isotherm to determine the K_d.[29,30]

In this thermodynamic NR dimer formation assay, if the K_d of dimer affinity were above 0.1 nM, only a small fraction of the maximum percent FRET would be observed from the small population of donor–acceptor dimers as an equal concentration of acceptor-labeled receptor is added to the 0.1 nM donor-labeled ER sample. However, if the NR dimer affinity has a K_d that is well below 0.1 nM, then nearly half of the maximum percent FRET would be observed from the large population of donor–acceptor dimers formed as an equal concentration of acceptor-labeled receptor is added to the 0.1 nM donor-labeled ER sample.

Under native conditions (without denaturant), the K_d for the ERα-LBD dimer affinity, with or without ligand, is <0.1 nM. This is below the sensitivity limit for our FRET-based method, which is limited by our ability to detect the fluorescence of fluorescein-labeled ER. However, we found that we were able to raise the K_d of the ER dimer to the low nanomolar range by

[29] S. Y. Tetin and T. L. Hazlett, *Methods* **20**, 341 (2000).
[30] S. Y. Tetin, C. A. Rumbley, T. L. Hazlett, and E. W. Voss, *Biochemistry* **32**, 9011 (1993).

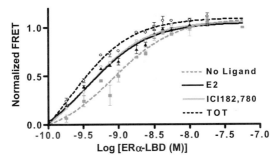

FIG. 3. Thermodynamic ERα-LBD dimer affinity. All reactions were conducted with 0.1 nM donor ER with 1 μM ligand (or vehicle), and titrating increasing concentration of acceptor ER in 2 M urea. The calculated equilibrium dissociation constant (K_d) value for no ligand was 1.0 ± 0.2 nM, for estradiol (E$_2$) was 0.33 ± 0.06 nM, for ICI 182,780 was 0.34 ± 0.06 nM, and for TOT was 0.27 ± 0.04 nM. The K_d values are the mean $\pm$ SD from four similar experiments.

the addition of a modest concentration of a denaturant (2 M urea). This concentration of denaturant causes less than a 3-fold reduction of estradiol-binding affinity. Under these equilibrium conditions, we could quantitate ligand-induced enhancement of dimer affinity (Fig. 3). The K_d values of ERα-LBD dimer affinity in 2 M urea (Fig. 3) are: in the absence of ligand (apo, 1.0 nM), with an agonist (E$_2$, 0.33 nM), a mixed agonist–antagonist [*trans*-hydroxytamoxifen (TOT), 0.27 nM], and a pure antagonist (ICI 182,780, 0.34 nM). Thus, there is a 3- to 4-fold increase in ERα-LBD dimer affinity with these three ligands of different pharmacological character.

Ligand Effects on the Kinetic Stability of ERα-LBD Dimers

To measure ERα-LBD dimer stability under native conditions (without denaturant), we used a standard *kinetic* FRET technique termed "monomer exchange."[31] Monomer exchange is conducted by mixing a preparation of donor-labeled dimers with a preparation of acceptor-labeled dimers, and monitoring the development of a FRET signal with time, as donor–acceptor dimers are formed. In our experiments, the rate at which a FRET signal (measured through a decrease in donor intensity) develops is governed by the rates at which "donor–donor homodimers" and "acceptor–acceptor homodimers" dissociate into monomers and then re-associate to form "donor–acceptor heterodimers." We have found, as expected, that monomer exchange kinetics do not change with increasing excess of acceptor-labeled receptor, which indicates that dimer dissociation (rather than monomer

---

[31] L. Erijman and G. Weber, *Photochem. Photobiol.* **57**, 411 (1993).

re-association) is the rate-determining step of this process.[31] Thus, the monomer exchange kinetics we observe with fluorescent receptor is an accurate measure of the rate of dimer dissociation, and hence represents the kinetic stability of the ERα-LBD dimer.

To maximize the FRET signal (in our experiments typically 35–50%), we typically use a 4:1 ratio of acceptor- to donor-labeled receptor, so that every donor-labeled monomer is more likely to find an acceptor-labeled monomer. To confirm that the FRET signal results from the formation of donor–acceptor dimers, we determined that we could abrogate the FRET signal by adding an excess of unlabeled ER, which would then undergo monomer exchange and disrupt the donor–acceptor dimers.

We conduct our NR monomer exchange assays in black 96-well Nunc polypropylene microtiter plates (Nalge Nunc, International Corporation), and monitor the reactions using a Molecular Devices Gemini XS (Sunnyvale, CA) fluorescence plate reader. The protocols for measuring ER dimer kinetic stability are listed below:

1. Prepare two separate stocks of 55 nM donor- or acceptor-labeled receptor in Tris–glycerol pH 8.0 buffer with 0.3 mg/ml chicken ovalbumin.

2. Place 45 μl of the donor-labeled ER stock into separate wells of a black 96-well microtiter plate (plate 1), and 225 μl of acceptor-labeled ER stock into separate wells of another 96-well microtiter plate (plate 2).

3. Add 5 μl of 10 μM ligand (or vehicle) into the wells of plate 1 containing donor-labeled ER, and 25 μl of 10 μM ligand (or vehicle) into the wells of plate 2 containing acceptor-labeled ER (both donor- and acceptor-labeled ERs will have a final concentration of 50 nM while incubating with 1 μM ligand for 1 hr in the dark to allow reactions to reach equilibrium).

4. With a multichannel pipette (8 or 12 channels), rapidly remove 200 μl of the acceptor-labeled solution from wells in plate 2 and place into donor-labeled wells in plate 1 that were incubated with the same ligand (yielding final concentrations of 10 nM donor-, 40 nM acceptor-labeled ERs, 1 μM ligand (or vehicle), and 0.3 mg/ml chicken ovalbumin in Tris–glycerol pH 8.0 buffer).

5. After mixing, immediately cover plate 1 with a clear polyolefin sealing tape (Nalge Nunc, International Corporation) to minimize evaporation, and monitor the decrease in the donor emission intensity (FRET signal) at 530 nm with time, while exciting at 485 nm, with a 515 nm cutoff filter in the excitation pathway and the chamber temperature controlled at 28°C.

We also include the following three controls along with each reaction plate: (1) reactions containing all components except for acceptor-labeled ER (to ensure that the decrease in donor intensity is due to a FRET signal resulting from donor–acceptor ER dimer formation, and not due to nonspecific quenching), (2) Tris–glycerol pH 8.0 buffer with 0.3 mg/ml chicken ovalbumin (as a blank to quantitate level of signal to background), and (3) a solution of 10 nM free fluorescein sample (to detect any changes in the light intensity of the instrument during the time course of the experiment).

To obtain the ERα-LBD dimer dissociation rates, we fit the rate of FRET signal development to a nonlinear regression, one-phase exponential decay function using Prism 3.00 (GraphPad Software, Inc., San Diego, CA). Changes in the rate of dimer dissociation reflect ligand-induced modulation of the ER dimer kinetic stability; therefore, we can obtain ligand structure–activity relationships (SARs) in terms of NR dimer stability.

We find that ligand binding affects the kinetic stability of ERα-LBD dimers. The rate of dimer dissociation for apo (no ligand) ERα-LBD is slow (half-life of 39 ± 3 min at 28°C), indicating that ERα-LBD dimers are stable in the absence of ligand. The rate of ER dimer dissociation becomes progressively slower (an indication of enhanced dimer kinetic stability) as receptor becomes occupied with most ligands and reaches a maximum at ligand saturation.

Titration experiments allowed us to estimate ligand concentrations that fully saturate the receptor (Fig. 4A). Here, the EC_{50} values reflect relative ligand affinity, whereas, the "efficacy" of dimer kinetic stability (i.e., maximum half-life under saturating ligand conditions) indicates the degree to which a particular ligand induces a conformation that stabilizes the dimer. Shown in Fig. 4A are both a high-affinity ligand [estradiol, relative-binding affinity (RBA) 100%] and a low-affinity ligand (genistein, RBA 0.013%). Using our convenient FRET-based dimer kinetic stability assay, we have systematically assessed over 30 natural and synthetic ligands for their effects (under saturating ligand conditions, 1 μM for most ligands, and 5 μM for very low-affinity ligands) on dimer dissociation of ERα-LBDs.

The ERα-LBD dimer dissociation half-lives with a selection of these ligands are summarized in Figs. 4B and C. Our studies show that not all agonists alter the ERα-LBD dimer dissociation kinetics to the same degree (Fig. 4B). Compared to the apo receptor, different agonist ligands affect the dimer dissociation kinetics by factors that range from 0.65 to 6.2-fold. Mixed agonist–antagonist ligands showed, in general, an even greater effect than agonists (4.5 to 7-fold), but within a narrower range. Compared with mixed agonist–antagonists, the pure antagonist ICI compounds had a somewhat lesser dimer stabilizing effect (3.5 to 4.5-fold relative to apo receptor) (Fig. 4C).

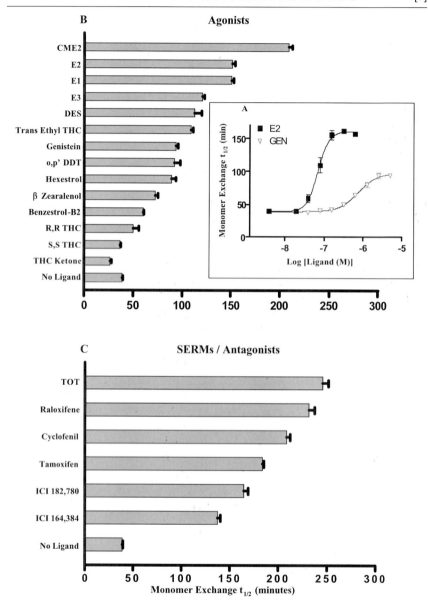

FIG. 4. (A) Ligand binding effects on ERα-LBD dimer kinetic stability. Dimer dissociation half-life increases and reaches a limiting value as a function of receptor ligand occupancy. Results are indicative of two similar experiments. GEN, genistein; E_2, estradiol. (B) The ERα-LBD dimer dissociation half-life values of apo and in the presence of saturating agonist ligand concentration. (C) The ERα-LBD dimer dissociation half-life values of apo and in the presence of saturating mixed agonist–antagonist and pure antagonist ligand concentrations.

A comparison of ligand affinity and dimer kinetic stability showed no correlation, indicating that dimer kinetic stability reflects ligand pharmacological character (i.e., agonist vs antagonist), not affinity. Furthermore, dimer dissociation rates do not appear to be a simple reflection of ligand dissociation rates (an additional measure of ligand affinity separate from RBA), which we have reported previously by both radiometric- and fluorescence-based assays.[19,32,33]

Some interesting SARs were observed from the ligands monitored for dimer kinetic stability effects on ERα-LBD (Fig. 4). An 11β substituent enhanced dimer kinetic stability compared to a similar ligand lacking a substituent at this position [11β-chloromethylestradiol (CME$_2$) vs E$_2$]. Also, it appears that antagonists stabilize the ERα-LBD dimer more than do agonists, a trend that is supported by an earlier report in which a chromatographic technique was used to monitor the rate of wild-type ERα-LBD monomer exchange.[26] The degree of dimer stabilization within each class of ligand character, however, covers a range of values.

A Functional Assay to Identify Ligand Character based on Selective Stabilization of Agonist-Bound ERα-LBD Dimers by Coactivator Peptides

Our FRET-based monomer exchange assay provides a convenient measure of modulation in NR dimer kinetic stability using a highly purified *in vitro* model system. In addition to monitoring ligand-induced modulation of NR dimer kinetic stability, our model system can also be used to monitor coregulator-induced modulation of this effect. Coactivators, corepressors, or other proteins thought to modulate NR dimer stability (i.e., heat shock proteins), can easily be added to our kinetic monomer exchange assay protocols outlined in the previous section.

Having obtained an initial SAR profile of ligand-induced ER dimer stability, we investigated whether coactivator peptides, which should bind only to agonist–ER complexes, would enhance and sharpen the SAR. Additionally, coactivators are present in the *in vivo* systems, so adding them to our *in vitro* model system makes it more similar to the *in vivo* situation. The presence of a coactivator peptide containing residues 629–831 of SRC-1 (encompassing LXXLL NR boxes 1–3) or of a 15-residue peptide containing only the NR-Box-2 of SRC-1 coactivator protein, caused up to a 2-fold increase in the kinetic stabilizing of E$_2$-ERα-LBD dimers (Fig. 5A). In terms of this effect, the longer SRC-1 fragment containing three NR

[32] A. C. Gee, K. E. Carlson, P. G. V. Martini, B. S. Katzenellenbogen, and J. A. Katzenellenbogen, *Mol. Endocrinol.* **13**, 1912 (1999).

[33] R. D. Bindal, K. E. Carlson, G. C. Reiner, and J. A. Katzenellenbogen, *J. Steroid Biochem.* **28**, 361 (1987).

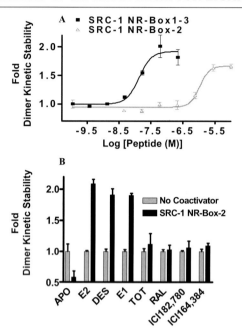

FIG. 5. Coactivator peptide selective enhancement in agonist–ER dimer stability. (A) The effects of increasing concentration of coactivator peptides on E_2-ERα-LBD dimer kinetic stability are illustrated. (B) The effect of 10 μM SRC-1-NR-Box-2 peptide on dimer kinetic stability of ERα-LBD with various ligands. All values are expressed as fold enhancement of dimer kinetic stability in the presence of coactivator peptide. Results are indicative of three similar experiments.

boxes has about 100-fold higher potency than the single NR-Box-2 peptide of SRC-1. Both coactivator peptides, as expected, failed to stabilize antagonist-bound ERα-LBD. Thus, the coactivator peptide-enhanced ER dimer stabilization is highly selective for agonist-bound receptor.

Our finding of coactivator-mediated NR dimer stabilization suggests that coactivator content of different tissues could play a role in the kinetic stabilization of transcriptionally active ER dimers, which, in turn, may offer an explanation for the complex tissue-selective pharmacology observed with ER ligands. We also find that conducting our ER monomer exchange experiments with a coactivator peptide concentration about 10-fold above the EC_{50} of E_2-ERα-LBD dimer stability, destabilizes apo dimers, stabilizes agonist-bound dimers, and has no effect on mixed agonist–antagonist and pure antagonist-bound dimers (Fig. 5C). Therefore, the addition of coactivator peptides to our ER dimer stability model system separates agonist from antagonist ligands, thus providing a definitive identity to the pharmacological character of any novel ER ligand.

Conclusion

We have successfully prepared site-specific fluorescent-labeled ERs that have native functional activity in terms of ligand binding and coactivator recruitment profiles. Using FRET and fluorescent-labeled ERα-LBDs, we have monitored the regulation of homodimer stability by ligands and coactivator peptides. This technique provides SARs of ligands in terms of ER dimer stability and can be used as a functional assay to identify the pharmacological character of novel synthetic or environmental ER ligands, as described herein. In addition, fluorescent-labeled ERβ-LBDs can be used to study homo- and heterodimers of both ER subtypes and the manner in which SERMs and ER subtype-selective ligands (both agonists and antagonists) modulate differential homo- and heterodimer stability. Site-specific fluorescent ERs can also be used for the development of other types of assays to characterize receptor conformation, conformational dynamics, and ligand or coregulator interactions.

Many steroid, nonsteroid, and orphan NRs have a low number of conserved cysteine residues in their LBDs (generally 3–5 cysteines per LBD).[27] Thus, the methodology that we have developed for ERα should be applicable, as well, to the study of ligand-induced effects on dimer affinity and monomer exchange dynamics in these other NR systems.

Acknowledgment

We would like to thank past members of our group who synthesized many of the ligands used in this report: Elio Napolitano, Ying R. Huang, Marvin J. Meyers, Deborah S. Mortensen, Donald A. Seielstad, Kwang Jim Hwang, and Shaun R. Stauffer. We would also like to thank Kathryn E. Carlson for excellent technical assistance. We are grateful for support of this work through a grant from the National Institutes of Health (PHS 5R37 DK15556).

[4] Cell-Free Ligand Binding Assays for Nuclear Receptors

By STACEY A. JONES, DEREK J. PARKS, and STEVEN A. KLIEWER

Introduction

The nuclear receptor (NR) superfamily of ligand-activated transcription factors includes the steroid, retinoid, and thyroid hormone receptors as well as many "orphan" receptors for which ligands have not yet been identified.[1,2] Members of this family play pivotal roles both during development and in adult physiology. NRs are also the molecular targets for many widely-used prescription drugs including those used to treat cancer, inflammation, osteoporosis, thyroid hormone deficiency, and diabetes.

Members of the NR family share a conserved ligand-binding domain (LBD) of ~250 amino acids in the carboxy-terminal region of the protein. As its name implies, the LBD is responsible for the selective binding of ligands. Biochemical, genetic, and structural studies have provided a great deal of insight into the mechanism whereby the binding of a ligand is converted into changes in gene transcription.[3] The docking of a ligand into the hydrophobic pocket of the LBD induces conformational changes in the protein, most notably in the position of the activation function 2 (AF-2) helix in the extreme carboxy terminus of the receptor. The repositioning of the AF-2 helix completes a hydrophobic cleft and permits the NR to interact with LXXLL motifs of coactivator proteins, which are required to activate transcription. In this way, a small lipophilic molecule can switch the NR into an active conformation.

Over the past 30 years, ligand-binding assays have played a central role in the discovery and characterization of the NR family. Early studies with radiolabeled steroids demonstrated that these proteins are localized within the nucleus of the cell, where they bind with high-affinity to sites in the chromatin.[4] Radioligand-binding assays and competition assays were subsequently critical for the identification of the NRs for the classic endocrine hormones, including the steroid and thyroid hormones. Recently, powerful ligand-binding assays that exploit receptor–coactivator interactions have taken center stage in the quest to discover both natural and synthetic ligands for the orphan NRs.[5] These assays are now also widely used in the hunt for novel drugs that mediate their therapeutic effects through NRs.

In this chapter we describe several of these ligand-binding assays, ranging from the more standard gel filtration and scintillation proximity assays (SPA) that employ radioligands to those that exploit coactivator interactions in either the time-resolved fluorescence energy transfer, fluorescence polarization, or luminescent proximity assay formats.

[1] V. Giguere, Orphan nuclear receptors: from gene to function, *Endocr. Rev.* **20**(5), 689 (1999).

[2] D. J. Mangelsdorf, C. Thummel, M. Beato, P. Herrlich, G. Schutz, K. Umesono, B. Blumberg, P. Kastner, M. Mark, P. Chambon, The nuclear receptor superfamily: the second decade, *Cell* **83**(6), 835 (1995).

[3] R. V. Weatherman, R. J. Fletterick, and T. S. Scanlan, Nuclear-receptor ligands and ligand-binding domains, *Annu. Rev. Biochem.* **68**, 559 (1999).

[4] E. V. Jensen and E. R. DeSombre, Mechanism of action of the female sex hormones, *Annu. Rev. Biochem.* **41**, 203 (1972).

[5] A. K. Shiau and P. Coward, Orphan nuclear receptors: from new ligand discovery technologies to novel signaling pathways, *Curr. Opin. Drag. Disc. Dev.* **4**, 575 (2001).

Radioligand-Binding Assays

In any experiment designed to detect the binding of a ligand to a receptor, it is important to ensure that strict attention is given to the assay conditions, in order to guarantee the data are interpreted appropriately. For radioligand binding and displacement studies such as gel filtration and SPA, it is important to choose receptor concentrations where the radioligand and the competing ligand are not depleted. Analysis of the data using the conventional equations cited in this article assumes that the free concentration of both radioligand and competing ligand are essentially the same as the concentrations added to the reaction, and that the amount bound by the receptor is small in comparison. It is also important to ensure the incubation period is long enough to reach equilibrium. There are many excellent primers that detail the different parameters to be considered when developing a binding assay.[6] It is up to the reader to ensure the conditions of their assay are considered when interpreting the data.

Gel Filtration

Assessment of ligand binding using gel filtration generally requires a radiolabeled ligand that binds to the receptor under study. Because of this requirement, typical displacement gel filtration studies are not possible for orphan NRs for which ligands have not been identified. Gel filtration is a relatively low throughput format for studying receptor–ligand interactions, but it is faster than screening in most cell-based assays. Gel filtration is attractive in some circumstances because it is inexpensive and does not require specialized equipment beyond a centrifuge and a scintillation counter. Gel filtration assays also do not require purification or labeling of the NR protein. The G25 Sephadex-packed Quick Spin Protein Columns sold by Boehringer Mannheim (Indianapolis, IN) work well for small numbers of samples. For filtration of samples in 96-well format MiniSpin Column Kits (The Nest Group, Ind., Southboro, MA) are available that allow a somewhat higher throughput than the individual columns.

Protein Preparation

1. Novagen (Madison, WI) BL21 (DE3) pLYS S competent cells are transformed according to the manufacturer's directions with 1 μl of plasmid DNA containing a T7 promoter driving expression of the LBD of the NR of interest.

---

[6] E. C. Hulme, "Receptor-Ligand Interactions. A Practical Approach," Oxford: IRL Press at Oxford University Press, 1992.

2. Plate the transformed cells on Luria Broth (LB) agar containing an appropriate antibiotic for the plasmid.
3. Incubate overnight at 37°C.
4. Pre-culture the *Escherichia coli* by picking a single colony from the plate and placing it in 3 ml LB broth with antibiotic and 3 μl 33 μg/μl chloramphenicol. Incubate in a shaking incubator overnight at 37°C.
5. Add the entire pre-culture sample to 250 ml LB broth containing antibiotic and 250 μl 33 μg/μl chloramphenicol in a 500 ml shaker flask (Corning, Corning, NY). Place the flask in a shaking 37°C incubator until the optical density at 600 nm (OD 600) reaches 0.5–1.0.
6. Add isopropyl-β-D-thiogalactopyranoside (IPTG) to a final concentration of 0.5 mM to induce transcription of the NR contained on the plasmid and return to the 37°C shaking incubator for 2–3 more hours.
7. Harvest the cells by centrifugation in a GSA rotor at 6000 $\times g$ for 15 min at 4°C.
8. Resuspend the bacterial pellets in Lysis buffer [50 mM Tris pH 8.0, 250 mM KCl, 1% Triton X-100, 10 mM dithiothreitol (DTT), 500 μM Pefabloc SC] using a pipettor. Do not vortex.
9. Freeze the lysate on dry ice to induce cell lysis.
10. Thaw the lysate in a 37°C waterbath. Remove the lysate from the waterbath immediately upon thawing.
11. Pour the viscous lysate into 11 $\times$ 34 mm polycarbonate centrifuge tubes (Beckman #343778, Fullerton, CA) and centrifuge for 20 min at 80 K in a TLA 100.2 rotor in a Beckman LG-100 tabletop ultracentrifuge.
12. Remove the supernatant and add glycerol to a final concentration of 10%.
13. Determine the total protein concentration according to manufacturer's directions using a kit such as the BioRad (Hercules, CA) Protein Assay Dye Reagent.
14. Aliquot and store the protein at −80°C.

Preparation of Quick Spin Protein Columns

1. Remove the top cap and bottom tip from one Quick Spin Column for each sample. Place each column over one of the supplied 1.5 ml collection tubes. Place the column + collection tube in a 15 ml polypropylene conical tube (Becton Dickinson, Franklin Lakes, NJ).
2. Centrifuge the tubes in a swinging bucket rotor at 1000 $\times g$ for 3 min at 4°C.

3. Remove the column from the conical tube and discard the eluate and the collection tube.
4. Arrange each column in a new 1.5 ml collection tube in a 15 ml polypropylene conical tube and store at 4°C until needed (up to one hour).

Gel Filtration Saturation Assay for Estrogen Receptor Alpha LBD

This protocol will produce a 10-point saturation curve covering a 500-fold range of radioligand concentrations. For each radioligand concentration there will be a sample for determination of total (T) binding (no competitor) and a sample for determination of nonspecific (NS) binding. Nonspecific binding is determined by adding a competing ligand at a sufficiently high concentration to displace all specific binding of the radioligand. The example below was optimized for binding of ^{3}H-estradiol to estrogen receptor alpha (ERα). It will be necessary to optimize radioligand and protein concentrations for other receptors.

1. Add 140 μl gel filtration assay buffer (10 mM potassium phosphate pH 7.0, 2 mM EDTA, 50 mM NaCl, 1 mM DTT, 2 mM CHAPS, 10% glycerol, 500 μM Pefabloc, 1 μM Leupeptin, 1 μM Pepstatin) into a microcentrifuge tube on ice for each sample.
2. Dilute protein lysate in gel filtration assay buffer to 50 ng total protein/μl. Add 35 μl of diluted protein lysate into each sample tube.
3. Add 35 μl 10 μM 17-β-estradiol into sample tubes to be used for determination of NS binding. Add 35 μl buffer into tubes to be used for determination of T binding.
4. Dilute [1,4,6,7,16,17-^{3}H(N)]estradiol (Perkin Elmer, Boston, MA) with gel filtration assay buffer to 100 nM. Make two-fold dilutions of the radioligand solution to generate 400 μl of each of the following 2$\times$ stock solutions: 100, 50, 25, 12.5, 6.25, 3.1, 1.6, 0.8, 0.4, 0.2 nM radioligand.
5. Add 175 μl of each diluted 2$\times$ stock radioligand solution into NS sample tube (containing 17-β-estradiol) as well as a T sample tube.
6. Mix gently and incubate on ice for one hour.
7. Remove 100 μl of each sample and pipet onto the resin bed of a prepared Quick Spin Protein Column. Triplicate columns can be run for each sample to determine experimental error.
8. Immediately centrifuge samples in a swinging bucket rotor at 1000 $\times g$ for 3 min at 4°C.
9. Transfer eluates into scintillation tubes and add scintillation fluid.
10. Count samples on scintillation counter.

Determination of Precise Radioligand Concentration in the Assay Samples

1. Add duplicate 50 μl aliquots of the 2$\times$ stock radioligand solutions generated in step 4 into scintillation fluid and count on a scintillation counter.
2. Calculate the exact ligand concentration (L) in each 100 μl assay sample by converting DPM to nM using the specific activity (SA; Ci/mmol) and Eq. (1)

$$L \text{ (n}M) = (\text{DPM}/100 \text{ } \mu\text{l})^*(\text{Ci}/2.2 \times 10^{12} \text{ DPM})$$
$$^*(\text{mmol}/\text{SA Ci})^*1 \times 10^{12} \tag{1}$$

Data Analysis

1. Calculate specific binding by subtracting NS from T for each radioligand concentration and plot specific DPM versus radioligand concentration (Fig. 1A).
2. Determine K_d by non-linear regression using software such as Prism (GraphPad Software, San Diego, CA).

Gel Filtration Displacement Assay for Estrogen Receptor Alpha LBD

1. Add 100 μl gel filtration assay buffer into a microcentrifuge tube on ice for each sample.
2. Dilute protein lysate in gel filtration assay buffer to 50 ng total protein/μl.

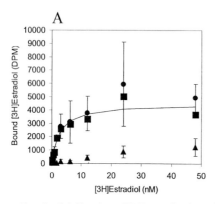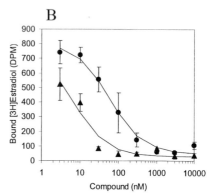

FIG. 1. Gel filtration. (A) Determination of K_d. Saturation binding assay for [³H]estradiol binding to ERα LBD. ● Total binding, ▲ nonspecific binding determined in the presence of 10 μM 17-β-estradiol, ■ specific binding is the difference between total and nonspecific binding. (B) Compound concentration response curves. Competition binding assay for [³H]estradiol binding to ERα. ● GI165638X, ▲ GI237604X.

3. Add 40 μl of diluted protein lysate to each sample.
4. Add 40 μl of test compound [10 × concentrate in assay buffer containing up to 0.1% dimethylsulfoxide (DMSO)] into sample tubes. For IC_{50} determination, run multiple concentrations spanning five orders of magnitude.
5. Add 220 μl 1.8 nM [1,4,6,7,16,17-^{3}H(N)]estradiol (Perkin Elmer) into each sample.
6. Mix gently and incubate on ice for one hour.
7. Remove 100 μl of each sample and transfer onto the resin bed of a prepared Quick Spin Protein Column. Triplicate columns can be run for each sample to determine experimental error.
8. Immediately centrifuge samples in a swinging bucket rotor at 1000 × g for 3 min at 4°C.
9. Transfer eluates into scintillation tubes and add scintillation fluid.
10. Count samples on scintillation counter.
11. Plot DPM values vs compound concentration (Fig. 1B). IC_{50} values can be generated using curve fitting software such as Prism.

Scintillation Proximity Assay

SPAs provide a tremendous increase in throughput compared to gel filtration. Because assays can be run in 96-well or 384-well plates, they are easily automated. The biggest advantage SPA provides compared to earlier methods is the elimination of the separation step: there is no need to separate bound radioligand from free radioligand. Binding of radioligand to NR that is captured on an SPA bead results in photon emission that can be detected by a scintillation counter. Radioligand molecules that do not bind the receptor do not come in close enough contact with the SPA bead to trigger photon emission. There are several capture methods available on the SPA beads including polylysine and polyethyleneimine for charge-based capture, glutathione beads for GST fusion protein capture, copper beads for His tag fusion protein capture and streptavidin beads for biotinylated protein capture.[7] Like gel filtration, SPA requires a radiolabeled ligand and thus is not feasible for orphan NR work. Also, like gel filtration, NR protein does not have to be purified for use in SPA. Crude lysate protein as detailed above can be used as illustrated in the following example. In these cases the amount of protein used to coat the bead must be determined empirically. For assays using purified protein, the amount of protein can be calculated based on bead capacity.

[7] J. S. Nichols, D. J. Parks, T. G. Consler, and S. G. Blanchard, *Anal. Biochem.* **257**, 112 (1998).

SPA Saturation Assay for Estrogen Receptor Alpha LBD

1. Add 10 μl SPA assay buffer (10 mM potassium phosphate pH 7.0, 2 mM EDTA, 50 mM NaCl, 1 mM DTT, 2 mM CHAPS, 10% glycerol, 500 μM Pefabloc, 1 μM Leupeptin, 1 μM Pepstatin) into wells in 96-well OptiPlate (Perkin Elmer) designated for determination of T binding.

2. Dilute protein lysate (see gel filtration protocol above) in SPA assay buffer to 33 ng total protein/μl. Add 15 μl of diluted protein lysate into each well.

3. Dilute polylysine yttrium silicate SPA beads (Amersham Pharmacia Biotech, Piscataway, NJ) in SPA assay buffer to 20 mg/ml. Place the bead suspension on a stir plate while pipetting to prevent the beads from settling out of suspension. Add 25 μl bead into each well.

4. Add 10 μl 10 μM 17-β-estradiol into wells designated for determination of NS binding.

5. Dilute [1,4,6,7,16,17-³H(N)]estradiol (Perkin Elmer) with SPA assay buffer to 100 nM. Make 2-fold dilutions of the radioligand solution to generate 400 μl of each of the following 2× stock solutions: 100, 50, 25, 12.5, 6.25, 3.1, 1.6, 0.8, 0.4, 0.2 nM radioligand.

6. Add 50 μl of each diluted 2× stock radioligand solution into NS and T wells.

7. Mix gently for one hour at room temperature then count on a TopCount (Perkin Elmer).

8. Determine precise radioligand concentration as in gel filtration protocol above.

9. Data analysis is performed as in gel filtration protocol (Fig. 2A).

SPA Displacement Assay for Estrogen Receptor Alpha LBD

1. Add 10 μl SPA assay buffer into wells of 96-well OptiPlate (Perkin Elmer) designated for determination of T binding.

2. Dilute protein lysate (see gel filtration protocol above) in SPA assay buffer to 50 ng total protein/μl. Add 10 μl of diluted protein lysate into each well.

3. Dilute polylysine yttrium silicate SPA beads (Amersham Pharmacia Biotech, Piscataway, NJ) in SPA assay buffer to 20 mg/ml. Place the bead suspension on a stir plate while pipetting to prevent the beads from settling out of suspension. Add 25 μl bead into each well.

4. Add 10 μl 10 μM 17-β-estradiol into wells designated for determination of NS binding.

5. Add 10 μl 10× stock test compounds into designated wells.

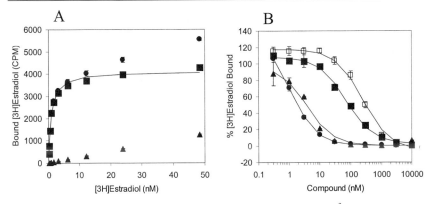

FIG. 2. SPA. (A) Determination of K_d. Saturation binding assay for [³H]estradiol binding to ERα LBD. ● Total binding, ▲ nonspecific binding determined in the presence of 10 μM 17-β-estradiol, ■ specific binding is the difference between total and nonspecific binding. (B) Compound concentration response curves. Competition binding assay for [³H]estradiol binding to ERα. ● 17-β-estradiol, ▲ ICI164384, ■ estrone, □ nafoxidine HCl.

6. Dilute [1,4,6,7,16,17-³H(N)]estradiol (Perkin Elmer) with SPA assay buffer to 1.8 nM. Pipet 55 μl into every well.
7. Mix gently for one hour at room temperature then count on a TopCount (Perkin Elmer).

Data Analysis

8. Mean CPM values for T (CPM_T) and NS (CPM_{NS}) are calculated for each plate and are used to calculate %[³H]Estradiol bound for each of the compound datapoints (CPM_{well}) using Eq. (2).
9. %[³H]Estradiol bound values for each compound datapoint are plotted against log[compound concentration]. Nonlinear regression is used to generate sigmoidal dose–response curves and obtain IC_{50} values (Fig. 2B).
10. K_i values are calculated from the IC_{50} values using the Cheng–Prosoff Eq. (3) and the K_d obtained from saturation binding studies.

$$\%[^3H]\text{Estradiol bound} = 100*[(CPM_{well} - CPM_{NS})/(CPM_T - CPM_{NS})] \tag{2}$$

$$K_i = IC_{50}/[1 + (\text{radioligand concentration}/K_d)] \tag{3}$$

Coactivator Interaction Assays

The discovery that coactivators and corepressors interact with NRs in a ligand mediated fashion[8] has provided new opportunities for NR assay development. Although this chapter will focus on coactivator–NR interaction assays, corepressors can be used as well. The mapping of the coactivator LXXLL motif interaction to the AF-2 region of the NR[9] has allowed utilization of purified NR LBD for these studies, which is advantageous because it can be difficult to express and purify full length NR. Several technologies have evolved that exploit the NR–coactivator interaction to detect ligand binding. Peptide sequences such as those containing the LXXLL motif have been utilized to discover natural and synthetic ligands for NRs.[10–12] Because coactivator interaction assays do not require a radiolabeled ligand it is possible to screen orphan NRs for which ligands have not yet been identified. However, it is optimal to purify the NR for each of these assay formats.[2]

As with radioligand binding assays, it is important to pay attention to assay conditions during development of coactivator interaction assays to ensure accurate interpretation of the results. To aid in the setup of these assays, the affinity of the NR–coactivator interaction ($Kd_{NR/C}$) should be determined when possible. If a known ligand for the NR is available, it is advantageous to also determine the $Kd_{NR/C}$ in the presence of a saturating amount of the ligand. In dose response studies for compounds it is desirable to use receptor and coactivator concentrations below the $Kd_{NR/C}$ for the receptor co-activator peptide complex. All of the assay methods described below can be used in various density formats, i.e., 96, 384, and 1536, and can be automated for high throughput screening campaigns.

Time-Resolved Fluorescence Resonance Energy Transfer (TR-FRET)

TR-FRET is a nonradioactive, homogenous proximity assay that utilizes the transfer of energy between two fluorescent probes. In the

[8] M. Tsai-Pflugfelder, S. M. Gasser, and W. Wahli, *Mol. Endo.* **12**(10), 1525 (1998).

[9] R. T. Nolte, G. B. Wisely, S. Westin, J. E. Cobb, M. H. Lambert, R. Kurokawa, M. G. Rosenfeld, T. M. Willson, C. K. Glass, and M. V. Milburn, *Nature* **395**, 137 (1998).

[10] G. Zhou, R. Cummings, Y. Li, S. Mitra, H. A. Wildinson, A. Elbrecht, J. D. Hermes, J. M. Schaeffer, R. G. Smith, and D. E. Moller, *Mol. Endo.* **12**, 1594 (1998).

[11] D. J. Parks, S. G. Blanchard, R. K. Bledsoe, G. Chandra, T. G. Consler, S. A. Kliewer, J. B. Stimmel, T. M. Willson, A. M. Zavacki, D. D. Moore, and J. M. Lehmann, *Science* **284**, 1365 (1999).

[12] M. Makishima, A. Y. Okamoto, J. J. Repa, H. Tu, R. M. Learned, A. Luk, M. V. Hull, K. D. Lustig, D. J. Mangelsdorf, and B. Shan, *Science* **284**, 1362 (1999).

following example, the coactivator peptide and NR are labeled with donor and acceptor fluorophores, respectively. Upon excitation of the donor fluorophore, energy transfer to the acceptor fluorophore occurs when the two are in sufficient proximity. Emission from the acceptor fluorophore can be detected in a time-resolved manner.[13] There are several donor–acceptor fluorophores in the market. Although expensive, the time-resolved component and sensitivity of lanthanide chelate-mediated energy transfer offer several advantages over traditional FRET-based assays. The incidence of compound interference is greatly reduced, sensitivity is greater and the background signal is usually lower than with other labels. There are many different ways to attach the donor and acceptor molecules to the coactivator peptide and the NR including antibody interactions and biotin–streptavidin. Kits for direct labeling with lanthanide chelates are also available (Perkin Elmer). Specialized plate readers such as a Perkin Elmer Victor with a time-resolved module are required for detecting these fluorophores.

TR-FRET Saturation Binding Assay for Farnesoid X-Activated Receptor (FXR) LBD

In coactivator interaction assays, the measurement of ligand affinity (Kd_L) is complicated by the coactivator–NR affinity. When attempting to determine the Kd_L, it is important to gather as much information as possible about the $Kd_{NR/C}$ of the NR–peptide interaction. Often $Kd_{NR/C}$ can be determined using other methods such as BIACore, Luminex, FP and AlphaScreen technologies. Some affinities between coactivator and NR in the absence of ligand are low. If a basal level of interaction exists, the determination of an IC_{50} for the interaction can be used to approximate the $Kd_{NR/C}$. An IC_{50} for the coactivator/NR interaction can also be determined in the presence of a known ligand.

In most cases it is optimal to label the coactivator peptide with a lanthanide chelate such as europium and the NR LBD with a suitable acceptor fluorophore such as allophycocyanin (APC), a fluorescent protein isolated from seaweed. Since the lanthanide chelates are expensive and at high concentrations will result in high background values, it is best to keep the peptide and europium at a lower, fixed concentration and titrate the APC labeled NR LBD. The following examples take advantage of the high affinity of a biotin streptavidin interaction by labeling both the purified biotinylated FXR and the biotinylated steroid receptor coactivator 1 (SRC1) peptide

[13] I. Hemmilä and S. Webb, Time-resolved fluorometry: an overview of the labels and core technologies for drug screening applications, *Drug Discov. Today* 2, 373–381 (1997).

(Biotin-CPSSHSSLTERHKILHRLLQEGSPS-CONH2) with APC labeled streptavidin and europium labeled streptavidin, respectively.

1. Add 162 μl of TR-FRET assay buffer (50 mM MOPS, pH 7.5, 50 mM NaF, 1 mg/ml fatty acid free BSA, 50 μM CHAPS, 5 mM DTT) into well Al of a deep well polypropylene 96 well plate (Beckman, Fullerton, CA).

2. Add purified biotinylated FXR LBD and streptavidin labeled APC (Molecular Probes, Eugene, OR) to give 5 μM of each in 350 μl volume. Gently mix by pipetting. Incubate at room temperature for one hour.

3. Add 3.5 μl of 10 mM biotin (Pierce Biotechnology, Inc., Rockford, IL) dissolved in DMSO to give a 20-fold molar excess of biotin. Gently mix and incubate at room temperature for 30 min.

4. Add 175 μl of TR FRET assay buffer into the remaining wells of row A. Serial dilute 2-fold across the plate by transferring 175 μl.

5. Add 1 μl of DMSO into row A, B, E and F of a black 96-well half area flat bottom assay plate (Corning Costar, Acton, MA).

6. Add 1 μl of a 500 μM solution of GW4064 into rows C and D of the assay plate.

7. Add 2.5 ml TR-FRET assay buffer into two separate polypropylene tubes for T and NS binding.

8. Add 2.5 μl 10 μM biotinylated SRC1 peptide in DMSO to the tube marked T, gently invert the tube to mix. Add 2.6 μl 9.5 μM europium labeled streptavindin (Perkin Elmer). Incubate 30 min then add 0.5 μl of 1 mM biotin in DMSO to block any remaining biotin binding sites.

9. Add 2.5 μl 10 μM non-biotinylated SRC1 peptide in DMSO to the tube marked NS, gently invert the tube to mix. Add 0.5 μl of 1 mM biotin in DMSO. Add 2.6 μl of 9.5 μM europium labeled streptavindin.

10. Transfer 25 μl of each concentration of the serially diluted biotinylated FXR–APC complex from the deep well plate to rows A, B, C, D, E, F of the assay plate (i.e., A1 of the deep well plate to A1, B1, C1, D1, E1, and F1 of the assay plate).

11. Add 25 μl of the tube marked T to rows A, B, C and D of the assay plate.

12. Add 25 μl of the tube marked NS to rows E and F of the assay plate.

13. Seal the plate to prevent evaporation.

14. Incubate at room temperature for 2 hr. Remove seal and count in a Victor 2V plate reader set up in a time resolved mode, collecting data at 620 and 665 nM.

Data Analysis

15. Divide the counts obtained at 665 nM by the counts obtained at 620 nM.

16. For each concentration of receptor, subtract the averaged NS values from the averaged T values obtained both in the presence and absence of ligand.

17. Plot these values versus FXR concentration for both liganded and unliganded receptor and fit to nonlinear regression (Fig. 3A).

TR-FRET Compound Concentration Response Assay
(EC_{50} Determination) for FXR LBD

1. Add 1 μl 50 × stock test compounds in DMSO into designated wells of a black 96-well half area flat bottom assay plate. Add 1 μl DMSO into wells designated for determination of basal (B) binding. Add 1 μl 250 μM GW4064 in DMSO into wells designated for determination of maximal (M) binding.

2. Add 3 ml of TR-FRET assay buffer into each of two polypropylene tubes.

3. Add 2.8 μl 21 μM biotinylated FXR LBD into tube 1 to give 20 nM FXR. Add 9.8 μl APC labeled streptavidin to 20 nM. Mix gently by inversion of the tube and incubate at room temperature for 30 min.

4. Add 0.6 μl 100 μM biotinylated SRC1 LCD2 peptide in DMSO into tube 2. Add 6.25 μl 9.6 μM europium-labeled streptavidin to give 20 nM. Mix gently by inversion of the tube and incubate at room temperature for 30 min.

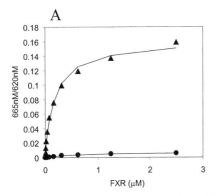 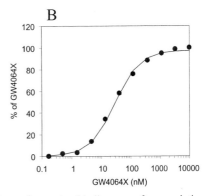

FIG. 3. TR-FRET. (A) Determination of $Kd_{NR/C}$. Saturation binding assay for association of SRC1 peptide to FXR LBD. ● without ligand, ▲ with 10 μM GW4064. (B) Compound concentration response curve. ● GW4064.

5. To both tubes add 1.2 μl 1 mM biotin in DMSO to give a 20-fold molar excess of biotin, incubate at room temperature for 15 min.
6. Mix the contents of tubes 1 and 2. Invert gently and add 50 μl to each well of the prepared assay plate.
7. Seal the plate and incubate at room temperature for 2 hr.
8. Remove seal and count in a Victor 2V plate reader set up in a time resolved mode, collecting data at 620 and 665 nM.

Data Analysis

9. Divide the counts obtained at 665 nM by the counts obtained at 620 nM.
10. Mean values for B and M are calculated for each plate.
11. %GW4064 response is calculated for each compound (C) concentration tested using Eq. (4).
12. %GW4064 Response values for each compound concentration are plotted against log[compound concentration]. Nonlinear regression is used to generate sigmoidal dose–response curves and obtain EC$_{50}$ values (Fig. 3B).

$$\% \text{ GW4064 Response} = 100*(C - B)/(M - B). \qquad (4)$$

Fluorescence Polarization (FP)

Fluorescence polarization is a technology that has been used for many years to detect ligand binding.[14] FP employs a fluorescently labeled molecule of relatively small molecular mass such as a coactivator peptide. Because the rate of rotation of a molecule in solution is related to its mass, binding of a small labeled molecule to a larger molecule, such as an NR, changes its rate of rotation. Instruments that are microtiter plate compatible have recently become available, making FP an attractive assay format. FP can be relatively inexpensive in comparison to other assay formats. The most common labels include fluoroscine, Rhodamine Green, carboxy-tetramethylrhodamine (TAMRA), and some of the Cy Dyes, which can be proprietary and expensive. There is potential for interference by the compounds being screened, so this should be considered when choosing a fluorescent label.

---

[14] J. R. Lakowicz, "Principles of Fluorescence Spectroscopy." Plenum Press, New York and London, 1983.

FP has been utilized in ligand identification campaigns for NRs[15] and in NR characterization assays.[16] In general, high concentrations of NR LBD are needed to obtain suitable quality control parameters for high throughput screens and structure activity relationship studies. If the supply of the NR protein of interest is not limiting, FP can be a useful screening technology.

FP Saturation Binding Assay for FXR LBD

It is important to gather as much information as possible about the $Kd_{NR/C}$ of the NR LBD–coactivator interaction. If a basal level of interaction exists, the determination of an IC_{50} for interaction can be used to approximate the $Kd_{NR/C}$. An IC_{50} for the interaction can also be determined in the presence of a known ligand.

1. Add 301 μl FP assay buffer (50 mM MOPS, pH 7.5, 50 mM NaF, 1 mg/ml fatty acid free BSA, 50 μM CHAPS, 5 mM DTT) into well Al of a deep well polypropylene 96-well plate. Add 49 μl 22 μM purified FXR LBD.

2. Add 175 μl FP assay buffer into the remaining wells of row A. Serially dilute 2-fold across the plate by transferring 175 μl.

3. Add 1 μl DMSO into rows A and B of a black 96-well half area flat bottom assay plate for determination of T binding.

4. Add 1 μl 500 μM GW4064 in DMSO into rows C and D.

5. Add 1 μl 10 mM unlabeled SRC1 peptide into column E and F for determination of NS binding.

6. Add 2.5 ml FP assay buffer into a polypropylene tube and add 2.5 μl 100 μM TAMRA-labeled SRC1 peptide solution in DMSO to give 100 nM.

7. Transfer 25 μl of each concentration of the serially diluted LBD from the deep well plate to rows A, B, C, D, E, and F of the assay plate.

8. Add 25 μl 100 nM TAMRA-labeled SRCl solution to rows A, B, C, D, E and F of the assay plate.

9. Seal the plate to prevent evaporation.

10. Incubate at room temperature for 2 hr. Remove seal and acquire mP values in an Acquest counter (LJL BioSystems, Inc., Sunnyvale, CA).

[15] J. R. Schultz, H. Tu, A. Luk, J. J. Repa, J. C. Medina, L. Li, S. Schwendner, S. Want, M. Thoolen, D. J. Mangelsdorf, K. D. Lustig, and B. Shan, *Genes Dev.* **14**, 2831 (2000).

[16] H. E. Xu, T. B. Stanley, V. G. Montana, M. H. Lambert, B. G. Shearer, J. E. Cobb, D. D. McKee, C. M. Galardi, K. D. Plunket, R. T. Nolte, D. J. Parks, J. T. Moore, S. A. Kliewer, T. M. Willson, and J. B. Stimmel, *Nature* **415**, 813 (2002).

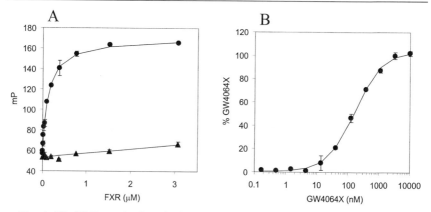

Fig. 4. FP. (A) Determination of $Kd_{NR/C}$. Saturation binding assay for association of SRC1 peptide to FXR LBD. ● without ligand, ▲ with 10 μM GW4064. (B) Compound concentration response curves. ● GW4064.

Data Analysis

11. For each concentration of receptor, subtract the averaged NS values from the averaged T and GW4064 values.

12. Plot the values versus LBD concentration for values obtained both for liganded and unliganded receptor and fit a curve using nonlinear regression (Fig. 4A).

FP Compound Concentration Response Assay for FXR LBD

1. Add 1 μl 50 × stock test compounds in DMSO into designated wells of a black 96-well half area flat bottom assay plate. Add 1 μl DMSO into wells designated for determination of basal (B) binding. Add 1 μl 250 μM GW4064 in DMSO into wells designated for determination of maximal (M) binding.

2. Add 6 ml FP assay into a polypropylene tube.

3. Add 54 μl 22 μM FXR LBD into the polypropylene tube. Mix gently by inverting the tube.

4. Add 3 μl 100 μM TAMRA labeled SRC1 in DMSO to the FXR solution. Invert gently to mix. Incubate at room temperature for 30 min.

5. Add 50 μl of the FXR SRC1 solution to each well of the prepared assay plate.

6. Seal or cover the plate and incubate at room temperature for 2 hr.

7. Remove seal and and acquire mP values in an Acquest counter.

Data Analysis

8. Mean values for B and M are calculated for each plate.
9. %GW4064 response is calculated for each compound (C) concentration tested using Eq. (3).
10. %GW4064 Response values for each compound concentration are plotted against log[compound concentration]. Nonlinear regression is used to generate sigmoidal dose–response curves and obtain EC_{50} values (Fig. 4B).

Amplified Luminescent Proximity Homogeneous Assay (ALPHAScreen)

ALPHAScreen is a relatively new technology that has recently been applied to NR coactivator recruitment assays.[11,17,18] In these assays the NR and the coactivator are coupled to derivatized polystyrene microbeads using typical techniques such as antibody capture, His-tag–nickel chelate or streptavidin–biotin. In the example that follows the coactivator is coupled to a "donor bead" whereas the NR is coupled to an "acceptor bead." Upon excitation of the "donor bead," singlet oxygen is released and will diffuse up to 200 nm before decaying. If the singlet oxygen encounters an "acceptor bead," the fluorophores encapsulated within the "acceptor bead" will emit a luminescent signal. Thus, ALPHAScreen is a nonradioactive homogenous proximity assay. ALPHAScreen is very sensitive, easy to use, easy to automate, and the detection kits are very reasonably priced. Specialized readers are required for reading the ALPHAScreen assays.

Because of the limitations imposed by the bead capacity, it is technically challenging to determine an $Kd_{NR/C}$ for NR–coactivator interactions in the traditional sense using ALPHAScreen. It is possible to estimate the $Kd_{NR/C}$ using uncoupled SRC1 peptide to generate an IC_{50} for the NR–coactivator interaction in the presence and absence of ligand. Technically this experiment is done in the same way as when generating an IC_{50} for a compound.

ALPHAScreen Assay for Estrogen Receptor Beta (ERβ)

1. Pipette 1 μl 50 × stock test compounds (or SRC1 peptide) in DMSO into designated wells of a black 96-well half area flat bottom assay plate.

[17] J. G. Glickman, X. Wu, R. Mercure, C. Illy, B. R. Bowen, Y. He, and M. Sills, *J. Biomol. Screen.* **7**, 3 (2002).

[18] R. K. Bledsoe, V. G. Montana, T. B. Stanley, C. J. Delves, C. J. Apolito, D. D. McKee, T. G. Consler, D. J. Parks, E. L. Stewart, T. M. Willson, M. H. Lambert, J. T. Moore, K. H. Pearce, and H. E. Xu, Crystal structure of the glucocorticoid receptor ligand binding domain reveals a novel mode of receptor dimerization and coactivator recognition, *Cell* **110**, 93 (2002).

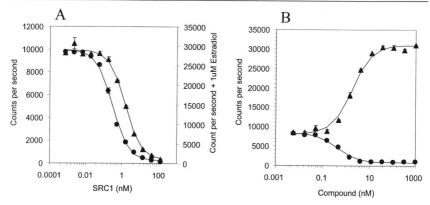

FIG. 5. ALPHAScreen. (A) Estimation of $Kd_{NR/C}$ for estrogen receptor beta LBD using SRC1 peptide ● without ligand, ▲ with 1 μM Estradiol. (B) Compound concentration response curves. ● Raloxifene, ▲ estradiol.

2. Pipette 12 ml assay buffer (50 mM MOPS, pH 7.5, 50 mM NaF, 1 mg/ml fatty acid free BSA, 50 μM CHAPS, 5 mM DTT) into a polypropylene tube.

3. Add 9.6 μl 5 mg/ml nickel chelate coated acceptor beads (Perkin Elmer) into the polypropylene tube. Add 1.2 μl 22 μM HIS-ERβ into the tube and mix gently by inversion.

4. Add 9.6 μl 5 mg/ml streptavidin coated donor beads (Perkin Elmer) to the tube. Add 3 μl 10 μM biotinylated SRC1 and mix gently by inversion.

5. Add 50 μl to each well of the prepared assay plate.

6. Seal plate and incubate at room temperature for 2 hr.

7. Collect data in counts per second on a Fusion or AlphaQuest (Perkin Elmer).

Data Analysis

8. Counts per second values for each compound datapoint are plotted against log[compound concentration]. Nonlinear regression is used to generate sigmoidal dose–response curves and obtain IC_{50} values (Fig. 5).

Summary

There has been tremendous progress in the development of ligand binding assays for NRs during the past several years. A major development has been the advent of homogeneous assay formats, including those that do not require a radioligand. These high throughput, low volume assay formats will be powerful tools for the identification and characterization of novel NR ligands, including both natural ligands for orphan NRs and new drugs that mediate their therapeutic effects through this class of receptors.

[5] Design and Synthesis of Receptor Ligands

By Hikari A. I. Yoshihara, Ngoc-Ha Nguyen, and Thomas S. Scanlan

Introduction

Nuclear Receptor Ligands

The members of the nuclear hormone receptor (NR) superfamily are generally regarded as ligand-activated transcriptional regulators that play important roles in modulating developmental and physiological processes.[1,2] Synthetic ligands of nuclear receptors have a number of properties that make them useful as pharmacological probes in the study of nuclear receptor action. These properties include: selective action, either in specifically binding to a receptor orthologue, isoform, or subtype; selective activation of specific hormone response elements; or antagonism, where a ligand competes with the natural hormone for binding to the receptor but does not activate it. In the case of orphan receptors for which a natural ligand has not been identified, a synthetic ligand may provide a means of artificially modulating the activity of the receptor and in so doing, reveal insight into the biological role of the protein.[3,4]

Nuclear hormone receptors play important roles in development and homeostasis and are implicated in a variety of diseases such as cancer, diabetes, and various endocrine disorders. The roles of the RXR, TR, LXR, FXR, and PPAR receptors in regulating lipid homeostasis[5] also make them potentially attractive targets to treat or prevent cardiovascular diseases.

[1] R. M. Evans, *Science* **240**, 885 (1988).

[2] D. J. Mangelsdorf, C. Thummel, M. Beato, P. Herrlich, G. Schutz, K. Umesono, B. Blumberg, P. Kastner, M. Mark, P. Chambon, and R. M. Evans, *Cell* **83**, 835 (1995).

[3] D. M. Kochhar, H. Jiang, J. D. Penner, A. T. Johnson, and R. A. S. Chandraratna, *Int. J. Dev. Biol.* **42**, 601 (1998).

[4] J. J. Repa, S. D. Turley, J. M. A. Lobaccaro, J. Medina, L. Li, K. Lustig, B. Shan, R. A. Heyman, J. M. Dietschy, and D. J. Mangelsdorf, *Science* **289**, 1524 (2000).

[5] A. Chawla, J. J. Repa, R. M. Evans, and D. J. Mangelsdorf, *Science* **294**, 1866 (2001).

Often, the medical use of the natural ligand, or compounds that closely mimic it, is less than ideal because the pleiotropic actions of the receptor can lead to undesirable side effects. Selective activators (or inhibitors) of nuclear receptors have the promise to reduce these side effects, thereby acting as safer and more useful drugs.

Biosynthesis

The known physiological ligands of NRs are of varied biosynthetic origin. Humans are able to synthesize some ligands, such as the steroids, from basic metabolites, while others require precursors obtained from the diet. The classic steroid hormones are derived from cholesterol by enzymatic modification.[6,7] Other cholesterol metabolites, including oxysterols and bile acids, serve as ligands for LXR[8,9] and FXR, respectively.[10–12] Retinoids are produced from the oxidative cleavage of β-carotene,[13] an essential nutrient. The respective PPARγ and PPARα ligands, 15-deoxy-$\Delta^{12,14}$-prostaglandin J$_2$[14,15] and leukotriene B$_4$,[16] are derived from arachidonic acid, which is produced from essential fatty acids. Thyroid hormone is unusual among NR ligands in that it is produced by the degradation of a precursor protein, thyroglobulin, which contains oxidatively coupled iodinated tyrosine residues.[17]

[6] D. B. Gower, in: H. L. J. Malkin (ed.), "Biochemistry of Steroid Hormones." Blackwell Scientific Publishers, Oxford, 1975.

[7] D. B. Gower and K. Fotherby, *in* "Biochemistry of Steroid Hormones" (H. L. J. Malkin, ed.). Blackwell Scientific Publishers, Oxford, 1975.

[8] B. A. Janowski, P. J. Willy, T. R. Devi, J. R. Falck, and D. J. Mangelsdorf, *Nature* **383**, 728 (1996).

[9] J. M. Lehmann, S. A. Kliewer, L. B. Moore, T. A. SmithOliver, B. B. Oliver, J. L. Su, S. S. Sundseth, D. A. Winegar, D. E. Blanchard, T. A. Spencer, and T. M. Willson, *J. Biol. Chem.* **272**, 3137 (1997).

[10] D. J. Parks, S. G. Blanchard, R. K. Bledsoe, G. Chandra, T. G. Consler, S. A. Kliewer, J. B. Stimmel, T. M. Willson, A. M. Zavacki, D. D. Moore, and J. M. Lehmann, *Science* **284**, 1365 (1999).

[11] H. B. Wang, J. Chen, K. Hollister, L. C. Sowers, and B. M. Forman, *Mol. Cell* **3**, 543 (1999).

[12] M. Makishima, A. Y. Okamoto, J. J. Repa, H. Tu, R. M. Learned, A. Luk, M. V. Hull, K. D. Lustig, D. J. Mangelsdorf, and B. Shan, *Science* **284**, 1362 (1999).

[13] W. S. Blaner and J. A. Olsen, *in* "The Retinoids: Biology, Chemistry, and Medicine" (M. B. Sporn, A. B. Roberts, and S. S. Goodman, eds.). Raven Press, New York, 1994.

[14] S. A. Kliewer, J. M. Lenhard, T. M. Willson, I. Patel, D. C. Morris, and J. M. Lehmann, *Cell* **83**, 813 (1995).

[15] B. M. Forman, P. Tontonoz, J. Chen, R. P. Brun, B. M. Spiegelman, and R. M. Evans, *Cell* **83**, 803 (1995).

[16] P. R. Devchand, H. Keller, J. M. Peters, M. Vazquez, F. J. Gonzalez, and W. Wahli, *Nature* **384**, 39 (1996).

[17] P. M. Yen, *Physiol. Rev.* **81**, 1097 (2001).

Chemical Features of NR Ligands

While natural ligands of NRs are produced by a variety of different pathways, the ligands themselves share common features. They are largely hydrophobic in character, consisting mostly of aliphatic, aromatic, or olefinic hydrocarbons. They have an elongated chemical structure with polar groups—oxo, carboxyl, or hydroxyl—at one or both ends. Additionally they have roughly the same Van der Waals volume—about 320 Å[3]—at least when considering thyroid hormones, steroid hormones, and retinoids.[18]

Structural Studies of Ligand–Receptor Interactions

X-ray crystal structures of the ligand-binding domains (LBDs) of various nuclear receptors show a common binding mode for their cognate ligands.[19–27] The bound ligand is buried completely within the protein, and forms a hydrophobic core of an internal subdomain. The ligand-binding pocket is complementary to the ligand with regard to steric and polar characteristics. Hydrophobic parts of the ligand are in contact with the side chains of hydrophobic residues, while the polar or charged functional groups at the ends of the ligand interact with polar or charged amino acid side chains. The tight fit between ligand and protein is reflected in the very strong conservation of ligand contact residues of each receptor type. Mutation of these residues can produce a receptor with altered ligand

[18] A. A. Bogan, F. E. Cohen, and T. S. Scanlan, *Nat. Struct. Biol.* **5**, 679 (1998).

[19] J. P. Renaud, N. Rochel, M. Ruff, V. Vivat, P. Chambon, H. Gronemeyer, and D. Moras, *Nature* **378**, 681 (1995).

[20] R. L. Wagner, J. W. Apriletti, M. E. McGrath, B. L. West, J. D. Baxter, and R. J. Fletterick, *Nature* **378**, 690 (1995).

[21] R. T. Nolte, G. B. Wisely, S. Westin, J. E. Cobb, M. H. Lambert, R. Kurokawa, M. G. Rosenfeld, T. M. Willson, C. K. Glass, and M. V. Milburn, *Nature* **395**, 137 (1998).

[22] A. M. Brzozowski, A. C. Pike, Z. Dauter, R. E. Hubbard, T. Bonn, O. Engstrom, L. Ohman, G. L. Greene, J. A. Gustafsson, and M. Carlquist, *Nature* **389**, 753 (1997).

[23] P. M. Matias, P. Donner, R. Coelho, M. Thomaz, C. Peixoto, S. Macedo, N. Otto, S. Joschko, P. Scholz, A. Wegg, S. Basler, M. Schafer, U. Egner, and M. A. Carrondo, *J. Biol. Chem.* **275**, 26164 (2000).

[24] P. F. Egea, A. Mitschler, N. Rochel, M. Ruff, P. Chambon, and D. Moras, *EMBO J.* **19**, 2592 (2000).

[25] N. Rochel, J. M. Wurtz, A. Mitschler, B. Klaholz, and D. Moras, *Mol. Cell* **5**, 173 (2000).

[26] R. E. Watkins, G. B. Wisely, L. B. Moore, J. L. Collins, M. H. Lambert, S. P. Williams, T. M. Willson, S. A. Kliewer, and M. R. Redinbo, *Science* **292**, 2329 (2001).

[27] H. E. Xu, T. B. Stanley, V. G. Montana, M. H. Lambert, B. G. Shearer, J. E. Cobb, D. D. McKee, C. M. Galardi, K. D. Plunket, R. T. Nolte, D. J. Parks, J. T. Moore, S. A. Kliewer, T. M. Willson, and J. B. Stimmel, *Nature* **415**, 813 (2002).

specificity. Among these altered receptors are those that do not bind the natural ligand tightly but instead prefer synthetic ligands that the wild-type receptor does not bind.[28–30] Similarly, mutation of variable residues in the ligand-binding pockets of receptor isoforms can alter the specificity of isoform-selective ligands.[31,32]

Synthetic ligands for NRs generally share many structural features with their natural counterparts. For high-affinity binding, it is important to retain functional groups on the ligand with correct orientation that allow important interactions, but do not create steric clashes with the protein.

Thyroid Hormone Receptor as Model System for Ligand Development

Among the NR superfamily members, the thyroid hormone receptor (TR) is relatively well-characterized from a structural standpoint, and is a useful system for the general design and synthesis of receptor ligands. Several thyroid hormone (3,3′,5-triiodo-L-thyronine, T_3, Fig. 1A) agonists and antagonists have been successfully developed recently[33–36] through the combined use of extensive structure–activity relationship (SAR) data[37,38] and the X-ray crystal structures of liganded TR complexes.[20,32,39] The SAR studies reveal the minimal requirements for a high-affinity TR ligand: (1) the

[28] D. J. Peet, D. F. Doyle, D. R. Corey, and D. J. Mangelsdorf, *Chem. Biol.* **5**, 13 (1998).

[29] D. F. Doyle, D. A. Braasch, L. K. Jackson, H. E. Weiss, M. F. Boehm, D. J. Mangelsdorf, and D. R. Corey, *J. Am. Chem. Soc.* **123**, 11367 (2001).

[30] Y. H. Shi and J. T. Koh, *Chem. Biol.* **8**, 501 (2001).

[31] M. Gehin, V. Vivat, J. M. Wurtz, R. Losson, P. Chambon, D. Moras, and H. Gronemeyer, *Chem. Biol.* **6**, 519 (1999).

[32] R. L. Wagner, B. R. Huber, A. K. Shiau, A. Kelly, S. T. C. Lima, T. S. Scanlan, J. W. Apriletti, J. D. Baxter, B. L. West, and R. J. Fletterick, *Mol. Endocrinol.* **15**, 398 (2001).

[33] T. S. Scanlan, H. A. Yoshihara, N. H. Nguyen, and G. Chiellini, *Curr. Opin. Drug Discov. Dev.* **4**, 614 (2001).

[34] H. A. I. Yoshihara, J. W. Apriletti, J. D. Baxter, and T. S. Scanlan, *Bioorg. Med. Chem. Lett.* **11**, 2821 (2001).

[35] G. Chiellini, N. H. Nguyen, J. W. Apriletti, J. D. Baxter, and T. S. Scanlan, *Bioorg. Med. Chem.* **10**, 333 (2002).

[36] J. D. Baxter, P. Goede, J. W. Apriletti, B. L. West, W. Feng, K. Mellstrom, R. J. Fletterick, R. L. Wagner, P. J. Kushner, R. C. Ribeiro, P. Webb, T. S. Scanlan, and S. Nilsson, *Endocrinology* **143**, 517 (2002).

[37] E. C. Jorgensen, *in* "Hormonal Proteins and Peptides" (C. H. Li, ed.). Vol. 6, p. 57. Academic Press, New York, 1978.

[38] E. C. Jorgensen, *in* "Hormonal Proteins and Peptides" (C. H. Li, ed.). Vol. 6, p. 107. Academic Press, New York, 1978.

[39] B. D. Darimont, R. L. Wagner, J. W. Apriletti, M. R. Stallcup, P. J. Kushner, J. D. Baxter, R. J. Fletterick, and K. R. Yamamoto, *Genes Dev.* **12**, 3343 (1998).

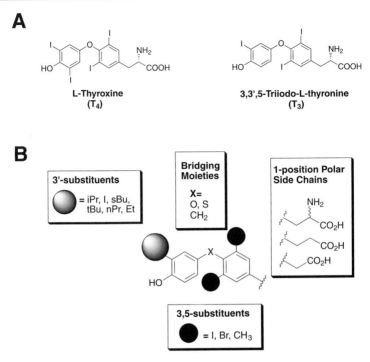

FIG. 1. (A) Chemical structures of natural thyroid hormones excreted by the thyroid gland. Deiodination of T_4 to T_3 in peripheral tissues gives the more active form of the hormone. (B) General profile of thyromimetic pharmacophore based on SAR data.

$4'$-hydroxyl- and carboxylic acid-containing side chain at the 1-position are essential for tight TR binding and thyromimetic activity, (2) halogen or hydrocarbon substitution at the 3, 5, $3'$-positions is required (Fig. 1B). This crude picture of the thyromimetic pharmacophore is further supported by the crystal structure, which shows the $4'$- and 1-positions of the thyronine structure are involved, respectively, in hydrogen bonding and electrostatic contacts with residues in the binding pocket. The 3,5- and $3'$-iodo substituents of T_3 reside in small hydrophobic pockets, and the ligand occupies almost 90% of the volume of the binding pocket.[20] Substituents significantly larger than iodine atoms hinder ligand binding and activity, presumably due to size limitations in the binding pocket of the receptor. These results provide important criteria required for developing high-affinity T_3 analogues.

Chemically, the thyronine structure is not an ideal scaffold upon which to design analogues. The iodine atoms of T_3 are highly susceptible to deiodination and therefore restrict the types of reagents that can be used in

FIG. 2. Chemical structures of thyromimetics DIMIT and GC-1.

subsequent chemical transformations. Synthesis of the biaryl ether is also a challenging problem, especially in the presence of the sterically bulky 3,5-diiodo substituents. Thus, it was necessary at the outset to design a chemically inert thyromimetic scaffold of high-binding affinity that would allow application of diverse synthetic methodologies for the preparation of agonist and antagonist analogues.

Design and Synthesis of Thyromimetic GC-1

The design of GC-1 was based on the structure of the halogen-free thyromimetic DIMIT (3,5-dimethyl-3'-isopropyl thyronine, Fig. 2). We incorporated changes that should not affect affinity adversely and should simplify the chemical synthesis. In GC-1, the DIMIT-like aryl–alkyl group substitution pattern is retained, but the biaryl ether oxygen is replaced with a methylene group, and the chiral L-alanine polar side chain is replaced with an achiral oxyacetic acid group (Scheme 1).

The alkyl groups at the 3,5- and 3'-positions provide the hydrophobic substitution requirement at these positions without the synthetic limitations imposed by the presence of aryl iodides. These substitutions do not affect affinity equally; the T_3 analogue with a 3'-isopropyl is slightly more active (104%) than T_3 itself, while the 3,5-diiodo to 3,5-dimethyl substitution in the context of thyronines results in a 150-fold reduction of activity.[38] Thus, DIMIT itself has 7% the activity of T_3 in the rat anti-goiter assay. However, these modifications were deemed a worthy compromise when considering the synthetic versatility gained by using the alkyl substitutions.

Another synthetic consideration was the ether bridge linking the two aromatic rings of thyronines. At the time these studies were performed, a good method for the preparation of biaryl ethers of the thyronine type was not available. Fortunately, new biaryl ether-forming reactions have since been reported,[40] and the Cu(II)-mediated coupling of aryl boronic acids with phenols developed by Evans et al.[41] has worked very well in our hands

[40] J. S. Sawyer, Tetrahedron 56, 5045 (2000).
[41] D. A. Evans, J. L. Katz, and T. R. West, Tetrahedron Lett. 39, 2937 (1998).

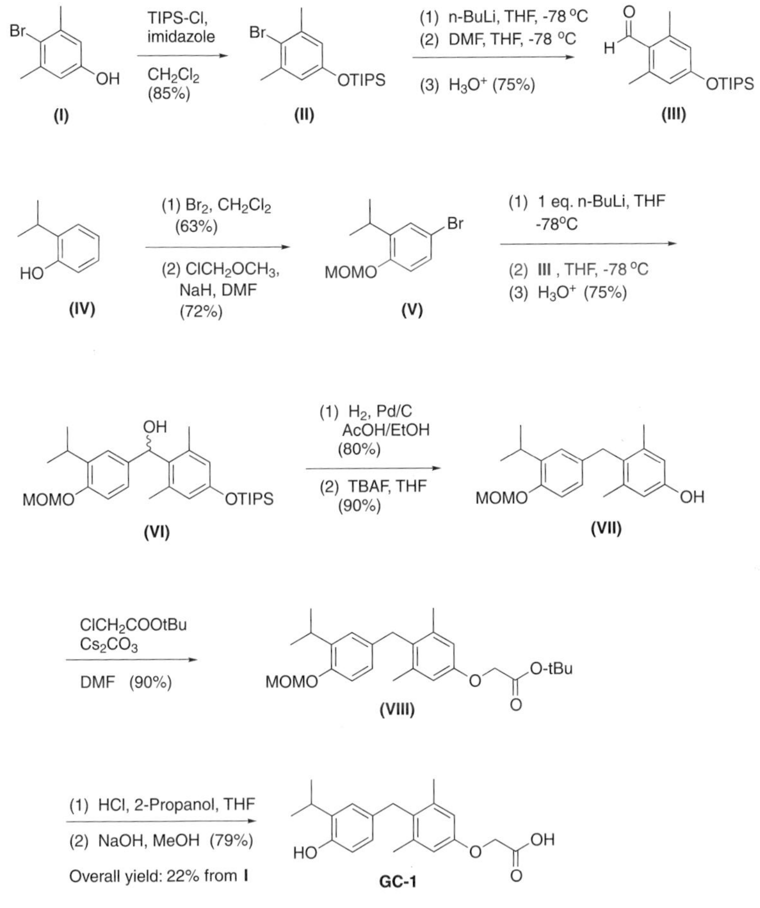

SCHEME 1. Synthetic route for the preparation of GC-1.

for the preparation of thyronine-like compounds. The methylene-bridged analogue of T_3, however, was a known analogue with high activity[42] and we therefore decided to incorporate the methylene bridge into our thyromimetic design. There are robust standard methods for forming the C–C bond in this system, and a methylene bridge provides an additional site for substitution unavailable with the biaryl ether.

While SAR studies of T_3 have shown that the alanine side chain is important for activity, the side chain length and the presence of a carboxylic acid group are the most important features. Chain lengths of two to three

[42] S. L. Tripp, F. B. Block, and G. Barile, *J. Med. Chem.* **16**, 60 (1973).

carbon atoms are well tolerated. For T_3, the chiral amino group can be in either the L or D configuration, and its analogue with a propionic acid side chain, lacking the amino group, is highly active. We chose the oxyacetic acid polar side chain as it is isosteric to propionic acid, and the aryl oxygen provides a convenient route to its preparation from a protected phenolic intermediate. The chemical synthesis of GC-1 is straightforward and improvements have been made to the original route that allow multigram quantities to be prepared in a standard chemical laboratory.[43,44]

Interestingly, GC-1 has a high affinity for TR—higher than was expected from its design and the available T_3 SAR data. The structural changes from DIMIT, while making the synthetic preparation easier, also resulted in a more potent thyromimetic. GC-1 binds to $TR\alpha_1$ with a K_d of 440 ± 120 pM and $TR\beta_1$ with a K_d of 67 ± 4 pM.[43] This selectivity in binding translates to different potencies for the $TR\alpha$ and $TR\beta$ isoforms in transient transfection assays.[43] *In vivo* effects of GC-1 in tadpoles[45] and rodents[46] are distinct from those of T_3 and are consistent with GC-1 behaving as a $TR\beta$-selective thyroid hormone agonist. The $TR\beta$-selectivity of GC-1 was not anticipated in its design, and illustrates both our limitations in predicting ligand-binding properties and how the understanding of thyroid hormone signaling can be improved by the development of new thyromimetic ligands.

The X-ray crystal structure of GC-1 bound to $hTR\beta1$-LBD (Fig. 3) shows that TR binds GC-1 in much the same way as it does T_3, DIMIT, and 3,3′,5-triiodothryoacetic acid (triac); however, the H-bond network of charged and polar residues around the GC-1 oxyacetic acid side chain is different from that observed with triac.[32] Asparagine 331, the one amino acid residue in the ligand-binding pocket that varies between $TR\beta$ and $TR\alpha$, participates in this H-bond network and provides a structural rationale for GC-1's selectivity. (The corresponding residue in $TR\alpha$ is Ser277.)

From the perspective of designing agonistic and antagonistic TR ligands, GC-1 is a good starting point. It is a shape/volume mimic of T_3 that contains all the essential molecular recognition components to ensure high-affinity binding to TR. It also fulfills the synthetic requirements of a thyromimetic scaffold, being chemically inert and easily synthesized,

[43] G. Chiellini, J. W. Apriletti, H. A. I. Yoshihara, J. D. Baxter, R. C. Ribeiro, and T. S. Scanlan, *Chem. Biol.* **5**, 299 (1998).

[44] G. Chiellini, N. H. Nguyen, H. A. I. Yoshihara, and T. S. Scanlan, *Bioorg. Med. Chem. Lett.* **10**, 2607 (2000).

[45] J. D. Baxter, W. H. Dillmann, B. L. West, R. Huber, J. D. Furlow, R. J. Fletterick, P. Webb, J. W. Apriletti, and T. S. Scanlan, *J. Steroid Biochem. Mol. Biol.* **76**, 31 (2001).

[46] S. U. Trost, E. Swanson, B. Gloss, D. B. Wang-Iverson, H. J. Zhang, T. Volodarsky, G. J. Grover, J. D. Baxter, G. Chiellini, T. S. Scanlan, and W. H. Dillmann, *Endocrinology* **141**, 3057 (2000).

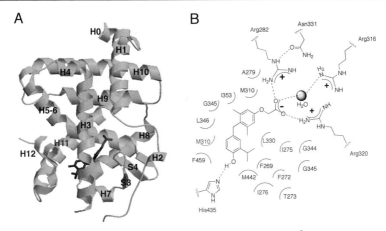

Fig. 3. (A) Cocrystal structure of GC-1 bound to hTRβ-LBD at 2.9 Å resolution. Like other liganded NRs, GC-1 is completely buried within the LBD. (B) Schematic diagram of residues lining the ligand-binding pocket that make contact with GC-1. Receptor–ligand contacts are like those observed with other thyromimetics and T_3. Figures were prepared with MolScript [P. J. Kraulis, *J. Appl. Crystallogr.* **24**, 946 (1991)] and Raster3D [E. A. Merritt and D. J. Bacon, *Methods Enzymol.* **277**, 505 (1997)].

and has a synthetic route amenable to scale-up and production of analogues.

Design of T_3 Antagonists: Extension Hypothesis

Our strategy for making TR antagonists incorporated the design principles of antagonist ligands of other NRs. As shown in Fig. 4, comparison of known antagonists across the NR superfamily reveals the same skeletal structures that allow tight binding to the receptors as are found in the corresponding agonists, but with a large extension protruding from the molecule to induce an abnormal fold relative to the agonist-liganded receptor. This "extension hypothesis" suggests that antagonistic ligands of NRs can be made by the addition of a largely hydrophobic appendage to the structure of an agonist ligand.[47]

Using GC-1 as the agonist core, we explored two potential sites to build extensions for developing antagonist ligands for TR: the methylene bridge center and the 5′-position (Fig. 5). We describe herein the design of antagonists HY-4 and NH-3 to illustrate the application of the "extension

[47] R. C. Ribeiro, J. W. Apriletti, R. L. Wagner, B. L. West, W. Feng, R. Huber, P. J. Kushner, S. Nilsson, T. Scanlan, R. J. Fletterick, F. Schaufele, and J. D. Baxter, *Recent Prog. Horm. Res.* **53**, 351 (1998).

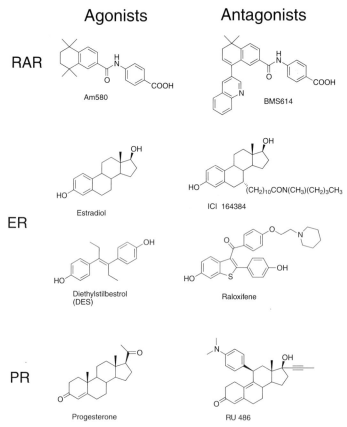

FIG. 4. Structures of selected agonist and antagonist ligands for NRs. Antagonists have similar agonist core structure, but also contain largely hydrophobic extension groups (highlighted in grey).

hypothesis." HY-4 is an analogue with a long alkylamide appendage attached to the bridging carbon of GC-1.[34] NH-3 is the GC-1 analogue with a 4-nitrophenylethynyl appendage at the 5'-position.[48]

TR Antagonist HY-4

Design

The design of HY-4 was conceived while inspecting the cocrystal structure of the estrogen receptor α ligand-binding domain (ERα-LBD) and

[48] N. H. Nguyen, J. W. Apriletti, S. T. Cunha Lima, P. Webb, J. D. Baxter, and T. S. Scanlan, J. Med. Chem. **45**, 3310 (2002).

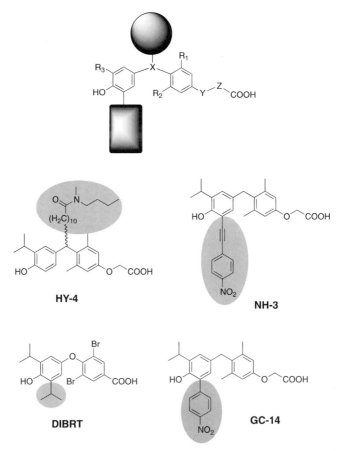

FIG. 5. Chemical structures of known thyroid hormone antagonists. HY-4, GC-14, and NH-3 are based on the halogen-free GC-1 agonist core structure; DIBRT is a thyronine-like compound containing biaryl ether. HY-4 contains an extension group stemming from the methylene bridging carbon; GC-14, NH-3, and DIBRT contain extensions at the 5′-position.

17β-estradiol (E_2).[22] The ER antagonist ICI-164,384 (ICI) consists of the steroid core of 17β-estradiol with a long alkylamide side chain substituent at the 7α-position (Fig. 4). Assuming ER binds the 17β-estradiol core in the same orientation for both 17β-estradiol and ICI, the side chain would project out toward helix 8 and the beta sheets (Fig. 6). This direction is opposite the orientation of the phenoxyethylpiperidine group of the selective estrogen receptor modulator raloxifene in its cocrystal structure with ERα-LBD.[22]

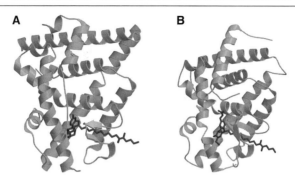

FIG. 6. Modeling of alkylamide side chain antagonists. (A) Superposition of ICI-164,384 modeled in place of estradiol and hERα. (B) Superposition of HY-4 (*R*-enantiomer) modeled in place of DIMIT and rTRα_1.

A manual superposition using SYBYL (Tripos, Inc.) of the DIMIT-TR-LBD and E_2-ER-LBD structures showed that the 7α-position of E_2 and the bridging ether DIMIT occupied approximately the same positions, suggesting that if a carbon atom replaced the ether oxygen, an alkyl chain attached to that carbon forming the *R*-enantiomer would point in a direction analogous to the antagonistic side chain of ICI (Fig. 5).

At the time this modeling was performed, a cocrystal structure of ER-LBD and ICI was not available. This structure, subsequently reported,[49] revealed that the E_2 steroid core in ICI adopted a flipped orientation and the antagonistic side chain protruded from the binding pocket to displace helix 12 (H12) in much the same way as the antagonistic side chain of raloxifene.[22]

This perturbation of H12, which is involved in the formation of a coactivator-binding surface,[39,50] appears to be a general mechanism of blocking NR transcriptional activity, as supported by the X-ray crystal structures of other antagonist-bound NRs (Fig. 6).[22,51,52] Furthermore, extension groups of antagonists can potentially form interactions with the receptor to stabilize an inactive conformation. In the case of raloxifene-bound ER, the extension makes extensive hydrophobic contacts and is anchored to the receptor by electrostatic interactions with its piperidine ring nitrogen.[22]

[49] A. C. Pike, A. M. Brzozowski, J. Walton, R. E. Hubbard, A. G. Thorsell, Y. L. Li, J. A. Gustafsson, and M. Carlquist, *Structure (Camb)* **9**, 145 (2001).

[50] W. Feng, R. C. Ribeiro, R. L. Wagner, H. Nguyen, J. W. Apriletti, R. J. Fletterick, J. D. Baxter, P. J. Kushner, and B. L. West, *Science* **280**, 1747 (1998).

[51] W. Bourguet, V. Vivat, J. M. Wurtz, P. Chambon, H. Gronemeyer, and D. Moras, *Mol. Cell* **5**, 289 (2000).

[52] A. K. Shiau, D. Barstad, P. M. Loria, L. Cheng, P. J. Kushner, D. A. Agard, and G. L. Greene, *Cell* **95**, 927 (1998).

Synthesis

HY-4 was prepared as a racemate from a biaryl methanol intermediate in the synthesis of GC-1. The phenolic hydroxyl groups were protected differently from those used for GC-1. The B-ring phenol was protected with a triisopropylsilyl (TIPS) ether, while the A-ring phenol was protected with a *tert*-butylmethoxyphenylsilyl (TBMPS) acetal.[34] TBMPS is resistant to acid hydrolysis, but the additional oxygen substituent on the silicon makes possible selective removal using a mild fluoride reagent while leaving the TIPS ether intact.[53,54] The long alkylamide side chain of HY-4 was installed as depicted in Scheme 2. An allyl group was placed at the bridging carbon using an efficient substitution reaction involving the presumed acidic solvolysis of the methyl ether intermediate.[55] The allyl group was then converted to a propionaldehyde by hydroboration and oxidation. The aldehyde, **XII**, was the key substrate for the Wittig reagent derived from 8-bromo-*N*-butyl-*N*-methyloctanamide (**XIV**). The synthesis of the alkylamide side chain was completed by catalytic hydrogenation of the resulting olefin (**XV**).

Properties

Compared to its parent, GC-1, HY-4 has a greatly reduced affinity for TR (K_d hTRα_1 112 ± 18 nM, hTRβ_1 148 ± 13 nM) and shows little selectivity.[34] In transactivation assays it reduces T_3-induced reporter gene activation in a dose-dependent fashion. This reduction can be overcome by the addition of T_3, which is consistent with the action of a competitive antagonist.

Substitution at the bridging carbon of GC-1 results in a large loss of affinity. The addition of the relatively small allyl group results in a 900-fold less potent agonist.[34] Extension of the allyl chain as with HY-4 does not significantly change the affinity for TR, but results in antagonism. The low affinity of HY-4 for TR makes it unsuitable for *in vivo* studies with *Xenopus* tadpoles, as the concentrations required to elicit an antagonistic response lead to toxic effects.[56] However, HY-4 serves as a proof of principle in the design of TR antagonists.

[53] J. W. Gillard, R. Fortin, H. E. Morton, C. Yoakim, C. A. Quesnelle, S. Daignault, and Y. Guindon, *J. Org. Chem.* **53**, 2602 (1988).

[54] Y. Guindon, R. Fortin, C. Yoakim, and J. W. Gillard, *Tetrahedron Lett.* **25**, 4717 (1984).

[55] H. A. I. Yoshihara, G. Chiellini, T. J. Mitchison, and T. S. Scanlan, *Bioorg. Med. Chem.* **6**, 1179 (1998).

[56] J. D. Furlow, Personal Communication.

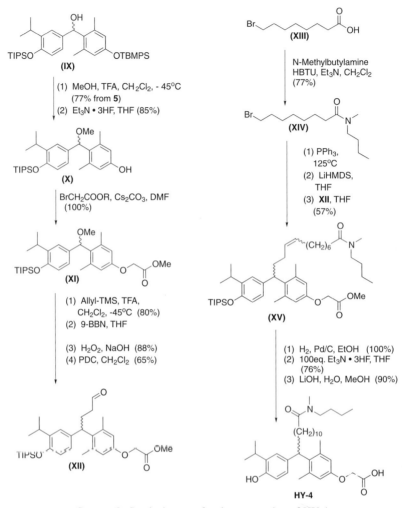

SCHEME 2. Synthetic route for the preparation of HY-4.

TR antagonist NH-3

Design

NH-3 is a second-generation antagonist based on a previous series of GC-1 analogues that targeted the 5′-position as an appendage site.[35] The original design of 5′-substituted antagonists stemmed from crystallographic studies revealing this position of the thyronine scaffold oriented in the direction of the loop between helices 11 and 12.[20] Since H12 is implicated as

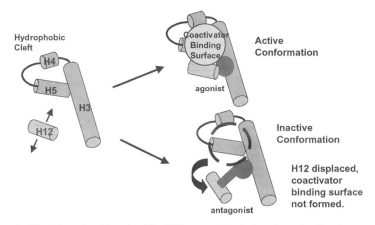

FIG. 7. The "Extension Hypothesis" of NR antagonist design. Extension-bearing antagonist ligands disrupt proper folding of H12 and formation of the coactivator-binding surface. Agonist ligands induce an active conformation of the NR LBD that allows formation of the putative coactivator-binding surface. Extension groups of antagonists protruding from the ligand-binding pocket can potentially disrupt formation of the active NR conformation and block coactivator binding.

the molecular switch in NR activation, 5′-extensions were designed to directly block proper packing of H12 resulting in disruption of the coactivator-binding surface. Numerous analogues containing 5′-alkyl and aryl substituents have been synthesized, but only a handful of compounds, including DIBRT and GC-14, exhibited antagonistic activity (Fig. 5).[35,36] The majority of 5′-analogues were agonists, suggesting that either the extensions employed were not large enough to perturb H12, or that the potential interactions between ligand and receptor that stabilize an inactive conformation had not been realized.

Previous studies showed that the antagonistic activity of GC-14, based on the GC-1 scaffold, required the 5′-aryl nitro (–NO₂) group in the *para*-position. Positional isomers of GC-14 in the *ortho*- and *meta*-positions, and isosteres of GC-14 with carboxylic acid and amino substitutions, exhibited agonistic properties, suggesting that the nitro group may be involved in key stereoelectronic interactions with H12 of the receptor. We postulated that the partial agonist activity of GC-14 might be due to flexibility of the 5′-aryl substitution that would still allow formation of the coactivator-binding surface. We therefore utilized an ethynyl group to rigidly link extensions to the agonist scaffold based on the hypothesis that decreasing the flexibility of the 5′-extension would allow more effective perturbation of H12.

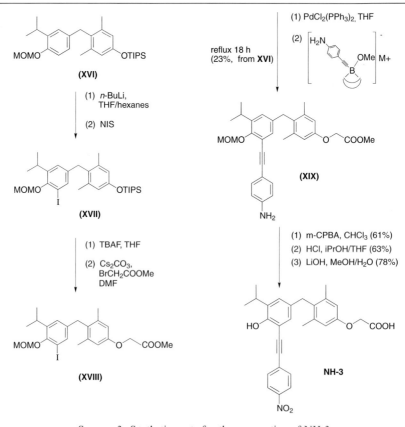

SCHEME 3. Synthetic route for the preparation of NH-3.

Synthesis

The GC-14 analogue, NH-3, was synthesized containing a 5'-*p*-nitrophenylethynyl extension (Scheme 3). Adapting the synthetic route of GC-1, the biarylmethane intermediate **XVI** was treated with butyl-lithium to achieve deprotonation at the 5'-position via methoxymethyl (MOM)-directed ortho-metalation[57,58] followed by *in situ* iodination with *N*-iodosuccinimide (NIS). Removal of the *O*-silyl protection with tetrabutylammonium fluoride (TBAF) and alkylation with methyl bromoacetate formed the methyl ester **XVIII**. The phenylethynyl moiety was attached by the palladium-catalyzed Suzuki–Miyaura coupling[59,60] of

[57] C. A. Townsend and L. M. Bloom, *Tetrahedron Lett.* **22**, 3923 (1981).
[58] R. C. Ronald and M. R. Winkle, *Tetrahedron* **39**, 2031 (1983).

XVIII with 4-amino-phenylethynyl boronate derivative, generated *in situ* under basic conditions with MeO-9-BBN and 4-ethynylaniline. Oxidation of the amine **XIX** to nitro with *m*-chloroperoxybenzoic acid (*m*-CPBA) followed by removal of the MOM protecting group and saponification of the methyl ester resulted in the desired product NH-3.

The critical step in this synthesis is formation of the 5′-iodinated intermediate **XVII**, which generally results in about 60% conversion to the iodinated species. This intermediate and the remaining starting compound cannot be separated by standard silica gel chromatography. Use of other iodinating agents such as iodine, iodine monochloride, and activated iodonium sources result in lower yields with decomposition of starting material. Alternative coupling methods were also explored to improve the yield, but were unsuccessful in our hands. Switching the iodide- and boronate-coupling partners, where a 5′-boronic acid GC-1 intermediate is coupled with iodoethynylbenzene using various palladium catalysts, gave no reaction. Various Sonogashira coupling conditions[61–64] were also tried, but generally resulted in low yields and decomposition of starting material. Despite the complication with the iodination step, the current synthetic route gives fairly good yields overall and NH-3 can be prepared in gram quantities.

Properties

Interestingly, NH-3 had improved binding affinity and antagonistic potency compared to GC-14; while GC-14 exhibited dual agonist/antagonist activity, NH-3 is essentially a full antagonist (Table I). Thus, the improved properties of NH-3 can be attributed directly to the internal ethynyl moiety. It has been proposed that liganded NRs exist in an equilibrium of active and inactive conformations and that the nature of the ligand can shift the equilibrium to favor one state over the other.[65–67] The 5′-nitrophenyl

[59] J. A. Soderquist, K. Matos, A. Rane, and J. Ramos, *Tetrahedron Lett.* **36**, 2401 (1995).

[60] N. Miyaura and A. Suzuki, *Chem. Rev.* **95**, 2457 (1995).

[61] R. Rossi, A. Carpita, and F. Bellina, *Org. Prep. Proc. Intl.* **27**, 127 (1995).

[62] S. Thorand and N. Krause, *J. Org. Chem.* **63**, 8551 (1998).

[63] K. Okuro, M. Furuune, M. Miura, and M. Nomura, *Tetrahedron Lett.* **33**, 5363 (1992).

[64] M. Alami, F. Ferri, and G. Linstrumelle, *Tetrahedron Lett.* **34**, 6403 (1993).

[65] A. L. Wijayaratne, S. C. Nagel, L. A. Paige, D. J. Christensen, J. D. Norris, D. M. Fowlkes, and D. P. McDonnell, *Endocrinology* **140**, 5828 (1999).

[66] R. E. Hubbard, A. C. Pike, A. M. Brzozowski, J. Walton, T. Bonn, J. A. Gustafsson, and M. Carlquist, *Eur. J. Cancer* **36 Suppl. 4**, S17 (2000).

[67] A. K. Shiau, D. Barstad, J. T. Radek, M. J. Meyers, K. W. Nettles, B. S. Katzenellenbogen, J. A. Katzenellenbogen, D. A. Agard, and G. L. Greene, *Nat. Struct. Biol.* **9**, 359 (2002).

TABLE I

BINDING AND TRANSCRIPTIONAL ACTIVATION DATA OF 5'-SUBSTITUTED GC-1 DERIVATIVES AT HUMAN TRα_1 AND TRβ_1

5'-Substituent	$K_d \pm$ SE (nM)[a]		% TRβ_1 activation[b]	TRβ_1 EC$_{50}$ (IC$_{50}$)[c] (nM)	% TRα_1 activation[b]	TRα_1 EC$_{50}$ (IC$_{50}$)[c] (nM)
	hTRβ_1	hTRα_1				
T$_3$[d]	0.10 ± 0.03	0.10 ± 0.03	100	2	100	2
GC-1[d] (H)	0.10 ± 0.02	1.8 ± 0.2	100	7	100	45
GC-13[d]	30 ± 13	170 ± 10	99	240	99	900
GC-14[d]	35 ± 12	200 ± 60	18	(680)[e]	35	(5000)[e]
GC-18[d]	21 ± 5	150 ± 80	100	550	n.d.	n.d.
NH-1[f]	37 ± 9	490 ± 100	70	500	n.d.	n.d.
NH-3[f]	20 ± 7	93 ± 29	3	(370)[e]	10	(950)[e]
NH-4[f]	90 ± 6	330 ± 60	11	n.d.	n.d.	n.d.

[a] The K_d and standard error (SE) values were measured by an equilibrium radioligand [^{125}I]T$_3$ displacement assay and calculated by fitting the competition data to the equations of Swillens and using the Graph-Pad Prism computer program (Graph-Pad Software, Inc.).

[b] HeLa cells were cotransfected with either hTRβ_1 or hTRα_1 expression vector and a TRE-luciferase reporter plasmid. Luciferase activity of 10^{-5} M analogue is expressed as a percent of the TRβ_1 or hTRα_1 response with 10^{-7} M T$_3$. Values are the mean $\pm$ SD for three separate experiments.

[c] The EC$_{50}$ value is the concentration of ligand required for half-maximum activation, whereas the IC$_{50}$ value is the concentration of ligand required for half-maximum inhibition in competition experiments with 10^{-9} M T$_3$. EC$_{50}$ and IC$_{50}$ values were calculated by nonlinear regression with the Graph-Pad Prism computer program (Graph-Pad Software, Inc.) using a sigmoidal dose response or single-site competition models, respectively. Values are the mean for three separate experiments.

[d] G. Chiellini, J. W. Apriletti, H. A. I. Yoshihara, J. D. Baxter, R. C. Ribeiro, and T. S. Scanlan, *Chem. Biol.* **5**, 299 (1998).

[e] These are IC$_{50}$ values for hTRβ_1 and hTRα_1, respectively.

[f] N. H. Nguyen, J. W. Apriletti, S. T. Cunha Lima, P. Webb, J. D. Baxter, and T. S. Scanlan, *J. Med. Chem.* **45**, 3310 (2002). NH-4 was not able to antagonize 10^{-9} M T$_3$-induced activation in a dose-dependent manner.

extension of GC-14 apparently is insufficient for complete perturbation of H12, thus allowing a significant concentration of a suboptimal active conformation and the observed partial agonism. With NH-3, H12 may be more effectively perturbed and stabilized in the inactive conformation as indicated by the inability of TR to interact with coactivators when bound with NH-3.[48] Rigidly extending the nitrophenyl group by two carbon atoms out of the ligand-binding pocket potentially allows additional and/or amplified stabilizing ligand–receptor interactions that shifts the equilibrium toward an inactive conformation.

Further support that the *p*-nitro group attached to the 5′-extension specifically dictates the antagonistic property, and not just the presence of any rigid 5′-phenylethynyl extension, is exemplified by analogues NH-1 and NH-4, corresponding to GC-13 and GC-18, respectively. Unlike the NH-3/GC-14 pair, addition of the ethynyl moiety to agonists GC-13 and GC-18 did not shift their activities toward antagonism, and in fact resulted in overall decreased binding affinity and agonist potency (Table I).

In contrast to GC-14 and NH-3, DIBRT does not contain a 4-nitrophenyl group but does exhibit antagonistic activity and weak partial agonist activity.[36] DIBRT is based on a thyronine scaffold and contains a 5′-isopropyl extension, which presents moderate hydrophobic steric bulk. This shorter 5′-extension is apparently sufficient to perturb H12, as suggested by the ability of DIBRT to block TR–coactivator interactions. Combined, these 5′-substituted analogues indicate that polar/electronic groups can be involved in key interactions with the receptor but are not required to confer antagonistic activity.

Molecular modeling via SYBYL was used to study the possible structural basis for antagonism of NH-3 (Fig. 8). Building the 5′-extension into the crystal structure of GC-1 bound to TRβ shows that the 4-nitrophenyl ethynyl group runs into significant steric clutter from residues of H5-6, H11, and H12, which presumably would alter the receptor conformation relative to the agonist-fold. Modeling suggests that the 5′-extension sterically perturbs His435, which is involved in H-bonding interactions with and recognition of the 4′-phenolic oxygen. This feature may account for the reduced binding affinity observed for NH-3 relative to GC-1; however, in accommodating the ligand extension, the receptor may adopt a conformation in which the nitro group is involved in stabilizing interactions. Knowing what residues are involved in ligand contacts in the agonist-bound TR crystal structure and surveying the secondary structure for nearby contacts[32] allows prediction of potential ligand–receptor interactions. The nitro group may be involved in electronic edge-to-face and/or pi-stacking interactions with proximal phenylalanine residues (Phe455 and Phe459) from H12 that lie along the rim of the ligand pocket

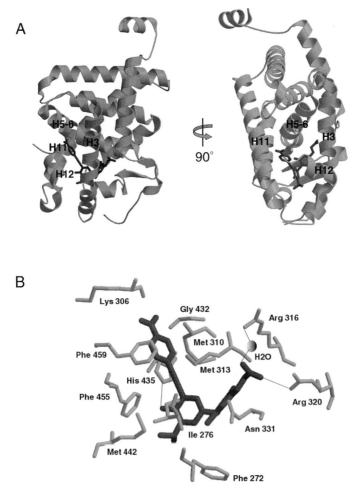

FIG. 8. Molecular modeling of NH-3 bound to hTRβ-LBD. (A) The 5′-*p*-nitrophenylethynyl extension protrudes from the ligand-binding pocket, resulting in steric clash with residues from helices H5-6, H11, and H12. (B) Receptor residues involved in ligand contact within 5 Å of NH-3 is shown. Modeling reveal residues 435, 459, 306, 432, and 310 lie within 3-Å proximity to the 5′-extension, suggesting that these residues must undergo some rearrangement in order to accommodate the ligand extension.

where the extension protrudes. Additionally, the strong dipole moment of the nitro group may participate in electrostatic interactions with Lys306 of H5-6. Although it is difficult to predict binding interactions for antagonists through modeling techniques, this method has been applied to various NRs through crystallographic comparison of available agonist- and antagonist-bound NR structures.[31,68] Unfortunately, no crystal structure of an antagonist-bound TR LBD has been solved, although efforts along this line are underway.

Conclusions

Currently, there is no clear-cut recipe for designing agonist and antagonist ligands for nuclear receptors. But as more SAR and structural information become available, important general considerations can be made to guide rational ligand design. We have had success using guiding principles from crystal structures combined with standard medicinal chemistry. Our main concern in these efforts has been synthetic accessibility of the ligands. For agonists, filling ligand-binding pocket volume and satisfying key polar interactions for molecular recognition are crucial for receptor-selectivity, ligand binding, and activity. For antagonists, although the "extension hypothesis" appears broadly applicable for antagonist design, the nature of chemical groups on the ligand extension is important in establishing specific interactions with receptor residues that stabilize an inactive receptor conformation. For example, with T_3 antagonists, we have found that simple steric blockade with large hydrophobic extensions is not sufficient to confer antagonist activity; instead correctly positioned groups of the appropriate chemistry are required. Nevertheless, without additional structural data, these putative interactions remain ill-defined and it is thus difficult to predict what ligand extension moieties will induce antagonism. Given what is known about NR function, antagonist ligands can potentially function by several mechanisms, including: (1) enhancing corepressor binding, (2) preventing coactivator recruitment, (3) disrupting NR dimerization, (4) disrupting NR interaction with DNA, (5) reducing NR half-life, or (6) a combination of these mechanisms.

---

[68] U. Egner, N. Heinrich, M. Ruff, M. Gangloff, A. Mueller-Fahrnow, and J. M. Wurtz, *Med. Res. Rev.* **21**, 523 (2001).

Section II

Structure/Function Analysis of Nuclear Receptors

[6] Bioinformatics of Nuclear Receptors

By MARC ROBINSON-RECHAVI and VINCENT LAUDET

Introduction

Nuclear hormone receptors are one of the most abundant classes of transcriptional regulators in metazoans (animals), in which they regulate functions as diverse as reproduction, development, metabolism, or homeostasis (for a review, see Ref. 1). They function as ligand-activated transcription factors, thus providing a direct link between signaling molecules that control these processes and transcriptional responses. Nuclear receptors form a superfamily of phylogenetically related proteins, which share a common structural organization: a variable N-terminal region (A/B domain) containing the AF-1 transcription activator, a central well-conserved DNA-binding domain (DBD, C domain), a nonconserved hinge (D domain), a long and moderately conserved ligand-binding domain (LBD, E domain), which also contains the AF-2 transcription activator, and eventually a C-terminal F domain (Table I).

There are 21 nuclear receptors in the complete genome of the fly *Drosophila melanogaster*,[2] 48 in humans,[3] but more than 250 in the nematode *Caenorhabditis elegans*.[4,5] The superfamily includes receptors for hydrophobic molecules such as steroid hormones (estrogens, glucocorticoids, progesterone, mineralocorticoids, androgens, vitamin D, ecdysone, oxysterols, bile acids, etc.), retinoic acids (all-*trans* and 9-*cis* isoforms), thyroid hormones, fatty acids, leukotrienes, and prostaglandins.[6] A large number of nuclear receptors have also been identified by homology with the

[1] V. Laudet and H. Gronemeyer, "The Nuclear Receptors Factsbook." Academic Press, London, 2002.

[2] M. D. Adams, S. E. Celniker, R. A. Holt, C. A. Evans, J. D. Gocayne, P. G. Amanatides, S. E. Scherer, P. W. Li, R. A. Hoskins, R. F. Galle, R. A. George, S. E. Lewis, S. Richards, M. Ashburner, S. N. Henderson, G. G. Sutton, J. R. Wortman, M. D. Yandell, Q. Zhang, L. X. Chen, R. C. Brandon, Y. H. Rogers, R. G. Blazej, M. Champe, B. D. Pfeiffer, K. H. Wan, C. Doyle, E. G. Baxter, G. Helt, C. R. Nelson, G. L. Gabor, J. F. Abril, A. Agbayani, H. J. An, C. Andrews-Pfannkoch, D. Baldwin, R. M. Ballew, A. Basu, J. Baxendale, L. Bayraktaroglu, E. M. Beasley, K. Y. Beeson, P. V. Benos, B. P. Berman, D. Bhandari, S. Bolshakov, D. Borkova, M. R. Botchan, and J. Bouck, *Science* **287**, 2185 (2000).

[3] M. Robinson-Rechavi, A.-S. Carpentier, M. Duffraisse, and V. Laudet, *Trends Genet.* **17**, 554 (2001).

[4] A. E. Sluder and C. V. Maina, *Trends Genet.* **17**, 206 (2001).

[5] A. E. Sluder, S. W. Mathews, D. Hough, V. P. Yin, and C. V. Maina, *Genome Res.* **9**, 103 (1999).

[6] H. Escriva, F. Delaunay, and V. Laudet, *Bioassays* **22**, 717 (2000).

TABLE I
DOMAIN LENGTH AND SEQUENCE CONSERVATION OF HUMAN NUCLEAR RECEPTORS

Subfamily	A/B Length	A/B id. (%)	DBD (C) Length	DBD (C) id. (%)	Hinge (D) Length	Hinge (D) id. (%)	LBD (E) Length	LBD (E) id. (%)	F Length
TR (NR1A)	87	34	58	90	32	78	208	84	0
RAR (NR1B)	82	37	58	93	18	61	220	80	36
PPAR (NR1C)	99	12	57	81	37	35	232	62	0
Rev-erb (NR1D)	121	40	59	95	167	26	206	65	0
ROR (NR1F)	63	50	58	88	141	39	231	48	47
LXR/FXR (NR1H)	118	15	60	69	43	22	233	32	0
VDR/PXR (NR1I)	35	28	57	57	51	27	242	34	0
HNF4 (NR2A)	47	36	58	91	17	76	201	79	102
RXR (NR2B)	183	24	58	90	49	45	182	90	0
TR2/4 (NR2C)	120	40	56	82	81	64	287	68	0
TLL1/PNR (NR2E)	42	7	60	71	22	38	248	40	0
COUPTF (NR2F)	79	15	58	86	64	32	190	75	0
ER (NR3A)	163	12	56	98	49	24	213	56	44
ERR (NR3B)	161	23	58	84	41	56	209	60	63
AR/GR (NR3C)	532	8	54	80	53	18	209	50	17
NR4A	263	32	56	93	53	61	180	61	0
NR5A	82	75	56	89	99	55	236	58	0
Average	134	29	57	85	60	45	219	61	20

Only human protein sequences were used; domains were defined on the alignment, relative to known structures. Length is the number of amino acids in the longest sequence of the group, "id." the proportion of identical sites between the most divergent pair of sequences of the group. Most F domains are not alignable, so identity proportions were not computed; we defined a F domain only when there are more than 10 amino acids in N-terminal of the LBD.

conserved DBD and LBD, but have no identified natural ligand, and are referred to as "nuclear orphan receptors." This diversity has been organized in a phylogeny-based nomenclature.[7]

As nuclear receptors bind small molecules which can easily be modified by drug design, and control functions associated with major diseases (cancer, osteoporosis, diabetes,...), they are promising pharmacological targets. The search of ligands for orphan receptors and the identification of novel signaling pathways has become a very active research field.[8,9] The importance of nuclear receptors has prompted the accumulation of rapidly increasing data from a great diversity of fields of research: sequences, expression patterns, 3D structures, protein–protein interactions, target

[7] Nuclear Receptors Nomenclature Committee, *Cell* **97**, 1 (1999).
[8] J. A. Gustafsson, *Science* **284**, 1285 (1999).
[9] S. A. Kliewer, J. M. Lehmann, and T. M. Willson, *Science* **284**, 757 (1999).

genes, physiological roles, mutations, etc. This has in turn highlighted the need for good bioinformatics, which are the subject of this review. For example, queries for the words "nuclear receptor" recover 44,497 hits in Medline (search over biomedical literature), and 126,000 hits in Google (search over the WWW), obviously not manageable manually.

The NCBI defines bioinformatics as "the field of science in which biology, computer science, and information technology merge to form a single discipline." In practice, this can include theoretical development of new algorithms, based on a biological question, as well as the standard use of a software. In this paper, we will try to review the information that biologists interested in nuclear receptors can get by using a computer, including general resources and those specialized in nuclear receptors, whether for access to data or analysis of data. We will not detail resources already well presented in a recent review on a similar subject,[10] and put emphasis on gaining time and knowledge in a dynamic and medically important field. Words followed by an asterisk are defined in Box 1.

Similarity Searches for Nuclear Receptors

Basic Use of BLAST

The strong conservation of the DBD of nuclear receptors (Table I) is a strong asset in the detection of new nuclear receptor sequences by all methods based on similarity, whether "wet" (PCR, Southern blot, . . .)[11] or "dry" (*in silico*). The main *in silico* similarity-based approach is BLAST,[12,13] although other algorithms, such as FASTA[14] or SSAHA,[15] may be advantageous in some situations.

BLAST is a heuristic* of the Smith and Waterman algorithm[16] for finding the optimum partial alignment between two biological sequences. BLAST finds local similarities based on "words," short conserved sequences. It does not seek to concatenate these words, thus allowing for domain shuffling or repetition, and for protein sequences it uses similarity

[10] M. Danielsen, *Methods Mol. Biol.* **176**, 3 (2001).
[11] H. Escriva, M. Robinson, and V. Laudet, *in* "Nuclear Receptors. A Practical Approach" (D. Picard, ed.). p. 1. Oxford University Press, Oxford, 1999.
[12] S. F. Altschul, W. Gish, W. Miller, E. W. Myers, and D. J. Lipman, *J. Mol. Biol.* **215**, 403 (1990).
[13] S. F. Altschul, T. L. Madden, A. A. Schaffer, J. Zhang, Z. Zhang, W. Miller, and D. J. Lipman, *Nucleic Acids Res.* **25**, 3389 (1997).
[14] W. R. Pearson and D. J. Lipman, *Proc. Natl. Acad. Sci. USA* **85**, 2444 (1988).
[15] Z. Ning, A. J. Cox, and J. C. Mullikin, *Genome Res.* **11**, 1725 (2001).
[16] T. F. Smith and M. S. Waterman, *J. Mol. Biol.* **147**, 195 (1981).

BOX 1
BIOINFORMATIC GLOSSARY

False negative	error in which an existing solution (a binding site, a gene, etc.) is not found; minimizing false negatives maximizes sensitivity
False positive	error in which a positive result (a binding site, a gene, etc.) is found whereas there is none; minimizing false positives maximizes specificity
Heuristic	approximate solution to a time-consuming problem, whose validity has not been formally demonstrated, but has experimentally been shown to work reasonably well with an economy of computation time and/or memory; a heuristic is not guaranteed to find the optimal solution
Homology	property of biological structures (organs, genes, amino-acid sites, . . .) which descend from a common ancestral structure; homology is often confused with similarity, yielding incorrect statements about "% homology": two structures are homologous or are not, whatever their % of similarity. See orthology and paralogy
Indel	insertion or deletion represented by a gap in a protein or a DNA alignment; it is often not possible to distinguish an insertion in some sequences from a deletion in the others, hence the acronym
Orthology	two homologous genes are orthologous if they diverged from their most recent common ancestor by speciation; this often implies the same function in different species (for example, AR in mouse and human), but not always (for example, vertebrate Rev-erbs and insect E75)
Paralogy	two homologous genes are paralogous if they diverged from their most recent common ancestor by gene duplication; this often implies different functions (for example, AR and PR). It should be noted that there can be paralogous genes in the same or different species; for example, human AR is paralogous to human PR, but also to mouse PR
Substitution	replacement of one nucleotide or amino acid by another during evolution
Synonymous	type of substitution or of mutation in which, due to degenerescence of the genetic code, the DNA sequence of a coding region is modified without modifying the encoded protein; codons which encode the same amino acid are said to be synonymous codons

matrixes which allow for conservative substitutions to be used even in the first step of the search; in addition, BLAST2[13] allows indels* in the local similarities. The output of BLAST includes a very useful statistic: the E-value. The E-value for a similarity score S is an estimate of the number of sequence hits with a score superior or equal to S, which would be found in a database of random sequences of similar size and amino acid or base composition as the database being searched. This score should not be taken as a classical "5% threshold" test, since the real distribution of sequences is not random and not known, the distribution of similarity scores may not be Normal, and in most cases many independent comparisons are made. But it remains the best way to measure the strength of a similarity search result (or "hit"), notably since it takes into account the size of the database searched.

BOX 2
USEFUL RESOURCES ON THE WORLD WIDE WEB

Nuclear receptor-specific resources

Nomenclature	http://www.ens-lyon.fr/LBMC/laudet/nomenc.html
Nuclear receptor resource	http://nrr.georgetown.edu/nrr/nrr.html
NucleaRDB	http://www.receptors.org/NR/
Nurebase	http://www.ens-lyon.fr/LBMC/laudet/nurebase.html

Bioinformatic methods

Blast	http://www.ncbi.nlm.nih.gov/BLAST/
Database list	http://www3.oup.co.uk/nar/database/a/
Online analyses	http://bioweb.pasteur.fr/intro-uk.html
Phylogeny links	http://evolution.genetics.washington.edu/phylip/software.html
Sequence alignment links	http://pbil.univ-lyon1.fr/alignment.html

Relevant biological resources

Classification of species	http://www.ncbi.nlm.nih.gov/Taxonomy/tax.html/
Drosophila genome	http://flybase.bio.indiana.edu/
Endocrine disruption	http://e.hormone.tulane.edu/
Gene interactions and pathways	http://www.genome.ad.jp/kegg/
Genome resources	http://www.ensembl.org/
Hormone levels	http://www.il-st-acad-sci.org//data2.html
Human genetic disorders	http://www3.ncbi.nlm.nih.gov/Omim/
Ligand binding	http://alto.rockefeller.edu/ligbase/
Nematode genome and biology	http://www.wormbase.org/
Pharmacology and nomenclature	http://www.iuphar.org/
Promoters	http://www.epd.isb-sib.ch/
Transcription factors (fly)	http://rana.lbl.gov/cis-analyst/
Transcription factors	http://transfac.gbf.de/TRANSFAC/

In practice, BLAST is available online or for download from the NCBI (see Box 2) in several flavors according to the type of sequences used.

The most straightforward use of BLAST on nuclear receptors is to search for homologues of a given protein in proteic databases (BLASTP). For example, a search (default options) with the complete human TRα against the nonredundant translation of GenBank yields more than 100 hits, all nuclear receptors, and of which the first 92 are annotated as thyroid hormone receptors. On the other hand, although most of the first hits for hTRα are TRα sequences, orthologues* of hTRα, and most of the next are TRβ sequences, paralogues* of hTRα, there is a frog sequence (A37952) with a lower E-value than eight TRβ sequences which is yet unambiguously a TRα. The results of this search also include heavy redundancy, with five hTRα and five hTRβ sequences retrieved. These results show that generalist databanks such as GenBank are redundant, and that BLAST scores are not a definitive measure of homology*. Still, a BLAST against such databases allows a rapid and easy search for all related nuclear receptors, and is clearly the first step of most other bioinformatic analyses, provided the results are treated with appropriate caution, including manual checking.

Data Mining in Nonannotated Sequences

Nonannotated sequences are the result of high-throughput sequencing programs, which provide a wealth of information if one knows to mine it. "Nonannotated" can signify that not even the number or the limits of genes are known. They include:

- High-throughput genome sequences (HTG) are high-quality sequences of long fragments (>100 kb) of genomic DNA; their assembly notably forms the human genome draft.[17] The best source for assembled and partially annotated versions of the draft human genome is the Sanger Institute Ensembl database[18] (Box 2). Ensembl has also released recently a draft version of the mouse genome, including 96% of euchromatin.
- Genome survey sequences (GSS) are often single-read short (≈ 1 kb) genomic DNA sequences; they are typically generated when sequencing an eukaryote genome by a shot-gun strategy. GSSs are available for a diversity of organisms, notably puffer fishes *Tetraodon negroviridis* and *Takifugu rubripes*. They are most easily accessed through the NCBI (Box 2) in the "GSS" division of GenBank.
- Expressed sequence tags (ESTs) are single-read short (<1 kb) cDNA sequences, from randomly amplified mRNAs; despite poor sequence quality, they give information on new genes, alternative transcripts, and the expression of known genes when the source of the cDNA library is known. Although ESTs are available for an increasing variety of species, most data still comes from human and mouse. Like GSSs, they form a division of GenBank accessible through NCBI.

---

[17] International Human Genome Sequencing Consortium, *Nature* **409**, 860 (2001).
[18] T. Hubbard, D. Barker, E. Birney, G. Cameron, Y. Chen, L. Clark, T. Cox, J. Cuff, V. Curwen, T. Down, R. Durbin, E. Eyras, J. Gilbert, M. Hammond, L. Huminiecki, A. Kasprzyk, H. Lehvaslaiho, P. Lijnzaad, C. Melsopp, E. Mongin, R. Pettett, M. Pocock, S. Potter, A. Rust, E. Schmidt, S. Searle, G. Slater, J. Smith, W. Spooner, A. Stabenau, J. Stalker, E. Stupka, A. Ureta-Vidal, I. Vastrik, and M. Clamp, *Nucleic Acids Res.* **30**, 38 (2002).

In these data redundancy is very high, there are sequencing errors, assembly and annotation errors, and contamination of libraries (by vectors, or by genomic DNA for ESTs). One should also take into account that the annotations of complete assembled genomes, such as those of *Caernorhabditis elegans* and *Drosophila melanogaster*, are very preliminary. Additional datamining can thus be necessary. For example, an ERR nuclear receptor gene has been identified in the fly,[19] but it is annotated only as "CG7404 gene product" in GenBank. *C. elegans* and *D. melanogaster* genome data are best accessed through specialized webpages (see Box 2).

Provided care is taken in interpreting results from "dirty" data, programs such as BLAST provide the most efficient way to exploit all these data. A notably useful option of BLAST in this case is TBLASTN, which compares a known protein sequence to all six possible translations of the nonannotated DNA sequence: three coding phases × two strands. As BLAST searches for local similarities, it can detect easily partial sequences and genes which are split among exons. For systematic searches of this type, it can be recommended to install the BLAST programs and the relevant databases locally, rather than passing repeatedly through webpages which are not optimized for such use. The NCBI (see Box 2) provides binaries for a wide variety of platforms, and computations can be done in a reasonable time for most applications concerning nuclear receptors on an average personal computer. Disk space for databanks can be more of a problem, since for example the FASTA format file of all EST sequences as of April 2002 has a size of 6.8 Go; standard personal computers do not manage files of this size, which can justify the use of Unix work stations.

An example, both of the care needed and of the information available, can be found in the human genome draft. Forty-eight human nuclear receptor genes were cloned and reported before the genome sequence,[1,7] of which two with no LBD. The two reports of draft human genome sequences[17,20] contained conflicting but much higher reports on the number of nuclear receptors: 60 nuclear receptors with both DBD and LBD (Fig. 39 in Ref. 17), and 47 nuclear receptor LBDs (Table 18 in Ref. 20), but 59 nuclear hormone receptors (Table 19 in Ref. 20). We reanalyzed the International Human Genome Sequencing Consortium HTGS data,[3] using TBLASTN as a first filter to find potential nuclear receptor genes, with a threshold E-value of 10^{-4}, which is in fact not very stringent given the strong

[19] J. M. Maglich, A. E. Sluder, X. Guan, Y. Shi, D. D. McKee, K. Carrick, K. Kamdar, T. M. Willson, and J. T. Moore, *GenomeBiology.com* **2**, research0029.1 (2001).
[20] J. C. Venter *et al.*, *Science* **291**, 1304 (2001).

conservation of nuclear receptor sequences. Treatment of the resulting hits included the use of the gene-prediction program GENSCAN,[21] manual alignment, phylogenetic reconstructions, further BLAST analyses against known nuclear receptor genes, and manual consideration of various factors such as chromosome position, divergence ratios, and the Celera version of the human genome. The final results[3,19] include confirmation of the 48 known nuclear receptor genes, three clear pseudogenes (ΨERR, ΨHNF4γ, ΨEAR2), of which two new, and most interestingly one ambiguous gene. In addition, a small region with high similarity to Rev-erb LBDs was found, but apparently not included in any functional nuclear receptor gene.[3] Thus most nuclear receptors were found by pregenomic techniques, but we now know the genomic sequence, complete with introns and flanking sequences, as well as the chromosome position of all human nuclear receptors.

The gene which was ambiguous from human genome data is a good example of the power and limits of bioinformatic approaches (Fig. 1). It appears to be a close paralogue* of FXR, a gene which has generated much interest for its affinity for bile acids (see Ref. 22, for a review), and has been called "FXRβ." Its protein sequence is typical of a vertebrate-specific paralogue, with high conservation of the DBD with FXR (77% identity), but several stop codons in the putative coding sequence.[3] Further experimental work has shown that it is in fact a new vertebrate nuclear receptor, which has become a pseudogene recently in primates.[23] This is the conclusion which was suggested by bioinformatic analysis alone,[3] but could only remain a suggestion without experimental work. Of note, the orthologue is indeed found by BLAST in the mouse genome, without stop codons, as well as in *Xenopus* (Fig. 1B). This has notably important consequences on the use of mouse and other mammalian models for the study of FXR receptor(s) in medicine.

Phylogeny, Classification and Evolution

Basic Animal Phylogeny

The phylogeny of nuclear receptors cannot be understood without reference to the phylogeny of the animals in which we find these proteins. It is typically by reference to this phylogeny that the investigator will choose a

[21] C. Burge and S. Karlin, *J. Mol. Biol.* **268**, 78 (1997).
[22] S. J. Karpen, *J. Hepatol.* **36**, 832 (2002).
[23] K. Otte, H. Kranz, I. Kober, *et al.*, *Mol. Cell. Biol.* **23**, 864 (2003).

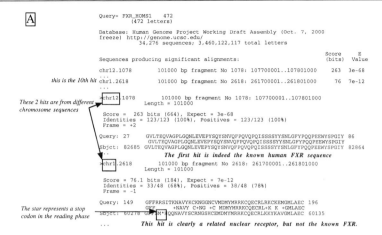

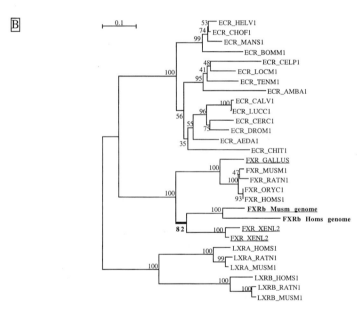

FIG. 1. Bioinformatic characterization of FXRβ. (A) Excerpts from the result of a TBLASTN search against the human genome sequence, with human FXR as a query; three dots represent deletion of text from the original file. In *italics*, comments added for the figure, which are not part of the original BLAST output. (B) Phylogeny of known FXR, LXR, and EcR sequences. Gene names follow NUREBASE[49] for cloned receptors; predicted genes from genomic sequences are in bold, sequences not available at the publication of the human genome are underlined. Phylogeny built by Neighbor-Joining with a Poisson model on aligned amino acid sequences, using only complete sites; figures at nodes are bootstrap proportions from 2000 repetitions; branch lengths are proportional to evolutionary distance, the measure bar representing 0.1 substitutions per site; the branch in bold supports a FXRβ group in the tree.

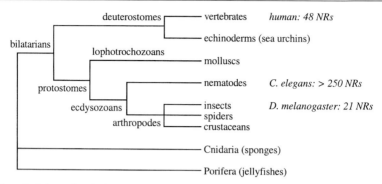

FIG. 2. Schematic phylogeny of animals (Metazoa). Based on published molecular phylogenies, notably using rRNA. Branch lengths are arbitrary. The number of nuclear receptor genes determined in sequenced genomes are shown in *italics* next to the relevant taxonomic group.

dataset, or detect a surprising result. The view of animal phylogeny based on morphology has been considerably reshuffled by the analysis of molecules, notably rRNA. We will describe here briefly this "new animal phylogeny"; for a more detailed review see Ref. 24. The major group of animals is bilaterians, divided between deuterostomes, including notably vertebrates and sea urchins (Echinoderma), and protostomes. Not included among bilaterians are notably sponges and jellyfishes. Protostomes are divided between Lophotrochozoans, including notably molluscs, and Ecdysozoans, including notably nematodes and arthropodes (insects, spiders, crustaceans, etc.). Thus it is expected to find more similarity, in sequence and function, between nuclear receptors from fly and nematode than between fly and human (Fig. 2).

Vertebrates, the most studied group of animals and including human, are a branch of chordates, the other branch being cephalochordates (amphioxus). The first splits among vertebrates separate jawless vertebrates (hagfishes and lampreys) from jawed vertebrates, and inside jawed vertebrates cartilaginous (sharks and rays) from bony (the others). Bony vertebrates are divided into two groups of similar size, ray-finned fishes, and "flesh-finned" Sarcopterygii, most of which are tetrapodes, the land vertebrates. It should be noted that the intuitive (and culinary) definition of "fish" is not valid in this context, since vertebrates originated in water and thus representatives of most groups still live there. A zebrafish is closer to us than to a shark, and a lung fish (Sarcopterygii) is closer to us than to a zebrafish.

[24] A. Adoutte, G. Balavoine, N. Lartillot, O. Lespinet, B. Prud'homme, and R. de Rosa, *Proc. Natl. Acad. Sci. USA* **97**, 4453 (2000).

Finally, it should be noted that phylogenies are statistical estimations based on experimental results, and are liable to be modified. Notably, basal animal phylogeny is still rather controversial,[24] whereas the basic outline of vertebrate phylogeny is not.[25] It is of course best to refer to the primary literature whenever possible, but the NCBI taxonomy webpage (Box 2) is a good source of information on classification and phylogeny for the nonspecialist.

Alignment and Phylogeny

Molecular phylogenetics is the estimation of a phylogenetic tree from aligned DNA or protein sequences. In the case of nuclear receptors, it is usually preferable to work on protein alignments, since synonymous* substitutions* are saturated beyond a divergence of the order of human–mouse (≈ 80 Myr).

The first step in a molecular phylogeny study is to choose sequences for the study. It is important to exclude alternative transcripts of the same gene in a phylogeny (for example, hTRα1 and hTRα2), since they are a mosaic of identical regions and nonhomologous* regions. Phylogenetic studies only allow the comparison of homologous structures. Also exclude very short sequences, as well as extremely divergent ones, unless they are indispensable in the analysis. The dataset should also include outgroup sequences. Indeed, all standard phylogenetic methods produce unrooted trees, that is to say trees which are not oriented relative to time. The only rigorous way to root a phylogeny is to include in the analysis one or several sequence(s) that are known to be external to the group of interest, yet homologous to the other sequences: they form the outgroup, and the root will be between these sequences and the others. Given the size and conservation of the nuclear receptor superfamily, it is usually easy to choose a few sequences from a subgroup close to that under investigation. For example, thyroid hormone receptor (TR) sequences can be used to root the phylogeny of retinoic acid receptors (RARs).

The next step is to align the sequences. This can be done automatically or by hand. The most wide-spread automatic alignment software is CLUSTALW,[26] which has the advantage of producing reasonably good alignments very quickly. CLUSTALX[27] is a version of CLUSTALW with

[25] N. D. Holland and J. Chen, *Bioassays* **23**, 142 (2001).

[26] J. D. Thompson, D. G. Higgins, and T. J. Gibson, *Nucleic Acids Res.* **22**, 4673 (1994).

[27] F. Jeanmougin, J. D. Thompson, M. Gouy, D. G. Higgins, and T. J. Gibson, *Trends Biochem. Sci.* **23**, 403 (1998).

a graphical interface, although it does not allow manual edition of the alignment. Recent developments include T-COFFEE,[28] which gives very good results but is both slower and memory consuming, and DbCLUSTAL,[29] which implements a new approach taking advantage of available online databases and BLAST results to improve CLUSTAL. In all cases, it is recommended to check the alignment in an editor; all automatic methods make mistakes, and these can sometimes be obvious to the experts' eye. There exist several manual editors for alignments (see sequence alignment links in Box 2). It is notably interesting to base the alignment on the 3D structure as much as possible,[30] a practice which should become standard with the increase of available structures and adapted software and databases.[31] It is standard phylogenetic procedure to exclude from the analysis all columns of the alignment with at least one gap, using only "complete sites"; this is a default option in some software, but not all. It explains why in some cases the inclusion of a single short sequence in the alignment dramatically diminishes the number of sites used in the final analysis.

To actually build the tree, we have to choose a method. Parsimony considers, for each possible tree and each site of the alignment, the minimum number of substitutions necessary, and each site is counted as supporting the tree for which this number is smallest; the tree supported by most sites is chosen. In theory, this method has the advantage of considering for each site all sequences together, and the disadvantage of using only part of the sites, those which support one tree over another ("informative sites"), as well as of not considering the probability of multiple substitutions. Indeed, during evolution several substitutions can occur at the same site, thus blurring the phylogenetic information; when the number of multiple substitutions is not too high, statistical models may recover the information correctly; when the number is too high, the sites are said to be saturated. The two other classes of methods use such statistical models, which will be briefly explained in the next paragraph. In practice, parsimony is known to give systematically the wrong answer in some cases, of which the most famous is "long-branch attraction,"[32] which will be detailed in a further paragraph. It remains a standard method in many phylogenetic studies.

[28] C. Notredame, D. G. Higgins, and J. Heringa, *J. Mol. Biol.* **302**, 205 (2000).

[29] J. D. Thompson, F. Plewniak, J. C. Thierry, and O. Poch, *Nucleic Acids Res.* **28**, 2919 (2000).

[30] J. M. Wurtz, W. Bourguet, J. P. Renaud, V. Viviat, P. Chambon, D. Moras, and H. Gronemyer, *Nat. Struct. Biol.* **3**, 87 (1996).

[31] E. Bettler, J. van Durme, S. Folkertsma, H.-J. Joosten, and G. Vriend, *in* "Nuclear Receptor Superfamily" (P. Chambon, D. J. Mangelsdorf, and B. S. Katzenellenbogen, eds.), p. 75. Keystone Symposia, Snowbird, Utah, USA, 2002.

[32] J. Felsenstein, *Syst. Zool.* **27**, 401 (1978).

The simplest model of protein evolution consists in assuming that all amino acid substitutions are equally probable. The probability of multiple substitutions at a same site can then easily be derived by a Poisson law, and thus a corrected estimate of the number of substitutions between protein sequences. It should be noted that however naive (or false) this model, the estimate thus obtained is much closer to the truth than the simple counting of differences between sequences, sometimes called "p-distance." Experience shows that in most cases this "Poisson" model is sufficient to reconstruct the phylogeny of groups of nuclear receptors, if these did not undergo specific evolutionary accelerations. More complex models often consist of matrices of probability of the substitution of one amino acid into another. Typically, it is more probable for a Leucine to be replaced by an Isoleucine than by a Tryptophane. These matrices may be based on theoretical considerations,[33] or on empirical comparison of many protein alignments,[34,35] and are also used in other bioinformatic applications such as BLAST.[13] In addition, it is possible to model the fact that the different sites of the protein do not evolve at the same rate. In nuclear receptors the DBD typically evolves slower than the LBD, and the other domains even faster; there are also more conserved sites inside each of these domains. The most widespread way to model this is to apply a gamma law to the distribution of rates among sites.[36] Although it can be much more time consuming, it is usually the best way to analyze very divergent nuclear receptors, either fast evolving or very distant in the tree. Typically, the alpha parameter of the gamma law varies between 0.5 and 1.5 for nuclear receptors, which represents medium heterogeneity along the protein.

The two other methods than parsimony are maximum likelihood (ML) and distances. Maximum likelihood is a statistical framework in which the phylogeny is chosen according to the complete alignment and the model.[37] As it takes all information into account in a rigorous manner, ML is the most robust phylogenetic method according to most comparative studies,[38] and the one to be prefered on theoretical grounds. Robustness is the property of giving mostly the correct answer when the model is false, and no model is absolutely true. In practice, ML has become the method of reference in phylogenetic studies, but it suffers from a major drawback,

[33] R. Grantham, *Science* **185**, 862 (1974).
[34] M. O. Dayhoff, R. M. Schwartz, and B. C. Orcutt, *in* "Atlas of Protein Sequence and Structure" (M. O. Dayhoff, ed.), p. 345. Natl. Biomed. Res. Fund, Washington DC, 1978.
[35] D. T. Jones, W. R. Taylor, and J. M. Thornton, *Comput. Appl. Biosci.* **8**, 275 (1992).
[36] Z. Yang, *Trends Ecol. Evol.* **11**, 367 (1996).
[37] J. Felsenstein, *J. Mol. Evol.* **17**, 368 (1981).
[38] S. Whelan, P. Li, and N. Goldman, *Trends Genet.* **17**, 262 (2001).

which is high computational times. Distance methods, on the other hand, are the fastest available methods and are to be used in priority when using very large datasets (more than 100 sequences). The distance between two sequences being defined as the number of substitutions estimated to have occurred since their divergence, the tree is chosen for which the distances between sequences, adapted to the tree, are the smallest. The disadvantage of this approach is considering sequences by pairs to estimate their distance, resulting in an impoverishment of the data. In practice, distance methods are not very robust (they work badly when the model is far from the truth), which is partly compensated by the huge choice of models which are implemented in various user-friendly programs, and by the computational speed which allows testing various models. "Neighbor-Joining" (NJ[39]) is a heuristic*, which has become to be the synonym of distance methods for most investigators. As both ML and distance methods use evolutionary models, which can modify the result, the model used should always be precised when presenting a tree from these methods.

Whatever the quality of your data, the phylogenetic method will give you a tree. It is important to know how well supported the conclusions you can draw from that tree are. The main method for this is the bootstrap,[40] which gives a support to each branch in the tree, based on N pseudoreplicates. Two points should be made about the bootstrap: it is not a real statistical test, thus setting a "5% threshold" not relevant; and the computational time involved is N times that is taken to make the original tree with the same method. It is a good procedure not to interpret branches with less than 50% bootstrap, but there is no consensus on what constitutes a "good" bootstrap score. Many other methods to evaluate phylogenies exist, and the most rigorous is probably comparing the likelihood of alternative phylogenies for the same data.[41]

Finally, a note on "long-branch attraction": when a sequence evolves much faster than the others, it tends to resemble them less, and many phylogenetic methods will then group this sequence with the outgroup rather than with its true sister sequences.[32] Long-branch attraction can lead to false conclusions with high bootstrap support. It should be detected by paying attention to branch lengths (not visible in parsimony) and to unstable results depending on evolution models. Although there is no magical solution to this problem, results can be improved by the use of more realistic evolutionary models (typically gamma law among sites) and incorporation of more sequences in the analysis. If these new sequences are

[39] N. Saitou and M. Nei, *Mol. Biol. Evol.* **4**, 406 (1987).
[40] J. Felsenstein, *Evolution* **39**, 783 (1985).
[41] N. Goldman, J. P. Anderson, and A. G. Rodrigo, *Syst. Biol.* **49**, 652 (2000).

evolutionarily close to the fast sequence, they will "break" the long branch by clustering with it, and diminish long-branch attraction. NJ under simplistic models and parsimony are known to be most sensitive to long-branch attraction, but it can be a problem to all methods.

To find programs which implement these various methods, we recommend passing through the website of Phylogeny links (see Box 2), where all software are listed according to methods and type of computer.

Evolution of Nuclear Receptors

It is feasible to align all nuclear receptors, DBDs and LBDs, although alignment of the LBD is less obvious. The alignment of these two domains provides 160 complete amino-acid sites to build a phylogeny across all receptors. A complete discussion of nuclear receptor evolution would be out of the scope of this paper, but a few points should be briefly made. Swapping between DNA-binding and ligand-binding domains during evolution has been proposed,[42,43] but is difficult to prove given the very poor resolution of phylogenetic trees based on the DBD alone.[6] Some extent of independent domain evolution did in any case exist, since we know of nuclear receptors which lack either DBD or LBD (the NR0 group of the nomenclature, see next paragraph), and the trithorax *Drosophila* gene contains a nuclear receptor-like DBD, but does not otherwise appear to be a nuclear receptor. The general case still appears to be evolution (duplication, divergence) of the whole nuclear receptor gene.

The wide diversity of unrelated ligands that are known for nuclear receptors is striking, given the high conservation of the structure and mode of action of the receptors. It would be very nice for the bioinformatic prediction of function if closely related nuclear receptors bound similar ligands, if there was for example, a clade of steroid receptors, a clade of retinoid receptors, a clade of orphans, etc. But it has been clear since the first studies that there is no such relation.[42] The observation of convergence toward similar ligands in different branches of the tree, and the presence of orphans in every subfamily, have led to the suggestion that the ancestral nuclear receptor may have been an orphan, with the capacity to gain ligand-binding specificity many times independently during evolution.[6,44] On the other hand, it should be noted that orthologue* receptors in different vertebrates usually have very similar, if not identical functions, as do the

[42] V. Laudet, C. Hänni, J. Coll, C. Catzeflis, and D. Stéhelin, *EMBO J.* **11**, 1003 (1992).

[43] J. W. Thornton and R. Desalle, *Syst. Biol.* **49**, 183 (2000).

[44] H. Escriva, R. Safi, C. Hänni, M.-C. Langlois, P. Saumitou-Laprade, D. Stehelin, A. Capron, R. Pierce, and V. Laudet, *Proc. Natl. Acad. Sci. USA* **94**, 6803 (1997).

"recent" duplicates which appeared at the origin of vertebrates such as estrogen receptors α and β, thyroid hormone receptors α and β, etc. Thus, a certain measure of functional prediction is possible from phylogenetic analysis and sequence comparison, but it should not be extended beyond its limits.

Classification of Nuclear Receptors

A standardized nomenclature of nuclear receptors has been proposed based on the phylogeny.[7] There is no outgroup to the complete phylogeny of nuclear receptors, so it is basically unrooted. Taking into account animal phylogeny and the function of the proteins, there are a limited number of reasonable positions for the rooting, which allow the definition and orientation of major subfamilies of nuclear receptors, supported by high bootstrap scores (>90% in distance analysis). Six such subfamilies are recognized in the nomenclature, and are further divided into groups and genes. Thus the official name of each gene is "NRxyz," where x is the subfamily (1–6), y is the group (A, B, C, . . .), and z is the gene (1, 2, . . .). For example, hTRα (thyroid hormone receptor) belongs to the first subfamily, to its first group, and is its first gene, thus its official nomenclature is NR1A1. The orthologues in different vertebrate species have the same nomenclature, as do the orthologues in different invertebrate species. The distinction is made because of the high number of duplicate nuclear receptor genes in vertebrates.[45] Official nomenclature of a nuclear receptor should always be quoted at least once in a scientific work, for example in the Introduction or the Materials section.

Subfamily 1 is the largest, comprising 11 groups of receptors, including thyroid hormone receptors, RARs, PPARs, Rev-erbs, ecdysone receptors, and vitamin D receptors, as well as several orphans. Subfamily 2 contains six groups, including major orphan receptors such as HNF4s, COUP-TFs, and TR2/4, but also retinoic X receptors. Subfamily 3 is the most homogenous subfamily, with three groups: estrogen receptors, estrogen-related receptors, and "classical" steroid receptors (AR, MR, PR, GR). Subfamily 4 is limited to the NGFIB group. Subfamily 5 contains two groups, FTZ-F1 and DHR39. Subfamily 6 is the smallest, with one known gene in its one group, GCNF1. In addition, a subfamily 0 was defined, which includes seven receptors lacking either the LBD (group 1, only known in invertebrates) or the DBD (group 2, only known in vertebrates).

---

[45] H. Escriva, V. Laudet, and M. Robinson-Rechavi, *J. Struct. Funct. Genomics* **3**, 177 (2003).

This classification, based on phylogeny of DBD + LBD, has been confirmed by an automatic classification procedure for proteins,[46] using only LBD sequences. The one exception is the huge diversity of nematode nuclear receptors,[4,5] which still await classification in the official nomenclature, and indeed disturbed this automatic procedure.[46]

Nuclear Receptors in Databases

Dedicated Databases

Three databases specialized in nuclear receptors are available (Box 2), and probably constitute together the major bioinformatic resource on nuclear receptors. The Nuclear Receptor Resource,[47] NRR, is the oldest. It consists of a network of webpages, specialized in different subfamilies of nuclear receptors, such as glucocorticoid receptors or androgen receptors, and maintained by experts in the field. The NRR features important information, such as a "Who's who?," in the field, meeting advertisements, educational resources, etc. On the other hand, it is not exhaustive, and it does not allow complex queries such as "all mouse orphan receptors." It can be considered as the best place to look for information first, especially for beginners in the field.

Two other databases have used a more classical approach, focusing on the sequences and related information, and relying on the official nomenclature to organize data. NucleaRDB[48] is accessible also through its own webpage, and includes all nuclear receptors, with alignments and trees per subfamily and group, as well as some information on dimerization and other properties of nuclear receptors. Sequence and related data in NucleaRDB is updated every two or three months. Limitations of NucleaRDB is that it does not allow queries, or manipulation of objects such as alignments, and the alignments and trees are mostly "as is" results of CLUSTALW.[26] Interesting features of NucleaRDB include data mining from MEDLINE for mutations, and the development of cross-references to a great diversity of other databases. As of 24th May 2002 there were 504 mutations listed in NucleaRDB, of which for 378 there was a single unambiguous Swissprot entry (noted "OK"), extracted from 134 Medline abstracts, which represents a serious gain of time of manual scanning of the literature.

---

[46] N. Wicker, G. R. Perrin, J. C. Thierry, and O. Poch, *Mol. Biol. Evol.* **18**, 1435 (2001).
[47] E. Martinez, D. D. Moore, E. Keller, D. Pearce, J. P. Vanden Heuvel, V. Robinson, B. Gottlieb, P. MacDonald, S. Simons, Jr., E. Sanchez, and M. Danielsen, *Nucleic Acids Res.* **26**, 239 (1998).
[48] F. Horn, G. Vriend, and F. E. Cohen, *Nucleic Acids Res.* **29**, 346 (2001).

The newest player in the field is NUREBASE,[49] which includes protein and DNA sequence data, alignments and phylogenetic trees reviewed by an expert. The protein entries are enriched by standardized names, nomenclature information, keywords including ligands, etc. NUREBASE is completed by automatically generated daily updates, stored in NUREBASE_DAILY. Both can be queried either by a webpage which allows complex queries, on keywords, taxonomy, etc., or by the graphical interface FamFetch,[50] which also allows interactive manipulation of trees and alignments. The strengths of NUREBASE for now are thus the complementarity between expertized data and automatic daily updates, and the flexibility of complex queries such as "all complete nuclear receptors known in mouse and fly."

In the future, further links between these different databases are desirable, while they will each continue to grow, hopefully. For example, information on transcripts (alternative splicing, expression levels, etc.) is being included in NUREBASE,[51] and several groups are working on integration of the 3D structure of nuclear receptors, which is notably of high interest for drug development. One of the advantages of all these databases is that each specialist in one domain can access data from other fields much more easily. For example, structural data for a developmentalist, or taxonomical and evolutionary information for a pharmacologist. Future developments should increase this trend.

Nuclear Receptors in Other Databases

GenBank, EMBL, and DDBJ are three complete DNA databanks, which contain all publicly available DNA sequences and exchange their data every day.[52–54] The major interest of these banks is their completedness, including preliminary genome fragments, ESTs, mutations, etc., but also a lot of redundancy, mistaken or incomplete annotations, etc. The most

[49] J. Duarte, G. Perriere, V. Laudet, and M. Robinson-Rechavi, *Nucleic Acids Res.* **30**, 364 (2002).

[50] G. Perriere, L. Duret, and M. Gouy, *Genome Res.* **10**, 379 (2000).

[51] J. Duarte, T. Ourjdal, G. Perriere, V. Laudet, and M. Robinson-Rechavi, in "Journées Ouvertes Biologie Informatique Mathématiques" (J. Nicolas and C. Thermes, eds.), Saint-Malo, France, 2002.

[52] D. A. Benson, I. Karsch-Mizrachi, D. J. Lipman, J. Ostell, B. A. Rapp, and D. L. Wheeler, *Nucleic Acids Res.* **30**, 17 (2002).

[53] G. Stoesser, W. Baker, A. van den Broek, E. Camon, M. Garcia-Pastor, C. Kanz, T. Kulikova, R. Leinonen, Q. Lin, V. Lombard, R. Lopez, N. Redaschi, P. Stoehr, M. A. Tuli, K. Tzouvara, and R. Vaughan, *Nucleic Acids Res.* **30**, 21 (2002).

[54] Y. Tateno, T. Imanishi, S. Miyazaki, K. Fukami-Kobayashi, N. Saitou, H. Sugawara, and T. Gojobori, *Nucleic Acids Res.* **30**, 27 (2002).

efficient way to access them for nuclear receptor-related information is through BLAST (see Similarity Searches for Nuclear Receptors), or by direct links from articles of interest.[55] Otherwise one risks drowning in a sea of data, and miss the relevant information, for example in GenBank (24 May 2002) there are 1571 entries for "nuclear receptor," of which not human ERα, which would be missed with these keywords. A good review of these databanks, and related resources, can be found in Ref. 10.

Maybe the most useful database for anyone interested in the function and structure of proteins is Swissprot.[56] It contains protein sequences with minimal redundancy, enriched by many annotations and links toward other databases such as posttranslational modifications, domains, nucleic sequences including alternative transcripts, 3D structures, diseases, etc. Moreover, Swissprot entries have standardized names, which make research of proteins much easier; for example, thyroid hormone receptors names all start by THA or THB. On the other hand, Swissprot is not up-to-date on sequences from nucleic databanks, due to the enormous work involved in this enrichment and standardization. It is thus completed by TrEMBL, the translation of EMBL-coding sequences, which ensures completeness if not high quality.

Another category of databanks of interest are those specialized in types of data other than sequences. Notably, the Protein Data Bank PDB[57] contains all experimentally solved protein 3D structures, as well as good-quality theoretical predictions. It can be queried by keywords, but "nuclear receptor" recovers also some unrelated molecules such as Transthyretin. The best way to access this data may be through links such as established in NucleaRDB, NUREBASE, or Swissprot. The latter possibility is especially rich, since you can find any type of protein (NR, cofactor, ...) by BLAST on Swissprot, and then follow the links to PDB and other databases. Also worthy of note is OMIM (see Box 2), a catalog of human genes and genetic disorders. There are 452 entries (24 May 2002) for "nuclear receptor," most of them relevant, but again it may be more efficient to follow links from other databases. OMIM is probably the best place to look not only for genetic disorders, but also for basic bibliography on each receptor.

Finally, nuclear receptors are ligand-dependent transcription factors, with many ligands being major hormones. There exists a variety of

[55] D. L. Wheeler, D. M. Church, A. E. Lash, D. D. Leipe, T. L. Madden, J. U. Pontius, G. D. Schuler, L. M. Schriml, T. A. Tatusova, L. Wagner, and B. A. Rapp, *Nucleic Acids Res.* **30**, 13 (2002).

[56] A. Bairoch and R. Apweiler, *Nucleic Acids Res.* **28**, 45 (2000).

[57] H. M. Berman, J. Westbrook, Z. Feng, G. Gilliland, T. N. Bhat, H. Weissig, I. N. Shindyalov, and P. E. Bourne, *Nucleic Acids Res.* **28**, 235 (2000).

bioinformatic resources (Box 2) which, although not dedicated to nuclear receptors, provide valuable information on transcription factors, endocrinology, or ligands and receptors. For example, LigBase[58] is a database of ligand-binding proteins aligned to structural templates, based on structures from the PDB. A simple search on "nuclear" and "receptor" provides 42 answers, all relevant, including natural and synthetic ligands. For each, a plot of the ligand–receptor interaction and links to other databases, including PDB, are provided. Transfac is a database on transcription factors, their genomic-binding sites and DNA-binding profiles,[59] which can be queried by keyword searches on specialized tables such as "genes," "factors," or "sites." A search for "nuclear receptor" on "factors" gives 158 entries, mostly relevant, although a search on "sites" gives a majority of artificial sequences. This is a very useful resource for anyone interested in regulation of nuclear receptors or by nuclear receptors.

Functional Bioinformatics

Exploiting the Structure

The obtention of increasing numbers of tridimensional structures of nuclear receptors, notably of LBDs, should ideally be exploited to understand and predict the ligand specificity of different proteins. For this, bioinformatic studies are needed which can compare different primary, secondary, and tertiary structures, and relate them to known properties such as ligand specificity, heterodimerization, etc. Such studies are still in their infancy, due to the limited number of available nuclear receptor structures, especially from orthologues* in different species, and to the complexity of manipulating 3D structures, compared to the relative ease of manipulating linear sequences.

Still some leads have been presented recently, for example to estimate the volume of many ligand-binding pockets in a comparable manner, despite different methods used to generate and describe data, and thus help comparison between nuclear receptor LBDs, and with potential ligands.[60] This is of course very important to search for new drugs, by "docking" strategies for example. Ligand docking is a computer search of compatibility

[58] A. C. Stuart, V. A. Ilyin, and A. Sali, *Bioinformatics* **18**, 200 (2002).

[59] E. Wingender, X. Chen, R. Hehl, H. Karas, I. Liebich, V. Matys, T. Meinhardt, M. Pruss, I. Reuter, and F. Schacherer, *Nucleic Acids Res.* **28**, 316 (2000).

[60] M. H. Lambert, S. P. Williams, and H. E. Xu, *in* "Nuclear Receptor Superfamily" (P. Chambon, D. J. Mangelsdorf, and B. S. Katzenellenbogen, eds.), Keystone Symposia, p. 159. Snowbird, Utah, USA, 2002.

between the 3D structures of a receptor and potential ligands.[61,62] Due to the complexity of the problem, including flexibility of protein folds, it is very time consuming, but it remains a powerful way of finding leads for new drugs *in silico*.[63] It has been applied for example to the vitamin D receptor of human, based on modeling of the 3D structure, using the experimentally solved structure of six other receptors.[64] In another work, paralogous* structures have been compared to find coevolving sites which may be important for conformational changes and dimer interactions of nuclear receptors.[65] In all cases, experimental verification remains necessary, notably to identify false positives*.

Promoter Prediction

The search for promoters in eukaryotic genomic DNA is a major aim of bioinformatics. Yet success is still very low, due to very diverse promoters and weak conservation.[66] Moreover, nuclear receptors are specific to animals, whose genomes are generally big, with large intergenic sequences in which to find the promoters, whereas most progress has been made in the more compact yeast *S. cerevisiae* genome. In tests we did, *ab initio* research for promoters on the 5' UTRs of mammalian nuclear receptor genes generated very little results. Methods based on similarity to known promoters were more successful, but with such a large proportion of false positives* as to make the results useless. False positives are indeed a general problem of such methods.[66] The most promising approach we tested was "phylogenetic footprinting," that is the comparison of large genomic regions between more or less closely related species. The most conserved noncoding sequences are well correlated with known promoters. From this point of view, the very recent release of the mouse genome (see Ensemble, Box 2) has high potential for functional bioinformatics of nuclear receptors and other genes.

The other way to characterize promoters is to use a database of promoters, and to search for the known sequences in 5' of a gene. It is important to start from a clean database, because many promoter

[61] Y. Z. Chen and D. G. Zhi, *Proteins* **43**, 217 (2001).

[62] R. M. Knegtel and M. Wagener, *Proteins* **37**, 334 (1999).

[63] R. Abagyan and M. Totrov, *Curr. Opin. Chem. Biol.* **5**, 375 (2001).

[64] K. Yamamoto, H. Masuno, M. Choi, K. Nakashima, T. Taga, H. Ooizumi, K. Umesono, W. Sicinska, J. VanHooke, H. F. DeLuca, and S. Yamada, *Proc. Natl. Acad. Sci. USA* **97**, 1467 (2000).

[65] A. I. Shulman, C. Larson, M. Makishima, R. Ranganathan, and D. J. Mangelsdorf, *in* "Nuclear Receptor Superfamily" (P. Chambon, D. J. Mangelsdorf, and B. S. Katzenellenbogen, eds.), p. 125. Keystone Symposia, Snowbird, Utah, USA, 2002.

[66] U. Ohler and H. Niemann, *Trends Genet.* **17**, 56 (2001).

annotations in generalist databanks contain errors, are ambiguous or are not up-to-date. The Eukaryotic Promoter Database (EPD)[67] notably collects promoter sequences in such an expertized manner. Although many promoter sequences are probably missing, we may thus hope to avoid the large amount of false positives* generated by most methods. The first use of a database, such as EPD, is to simply find all characterized promoters in front of a gene of interest, if it is in the database, without a time-consuming search of the literature. There are for now three nuclear receptors in EPD, which can be found easily through the use of the keywords "nuclear receptor": *Drosophila* E78 (two characterized promoters), human ERα, and PR. Then similar promoters may be found in other genes by BLAST for example, although the risk exists again of false positives.

The search for promoters can also be considered the other way around: since nuclear receptors are transcription factors, it is a legitimate question to look for genes which possess binding sites to these factors in their promoter, and are thus susceptible of being regulated by nuclear receptors. With the human genome sequence,[17] as well as the improvement of bioinformatic and high-throughput experimental methods, systematic searches of this kind become feasible. For example, a search for all HNF4α binding sites in the human genome was recently presented, in which experimental verification allowed confirmation of many of these sites.[68] In the forseeable future, these approaches will not replace experimental characterization of functionally important sites, but they hold great potential to guide biologists among the mass of data, indicating the most promising sequences to test further.

Data-Mining in Functional Genomics

Functional genomics include a variety of methods which have in common the production of large quantities of data per experiment ("high throughput"), which should then be sieved for the relevant information. The different types of data one can obtain in this manner, and their production, are out of the scope of this review, but bioinformatics must be used to make sense of the results. The most popular functional genomic approaches up to now are concerned with the transcriptome. ESTs can notably be used to characterize the expression pattern of nuclear receptors,[51] and to discover

[67] V. Praz, R. Perier, C. Bonnard, and P. Bucher, *Nucleic Acids Res.* **30**, 322 (2002).
[68] K. Ellrot, C. Yang, F. M. Sladek, and T. Jiang, *Bioinformatics* **18**(Suppl. 2), S100 (2002).

new alternative transcripts.[69] Chen *et al.*,[70] for example, discovered two new transcripts related to ERRβ by a bioinformatic analysis of EST data, and showed that, in fact, one previously described ERR gene was erroneously annotated. Such work concerning nuclear receptors, using ESTs or DNA chips, is scarce for now, but is bound to increase in the near future, making the associated bioinformatic methods more relevant to the field.[71,72]

An important notion in functional genomics is that of "guilt by association": if two genes, or proteins, have similar expression patterns, or bind to the same cofactors, or are induced by the same stimulants, it is worth investigating whether they are share functions or regulatory pathways. Such approaches can be very fruitful to find nuclear receptor cofactors and target genes, as well as to improve our understanding of the regulation of orphan receptors. For example, LION Biosciences has recently reported use of automated yeast two-hybrid methods coupled with bioinformatic analysis, a dedicated database, and functional tests, to determine cofactor interactions for nuclear receptors.[73]

Conclusion

The present of bioinformatics for nuclear receptors is mostly composed of three specialized databases, which allow faster and easier access to relevant information, and of more or less easy data-mining from other sources. The future lies in "postgenomic" bioinformatics: through the integration of experimental data into bioinformatic bases and the analysis of nonannotated sequences, we should be able to ask simple or complex queries on any aspect of nuclear receptor biology and biochemistry, and get an understandable, fast, complete, and relevant answer. This answer will in some cases be no more than a lead toward new experimental work, and in others hopefully be almost definitive, as can already be done for phylogenies for example. In any case, this necessitates integration of data as diverse as expression patterns, phenotypes, 3D structures, natural and artificial ligands, target genes, etc., plus methods such as promoter characterization, alignment, structure comparison, microarray analysis, etc. These challenges

[69] B. Modrek and C. Lee, *Nat. Genet.* **30**, 13 (2002).

[70] F. Chen, Q. Zhang, T. McDonald, M. J. Davidoff, W. Bailey, C. Bai, Q. Liu, and C. T. Caskey, *Gene* **228**, 101 (1999).

[71] S. Fields, Y. Kohara, and D. J. Lockhart, *Proc. Natl. Acad. Sci. USA* **96**, 8825 (1999).

[72] K. P. White, *Nat. Rev. Genet.* **2**, 528 (2001).

[73] M. Albers, C. Kaiser, I. Kober, H. Kranz, C. Kremoser, and M. Koegl, *in* "Nuclear Receptor Superfamily," (P. Chambon, D. J. Mangelsdorf, and B. S. Katzenellenbogen, eds.), Keystone Symposia, p. 99. Snowbird, Utah, USA, 2002.

will be answered in part by dedicated work to nuclear receptors, probably by extension of the existing databases, and in part by integration of new bioinformatic methods into the everyday habits of the nuclear receptor community. We hope that in some cases nuclear receptors will, in fact, be the model which will allow the development of new bioinformatic methods, then to be generalized to other proteins of interest. In any case, in the future much more will be accessible in a "click" of the computer mouse, thanks to high-throughput experiments, bioinformatics, and the sharing of knowledge.

[7] Application of Random Peptide Phage Display to the Study of Nuclear Hormone Receptors

By CHING-YI CHANG, JOHN D. NORRIS, MICHELLE JANSEN, HUEY-JING HUANG, and DONALD P. MCDONNELL

Introduction

The steroid–nuclear hormone receptors are ligand-activated transcription factors involved in the regulation of a variety of processes ranging from reproduction to cholesterol metabolism. One of the research focuses in our laboratory has been to understand the mechanisms underlying the pharmacology of estrogen receptor (ER) ligands. The classical models of ER pharmacology held that the receptor exists in either an active or an inactive state within target cells. According to this model, the function of an agonist is to transform a receptor from an inactive one to an active one. Antagonists, on the other hand, were believed to competitively inhibit estrogen binding and freeze the receptor in an inactive state. This simple model was challenged when it was determined that tamoxifen, an antiestrogen used to oppose estrogen action in ER-positive breast tumors, could function as an agonist in some tissues including the bone, uterus and the cardiovascular system. Another compound, raloxifene, has been shown to function as an antagonist in the breast and uterus, while functioning as an estrogen in the bone and cardiovascular system. Reflecting these properties, tamoxifen, raloxifene, and other mechanistically similar compounds have been reclassified as Selective Estrogen Receptor Modulators (SERMs). It has been difficult to reconcile SERM action with the classical models of receptor pharmacology. However, a more complex model which more adequately describes these results has recently emerged. This is based on observations that the shape of ER is regulated by the nature of the bound ligand and the ability of cells to distinguish between different receptor conformations. Specifically, it was demonstrated that the ER can adopt multiple conformations upon binding different ligands.[1,2] The impact of such conformational changes was further revealed when steroid receptor coactivator-1 (SRC-1), and subsequently other cofactor proteins, coactivators and corepressors, were isolated.[3,4] The cue for recruitment of coactivators and corepressors to the ER is provided by the conformational changes induced by ligand binding to the receptor. Analysis of the crystal structure of the ER ligand-binding domain revealed that the activation function 2 (AF-2) pocket, when bound by an agonist, undergoes a conformational change which allows the docking of a conserved leucine-rich LxxLL (L = leucine, x = any amino acid) motif present in all p160 coactivator proteins. Binding of an antagonist, conversely, alters the AF-2 structure so that it is incompatible with coactivator docking.[5,6] It is thus believed that the pharmacology of a SERM is determined by both the receptor conformation it induces, and the relative expression levels of cofactor proteins in the target tissue. Therefore, it was clear several years ago that defining the impact of conformation on receptor–cofactor interactions would be instructive in understanding SERM action.

In pursuing the concept that protein–protein interactions are important determinants of nuclear receptor transcriptional activity and that such interactions are dictated by receptor conformation, many investigators have used the yeast two-hybrid screen, expression cloning, proteomic analysis, crystallography, and various biochemical methods to identify receptor-interacting proteins and to define cofactor–receptor interfaces. To complement these other studies, we have applied a random peptide phage display approach to define and determine the protein–protein interaction surfaces on the ER. Phage display has been widely used as a method to map protein–protein interactions, identify peptide ligands for cell surface receptors, and map antibody–antigen epitopes.[7] Using phage display,

[1] J. M. Beekman, G. F. Allan, S. Y. Tsai, M.-J. Tsai, and B. W. O'Malley, *Mol. Endocrinol.* **7**, 1266 (1993).

[2] D. P. McDonnell, D. L. Clemm, T. Hermann, M. E. Goldman, and J. W. Pike, *Mol. Endocrinol.* **9**, 659 (1995).

[3] N. J. McKenna and B. W. O'Malley, *Nat. Med.* **6**, 960 (2000).

[4] N. J. McKenna, R. B. Lanz, and B. W. O'Malley, *Endocr. Rev.* **20**, 321 (1999).

[5] A. K. Shiau, D. Barstad, P. M. Loria, L. Cheng, P. J. Kushner, D. A. Agard, and G. L. Greene, *Cell* **95**, 927 (1998).

[6] A. C. Pike, A. M. Brzozowski, R. E. Hubbard, T. Bonn, A. G. Thorsell, O. Engstrom, J. Ljunggren, J. A. Gustafsson, and M. Carlquist, *EMBO J.* **18**, 4608 (1999).

[7] R. Cortese, P. Monaci, A. Luzzago, C. Santini, F. Bartoli, I. Cortese, P. Fortugno, G. Galfre, A. Nicosia, and F. Felici, *Curr. Opin. Biotechnol.* **7**, 616 (1996).

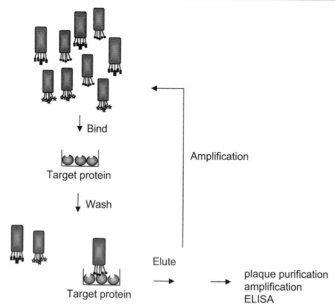

FIG. 1. Affinity selection of target protein binding peptides using phage display. M13 phage-based random peptide libraries are incubated with target protein immobilized on a solid support. After incubation, a washing step is used to remove unbound phage and the target protein binding phage are retained. Bound phage are eluted using a low-pH buffer, amplified in bacteria and subjected to subsequent rounds of selection. The selection process is repeated 3–4 times to enrich for target binding phage. Individual phage are then plaque purified, amplified, and their binding characteristics examined by ELISA. Phage that interact specifically with target proteins are selected and the sequences of the displayed peptides are deduced by DNA sequencing.

random peptide libraries can be generated and specific target-interacting peptides can be identified. The random peptide libraries are created by inserting short random oligonucleotides within the coding sequence of the M13 bacteriophage capsid protein, with the subsequent display of random peptides on the outer surface of the phage. Using routine molecular biology techniques, phage libraries containing billions of random peptide inserts can be easily constructed for use in affinity selection for target protein binders. Since the peptide is physically linked to the phage particle, target-binding phage can be recovered and amplified after affinity selection, and the oligonucleotide insert sequence can be determined by DNA sequencing to identify the corresponding peptide sequence (Fig. 1). Consequently, only a modest amount of time, effort and material are required to survey a vast number of peptides for their ability to bind to the target protein of interest.

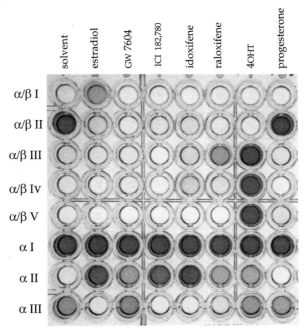

FIG. 2. Effect of ligands on ERα conformation. Biotinylated vitellogenin EREs (2 pmol) were immobilized on a 96-well plate coated with streptavidin. Subsequently, ERα (3 pmol) was immobilized on the ERE and incubated with ligand (1 μM) for 5 min before the addition of phage expressed peptides. Phage were incubated for 30 min at room temperature, and washed five times to remove unbound phage. The bound phage were detected using an anti-M13 antibody coupled to HRP and developed in ABTS and hydrogen peroxide. [Reprinted with permission from A. L. Wijayaratne, S. C. Nagel, L. A. Paige, D. J. Christensen, J. D. Norris, D. M. Fowlkes, and D. P. McDonnell, Comparative analyses of the mechanistic differences among antiestrogens. *Endocrinology* **140**, 5828–5840 (1999). Copyright The Endocrine Society.]

We have successfully used phage display to probe the various conformations of ER induced by binding to different SERMs (Fig. 2).[8,9] Characterization of ligand-induced conformational changes serves three purposes: (A) it enables the establishment of a link between the receptor conformation and biological activity, (B) it will lead to the identification of specific protein–protein interaction surfaces which may be important for receptor function, and (C) it allows the identification of novel ligands with altered peptide-binding specificities–biological functions. For example, one

[8] A. L. Wijayaratne, S. C. Nagel, L. A. Paige, D. J. Christensen, J. D. Norris, D. M. Fowlkes, and D. P. McDonnell, *Endocrinology* **140**, 5828 (1999).
[9] J. D. Norris, L. A. Paige, D. J. Christensen, C.-Y. Chang, M. R. Huacani, D. Fan, P. T. Hamilton, D. M. Fowlkes, and D. P. McDonnell, *Science* **285**, 744 (1999).

can imagine that a common biological response shared by all ER ligands is expected to result from the interaction of a common cofactor with a surface on ER that is formed regardless of which compound is used as an activating ligand (for instance, the αII peptide-binding site). On the other hand, a unique conformation created on the surface of tamoxifen-activated ER (for example the α/β V peptide-binding site) may be the surface used by tamoxifen–ER complex to recruit a novel cofactor that is specific to only this receptor–ligand complex, and which may be responsible for the partial agonist activity manifest by tamoxifen. This hypothesis was confirmed by the demonstration that the α/β V peptide was able to disrupt tamoxifen-induced ERα transcriptional activity but had no effect on the activity induced by estradiol, suggesting that the mechanism by which tamoxifen and estradiol manifest their activity may not be the same (Fig. 3). Regardless, by defining these receptor–cofactor interaction surfaces, it now appears possible to predict the biological activity of an unknown compound based on the conformation it induces within the receptor. Furthermore, since the interaction between receptor and cofactor ultimately determines the biological output of a compound–receptor complex, one should be able to generate novel classes of antagonists which target specific protein–protein interactions.

In addition to using phage display to study receptor conformation and SERM pharmacology, we have used this powerful technique to develop highly specific peptide antagonists which block receptor function by interfering with required protein–protein interactions. This approach has been particularly useful in the studies of ERα and ERβ where specific small molecule modulators are not yet available. We were able to find a small peptide that binds to ERβ but not ERα using phage display. Importantly, when introduced into cells this peptide specifically disrupted ERβ transcriptional activity without affecting the activity of the ERα protein (Fig. 4). In this post-genome era, it is also possible to use peptide sequences obtained from phage display screens to identify *bona fide* cofactor proteins and define unexpected interactions between known proteins by sequence analysis.[10,11] It is worth emphasizing that the phage display of random peptides is a powerful technology which clearly has utility in the study of most protein–protein interactions.

Although the *in vitro* phage display screen approach allows for the identification of small peptides that can probe the surface of the receptor, the biological significance of these surfaces cannot be inferred

[10] X. Li, E. A. Kimbrel, D. J. Kenan, and D. P. McDonnell, *Mol. Endocrinol.* **16**, 1482 (2002).
[11] X. Li and D. P. McDonnell, *Mol. Cell. Biol.* **22**, 3663 (2002).

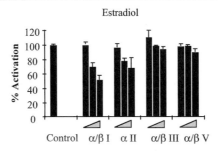

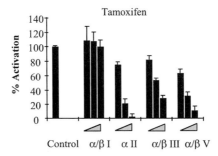

FIG. 3. Disruption of ERα transcriptional activity using cofactor-specific peptides. Cofactor-specific peptides can be used to selectively disrupt estradiol- or tamoxifen-activated ERα transcriptional activity. HepG2 cells were transfected with the estrogen-responsive C3-Luc reporter gene along with an ERα expression vector. Also included in the transfections were increasing amounts of different peptide expression vectors (as Gal4DBD fusions) as indicated. Cells were treated with 10 nM estradiol or 10 nM 4-hydroxytamoxifen. Control represents the transcriptional activity of estradiol-activated ER (or 4-hydroxytamoxifen-activated ER) in the presence of the Gal4DBD alone and is set at 100% activity. Increasing amounts of each Gal4DBD-peptide fusion expression vectors were introduced into cells (triangle) and the resulting transcriptional activity presented as percent activation of control is shown. The estradiol–ER binding peptide, α/β I, specifically blocked the interaction between ER and coactivator(s) recruited by the estradiol–ER complex leading to disruption of estradiol-induced ER transcriptional activity without affecting the activity induced by tamoxifen. The reverse is also true for tamoxifen–ER binding phage, α/β III and α/β V. Thus tamoxifen and estradiol do not manifest activity in the same manner. [Reprinted with permission from J. D. Norris, L. A. Paige, D. J. Christensen, C.-Y. Chang, M. R. Huacani, D. Fan, P. T. Hamilton, D. M. Fowlkes, and D. P. McDonnell, Peptide antagonists of the human ER. *Science* **285**, 744–746 (1999). Copyright 1999 American Association for the Advancement of Science.]

based solely on *in vitro* binding studies. A cell-based analysis is therefore required, and has been implemented to complement the *in vitro* approach. The cell-based approach is the modification of a commonly used mammalian two-hybrid assay to assess the interaction between the isolated peptides and the receptor(s) of interest inside cells and will be discussed later in this chapter.

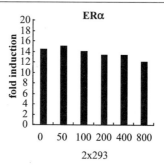

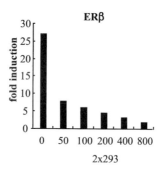

Fig. 4. An ER subtype-specific peptide selectively disrupts ERβ-dependent reporter gene expression without affecting ERα-mediated transcription when expressed in target cells. HeLa cells were transfected with an ER-responsive 3xERE-TATA-luc reporter gene alone with either an ERα or ERβ expression plasmid. Increasing amounts of the ERβ-selective peptide #293 (as Gal4DBD fusion) was also included in the transfection as indicated. Fold induction represents the ratio of the activity induced by estradiol versus no-hormone control for each transfection. [Reprinted with permission from C.-Y. Chang, J. D. Norris, H. Grøn, L. A. Paige, P. T. Hamilton, D. J. Kenan, D. M. Fowlkes, and D. P. McDonnell, Dissection of the LXXLL nuclear receptor-coactivator interaction motif using combinatorial peptide libraries: discovery of peptide antagonists of ERs alpha and beta. *Mol. Cell. Biol.* **19**, 8226–8239 (1999). Copyright 1999 American Society for Microbiology.]

This article is organized into four sections: the construction of a peptide phage display library, a step-by-step protocol for biopanning (the affinity selection process), the discussion of a mammalian cell-based screening technology to validate and complement the *in vitro* phage display approach, and finally the use of phage display selected peptides as antagonists of nuclear receptor function.

Construction of Random Peptide Libraries in Bacteriophage M13

Several types of bacteriophage and different coat proteins have been used to construct random peptide libraries, among which the filamentous

M13 phage and its capsid protein pIII are the most frequently used. Due to space limitations, we will discuss only the construction and biopanning of this type of random peptide library. A detailed discussion of other library formats can be found elsewhere.[12]

Several random peptide libraries are available commercially through New England Biolabs (Beverley, MA). However, they exist in limited formats and for most applications, the investigators will have to generate their own libraries. Construction of the random peptide library involves three steps: (1) generation of double stranded oligoucleotides, (2) ligation of oligonucleotides into the M13 phage genome, and (3) transformation of ligated DNA into *Escherichia coli* cells.

Materials

All the standard buffers and solutions can be found in Current Protocols in Molecular Biology.[13]

Common chemicals are obtained from Sigma-Aldrich (http://www.sigma-aldrich.com).

SDS-PAGE purified, degenerate oligonucleotides (Life Technologies, http://www.lifetech.com).

*Xho*I, *Xba*I and Klenow polymerase (Boehringer Mannheim Corp., Indianapolis, IN).

T4 DNA ligase (New England Biolabs, http://www.neb.com).

mBAX vector (a gift from Dr. Daniel Kenan, Duke University, Durham, NC). A similar vector is available from New England Biolabs.

15% nondenaturing polyacrylamide gel:

5 ml 30% acrylamide solution (29:1) (Bio-Rad Laboratory, Hercules, CA)
1 ml 10 × TBE
4 ml H_2O
100 μl 10% ammonium persulfate
10 μl N,N,N',N'-tetramethylethylenediamine (TEMED)

Phenol–chloroform–isoamyl alcohol (25:24:1) (Life Technologies, http://www.lifetech.com).

3 *M* sodium acetate (pH 5.2)
Dissolve 408 g sodium acetate · $3H_2O$ in 800 ml H_2O.
Add H_2O to 1 liter.
Adjust pH to 5.2 with acetic acid.

[12] B. K. Kay, J. Winter, and J. McCafferty, "Phage Display of Peptides and Proteins: A Laboratory Manual," p. xxii. Academic Press, San Diego, 1996.
[13] F. M. Ausubel, "Current Protocols in Molecular Biology." John Wiley, New York, 2001.

2xYT medium
    Dissolve 31 g 2xYT broth (Life Technologies) in 900 ml H_2O.
    Adjust volume to 1 liter with H_2O.
    Sterilize by autoclaving.

2xYT plates
    Dissolve 31 g 2xYT broth (Life Technologies) in 900 ml H_2O.
    Add 15 g bacto-agar (Life Technologies), adjust volume to 1 liter with H_2O.
    Sterilize by autoclaving.
    Let the agar cool to around 55 or 56°C before pouring plates.

Top Agar
    Dissolve 31 g 2xYT broth (Life Technologies) in 900 ml H_2O.
    Add 8 g bacto-agar (Life Technologies), adjust volume to 1 liter with H_2O.
    Sterilize by autoclaving.

SOC (Life Technologies)
30% (w/v) PEG 8000–1.6 M NaCl
    Dissolve 300 g of PEG 8000 and 93.6 g NaCl in H_2O to a final volume of 1 liter. Filter sterilize.

Methods

Generation of Double-Stranded DNA Inserts

Oligonucleotides can be custom synthesized by any vendor; however, the codon schemes used to generate random peptide inserts need to be carefully considered before placing the order. For example, NNN, where N is an equimolar mix of all four bases, will produce all four possible codons that encode all 20 amino acids. Unfortunately, this codon scheme also encodes the three stop codons, leading to generation of non-productive clones containing pre-matured stop codons. The use of either the NNK (K = G or T) or the NNS (S = G or C) codon scheme is a partial solution to this problem, because both of them use 32 codons to encode all 20 amino acids and one stop codon (TAG), significantly reducing the occurrence of a stop codon in the insert. Other alternative synthesis approaches have been used to encode all the amino acids with no stop codon, but they require more sophisticated and expensive technologies.[14–16]

[14] S. M. Glaser, D. E. Yelton, and W. D. Huse, *J. Immunol.* **149**, 3903 (1992).
[15] J. Sondek and D. Shortle, *Proc. Natl. Acad. Sci. USA* **89**, 3581 (1992).
[16] B. P. Cormack and K. Struhl, *Cell* **69**, 685 (1992).

5'-AGTGTGTGC<u>CTCGAG</u>A(NNK)₇CTG(NNK)₂CTGCTG(NNK)₇<u>TCTAGA</u>CTGTGCAGT-3'   **Top strand**
      Xho I   **Xba I**

◀┈┈ 3'-AGATCTGACACGTCA-5'   **Bottom strand**

FIG. 5. The design of the oligonucleotides used in the construction of the "LxxLL" library. The top strand oligonucleotide was designed to contain an *Xho*I site (CTCGAG) at the 5'-end and an *Xba*I site (TCTAGA) at the 3'-end, which are compatible with the cloning site in the M13 phage vector mBAX and will generate inframe fusions with the pIII capsid protein. A shorter oligonucleotide complementary to the 3' sequences of the top strand is synthesized and annealed to the top strand oligonucleotide. The resulting DNA complex is extended with Klenow polymerase to generate double stranded DNA and subsequently digested with *Xho*I and *Xba*I to ligate into mBAX vector, also digested with the same enzymes. N: any nucleotide, K: G or T.

For instructional purposes, we will discuss the construction of a library in the format of -(X)₇LXXLL(X)₇- (X = A, C, G, or T) using the NNK codon scheme.

The top strand oligonucleotide (Fig. 5) was designed to contain an *Xho*I site (CTCGAG) at the 5'-end and an *Xba*I site (TCTAGA) at the 3'-end, which are compatible with the cloning site in the M13 phage vector mBAX and will generate inframe fusions with the pIII capsid protein. A shorter oligonucleotide complementary to the 3' sequences of the top strand was synthesized and annealed to the top strand oligonucleotide. The resulting DNA complex was extended with Klenow polymerase to generate double stranded DNA and subsequently digested with *Xho*I and *Xba*I to ligate into mBAX vector, also digested with the same enzymes. The ligated DNA was electroporated into *E. coli* JS-5 cells to generate phage libraries.

1. Combine 400 pmol top strand oligonucleotide
                400 pmol bottom strand oligonucleotide
                20 μl 10 × Klenow buffer (final = 1 ×)
                Adjust with H₂O to final volume of 200 μl
2. Incubate the mixture at 75°C for 15 min. Allow the reaction to cool slowly to room temperature.
3. Add   4.0 μl 10 mM dNTPs   (final = 200 μM)
                2.0 μl acetylated BSA (10 mg/ml) (final = 0.1 μg/μl)
                2.0 μl 100 mM DTT   (final = 1 mM)
                8.0 μl Klenow polymerase (2 unit/ul)
                Incubate at 37°C for 1 hr.
4. Heat inactivate Klenow polymerase at 65°C for 1 hr.
5. Combine the double stranded DNA with
         40 μl 10 × restriction enzyme buffer   (final = 1 ×)
         4 μl acetylated BSA (10 mg/ml)   (final = 0.1 μg/μl)

4 μl 100 mM DTT   (final $= 1$ mM)
Adjust with H_2O to final volume of 400 μl
Save 10 μl as no-enzyme control. Split the remaining mix into two tubes. To one tube add 100 units of *Xho*I and to the other 100 units of *Xba*I. Save 10 μl from each single enzyme digest and combine the remaining solutions into one tube to generate double digested insert. Incubate at 37°C for 3 hr.

6. Load the no-enzyme control, single-enzyme digested and double-digested samples on a preparative 15% nondenaturing polyacrylamide gel. Run the gel at 100 V until the bromophenol blue dye reaches the bottom. After electrophoresis, stain the gel with 0.2 μg/ml ethidium bromide and visualize under a UV light box to determine if the enzyme digest is complete.

7. Upon completion of the enzyme digestion, load the double digested DNA onto a 15 cm × 15 cm × 1 mm 15% polyacrylamide gel to separate any undigested or single-cut DNA from double digested DNA. Stain the gel as described above and use a razor blade to isolate gel pieces containing the double digested DNA. Collect all the gel pieces in 2–3 microcentrifuge tubes.

8. Recover the DNA insert by crushing the gel slices against the wall of the microcentrifuge tube using a pipet tip. Elute the DNA by adding 3.0 ml of 0.5 M ammonium acetate and incubate the tube at 37°C overnight with end-to-end rocking. Pellet the gel pieces by centrifugation. Save the supernatant, which contains eluted DNA. Reduce the volume to less than 500 μl by repeated extraction with 1 volume of 1-butanol. The DNA should remain in the aqueous phase (bottom layer).

9. Extract the eluted DNA solution with an equal volume of phenol–chloroform–isoamyl alcohol (25:24:1). The DNA should stay in the aqueous phase (top layer). Precipitate the extracted DNA with 0.1 volume of 3 M sodium acetate (pH 5.2) and 2.5 volume of ice-cold 100% ethanol. Pellet the DNA by centrifugation at 14,000 rpm for 20 min. Discard the ethanol and wash the pellet with 0.5 ml ice-cold 80% ethanol. Air dry the pellet and dissolve the DNA in 200 μl TE buffer. The DNA can be stored at −80°C or taken to the ligation step.

Preparation of Vector

Several M13 vectors have been constructed to accept insert DNA for phage display. The vector we use (mBAX) contains a TAG stop codon within its parental insert. The parent vector can only be propagated in a bacterial strain that carries suppressor tRNAs (*SupE* or *SupF*), such as DH5αF′ or TG-1, but not a strain that lacks suppressor tRNA, such as JS5. Libraries constructed with this vector will eliminate the production of non-recombinant phage and select for only phage containing the inserts when propagated in JS-5 cells. Detailed descriptions of other types of M13 cloning vectors can be found elsewhere.[12]

1. Combine 250 μg mBAX DNA
            100 μl restriction enzyme buffer
            10 μl acetylated BSA (10 mg/ml)
            10 μl 100 mM DTT
   Adjust with H_2O to final volume of 1000 μl.
2. Add 300 units of *Xho*I and 300 units of *Xba*I as described in preparation of inserts. Incubate at 37°C for 3 hr.
3. Check the completeness of digestion by running samples on an 0.8% agarose gel.
4. Extract and precipitate DNA as described in the preparation of insert DNA.
5. Dissolve the vector DNA in 500 μl TE buffer.

Ligation of Insert and Vector DNA

1. In a microcentrifuge tube,
   combine 100 μg vector DNA
            appropriate amount of insert DNA
            400 μl of 10 × T4 DNA ligase buffer
            25 μl of T4 DNA ligase (125 Weiss unit total)
   Adjust with H_2O to final volume of 4000 μl.
   Incubate at 15°C overnight.
   Note. The amount of insert DNA to be used in the ligation should be pre-determined in a small-scale pilot experiment. Prepare the ligations with different insert to vector ratios and transform the ligated product into JS-5 cells as described below. Use the insert:vector ratio that produces the highest number of recombinant plaques for library construction.
2. Extract the ligation reaction twice with an equal volume of phenol–chloroform–isoamyl alcohol (25:24:1). DNA stays in the top layer. Precipitate the extracted DNA with 0.1 volume of 3 M sodium acetate (pH 5.2) and 2.5 volume of ice-cold 100% ethanol. Pellet DNA by centrifugation at 14,000 rpm for 20 min. Discard ethanol and wash the pellet with 0.5 ml ice-cold 80% ethanol. Air dry the pellet and dissolve the DNA in 200 μl TE buffer.

The DNA can be stored at −80°C or taken directly to the transformation steps.

Transformation of Ligated DNA into Bacteria to Produce Phage Libraries

Electroporation competent JS-5 cells can be purchased from Stratagene (http://www.stratagene.com) or Bio-Rad (http://www.biorad.com), or prepared following the protocol described in Current Protocols in Molecular Biology.[13]

1. Aliquot 100 μl/tube of electroporation competent JS-5 cells into 50 pre-chilled microcentrifuge tubes.
2. Add 4 μl of ligated DNA to each tube.
3. Transfer the contents of each tube into a pre-chilled electroporation cuvette (0.2 cm pathlength). Use a kimwipe to remove any condensation on the outside of the cuvette to be electroporated. Place it in the electroporation chamber and electroporate ($V = 2.0$ kV, $C = 25$ μF, $R = 400$ W).
4. Immediately add 1 ml of SOC to the cells.
5. Split each electroporation into two 6-ml Falcon tubes. Add 3 ml of 42°C Top Agar to each tube and immediately pour the entire contents onto a pre-warmed (37°C), 10-cm 2xYT plate. Incubate the plates first at room temperature for 10 min, allowing the top agar to solidify. Transfer and incubate the plates upside down in a 37°C incubator for 8 hr.
6. Elute the phage by adding 5 ml sterile PBS to each plate with gentle rocking at 4°C for 2–4 hr.
7. Collect the PBS (containing eluted phage) from each plate, combine them and centrifuge at 6000 rpm at 4°C for 10 min to spin out bacterial debris.
8. Transfer the supernatant to a clean tube, add 0.2 volume of 30% PEG 8000–1.6 M NaCl. Mix well. Incubate at 4°C for 1 hr to precipitate phage.
9. Pellet the phage by centrifugation at $10,000 \times g$ for 20 min at 4°C. Discard supernatant and centrifuge again to remove residual supernatant.
10. Resuspend phage pellet in 20 ml PBS–20% glycerol. Centrifuge at $6000 \times g$ for 10 min to remove any insoluble debris. Transfer the supernatant to a clean tube.
11. Dispense the phage library into 100–500 μl aliquots. Flash freeze in liquid nitrogen and store at −80°C.

Affinity Selection of Nuclear Receptor Binding Peptides

Selection of peptides that bind to the target proteins of interest is achieved by incubating phage libraries containing small random peptide inserts with the target proteins that are immobilized on a solid support. Ninety-six-well microtiter plates are the most commonly used solid support for target immobilization. While paramagnetic beads, sepharose, immunotubes and other solid supports have been used successfully in phage display, 96-well microtiter plates afford easy handling of a large number of samples; thus they have always been the first choice in our laboratory. Most target proteins can be immobilized on the solid support using a basic solution (NaHCO$_3$, pH 8.5 or Tris–HCl, pH 8.5). The high pH promotes hydrophobic interactions between the target protein and the plastic. A very good alternative to immobilizing steroid–nuclear hormone receptors and other DNA-binding transcription factors is by using biotinylated oligonucleotides (corresponding to their cognate response element) pre-captured on streptavidin-coated plates. We will discuss both direct coating (using a basic solution) and DNA-mediated coating of target proteins in this chapter.

Materials

Anti-M13 antibody conjugated with horseradish peroxidase (Pharmacia)
Streptavidin (Sigma-Aldrich)
Adhesive lid (USA Scientific)
ABTS (2′-2′-azino-bis-ethylbenzthiazoline-6-sulfonic acid) solution. Dissolve 10.5 g citric acid monohydrate in 1.0 liter sterile deionized water. Adjust pH to 4.0 with approximately 6 ml of 10 M NaOH. Add 220 mg ABTS.
Filter sterilize and store at 4°C, protecting from light.
Immediately before use, add 30% H_2O_2 to 0.05%.
Stock of TG-1 or DH5αF′ (Promega, Life Technologies, or Stratagene)
40% glycerol
Dilute 4 ml of glycerol with 6 ml deionized water.
Sterilize by autoclaving.
0.1 M HCl
Dilute 1 ml of concentrated HCl in 99 ml deionized water.
Sterilize by filtration.
1 M Tris–HCl, pH 7.4
Dissolve 74.5 g of Tris base in 800 ml H_2O.
Adjust the pH to 7.4 with HCl. Bring the final volume to 1000 ml with H_2O.
Sterilize by autoclaving.

2% Isopropylthio-β-D-galactoside (IPTG)
  Dilute 20 mg of IPTG with 1 ml of H_2O.
  Filter sterilize.
  Store at $-20°C$.
100 mM NaHCO$_3$, pH 8.5
  Dissolve 4.2 g of NaHCO$_3$ in deionized water to ~ 400 ml.
  Adjust the pH to 8.5 if necessary.
  Adjust the volume to 500 ml.
  Filter sterilize, store at room temperature. Good for ~ 2 weeks.
1% BSA stock
  Dissolve 100 mg of BSA in 10 ml deionized water.
  Filter sterilize.
  Dispense 1 ml aliquots into sterile microcentrifuge tubes.
  Store at $-20°C$.
2% X-Gal, (5-bromo-4-chloro-3-indoyl-β-D-galactoside)
  Dissolve 20 mg of X-Gal in 1 ml of dimethylformamide in glass or
  polypropylene tube.
  Store wrapped in foil at $-20°C$.

Methods

Immobilization of Targets onto High-Binding Microtiter Plates

Method 1: Direct Adsorption of Proteins to the Plastic Wells.

1. Prepare the protein solutions to be used for immobilization by
   diluting your protein in 100 mM NaHCO$_3$ (pH 8.5) to a final
   concentration of 2.5 μg/ml immediately before use. Also prepare
   100 μl of 2.5 μg/ml positive control protein solution in the same
   buffer.
   Note. The quality of the target protein is very important for a
   successful screen. Before the screen, steps need to be taken to make
   sure that the target proteins are not denatured during the purification
   steps or immobilization process, and that they are functionally
   active. Protein targets should be as pure as possible, free of any
   contaminants and affinity tags (6-His, GST, thioredoxin, etc.).
   Because of the strong selection power, many have noticed that if
   target proteins contain an affinity tag, peptides that bind to the
   affinity tag will be co-selected in the panning process, despite
   attempts to pre-block phage with reagents containing the affinity tag
   (see Ref. 17 and our own observation). If using an affinity tag is

[17] K. K. Murthy, I. Ekiel, S. H. Shen, and D. Banville, *Biotechniques* **26**, 142 (1999).

unavoidable, extra efforts will be required to weed out the tag binders from the target protein binding phage.

2. Obtain an Immulon 4 or equivalent high-binding microtiter plate. Mask every other row using narrow lab tape. It is important to skip wells in order to prevent cross contamination. Also, label one position to be used for the positive control. We frequently use estradiol-activated ERβ (Panvera, http://www.panvera.com) as a positive control in our screens.

   Note. Immulon 4 (Dynateck) plates provide higher protein-binding affinity. We have found, however, that certain phage appear to have high background binding to this plastic despite pre-blocking the wells with BSA and milk. We found that the Costar 96-well cell culture plates can work just as well for protein immobilization and appear to have the least background phage binding.

3. Add 100 μl of your target protein solution to each well to be used in panning. Add 100 μl of the positive control protein solution to the control well.

4. Seal the plate with an adhesive lid to avoid evaporation, and incubate the plate overnight at 4°C or room temperature for 2 hr.

5. After coating wells with target protein, add 150 μl of BSA (0.1% in PBS) or milk (nonfat dry milk, 2% in PBS) to each well to block nonspecific binding. Let incubate at room temperature for 1 hr. Wash wells five times with 300 μl of PBST (PBS + 0.1% Tween-20). The plate is now ready for panning.

Method 2: Tethering Target Protein to the Wells using Biotinylated Oligonucleotide. Transcription factors can bind to and form stable complexes with their DNA response elements, thus this class of proteins can be immobilized on a solid support through binding to oligonucleotide-coated wells. Synthetic biotinylated double-stranded DNA containing the response element is first immobilized on a streptavidin-coated plate (through a biotin–streptavidin interaction), then used to capture the target protein. Target protein immobilized using this method, presumably, resembles more closely the conformation it adopts in cells. We have also found in the ELISA assay that the conformation of ER is slightly different when bound to DNA. DNA bound ER allows more efficient recruitment of corepressor-like peptides than ER immobilized directly on the plastic.[18]

1. Prepare 10 μg/ml streptavidin in 100 mM NaHCO$_3$, pH 8.5 immediately before use.

---

[18] H.-J. Huang, J. D. Norris, and D. P. McDonnell, Mol. Endocrinol. 16, 1778 (2002).

2. Add streptavidin solution to 96-well microtiter plates, seal the plate with an adhesive lid and incubate overnight at 4°C.
3. Block nonspecific binding sites with 150 μl of 2% milk (in PBS) or 0.1% BSA (in PBS). Incubate at room temperature for 1 hr.
4. Wash plate five times with 300 μl of PBST (PBS + 0.1% Tween-20).
5. Add 2 pmol of biotinylated double-stranded DNA (containing appropriate binding sequence for your target protein) in 100 μl of PBST to the wells. Incubate at room temperature for 1 hr.
6. Wash wells five times with 300 μl of PBST.
7. Add 2 pmol/well of your target protein (or 4 pmol, if your target protein binds as a dimer) diluted in 100 μl of PBST to the wells. Incubate at room temperature for 1 hr.
8. After incubation, wash wells five times with 300 μl of PBST. The plate is now ready for panning.

Affinity Partitioning of Binding Phage

In this step, phage containing random peptide libraries are incubated with the immobilized target. Nonspecific binding phage are removed by multiple washes and target-binding phage recovered by eluting with low pH buffer. The eluted phage are amplified in *E. coli* containing the F conjugative plasmid (F pili is required for M13 phage infection), such as DH5αF' or TG-1. This process is repeated several times to enrich for target-binding peptides.

1. Start an overnight culture of DH5αF' in 2xYT media from a single colony the day before panning. On the day of panning, dilute the overnight culture 1:100 into sufficient volume of 2xYT media and grow to log phase (OD = 0.5–1.0) to amplify the eluted phage from the day's panning.
2. Remove the blocking solution from target-coated wells by flicking the solution into the sink and "slapping" the inverted plate onto a stack of dry paper towels. Wash wells five times with 300 μl of PBST. Do not allow the wells to dry out completely.
3. Add 25 μl of random peptide library phage (> 10^{10} pfu) in 125 μl of PBST to the corresponding labeled well. Seal the plates and incubate at room temperature for 2 hr.
   Note. We have found that pre-blocking the phage libraries with milk or BSA before applying them to the targets significantly reduces background binding phage in the panning process. Simply add milk (2%) or BSA (0.1%) to the phage aliquots in a microcentrifuge tube and incubate on ice for 1 hr before applying them to the target wells.

Note. To avoid contamination, frequent changes of gloves is recommended. Also, aerosol resistant pipet tips should be used for all pipetting involving phage.

4. Remove nonbinding phage with five washes of PBST, then another five washes of PBS.

5. Elute the bound phage by adding 100 μl of 0.1 M HCl to the well and incubate for 10 min at room temperature. Neutralizing the eluted phage immediately with 50 μl of 1 M Tris–HCl (pH 7.4) and proceed directly to amplification, or store at 4°C for up to 2 weeks.

6. Save 25–50 μl of eluted phage to determine the elution titer. Mix 5 ml of log-phase DH5αF$'$ (or TG-1) with the rest of the eluted phage. Incubate first in 37°C water bath for 20 min without shaking, then in a 37°C shaking incubator for 3–8 hr.
   Note. The amplification is ideally for no longer than 8 hr to minimize the chance of proteolytic degradation of displayed peptides.
   Depending on the titer of the eluted phage, typically a 5-hr amplification should be sufficient.

7. Spin down cells at $4000 \times g$ for 10 min, then transfer the supernatant (containing amplified phage) to a new tube. Heat at 65°C for 2 min to pasteurize the supernatant. Store at 4°C until use. Also make aliquots of frozen stock for long-term storage and for archival purposes in case there is a need to re-pan or analyze the isolates from this selection (mix 1 volume of phage supernatant with 1 volume of sterile 40% glycerol, flash freeze in liquid nitrogen, and store at -80°C). Titer the amplified phage as described below.

8. Prepare target-coated plates as described above and repeat the affinity selection using 10^9–10^{12} pfu of the amplified phage. Repeat this process 2–3 times to enrich for target-binding phage. An increase in the elution titers following subsequent rounds of panning is an indication of enrichment. The panning stringency can be increased by lowering the input phage to as low as 10^9 pfu, or by decreasing the target protein coated on the well.

Determine Phage Titer

1. Pre-warm a sufficient number of 2xYT plates in 37°C incubator.
2. Make serial 10-fold dilutions of phage supernatant in PBST.
3. Mix 30 μl 2% X-Gal, 30 μl 2% IPTG, 200 μl log-phase DH5αF$'$ and 100 μl phage dilutions in a 10 ml tube. Incubate for 20 min at 37°C without shaking.

4. Add 3 ml of melted top agar (kept at 50°C) to each tube, mix well and spread the contents evenly on a 2xYT plate. Incubate the plates at 37°C for 8 hr to overnight until blue plaques are visible and can be counted.

Determination of Binding Activities within Phage Pools by ELISA

1. Coat target proteins on microtiter plates for ELISA the same way as for panning except that one well of the target must be prepared for each round of selection for each library that was panned. Typically, pool ELISAs are performed after 3–4 rounds of panning. Include an appropriate number of wells for the positive and negative controls.
2. Block and wash the wells as before.
3. Add 50 μl of PBST to each well containing immobilized protein to prevent wells from drying out, then add an equal number of phage (10^9–10^{12} pfu/well) from the phage stock representing each pool to the appropriate wells. Seal the plates and incubate at room temperature for 1 hr.
4. Remove nonbinding phage by washing the wells five times with PBST.
5. Dilute horseradish peroxidase-conjugated anti-phage antibody 1:5000 in PBST. Add 100 μl of the diluted antibody to each well. Seal the wells and incubate the plate at room temperature for 1 hr. Wash the wells five times with PBST.
6. Add 100 μl ABTS reagent containing 0.05% H_2O_2 to each well. Incubate the plate at room temperature for 10 min. Measure the absorbance at 405 nm with a microtiter plate reader.
   Note. The ELISA signal from pools of a particular library should show a steady increase for subsequent rounds (Fig. 6). If enrichment is not observed after four rounds, then there is a good chance that the target is denatured or inactive. Consider an alternative method of immobilization.

Isolation and Propagation of Affinity-Purified Phage Clones

1. Melt top agar in a microwave and keep it at 50°C until ready for use. Pre-warm 2xYT plates in a 37°C incubator. Plan ahead to make sure that a fresh log-phase culture of an appropriate host strain (DH5αF' or TG-1) is ready for use.
2. In a 96-well plate, perform serial 10-fold dilutions of phage stocks which gave best results in the ELISA assay.

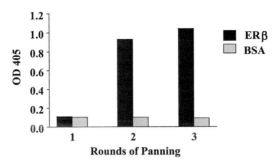

Rounds of Panning

Fig. 6. Enrichment of target binding phage during panning process. A CoRNR box library in the format of [X7-LXX(H/I)IXXX(I/L)-X7] corresponding to the receptor interacting domain of the NCoR and SMRT corepressors was constructed and used to screen for peptides that bind ERβ in the presence of ICI 182,780.[18] To prepare the target for these screens, a biotinylated ERE was first immobilized on a 96-well microtiter plate precoated with streptavidin. ERβ was then immobilized on the coated EREs in the presence of ICI 182,780 (1 μM). Panning was performed by incubating 10^{10} pfu of library with the target overnight at 4°C. After washing, the bound phage were eluted, amplified and subjected to two additional rounds of panning. Enrichment of ERβ specific binding phage was assessed by measuring the amount of phage bound to ERβ-coated versus BSA-coated wells using an anti-phage ELISA.

3. Mix 30 μl 2% X-Gal, 30 μl 2% IPTG, 200 μl log-phase DH5αF′ and 100 μl phage dilution in a 10 ml tube. Incubate for 20 min at 37°C without shaking.

4. Add 3 ml of melted top agar to each tube, mix well, and spread the contents evenly on a 2xYT plate. Allow the plates to sit undisturbed for 10 min until the top agar hardens. Incubate the plates at 37°C for 8 hr to overnight, until the blue plaques are easily identifiable. By this time, some of the dilutions should have produced isolated single plaques. Store plates at 4°C until ready to pick up the isolated phage clones.

5. For each isolated plaque to be propagated, add 3 ml of log-phase DH5αF′ to a 15 ml tube. At least 48 plaques should be picked from each pool. Amplification can be done in 96-well, deep well plates (2 ml volume). Grow a 1-ml culture for each individual plaque isolated.

6. Pick (touch and twist) and inoculate a blue isolated plaque into each of the 15-ml tubes (or 96-well deep well plates). Incubate at 37°C with vigorous agitation for 6 hr.

7. Pellet the bacterial cells by centrifugation at 4°C, 4000 × g for 10 min. Transfer phage supernatant into a new tube and store at 4°C until ready to use. The bacterial pellet can be used for plasmid preparation for sequencing.

Confirmation of Binding Activity of Individual Phage Clones by ELISA

1. Coat a microtiter plate with target protein for ELISA as described above. For each phage clone to be tested, prepare one well coated with target protein and another well coated with a negative control protein, such as the bacterial fusion partner (i.e., GST, maltose-binding protein, etc.), milk or BSA. It is not necessary to skip wells at this time. An appropriate positive control for the ELISA should also be included.

2. Block and wash the wells as described before.

3. Add 50 μl of PBST to each well containing immobilized protein to prevent wells from drying out. Add 50 μl of each phage supernatant to be tested to each set (target and negative controls) of wells. Seal the plates and incubate at room temperature for 1 hr.

4. Wash the wells five times with PBST.

5. Dilute horseradish peroxidase-conjugated anti-phage antibody 1:5000 in PBST. Add 100 μl of the diluted antibody to each well. Seal the wells and incubate the plate at room temperature for 1 hr. Wash the wells five times with PBST.

6. Add 100 μl ABTS reagent containing 0.05% H_2O_2 to each well. Incubate at room temperature for 10 min. Quantify the reaction by measuring the absorbance at 405 nm with a microtiter plate reader. Optical density (OD) values in the range of 0.5–3.0 generally constitute positive signals, while negative signals are typically in the range of 0.05–0.3, although your individual experience may differ.

7. Re-plaque and re-ELISA the positive phage prior to subsequent analysis. Only those isolates whose activity is confirmed by ELISA should be carried forward for subsequent analysis (sequencing and mammalian two-hybrid assays).

Validation of Receptor-Interacting Peptides in Mammalian Cells

Although an *in vitro* ELISA assay can verify the interaction between the isolated peptide and the target receptor, a secondary approach is often required to eliminate peptides that may be binding to a minor contaminant in the protein preparation or to the fusion tag. There is also concern that the purified receptor, produced either in bacterial or insect cells, may not have the same conformation and/or post-translational modification(s) as the proteins expressed in mammalian cells. We have therefore implemented a cell-based assay to further validate the peptides obtained from phage display screens.

The cell-based assay is a modification of a commonly used mammalian two-hybrid analysis. Isolated peptides are made as fusion proteins to the Gal4-DBD and the receptor of interest is made as a fusion partner of the VP16 acidic transactivation domain. The assay consists of co-transfection of the Gal4DBD-peptide and VP16-receptor plasmids into mammalian cells together with a luciferase reporter gene containing a Gal4 response element. If the DNA-bound peptide can interact with the receptor in target cells, the VP16 activation domain, via its fusion to the receptor, will be brought to the DNA and enable the expression of the reporter gene. An added advantage of moving the validation step into mammalian cells is the ability to cross screen multiple nuclear receptors without having to purify individual receptors for *in vitro* assays. We have used this approach to test the receptor-binding specificity of phage identified using ERα and ERβ as bait in our primary screens. With ease we evaluated the ability of over 50 peptides to interact with 10 different receptors using this approach.[19,20] ER subtype-specific peptides were identified and have proven to be very useful in dissecting ERα and ERβ signaling. Non-discriminating peptides that bind multiple receptors have also been useful for studying the mechanism of action of those receptors. For example, peptides identified in ERα screens have been used to study the mechanisms of action of the androgen receptor (AR),[21] retinoic receptor-related orphan receptor-alpha (RORα),[22] and the Vitamin D3 receptor (VDR)[23] without having to initiate a primary screen against these receptors.

Materials

pM and pVP16 vectors can be obtained from Clontech (http://www.clontech.com).

pM-peptide and pVP16-receptor constructs are made using regular cloning techniques.

Lipofectin, cell culture media (Life Technologies).

---

[19] C.-Y. Chang, J. D. Norris, H. Grøn, L. A. Paige, P. T. Hamilton, D. J. Kenan, D. Fowlkes, and D. P. McDonnell, *Mol. Cell. Biol.* **19**, 8226 (1999).

[20] J. M. Hall, C.-Y. Chang, and D. P. McDonnell, *Mol. Endocrinol.* **14**, 2010 (2000).

[21] C.-Y. Chang and D. P. McDonnell, *Mol. Endocrinol.* **16**, 647 (2002).

[22] C. D. Kane and A. R. Means, *EMBO J.* **19**, 691 (2000).

[23] P. Pathrose, O. Y. Barmina, C.-Y. Chang, D. P. McDonnell, N. K. Shevde, and J. W. Pike, *J. Min. Bone Res.* **17**, 2196 (2002).

Method

1. Human hepatocarcinoma cells HepG2 are split into 24-well plates the day before transfection. We have performed this assay in a number of other cell lines with similar success.
2. For triplicate wells using lipofectin-mediated transfection, we use

    400 ng p*M*-peptide plasmid
    400 ng pVP16-receptor plasmid
    200 ng pCMV-βgal
    2000 ng 5 × Gal4-Luc3 reporter plasmid

    The amounts of DNA used vary with different transfection media. One can modify the input DNA amounts to accommodate the requirements of that particular transfection protocol.
3. Perform transfection using your chosen transfection protocol.
4. After transfections, add the appropriate hormone to the cells if the receptor–peptide interaction is expected to be ligand-dependent. Incubate the cells for 24 hr. Perform luciferase and β-galactosidase assays according to manufacturer's instruction.

Use Receptor-Interacting Peptides to Inhibit Receptor Transcriptional Activity

Since cofactor–receptor interactions are required for nuclear receptors to fully manifest transcriptional activity, disruption of such interactions is expected to have an inhibitory effect on receptor activity. Many of the peptides identified in the phage display screens have been demonstrated to bind to important protein–protein interaction surfaces on the receptors. For example, several of the peptides we identified contain an LxxLL motif that mimics the receptor interaction domain(s) of the p160 class of coactivators.[9,19] Two major concerns that may limit the use of these peptides as antagonists of nuclear receptors are: (1) the selected peptides may not possess high enough affinity to disrupt receptor–cofactor interactions, and (2) the peptides obtained may bind to multiple nuclear receptors. To address the first concern, we have used fluorescence polarization assays to measure the affinity of some of our ER-binding peptides and found that most bind in the 100 nM range. One of the peptides identified in our primary screen, however, possesses an affinity of 60 nM for ERβ, similar to the affinity of coactivator SRC-1 to ERα.[24,25] Not surprisingly, when introduced into cells,

[24] M. Jansen, personal communication.
[25] E. Margeat, N. Poujol, A. Boulahtouf, Y. Chen, J. D. Muller, E. Gratton, V. Cavailles, and C. A. Royer, *J. Mol. Biol.* **306**, 433 (2001).

these peptides efficiently blocked ERα and ERβ transcriptional activity.[19,20] In addition, peptides that demonstrate receptor-specific binding characteristics have also been identified, eliminating our second concern. We found two peptides which bind specifically to ERβ but show no interaction with any other nuclear receptors tested.[20] These peptides are powerful tools to dissect the pharmacology of the closely related ERα and ERβ isoforms, where no ER subtype-specific ligands are available. We believe that the same approach can be applied to other nuclear receptors, particularly the orphan nuclear receptors where there is no known ligand to modulate their activity.

The simplest way of introducing these peptides into mammalian cells is to transfect plasmids encoding the peptide (or peptide fusion proteins, i.e., Gal4-DBD fusion used in the mammalian two-hybrid assay) using transient transfection. We have also had success expressing peptides in cells and in whole animals using an adenoviral delivery system. Furthermore, we have found stable cell lines expressing peptides under the control of a regulated promoter to be useful in our studies. Also available are several emerging technologies which allow the introduction of synthetic peptides or *in vitro* purified recombinant protein–peptides into cells.[26–32]

In conclusion, we have described in this chapter the use of a combinatorial peptide approach to dissect the cofactor–nuclear receptor interface. We have also discussed the use of these peptides in the study of nuclear receptor pharmacology. We believe that the same approach can be applied to other cellular proteins whose activity is modulated by protein–protein interactions. For example, phage display has been used to identify peptides which disrupt the interaction between htm2 and p53, preventing htm2-mediated p53 degradation, leading to subsequent stabilization and activation of p53 in cells.[33,34] The full potential of using phage display to study protein–protein interactions we believe has yet to be realized. We hope that the examples presented here, where it has been used to study

[26] E. Vives, P. Brodin, and B. Lebleu, *J. Biol. Chem.* **272**, 16010 (1997).

[27] S. Fawell, J. Seery, Y. Daikh, C. Moore, L. L. Chen, B. Pepinsky, and J. Barsoum, *Proc. Natl. Acad. Sci. USA* **91**, 664 (1994).

[28] S. R. Schwarze, A. Ho, A. Vocero-Akbani, and S. F. Dowdy, *Science* **285**, 1569 (1999).

[29] D. Derossi, A. H. Joliot, G. Chassaing, and A. Prochiantz, *J. Biol. Chem.* **269**, 10444 (1994).

[30] G. Elliott and P. O'Hare, *Cell* **88**, 223 (1997).

[31] P. A. Wender, D. J. Mitchell, K. Pattabiraman, E. T. Pelkey, L. Steinman, and J. B. Rothbard, *Proc. Natl. Acad. Sci. USA* **97**, 13003 (2000).

[32] D. J. Mitchell, D. T. Kim, L. Steinman, C. G. Fathman, and J. B. Rothbard, *J. Pept. Res.* **56**, 318 (2000).

[33] A. Bottger, V. Bottger, A. Sparks, W. L. Liu, S. F. Howard, and D. P. Lane, *Curr. Biol.* **7**, 860 (1997).

[34] V. Bottger, A. Bottger, S. F. Howard, S. M. Picksley, P. Chene, C. Garcia-Echeverria, H. K. Hochkeppel, and D. P. Lane, *Oncogene* **13**, 2141 (1996).

nuclear receptor function, will encourage other investigators to utilize this technology.

Acknowledgment

This work was supported by grants to D. P. M. from the NIH (DK48807, DK50494) and NCI (CA90645), and postdoctoral fellowships to C.-Y. C (DAMD17-99-1-9173) and M. J. (CA92984) from the US Army Medical Research and Material Command and NIH, respectively. We thank Dr. D. Kenan (Duke University, Durham, NC) and Karo-Bio, USA (RTP, NC) for sharing reagents and technical expertise.

[8] Methods for Detecting Domain Interactions in Nuclear Receptors

By Elizabeth M. Wilson, Bin He, and Elizabeth Langley

Amino- and carboxyl-terminal (N–C) domain interactions in the human androgen receptor (AR) were initially demonstrated in studies on the dissociation rate of bound ligand. Deletion of the AR NH_2-terminal region to produce a truncated AR containing the DNA and ligand-binding domains increased the dissociation rate of a bound synthetic androgen (1881) some 3- to 5-fold.[1] Since there was no change in the apparent equilibrium binding affinity as determined by Scatchard plot analysis, this result suggested that the association rate of [3H]R1881 binding was increased in the deletion mutant. Further, the data indicated that the AR NH_2-terminal domain stabilized the binding of androgen to the ligand-binding domain located at the C-terminus.

An N–C interaction in the AR was supported also by the finding that NH_2- and carboxyl-terminal fragments of the receptor each containing the DNA-binding domain, exhibited androgen-dependent DNA binding and dimerization.[2] This result was in contrast to that obtained with a C-terminal fragment containing the DNA and ligand-binding domains, which bound DNA equally well in the presence and absence of androgen. These findings raised the possibility that the dimer formed between the NH_2- and carboxyl-terminal fragments was structurally different from that formed by the carboxyl-terminal domain alone.

[1] Z. X. Zhou, M. V. Lane, J. A. Kemppainen, F. S. French, and E. M. Wilson, *Mol. Endocrinol.* **9**, 208 (1995).
[2] C. I. Wong, Z. X. Zhou, M. Sar, and E. M. Wilson, *J. Biol. Chem.* **268**, 19004 (1993).

A direct demonstration of an N–C interaction was obtained using a mammalian two-hybrid assay.[3] It was known at this time that the DNA-binding domains of steroid receptors dimerize, thus an assay was established to score for N–C interactions using AR proteins that lacked the DNA-binding domain. To accomplish this goal, we constructed two expression plasmids, GALD-H (GALAR624-919), which encoded the C-terminal domain of AR encompassing the hinge region and ligand-binding domain (624-919), and VPA1 (VPAR1-503), which encoded the NH$_2$-terminal domain of AR (residues 1–503).[4] The starting plasmids for these constructs were pGAL0 and pNLVP16, respectively. The GALD-H plasmid expresses the *Saccharomyces cerevisiae* GAL4 DNA-binding domain (residues 1–147) as a fusion protein with the AR ligand-binding domain. The D–H designation of this plasmid refers to AR exons D–H[5] or exons 4–8[6] that code for the ligand-binding domain. VPA1 specifies the herpes simplex virus VP16 protein activation domain (residues 411–456) as a fusion protein with the AR NH$_2$-terminal domain. A reporter plasmid, G5E1bLuc, was obtained, which contained five DNA-binding sites for the yeast GAL4 transcription factor linked to the firefly luciferase reporter sequence.

Two-hybrid assay results obtained with these constructs indicated a strong androgen-dependent N–C interaction, with as much as a 100-fold induction of luciferase activity when both domains were present.[3] We determined that the N–C interaction was specifically induced by androgens and that it was inhibited by antiandrogens such as hydroxyflutamide[3,7] and casodex.[8] A striking aspect of the N–C interaction was its high degree of specificity for androgens and inhibition by antiandrogens. In contrast, there is a relative lack of androgen specificity in transient transfection studies in which AR nuclear transport and induction of MMTV-luciferase reporter activity are monitored.[9] Androgen specificity and antagonism by anti-androgens indicated that the N–C interaction was physiologically relevant. In addition, the use of androgen insensitivity mutants and AR deletion mutants supported the hypothesis that the androgen-bound AR dimer is

[3] E. Langley, Z. X. Zhou, and E. M. Wilson, *J. Biol. Chem.* **270**, 29983 (1995).

[4] D. B. Lubahn, D. R. Joseph, M. Sar, J. A. Tan, H. N. Higgs, R. E. Larson, F. S. French, and E. M. Wilson, *Mol. Endocrinol.* **2**, 1265 (1988).

[5] D. B. Lubahn, T. R. Brown, J. A. Simental, H. N. Higgs, C. J. Migeon, E. M. Wilson, and F. S. French, *Proc. Natl. Acad. Sci. USA* **86**, 9534 (1989).

[6] C. A. Quigley, A. De Bellis, K. B. Marschke, M. K. El-Awady, E. M. Wilson, and F. S. French, *Endocr. Rev.* **16**, 271 (1995).

[7] J. A. Kemppainen, E. Langley, C. I. Wong, K. Bobseine, W. R. Kelce, and E. M. Wilson, *Mol. Endocrinol.* **13**, 440 (1999).

[8] C. W. Gregory and E. M. Wilson, unpublished studies.

[9] J. A. Kemppainen, M. V. Lane, M. Sar, and E. M. Wilson, *J. Biol. Chem.* **267**, 968 (1992).

organized in an antiparallel orientation in which the N and C termini interact.[3,10] More recent studies indicate that the AR N–C interaction is required for androgen responsiveness of the naturally occurring enhancer–promoter regions of the prostate-specific antigen and probasin genes.[11]

A direct androgen-dependent interaction between the NH_2- and carboxyl-terminal domains of the human AR was confirmed by glutathione-S-transferase (GST) affinity matrix assays.[12] Studies then progressed to determine the molecular basis of the N–C interaction. We focused our efforts on several AR mutations in the ligand-binding domain that caused androgen insensitivity syndrome and that did not diminish the apparent equilibrium binding affinity of the receptor for [³H]R1881. We reasoned that this approach would more faithfully resemble a physiologically relevant process, since AR mutations that cause androgen insensitivity inactivate the AR *in vivo* resulting in partial or complete inhibition of male sexual development.[6] These and other ligand-binding domain mutations helped to define the activation function 2 (AF2) domain as the interaction site in the C-terminal region that participates in the N–C interaction.[12] We also established that the N–C interaction site in the N-terminus maps to an FXXLF motif (23-FQNLF-27) and a WXXLF motif (433-WHTLF-437[13]). AF2 forms a hydrophobic binding surface for the LXXLL motifs of the p160 family of coactivators, raising the possibility that competition between these binding motifs might occur. Additional studies suggested that the FXXLF motif is specific for AR AF2, and that these amino acids mediate AR interactions with several coregulators.[14,15] Studies from other laboratories confirmed the N–C interaction in the AR[16–22]

[10] E. Langley, J. A. Kemppainen, and E. M. Wilson, *J. Biol. Chem.* **273**, 92–101 (1998).

[11] B. He, L. W. Lee, J. T. Minges, and E. M. Wilson, *J. Biol. Chem.* **277**, 25631 (2002).

[12] B. He, J. A. Kemppainen, J. J. Voegel, H. Gronemeyer, and E. M. Wilson, *J. Biol. Chem.* **274**, 37219 (1999).

[13] B. He, J. A. Kemppainen, and E. M. Wilson, *J. Biol. Chem.* **275**, 22986 (2000).

[14] B. He, J. T. Minges, L. W. Lee, and E. M. Wilson, *J. Biol. Chem.* **277**, 10226 (2002).

[15] Z. X. Zhou, B. He, S. H. Hall, E. M. Wilson, and F. S. French, *Mol. Endocrinol.* **16**, 287 (2002).

[16] P. Doesburg, C. W. Kuil, C. A. Berrevoets, K. Steketee, P. W. Faber, E. Mulder, A. O. Brinkmann, and J. Trapman, *Biochemistry* **36**, 1052 (1997).

[17] T. Ikonen, J. J. Palvimo, and O. A. Jänne, *J. Biol. Chem.* **272**, 29821 (1997).

[18] C. A. Berrevoets, P. Doesburg, K. Steketee, J. Trapman, and A. O. Brinkmann, *Mol. Endocrinol.* **12**, 1172 (1998).

[19] C. L. Bevan, S. Hoare, F. Claessens, D. M. Heery, and M. G. Parker, *Mol. Cell. Biol.* **19**, 8383 (1999).

[20] T. Slagsvold, I. Kraus, T. Bentzen, J. Palvimo, and F. Saatcioglu, *Mol. Endocrinol.* **14**, 1603 (2000).

[21] J. Thompson, F. Saatcioglu, O. A. Jänne, and J. J. Palvimo, *Mol. Endocrinol.* **15**, 923 (2001).

[22] A. Bubulya, S. Y. Chen, C. J. Fisher, Z. Zheng, X. Q. Shen, and L. Shemshedini, *J. Biol. Chem.* **48**, 44704 (2001).

and showed that similar binding events take place in other nuclear receptors.[23–26]

In addition to binding the FXXLF motif in the N-terminus of AR, the AF2 sequence also binds FXXLF motifs in several AR coactivators[14] and LXXLL motifs in the p160 coactivators.[27] We demonstrated that the androgen-induced N–C interaction inhibits AR activation by p160 coactivators through AF2, most likely by competition for the shared AF2-binding site.[27] The androgen-dependent, N–C interaction-induced inhibition may limit AR activation by p160 coactivators under normal physiological conditions of low p160 coactivator expression. Immunoblotting and immunohistochemical results indicate that the levels of SRC1 and TIF2, two members of the p160 coactivator family, are very low in the hyperplastic prostate. In contrast, the majority of advanced recurrent prostate cancer samples express markedly increased levels of these coactivators.[28] We postulated that one mechanism for AR activation in the androgen-deprived patient with recurrent prostate cancer involves AR activation by abundant p160 coactivators through the AF2 domain.[28]

Methods for Determining Nuclear Receptor Domain Interactions

Detection methods for the AR N–C interaction have changed over the years to optimize transfection efficiency and signal intensity, and to facilitate tests for specific interacting sequences. Instead of the Chinese hamster ovary (CHO) cells used originally in the assay,[3,10] we currently employ human epithelioid cervical carcinoma (HeLa) or human hepatocellular carcinoma (HepG2) cells. The Effectene[TM] reagent from Qiagen is used for transient transfection[14] rather than DEAE dextran.[3,10] The G5E1bLuc reporter vector[3,10] is replaced by 5XGAL4Luc3, which yields a stronger signal.[29] The 5XGAL4Luc3 reporter plasmid is in the pGL3-basic vector, which has a cryptic androgen response element that is active in HeLa cells but not

[23] W. L. Kraus, E. M. McInerney, and B. S. Katzenellenbogen, *Proc. Natl. Acad. Sci. USA* **92**, 12314 (1995).

[24] D. Shao, S. M. Rangwala, S. T. Bailey, S. L. Krakow, M. J. Reginato, and M. A. Lazar, *Nature* **396**, 377 (1998).

[25] M. J. Tetel, P. H. Giangrande, S. A. Leonhardt, D. P. McDonnell, and D. P. Edwards, *Mol. Endocrinol.* **13**, 910 (1999).

[26] R. Métivier, G. Penot, G. Flouriot, and F. Pakdel, *Mol. Endocrinol.* **15**, 1953 (2001).

[27] B. He, N. T. Bowen, J. T. Minges, and E. M. Wilson, *J. Biol. Chem.* **276**, 42293 (2001).

[28] C. W. Gregory, B. He, R. T. Johnson, O. H. Ford, J. L. Mohler, F. S. French, and E. M. Wilson, *Cancer Res.* **61**, 4315 (2001).

[29] C. Chang, J. D. Norris, H. Gron, L. A. Paige, P. T. Hamilton, D. J. Kenan, D. Fowlkes, and D. P. McDonnell, *Mol. Cell. Biol.* **19**, 8226 (1999).

HepG2 cells, presumably reflecting a different complement of transcription factors in these two cell lines. To minimize the contribution of this cryptic element, assays in HeLa cells use only 0.01 μg of AR expression vector (pCMVhAR) per well of a 12-well tissue culture plate. There is little background activity with the AR NH_2-terminal and DNA-binding domain fragment VPAR1-660 in HeLa cells, suggesting that full-length AR is required for activation of the cryptic androgen response element in the 5XGAL4Luc3 reporter construct.

Another change in the current N–C assay is the use of VPAR1-660, which encodes both the AR NH_2-terminal and DNA-binding domains. VPAR1-660 replaces the VP-A1 plasmid, which encodes just the AR NH_2-terminal domain (residues 1–503). Although the AR DNA-binding domain is not required for the N–C interaction,[3] its presence leads to greater androgen-induced luciferase activity in the assay and hence increased sensitivity. The DNA-binding domain of the progesterone receptor is reported to stabilize the NH_2-terminal region[30] and a similar stabilization may occur in the AR.

Once the binding motifs in the AR NH_2-terminal region were identified, it became clear that an assay to measure the interaction of peptide sequences was required to precisely identify interacting amino acids. Since mammalian two-hybrid assays using GAL4-peptides and a VP16-estrogen receptor α fusion protein were shown to be robust,[29] we adopted a similar approach using VPAR507-919 to address the comparative binding of FXXLF motifs of various AR coregulators and the LXXLL motifs of p160 coactivators.[14] Studies with these reagents revealed that protein sequences within and flanking the FXXLF motif are crucial in determining the strength and specificity of the interaction with the AF2 region of the AR ligand-binding domain.[31]

An additional method to confirm protein–protein interaction sites among nuclear receptor sequences involves determining the dissociation rate of bound radiolabeled ligand from chimeric receptor proteins.[14,27] For this assay, DNA encoding the interaction sequence of interest, e.g., sequences containing FXXLF or LXXLL motifs, is cloned into a nuclear receptor cDNA specifying the NH_2-terminal region, expressed, and then dissociation rate kinetics of bound radiolabeled agonist are determined. The degree of interaction between the NH_2-terminal motif and AF2 in the ligand-binding domain is reflected by the extent of change in dissociation half-time. A greater ligand dissociation half-time reflects stabilization of the ligand

[30] D. L. Bain, M. A. Franden, J. L. McManaman, G. S. Takimoto, and K. B. Horwitz, *J. Biol. Chem.* **275**, 7313 (2000).
[31] B. He and E. M. Wilson, *Mol. Cell. Biol.* **23**, 2135 (2003).

in the binding pocket. We used this approach to demonstrate that engineering an artificial N–C interaction into a TIF2-glucocorticoid receptor chimera led to a dramatic effect on ligand binding. Previous studies showed that an N–C interaction did not occur in the glucocorticoid receptor because the dissociation rate of bound [^{3}H]dexamethasone was unaffected by deletion of the NH$_2$-terminal region;[1] however, when the LXXLL motif region of the p160 coactivator TIF2 was introduced into the NH$_2$-terminal region of the glucocorticoid receptor, the dissociation half-time of [^{3}H]dexamethasone was 5-fold slower.[27] In fact, the dissociation half-time of [^{3}H]dexamethasone bound to the TIF2-glucocorticoid receptor chimera slowed to that observed for [^{3}H]R1881 dissociation from the normal AR, the latter reflecting the inherent stabilizing effect of the N–C interaction in AR.

Another approach to demonstrate protein–protein interactions is GST affinity matrix or pull-down assays. In the case of the androgen receptor, the AR hinge region and ligand-binding domain (residues 624–919) expressed as a fusion protein with GST is not readily soluble. This problem is remedied to some extent by the growth of bacterial cells at 15°C; however, protein recovery is not ideal. To overcome these limitations, we express the AR NH$_2$-terminal region as a fusion protein with GST, and produce [^{35}S]-labeled AR ligand-binding domain using Promega TNT coupled Transcription/Translation of mRNAs in rabbit reticulocyte lysates. Using these two reagents, we confirmed a direct interaction between the AR NH$_2$-terminal and ligand-binding domains[12] and demonstrated that the FXXLF and WXXLF motifs were required for the NH$_2$-terminal domain interaction with AF2 in the C-terminal domain.[13] GST affinity matrix assays demonstrate protein interactions, but it must be kept in mind that nonspecific binding can occur when high concentrations of protein are used. In studies with nuclear receptors, this potential complication is minimized by demonstrating a hormone dependence for the interaction.

N–C Two-Hybrid Interaction Assay

Procedure

1. HeLa or HepG2 cells are obtained from the American Type Culture Collection. HeLa cells are cultured in minimal essential medium with Earle's salts and L-glutamine (MEM, GibcoBRL) containing 10% (v/v) fetal bovine serum (Hyclone), 2 mM L-glutamine (GibcoBRL), and 1:100 dilution of penicillin–streptomycin (GibcoBRL). HepG2 cells are cultured in MEM containing 10% fetal bovine serum, 0.1 mM MEM nonessential amino acids

(GibcoBRL), 1 mM MEM sodium pyruvate (GibcoBRL), 2 mM L-glutamine, and penicillin–streptomycin as above. The lysis buffer contains 2 mM EDTA, 1% (v/v) Triton X-100, and 25 mM Trizma phosphate, pH 7.8. The plasmids used are GALAR624-919, VPAR1-660, and 5XGAL4Luc3. Transient transfections are performed using the Effectene reagent (Qiagen). Plasmid DNAs are propagated in bacteria and isolated using the QIAfilter Plasmid Maxi protocol (Qiagen).

2. One or more days before the experiment, expression and reporter plasmid DNAs are aliquoted into 14 ml polystyrene round-bottom (17 × 100 mm) Falcon tubes, which are then stored at −20°C. DNAs are aliquoted together for two or more tissue culture wells in sufficient number for replicates and for determinations in the absence and presence of 10 nM dihydrotestosterone (DHT) for HeLa cells, or 10 nM R1881 (methyltrienolone) for HepG2 cells. For each well of a 12-well tissue culture plate, 0.15 μg GALAR624-919, 0.15 μg VPAR1-660, and 0.1 μg 5XGAL4Luc3 are added.

3. HeLa cells are plated at 0.1 × 10^6 cells/well in 12-well tissue culture plates in 2 ml medium containing serum, phenol red, and additives. HepG2 cells are plated at 0.2 × 10^6 cells/well. The next day, the medium is exchanged with 0.8 ml fresh medium containing serum and phenol red. Effectene transfection reagents (50 μl EC buffer and 1 μl Enhancer per well) are added to the Falcon tubes containing the pre-aliquoted DNAs, and the tubes are incubated for 5 min at room temperature followed by the addition of 1 μl Effectene per well and vortexing for 10 sec. Incubate for 10 min and then add 0.4 ml medium per well containing serum and phenol red; mix well.

4. Add 0.4 ml of DNA-media mix to each well containing 0.8 ml medium and incubate overnight at 37°C in a 5% CO$_2$ incubator. Aspirate the medium and wash with 2 ml phosphate-buffered saline. Add 2 ml/dish serum-free, phenol red-free medium and hormones dissolved in ethanol (10 nM DHT for HeLa cells or 10 nM R1881 for HepG2 cells, final concentrations).

5. Next day the medium is aspirated and the cells are washed once with 2 ml phosphate-buffered saline. The saline is aspirated to dryness and the cells harvested in 0.22 ml lysis buffer/well. Samples are analyzed by assaying 0.1 ml aliquots for luciferase activity in an automated luminometer. Typical induction in the presence of 10 nM DHT or 10 nM R1881 is ∼50-fold relative to the no-hormone control. Results are similar using HeLa or HepG2 cells, but note the use of different androgens for the two cell lines.

GAL4-Peptide Two-Hybrid Interaction Assay

Procedure

1. HepG2 cells are cultured as above. The plasmids used are pCMVhAR or VPAR507-919, GAL-peptide, and 5XGAL4Luc3. The Effectene transfection reagent kit (Qiagen) is used. DNAs for two or more wells for replicates and for determinations in the absence and presence of hormone (10 nM R1881) are aliquoted into Falcon tubes one or more days before the experiment as described earlier. Plasmids (amount per well) for 12-well tissue culture plates include 0.05 μg pCMVhAR (0.01 μg pCMVhAR for HeLa cells) or 0.15 μg VPhAR507-919, together with 0.05 μg GAL-peptide and 0.1 μg 5XGAL4Luc3.

2. HepG2 cells are plated at 0.2×10^6 cells/well in 12-well plates. The next day, the medium is replaced with 0.8 ml fresh medium containing serum and phenol red. To Falcon tubes containing the DNAs, Effectene transfection reagents (50 μl EC buffer and 1 μl Enhancer/well) are added and incubated for 5 min at room temperature. Effectene reagent (1 μl/well) is added and the tubes vortexed for 10 sec. After a 10 min incubation, 0.4 μl per well of medium containing serum and phenol red is added and the solution is mixed well.

3. Add 0.4 ml of the DNA-medium mix to each well of cells containing 0.8 ml of medium and incubate overnight at 37°C in a 5% CO_2 incubator. Aspirate the medium, and wash cells with 2 ml phosphate-buffered saline. Add 2 ml/well serum-free, phenol red-free medium, and hormone dissolved in ethanol (10 nM R1881, final concentration).

4. The next day, aspirate the medium and wash the cells with 2 ml phosphate-buffered saline. Harvest cells in 0.22 ml lysis buffer and analyze 0.1 ml for determination of luciferase activity using an automated luminometer. Induction of luciferase activity in the presence of 10 nM R1881 depends on the peptide tested and has been up to 200-fold using pCMVhAR and 50-fold with VPAR507-919.

Dissociation Kinetics of Bound Ligand from Nuclear Receptor Chimeras

Procedure

1. Monkey kidney COS-1 cells are obtained from the American Type Culture Collection. The tissue culture medium is Dulbecco's MEM (DMEM), supplemented with 20 mM HEPES, pH 7.2, 2 mM

L-glutamine, 10% (v/v) bovine calf serum (Hyclone), and penicillin–streptomycin as earlier. TBS buffer is 0.14 M NaCl, 3 mM KCl, 1 mM CaCl$_2$, 0.05 mM MgCl$_2$, 0.9 mM NaH$_2$PO$_4$, and 25 mM Tris–HCl, pH 7.4. The chloroquine medium is 5 mg chloroquine–100 ml in DMEM with 10% bovine calf serum. The glycerol medium is 15% (v/v) glycerol and 10% bovine calf serum in DMEM. The labeling medium is 5 nM [^{3}H]R1881 (70–90 Ci/mmol) for AR, and 8 nM [^{3}H]dexamethasone (84 Ci/mmol) for the glucocorticoid receptor, both in serum-free, phenol red-free DMEM–HEPES medium, as earlier, with or without a 100-fold excess of unlabeled hormone. Cells are harvested in a buffer containing 2% (w/v) SDS, 10% (v/v) glycerol, and 10 mM Tris–HCl, pH 6.8.

2. One or more days before the experiment, DNAs are aliquoted into Falcon tubes and stored as described earlier, using 2 μg DNA/well for 6-well tissue culture plates and normal or chimeric receptor DNAs. Sufficient wells are set up for replicates and multiple time points, and for determining specific and nonspecific binding. The latter binding is determined in the presence of a 100-fold excess of unlabeled hormone added at the beginning of the 2-hr labeling period.

3. Place 0.4×10^6 COS-1 cells/well in 6-well tissue culture plates and add 3 ml/well DMEM containing serum and phenol red. The next day, add 0.95 ml of $1.08 \times$ TBS and 0.11 ml of 500 mg DEAE dextran/well to each Falcon tube containing plasmid DNAs to be expressed. The cell culture medium is aspirated and 1 ml of the DNA mix is added followed by incubation for 30 min at 37°C. The medium is aspirated and 3 ml/well of chloroquine medium is added and cells are incubated for 3 hr at 37°C. This medium is aspirated and cells are incubated for 4 min at room temperature with 1 ml glycerol medium. The glycerol medium is aspirated and cells are washed once with 3 ml TBS. Cells are overlayed with 3 ml of prewarmed (37°C) DMEM containing 10% (v/v) bovine calf serum and placed in a 5% CO$_2$ incubator.

4. After 48 hr, the medium is aspirated, 0.6 ml labeling medium is added and the cells are incubated for 2 hr at 37°C. [^{3}H]-Ligand dissociation is initiated by the addition of a 10,000-fold molar excess of unlabeled ligand in 0.1 ml serum-free, phenol red-free medium to all wells except for the zero time point controls. At time intervals after the start of ligand dissociation, cells are carefully washed once with 3 ml phosphate-buffered saline and collected in 0.5 ml harvest buffer. Radioactivity is measured by scintillation counting.

GST Affinity Matrix Assay

Procedure

1. XL1-Blue *Escherichia coli* cells are obtained from Stratagene. For GST-AR1-660 expression, the LB media (BIO 101, Inc.) contains 0.5 ml/liter of 100 mg/ml ampicillin (Sigma) and (isopropyl-1-thio-β-D-galactopyranoside (IPTG, Sigma) as indicated. Glutathione agarose beads (Amersham Pharmacia Biotech) are washed twice in Sonication Buffer and resuspended 1:1 in Sonication Buffer containing 0.5% (v/v) Nonidet P-40, 1 mM EDTA, 0.1 M NaCl, 0.02 M Tris–HCl, pH 8.0. The SDS Buffer contains 2% (w/v) SDS, 10% (v/v) glycerol, and 10 mM Tris–HCl, pH 6.8. [^{35}S]Methionine (1175 Ci/mmol) is from NEN Life Science Products. The plasmids are GST-AR1-660 in pGEX-5X-1 (Amersham Pharmacia Biotech), and pcDNA3HA-AR624-919 for *in vitro* translation. The TNT T7 Quick-coupled Transcription–Translation System (Promega) reactions contain 40 μl TNT T7 Quick master mix, 2.5 μl [^{35}S]methionine ($\geq$ 1000 Ci/mmol) at 10 mCi/ml, and 1 μg pcDNA3HA-AR624-919 in a final volume of 50 μl, and are incubated for 60 min at 30°C according to a protocol supplied by the manufacturer.

2. A single bacterial colony is grown overnight in 10 ml ampicillin medium at 37°C with shaking. The following day, make a 1:10 dilution of this culture in 100 ml of LB medium with ampicillin and grow for 3 hr at 37°C. Add IPTG to 0.5 mM (less for low solubility proteins; more for high solubility proteins; range 0.1–1 mM IPTG), grow for 3 hr at 30°C in a shaking incubator, collect the bacteria by centrifugation for 10 min at 3000 rpm in a Sorvall KA12 rotor. Sonicate the bacterial pellet in 4 ml ice cold Sonication Buffer using a microprobe at the highest setting; pulse the slurry three times for 15 sec each on ice.

3. Centrifuge the lysate for 10 min at 12,000 rpm in a microfuge at 4°C and decant the supernatant to a fresh tube. Incubate 30 μl washed glutathione agarose beads in 0.5 ml bacterial supernatant containing GST-AR1-660 for 1 hr with gentle rocking at 4°C. Centrifuge the slurry for 2 min at 3000 rpm, aspirate the supernatant and wash beads with 0.5 ml ice cold Sonication Buffer, repeat wash procedure three times and discard the supernatant.

4. Combine washed beads with 0.2 ml ice cold Sonication Buffer and 10 μl of *in vitro* translation mix containing [^{35}S]methionine-labeled human AR624-919 (prepared in stage 1 above); incubate 2 hr with rocking at 4°C in the absence and presence of 0.2 μM DHT. Pellet the beads by centrifugation at 3000 rpm for 2 min in a microfuge,

wash beads with 0.5 ml ice cold Sonication Buffer, repeat wash procedure three times.

5. Bound proteins are released from the beads by boiling for 10 min in 50 μl SDS Buffer, and the sample is analyzed by electrophoresis through 12% polyacrylamide gels containing SDS. An aliquot (1 μl) of *in vitro* translation mix is analyzed in parallel on the gel, representing 10% of the input radioactivity. The gels are dried and exposed to X-ray film to visualize the bound C-terminal fragment of AR.

Acknowledgment

We gratefully acknowledge the excellent and dedicated technical assistance provided at different times over the years by Jon A. Kemppainen, Malcolm V. Lane, K. Michelle Cobb, Natalie T. Bowen, John T. Minges, Lori W. Lee, and De-Ying Zang. We also acknowledge the scientific contributions of Zhoug-xun Zhou, Choi-iok Rebecca Wong, Christopher W. Gregory, and Philip D. Reynolds. Plasmid vectors were kindly provided by Gordon Tomaselli, Hinrich Gronemeyer, and Donald P. McDonnell. We thank Frank S. French for his encouragement and critical discussions. The work was supported by Public Health Service grant HD 16910 from the National Institutes of Child Health and Development, by cooperative agreement U54-HD35041 as part of the Specialized Cooperative Centers Program in Reproductive Research of National Institutes of Health, by the United States Army Medical Research and Material Command Grant DAMD17-00-1-0094, and by the International Training and Research in Population and Health Program supported by the Fogarty International Center and National Institutes of Child Health and Development, National Institutes of Health.

[9] The Orphan Receptor SHP and the Three-Hybrid Interference Assay

By Yoon-Kwang Lee and David D. Moore

Introduction

SHP (Small Heterodimer Partner, NR0B2) is an unusual member of the nuclear hormone receptor superfamily that completely lacks a DNA-binding domain. It was initially isolated using yeast two-hybrid screening with CAR, another orphan nuclear hormone receptor, as a bait,[1] and was also isolated independently in two similar screens for PPAR

---

[1] W. Seol, H. S. Choi, and D. D. Moore, An orphan nuclear hormone receptor that lacks a DNA binding domain and heterodimerizes with other receptors, *Science* **272**(5266), 1336–1339 (1996).

interacting proteins.[2,3] A variety of studies have demonstrated that SHP can interact directly with many activated nuclear receptors and inhibit transactivation.[3–7]

Initial results suggested that SHP could inhibit transactivation by RAR–RXR heterodimers at the level of DNA binding.[1] However, additional studies revealed an autonomous transcriptional repression domain in the C-terminal region, suggesting that SHP could exert a more direct repression function if tethered to DNA through protein–protein interaction.[4]

In contrast to the conventional receptor–receptor dimeric interactions originally assumed to account for the interaction of SHP with other receptors, two independent studies showed that it binds to the AF-2 coactivator-binding site. Thus, Johansson et al. showed that SHP competes with coactivators for binding the ligand-activated estrogen receptor.[3] Similarly, Lee et al. found that the helix 12 AF-2 core motif is critical for SHP interaction with either the constitutively active orphan receptor HNF-4, or retinoid-activated RXR and developed a novel three-hybrid interference assay to characterize the in vivo competition between SHP and coactivators.[6] Based on these results, a two-step model for SHP repression was proposed in which the effects of the initial step, competition with coactivators, are amplified by the subsequent effects of the SHP autonomous repression function.[6,8]

This chapter describes the application of the three-hybrid interference assay to study of SHP interaction with nuclear receptor targets and its broader utility in studying other protein–protein interactions.

[2] N. Masuda et al., An orphan nuclear receptor lacking a zinc-finger DNA-binding domain: interaction with several nuclear receptors, Biochim. Biophys. Acta 1350(1), 27–32 (1997).

[3] L. Johansson et al., The orphan nuclear receptor SHP inhibits agonist-dependent transcriptional activity of estrogen receptors ERalpha and ERbeta, J. Biol. Chem. 274(1), 345–353 (1999).

[4] W. Seol, M. Chung, and D. D. Moore, Novel receptor interaction and repression domains in the orphan receptor SHP, Mol. Cell. Biol. 17(12), 7126–7131 (1997).

[5] W. Seol et al., Inhibition of estrogen receptor action by the orphan receptor SHP (short heterodimer partner), Mol. Endocrinol. 12, 1551–1557 (1998).

[6] Y. K. Lee et al., The orphan nuclear receptor SHP inhibits hepatocyte nuclear factor 4 and retinoid X receptor transactivation: two mechanisms for repression, Mol. Cell. Biol. 20(1), 187–195 (2000).

[7] L. Johansson et al., The orphan nuclear receptor SHP utilizes conserved LXXLL-related motifs for interactions with ligand-activated estrogen receptors, Mol. Cell. Biol. 20(4), 1124–1133 (2000).

[8] Y. K. Lee and D. D. Moore, Dual mechanisms for repression of the monomeric orphan receptor liver receptor homologous protein-1 by the orphan small heterodimer partner, J. Biol. Chem. 277(4), 2463–2467 (2002).

A Three-Hybrid Assay for SHP Interference

The three-hybrid interference assay is based on the mammalian version of the two-hybrid assay as described by Liu and Green[9] and Forman et al.[10] In this variation on the original yeast strategy,[11,12] the interaction of a protein or protein segment fused to a DNA-binding domain with a second hybrid peptide fused to a transcriptional activation domain is detected by the effects of the recruitment of this second hybrid protein to an appropriate reporter. The yeast transcription factor GAL4[13] provides the DNA-binding domain for the mammalian two-hybrid assay; the transcriptional activation domain is from the Herpes simplex virus VP16 protein.[14]

The adaptation of this assay to study the competition between SHP and coactivators for binding activated nuclear receptors is quite similar to a previously described modification of the yeast two-hybrid assay in which a third protein functionally disrupts the interaction between the DNA-binding domain and transcriptional activation domain fusions.[15] However, such interference assays are quite different from another class of three-hybrid assay in which a third chimeric molecule, such as a bivalent small molecule[16] or RNA,[17] promotes assembly of the DNA binding and transcriptional activation domain hybrids.

The potent autonomous repression function of SHP prevents the simple application of the interference strategy to study its effects on coactivator

[9] F. Liu and M. R. Green, A specific member of the ATF transcription factor family can mediate transcription activation by the adenovirus E1a protein, Cell 61(7), 1217–1224 (1990).

[10] B. M. Forman et al., Unique response pathways are established by allosteric interactions among nuclear hormone receptors, Cell 81, 541–550 (1995).

[11] S. Fields and O. Song, A novel genetic system to detect protein-protein interaction, Nature 340, 245–246 (1989).

[12] J. Gyuris et al., Cdi1, a human G1 and S phase protein phosphatase that associates with Cdk2, Cell 75, 791–803 (1993).

[13] L. Keegan, G. Gill, and M. Ptashne, Separation of DNA binding from the transcription-activating function of a eukaryotic regulatory protein, Science 231(4739), 699–704 (1986).

[14] S. J. Triezenberg, K. L. LaMarco, and S. L. McKnight, Evidence of DNA: protein interactions that mediate HSV-1 immediate early gene activation by VP16, Genes Dev. 2(6), 730–742 (1988).

[15] C. R. Geyer, A. Colman-Lerner, and R. Brent, "Mutagenesis" by peptide aptamers identifies genetic network members and pathway connections, Proc. Natl. Acad. Sci. USA 96(15), 8567–8572 (1999).

[16] E. J. Licitra and J. O. Liu, A three-hybrid system for detecting small ligand-protein receptor interactions, Proc. Natl. Acad. Sci. USA 93(23), 12817–12821 (1996).

[17] D. S. Bernstein et al., Analyzing mRNA-protein complexes using a yeast three-hybrid system, Methods 26(2), 123–141 (2002).

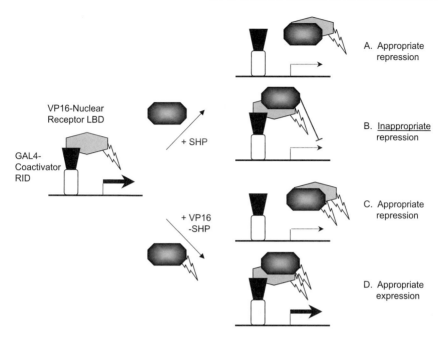

FIG. 1. The three-hybrid interference assay for analysis of protein–protein interference. In the standard two-hybrid assay, an activated nuclear receptor ligand-binding domain interacts strongly with the RID of a coactivator, as indicated on the left. Four possible outcomes from coexpression of intact SHP or a VP16-SHP hybrid are indicated. If SHP competes with the coactivator for binding the receptor LBD, the expression of the reporter gene would be decreased similarly for both SHP and VP16-SHP, and the conclusion that the receptor–coactivator interaction was lost would be correct (A and C). However, if SHP were to be recruited to the receptor–coactivator complex via another interaction, it would decrease reporter expression due to its direct repression function. This would lead to an incorrect conclusion that receptor–coactivator interaction was decreased (compare B with D).

binding to its target receptors. As diagrammed in Fig. 1, any recruitment of SHP to the DNA-binding hybrid, via either a direct or indirect interaction, would decrease transcriptional output. This decrease could be inappropriately interpreted as an inhibition of the direct binding of the two-hybrid proteins. Based on previous work demonstrating that the VP16 activation domain could counteract the SHP repression function,[4] we chose to fuse this activation domain to SHP in a third hybrid protein. As indicated in Figs. 1A and C, this additional fusion does not affect the ability of SHP to compete for coactivator binding if that mechanism is active. However, it does prevent the inappropriate conclusion that coactivator binding is decreased if SHP is recruited to the DNA by indirect interactions (Fig. 1B vs D).

Experimental Procedure

Plasmids

These studies focus on the coactivator SRC-3. Since this is a relatively big protein with several protein interaction domains, only the receptor interaction domain (RID) was used to eliminate possible unexpected interactions. In order to construct GAL4-SRC-3(RID), the human SRC-3 RID (amino acids 601–761) containing three LXXLL motifs[18] was amplified by PCR and inserted into pCMXGAL4. VP16-RXR(L) contains the human RXRα ligand-binding domain (amino acids 203–462). VP16-SHP was constructed by insertion of full length of SHP into pCMXVP16 plasmid. The luciferase reporter construct used in this method contains four copies of the consensus GAL4-binding site, as described.[10]

Transfection

For the results described here, HepG2 cells (human hepatoma cell line) were maintained in DMEM supplemented with 10% fetal bovine serum, penicillin (100 U/ml), and streptomycin (100 μg/ml). Twenty-four hours before transfection, the cells are trypsinized and seeded in a density of 1×10^5 cells per well to 24-well cell culture cluster (Costar) resulting in approximately 90% confluency at the time of transfection. Cells were transfected using calcium phosphate precipitation as described.[19] Transfections were carried out in triplicate with the indicated amounts (per well) of pCMXVP16-SHP (0.1, 0.4, and 1 μg) plus appropriate amounts of pCMXVP16 empty vector to maintain the total amount of DNA at 1 μg, 50 ng of pCMXVP16-RXR(L) or pCMXVP16, 100 ng of GAL4-SRC3(RID), 200 ng of GAL4TKluc reporter plasmid, and 200 ng of TKGH internal control plasmid. After transfection the cells were further incubated at 37°C for 20 hr and provided with fresh DMEM containing 10% charcoal stripped FBS and 1 μM 9-*cis*-retinoic acids or vehicle (ethanol) alone, and harvested for analysis of reporter activities using standard procedures as described.[6]

---

[18] H. Li, P. J. Gomes, and J. D. Chen, RAC3, a steroid/nuclear receptor-associated coactivator that is related to SRC-1 and TIF2, *Proc. Natl. Acad. Sci. USA* **94**(16), 8479–8484 (1997).

[19] F. M. Ausubel *et al.* (eds.), "Current Protocols in Molecular Biology," John Wiley & Sons, New York, 2003.

To extend these results with SHP to other transcriptional corepressors, several issues should be kept in mind. The first is that it is important to confirm that the transcriptional repression function of the VP16-repressor fusion is lost. This could be accomplished using an assay specific to the repressor protein of interest, or more generally by fusion of the chimeric protein to GAL4. The second is that it is important to incorporate a number of controls to confirm the specificity of any effects observed. These include combinations of the GAL4 alone and VP16 alone expression vectors with the hybrid expression vectors of interest, as well as comparisons of the VP16-repressor fusion to the activity of the repressor alone. In the case of the effects described here, the absence of a difference between the results with SHP alone and VP16-SHP added support for the conclusion that this orphan directly competes with coactivators for binding activated receptor targets.[6,8]

Results

A variety of results are anticipated depending on three presumptive roles of the third hybrid. If VP16-SHP does not interact with either of the two fusion proteins, the reporter activity will not change unless VP16-SHP is strongly overexpressed, resulting in nonspecific squelching. To avoid such nonspecific effects, it is important to use appropriate amounts of expression vectors, reporters, and internal control reporter determined in dose response studies, and to include a number of controls, particularly expression vectors for GAL4 alone and VP16 alone. As depicted in Fig. 1, if VP16-SHP competes with GAL4-SRC3 to bind to VP16-RXR, the reporter activity will decrease with increased expression of VP16-SHP. If VP16-SHP does interact with either of the two fusion proteins, but does not interfere with their binding to each other, the reporter activity will either remain level or possibly increase with overexpression of VP16-SHP.

As indicated in Fig. 2, the very strong interaction of GAL4-SRC3(RID) and VP16-RXR(L) in the presence of 9-*cis*-retinoic acid was strongly decreased by increasing amounts of coexpressed VP16-SHP. In an additional control experiment, neither SHP nor VP16-SHP had any effect on SF-1–coactivator interaction. This is expected from the absence of a functional interaction between the two orphans. The conclusion from these results that SHP competes with coactivators for binding-activated receptors has been strongly supported by direct biochemical and other studies, both in this laboratory[6,8] and others.[3,7]

Overall, we conclude that this three-hybrid interference assay provides a novel strategy to address the complications associated with the potential

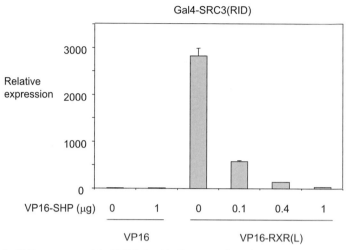

Fig. 2. SHP competes with SRC-3 for binding to activated RXR. A GAL4-SRC3 RID chimera was cotransfected with a VP16-RXR ligand-binding domain chimera and a GAL-responsive reporter in the presence of 1 mM 9-*cis*-retinoic acid, as described.[6]

contribution of a transcriptional repressor to the reporter gene activity in a conventional two-hybrid assay. The recent application of this strategy to address such issues in the competitive binding of the corepressor SMRT and the coactivator GRIP1/SRC-2 to the orphan receptor HNF-4 supports this conclusion.[20]

Summary

The three-hybrid interference method described here is similar to previous interference assays based on the two-hybrid system. In these approaches the presence of a third protein disrupts the interaction between the two different hybrid proteins. The three-hybrid interference assay circumvents a potential problem that arises when the third protein is itself a direct transcriptional repressor by preventing the decrease in transcriptional readout that could occur if the third protein is recruited indirectly to the two-hybrid protein complex.

This three-hybrid method should be applicable to any third protein that is a transcriptional repressor. To use this method to explore the effects of

[20] M. D. Ruse, Jr., M. L. Privalsky, and F. M. Sladek, Competitive cofactor recruitment by orphan receptor hepatocyte nuclear factor 4alpha1: modulation by the F domain, *Mol. Cell. Biol.* **22**(6), 1626–1638 (2002).

such repressors on protein–protein interactions, however, it is important to confirm that the VP16-repressor fusion does not retain active repressor function. It is also appropriate to compare the effects of the VP16-repressor fusion to those with the repressor alone. Finally, it is essential to confirm any conclusions from this or any other single method to study protein–protein interactions with alternative independent approaches.

[10] Regulation of Glucocorticoid Receptor Ligand-Binding Activity by the hsp90/hsp70-based Chaperone Machinery

By Kimon C. Kanelakis and William B. Pratt

Introduction

Nearly 100 proteins are known to be regulated by hsp90.[1] Most of these substrates, or "client" proteins, are involved in signal transduction, and they are assembled into heterocomplexes with hsp90 by a multiprotein hsp90/hsp70-based chaperone machinery. Included among the hsp90 substrates are a number of nuclear receptors, some of which must be in a complex with hsp90 to have ligand-binding activity (e.g., glucocorticoid (GR), mineralocorticoid, and dioxin receptors). The receptor-hsp90 heterocomplexes can be assembled by incubating immunoadsorbed receptors that have been salt stripped of hsp90 with rabbit reticulocyte lysate.[2,3] The hsp90-free GR has no steroid-binding activity, but when it is incubated with reticulocyte lysate, steroid-binding activity is generated in direct proportion to the number of GR-hsp90 heterocomplexes that are assembled.[4] Generation of steroid-binding activity with GR-hsp90 heterocomplex assembly provides an *in vitro* assay of hsp90 action that is clearly relevant to hsp90 action *in vivo*.

The Five-Protein hsp90 Heterocomplex Assembly System

Receptor-hsp90 heterocomplex assembly requires ATP, Mg^{++}, and a monovalent cation (e.g., K^+).[4,5] The heterocomplex assembly system of

[1] W. B. Pratt and D. O. Toft, *Exp. Biol. Med.* **228**, 111 (2003).
[2] D. F. Smith, D. B. Showalter, S. L. Kost, and D. O. Toft, *Mol. Endocrinol.* **4**, 1704 (1990).
[3] L. C. Scherrer, F. C. Dalman, E. Massa, S. Meshinchi, and W. B. Pratt, *J. Biol. Chem.* **265**, 21397 (1990).
[4] K. A. Hutchison, M. J. Czar, L. C. Scherrer, and W. B. Pratt, *J. Biol. Chem.* **267**, 14047 (1992).
[5] D. F. Smith, B. A. Stensgard, W. J. Welch, and D. O. Toft, *J. Biol. Chem.* **267**, 1350 (1992).

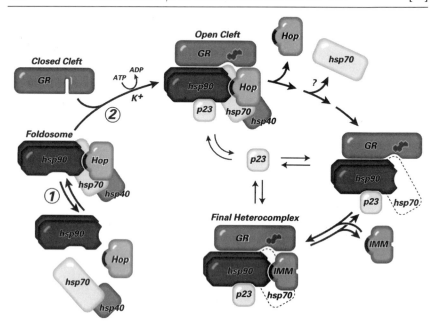

FIG. 1. Model of steroid receptor-hsp90 heterocomplex assembly. The hsp90/hsp70-based chaperone machinery, or *foldosome*, converts the GR LBD from a folded conformation in which the steroid-binding cleft is closed and not accessible to hormone to an open-cleft conformation that can be accessed by steroid. Hop binds via independent TPR domains to hsp90 and hsp70 to form the foldosome machinery. Hsp90 functions as a dimer and one Hop is bound to an hsp90 dimer. Hsp40 is an hsp70 bound cochaperone that is also present in the machinery that carries out the ATP/Mg^{++}-dependent and K^{+}-dependent opening of the steroid-binding cleft. During cleft opening, hsp90 is converted to its ATP-dependent conformation, which is dynamically stabilized by p23, and Hop is released from hsp90. Thus, the complex shown at the end of step 2 represents a composite in which several changes are occurring in dynamic fashion. The immunophilins (IMM) also bind to hsp90 via TPR domains (indicated by the solid black crescents), and when Hop exits the complex, an immunophilin can bind to the single TPR acceptor site on the receptor-bound hsp90 dimer. Some hsp70 also leaves the intermediate complex, and the broken line for hsp70 in the *final heterocomplex* indicates that it is present at substoichiometric levels with respect to the receptor.

reticulocyte lysate has been reconstituted,[6] and we now assemble receptor hsp90 heterocomplexes with a minimal system of five purified proteins—hsp90, hsp70, Hop, hsp40, and p23.[7,8] An overview of the assembly process is provided in Fig. 1.

[6] K. D. Dittmar, K. A. Hutchison, J. K. Owens-Grillo, and W. B. Pratt, *J. Biol. Chem.* **271**, 12833 (1996).
[7] K. D. Dittmar, M. Banach, M. D. Galigniana, and W. B. Pratt, *J. Biol. Chem.* **273**, 7358 (1998).

Hop (Hsp organizing protein) is a protein with two tetratricopeptide repeat (TPR) domains that bind independently to hsp90 and hsp70 to bring the two chaperones together in an hsp90-Hop-hsp70-hsp40 machinery that is labeled a "foldosome" in Fig. 1. All of the Hop in reticulocyte lysate is present in these complexes,[9] which can be immunoadsorbed from lysate with an antibody against Hop.[10,11] The hsp90-Hop-hsp70-hsp40 complex is formed spontaneously on mixing of the purified proteins, and both the immunoadsorbed purified protein complex and the native complex from reticulocyte lysate convert the GR ligand-binding domain (LBD) to a steroid-binding state.[7,11,12]

In Fig. 1, the receptor that interacts with the machinery is shown in a folded state in which the hydrophobic steroid-binding cleft is closed and not accessible to steroid. Both hsp90 and hsp70 bind to the LBD of the receptor, and receptor-bound hsp90 must achieve its ATP-bound conformation for the binding cleft to be open to access by steroid.[13] Although hsp90, hsp70, Hop, and hsp40 act together to open the steroid-binding cleft, the GR-hsp90 heterocomplexes that are produced are unstable and rapidly disassemble unless p23 is present. p23 is not a component of the foldosome machinery, but when receptor-bound hsp90 assumes its ATP-dependent conformation, p23 binds dynamically to hsp90 to stabilize it in that conformation, thus stabilizing the GR-hsp90 heterocomplex.[11]

At this point, Hop and variable amounts of hsp70 exit the intermediate complex formed in step 2 (Fig. 1). Exit of Hop from the intermediate complex frees the TPR acceptor site on hsp90 to bind the TPR domain of immunophilins, such as the FK506-binding proteins FKBP51 and FKBP52, the FKBP homolog protein phosphatase 5, or the cyclosporin A-binding protein Cyp-40.[1] Thus, when receptor-hsp90 heterocomplexes are assembled in cells or in reticulocyte lysate, a variety of complexes are formed, with dynamic exchange of immunophilins taking place.

[8] H. Kosano, B. Stensgard, M. C. Charlesworth, N. McMahon, and D. O. Toft, *J. Biol. Chem.* **273**, 32973 (1998).

[9] P. J. M. Murphy, K. C. Kanelakis, M. D. Galigniana, Y. Morishima, and W. B. Pratt, *J. Biol. Chem.* **276**, 30092 (2001).

[10] D. F. Smith, W. P. Sullivan, T. N. Marion, K. Zaitsu, B. Madden, D. J. McCormick, and D. O. Toft, *Mol. Cell. Biol.* **13**, 869 (1993).

[11] K. D. Dittmar, D. R. Demady, L. F. Stancato, P. Krishna, and W. B. Pratt, *J. Biol. Chem.* **272**, 21213 (1997).

[12] K. D. Dittmar and W. B. Pratt, *J. Biol. Chem.* **272**, 13047 (1997).

[13] J. P. Grenert, B. D. Johnson, and D. O. Toft, *J. Biol. Chem.* **274**, 17525 (1999).

Essential versus Nonessential Chaperones

The purified five-protein system has allowed us to define essential *versus* nonessential chaperones for opening the steroid-binding cleft and for "stable" GR-hsp90 heterocomplex assembly.[14] In the experiment of Fig. 2, stripped GR (*bar 2*) was incubated with the complete five-protein system [hsp90, hsp70, Hop, YDJ-1 (the yeast hsp40 ortholog), and p23] (*bar 3*) or with all components of the system except one protein. At the end of the incubation, the immune pellets were washed and incubated with [³H]triamcinolone acetonide to determine steroid-binding activity. Elimination of hsp90 (*bar 4*) or hsp70 (*bar 5*) reduced receptor activation to levels near that of the stripped receptor, but some activation occurred in the absence of Hop (*bar 6*) or YDJ-1 (*bar 7*). Thus, Hop and YDJ-1 are not

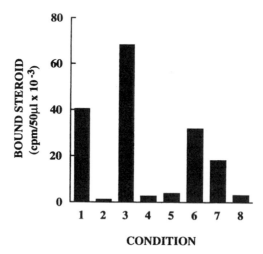

FIG. 2. Reconstitution of steroid-binding activity with the purified five-protein assembly system. Stripped GR immune pellets were incubated for 20 min at 30°C with the purified assembly system (hsp90, hsp70, Hop, YDJ-1, and p23) in the presence of sodium molybdate. The immune pellets were then washed and incubated with [³H]triamcinolone acetonide to determine steroid-binding activity. *Bar 1*, mouse GR immune pellet prepared from Sf9 cytosol; *bar 2*, stripped GR pellet; *bars 3–8*, stripped GR incubated with all five proteins (*bar 3*), all but hsp90 (*bar 4*), all but hsp70 (*bar 5*), all but Hop (*bar 6*), all but YDJ-1 (*bar 7*), and all but the heterocomplex stabilizers p23 and molybdate (*bar 8*). It should be noted that the hsp70 and hsp90 used in this experiment have been purified free of all Hop and hsp40. [Data from Y. Morishima, K. C. Kanelakis, A. M. Silverstein, K. D. Dittmar, L. Estrada, and W. B. Pratt, *J. Biol. Chem.* **275**, 6894 (2000). Reprinted with permission.]

[14] Y. Morishima, K. C. Kanelakis, A. M. Silverstein, K. D. Dittmar, L. Estrada, and W. B. Pratt, *J. Biol. Chem.* **275**, 6894 (2000).

essential components of the assembly system. These results are consistent with studies in yeast showing that deletion of *STI1*, which encodes the yeast ortholog of Hop, or deletion of *YDJ-1*, reduces but does not eliminate GR steroid-binding activity.[15,16]

Although it appears in Fig. 2 that p23 may be essential for cleft opening (*bar 8*), this is not the case. Here, we have assembled GR-hsp90 hetero-complexes at 30°C and then incubated on ice with steroid to determine steroid-binding activity. This procedure assays the formation of stable heterocomplexes that remain intact during subsequent incubation with steroid, and when p23 is omitted from the purified assembly system, no steroid-binding activity is generated in this assay.[8,12,14] However, opening of steroid-binding clefts in the absence of p23 can be detected by having steroid present during the assembly reaction at 30°C. Under these conditions, as soon as the steroid-binding cleft is opened, the steroid enters, and generation of steroid binding shows that the appropriate conformational change has occurred in the absence of p23.[11] Thus, p23 is also a nonessential component of the assembly system for steroid-binding cleft opening, but it is essential for the formation of stable heterocomplexes. Again, this finding is consistent with studies in yeast showing that deletion of *SBA1*, which encodes the yeast p23 ortholog, does not affect dexamethasone-dependent activation of transcription.[17,18]

Figure 3 shows an experiment in which the GR was incubated at 30°C with highly purified proteins in the presence of [^{3}H]steroid. It can be seen that no steroid-binding activity is generated by hsp70 or hsp90 alone, but the two chaperones together yield some ATP-dependent activation. When GR-hsp90 heterocomplex disassembly is retarded by the presence of p23 and molybdate, more steroid binding is detected. Thus, hsp90 and hsp70 work together as the only two essential components of the five-protein system for converting the GR to the steroid-binding state, and Hop, hsp40 (YDJ-1), and p23 function as nonessential cochaperones that optimize the process.

The Mechanism of Cleft Opening

As shown in Figs. 2 and 3, cleft opening occurs without Hop, but we have shown that the rate of GR activation is faster when Hop is present.[14]

[15] H. J. Chang, D. F. Nathan, and S. Lindquist, *Mol. Cell. Biol.* **17**, 318 (1997).
[16] A. E. Fliss, J. Rao, M. W. Melville, M. E. Cheetham, and A. J. Caplan, *J. Biol. Chem.* **274**, 34045 (1999).
[17] S. Bohen, *Mol. Cell. Biol.* **18**, 3330 (1998).
[18] Y. Fang, A. E. Fliss, J. Rao, and A. J. Caplan, *Mol. Cell. Biol.* **18**, 3727 (1998).

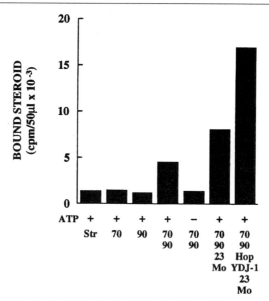

FIG. 3. Some GR activation is obtained with just hsp90 and hsp70. Immunoadsorbed GR pellets were stripped with salt, and the pellets were then washed and incubated 20 min at 30°C with (+) or without (−) an ATP-regenerating system in the presence of 100 nM [^{3}H]triamcinolone acetonide and highly purified (hsp40-free/Hop-free) hsp70, or hsp90, or both hsp70 and hsp90 with additions as noted under each bar in the graph. *Mo* indicates the presence of 20 mM molybdate. *Str*, stripped receptor. [Data from Y. Morishima, K. C. Kanelakis, A. M. Silverstein, K. D. Dittmar, L. Estrada, and W. B. Pratt, *J. Biol. Chem.* **275**, 6894 (2000). Reprinted with permission.]

Thus, the system works more efficiently when Hop brings hsp90 and hsp70 together into a machinery.[14] However, the assembly machinery does not have to be formed prior to contact with the receptor; it can be assembled in stepwise fashion on the receptor.[19] Such stepwise assembly experiments with purified proteins are beginning to separate the cleft opening process into an ordered series of events as diagrammed in Fig. 4.[19–21]

In the first step, immunoadsorbed GR is incubated with purified hsp70 and hsp40 (YDJ-1) in the presence of ATP. This ATP-dependent step produces a GR-hsp70-hsp40 complex that can be washed free of unbound hsp70 and hsp40 and then incubated with purified hsp90, Hop, and p23.

[19] Y. Morishima, P. J. M. Murphy, D.-P. Li, E. R. Sanchez, and W. B. Pratt, *J. Biol. Chem.* **275**, 18054 (2000).
[20] Y. Morishima, K. C. Kanelakis, P. J. M. Murphy, D. S. Shewach, and W. B. Pratt, *Biochemistry* **40**, 1109 (2001).
[21] M. P. Hernandez, A. Chadli, and D. O. Toft, *J. Biol. Chem.* **277**, 11873 (2002).

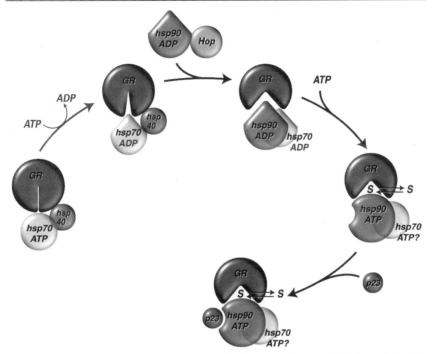

FIG. 4. Mechanism by which hsp70 and hsp90 open the steroid-binding cleft. The mechanism is derived from stepwise assembly experiments in which hsp70 binds to the GR in an ATP-dependent and hsp40-dependent step to form a "primed" GR-hsp70 complex that can then bind Hop and hsp90. After hsp90 binding, there is a second ATP-dependent step (or steps) that is rate limiting and leads to opening of the steroid-binding cleft, which enables access by the steroid. The hsp90 is now in its ATP-dependent conformation and can be bound by p23, which stabilizes the chaperone in that conformation, preventing disassembly of the GR-hsp90 heterocomplex. The hsp40 and Hop components of the five-protein system have been omitted from later steps for simplicity.

In the first reaction the receptor is "primed" to bind hsp90 and then be activated in the second reaction.[22] The second step with hsp90 is also ATP-dependent, and steroid-binding activity is generated only during the second step. The priming step and subsequent binding of hsp90 to the primed receptor are rapid, and the rate-limiting step in the overall process is the ATP-dependent opening of the steroid-binding cleft.[22] The reader is referred to Pratt and Toft[1] for a more detailed discussion of the mechanism.

[22] K. C. Kanelakis, D. S. Shewach, and W. B. Pratt, *J. Biol. Chem.* **277**, 33698 (2002).

Method of GR-hsp90 Heterocomplex Assembly

Source of GR

We have used two sources of mouse GR for GR-hsp90 heterocomplex assembly with the five-protein system. GR may be immunoadsorbed from the $100,000 \times g$ supernatant fraction (cytosol) of L929 mouse fibroblasts prepared by suspension of cells in 1.5 volumes of HE buffer (10 mM HEPES, 1 mM EDTA, pH 7.35) with 20 mM molybdate, followed by Dounce homogenization. Cytosols prepared from these cells or other cultured cells or tissues with reasonable levels of receptor can be flash frozen and stored at $-70°$C. When stripped of hsp90 and incubated with either reticulocyte lysate or the purified protein system, 80–100% of the receptors in a preparation of this type can be activated to the steroid-binding state.

We have described previously the construction of a recombinant baculovirus for overexpression of mouse GR in Sf9 cells.[19] Sf9 cells infected in log-phase of growth with recombinant baculovirus at a multiplicity of infection of 3.0 are supplemented with 1.0% glucose at infection and 24-h postinfection as described by Srinivasan *et al.*[23] The Sf9 cells are harvested 48-h postinfection by suspension in 1.5 volumes HE buffer with 20 mM sodium molybdate, 1 mM phenylmethylsulfonyl fluoride, and one tablet of Complete-Mini protease inhibitor mixture (Roche Molecular Biochemicals, Mannheim, Germany) per 7 ml of buffer, followed by Dounce homogenization. The GR in the Sf9 cytosol is present in large aggregates, and only 12–15% of it is bound by insect hsp90 and has steroid-binding activity. When the GR is immunoadsorbed from this cytosol, the receptors can be stripped of hsp90 and then reactivated to the steroid-binding state with reticulocyte lysate or the five-protein system; however, we are not able to activate $\sim 85\%$ of the GR that is in aggregated form. Despite the loss to aggregation, use of Sf9 cell-produced receptor is time saving, inexpensive, and convenient. The overexpressed receptor, however, cannot be used for measurement of chaperone–substrate stoichiometry during the various stages of assembly.

Immunoadsorption of GR

GR is immunoadsorbed from 100-μl aliquots of L cell cytosol or 50-μl aliquots of Sf9 cytosol by rotation for 2 hr at 4°C with 14 μl of protein A-Sepharose precoupled to 7 μl of FiGR ascites suspended in 200 μl of TEG buffer (10 mM TES, pH 7.6, 50 mM NaCl, 1 mM EDTA, 10% glycerol).

[23] G. Srinivasan, J. F. M. Post, and E. B. Thompson, *J. Steroid Biochem. Mol. Biol.* **60**, 1 (1997).

The immunoadsorbed receptors are stripped of associated hsp90 by incubating the immunopellet an additional 2 hr at 4°C with 300 μl of 0.5 M NaCl in TEG. The pellets are then washed once with 1 ml of TEG followed by a second wash with 1 ml of HEPES buffer (10 mM HEPES, pH 7.35).

The FiGR antibody is a mouse monoclonal IgG directed against the same epitope on the GR as the BuGR2 antibody. The FiGR ascites was generously provided to us by Dr. Jack Bodwell (Department of Physiology, Dartmouth Medical School). The BuGR2 monoclonal IgG against the GR (Affinity Bioreagents, Golden, CO) also may be used for GR immuno-adsorption and reconstitution experiments. In this case, the BuGR2 antibody is prebound to protein A-Sepharose pellets by incubating 14 μl of protein A-Sepharose for 1 hr at 4°C with 20 μl of antibody at a concentration of 100 μg/ml and 150 μl of TEG buffer, followed by centrifugation and washing as above.

Note. The GR immunoadsorption procedure described above is the protocol for preparing immunopellets for activation of steroid-binding activity. When preparing immunopellets for analysis of GR-hsp90 heterocomplex assembly by Western blotting, we immunoadsorb receptor from 100-μl aliquots of Sf9 cytosol and 300-μl aliquots of L cell cytosol.

GR Heterocomplex Reconstitution

Immunopellets stripped of chaperones are incubated with 50 μl of rabbit reticulocyte lysate or with the five-protein mixture (20 μg of purified hsp90, 15 μg of purified hsp70, 0.6 μg of purified Hop, 6 μg of purified p23, 0.4 μg of purified YDJ-1), and adjusted to 50 μl with HKD buffer (10 mM HEPES, 100 mM KCl, 5 mM dithiothreitol, pH 7.35) containing 20 mM sodium molybdate and 5 μl of an ATP-regenerating system (50 mM ATP, 250 mM creatine phosphate, 20 mM magnesium acetate, and 100 units/ml creatine phosphokinase). The assay mixtures are incubated for 20 min at 30°C with suspension of the pellets by shaking the tubes every 2 min. At the end of the incubation, the pellets are washed twice in 1 ml of ice-cold TEGM buffer (TEG with 20 mM sodium molybdate), and assayed for steroid-binding capacity or for receptor and associated proteins.

Note. Trivial as it may seem, the resuspension by shaking (actually we flick the bottom of the incubation tube several times with the tip of a finger) is crucial for efficient heterocomplex reconstitution.

Assay of Steroid-Binding Capacity

Immunopellets to be assayed for steroid-binding activity are incubated overnight at 4°C in 50 μl of HEM buffer (HE with 20 mM

sodium molybdate) plus 50 nM [^{3}H]triamcinolone acetonide (or [^{3}H]dexamethasone). Samples are then washed three times with 1 ml of TEGM and counted by liquid scintillation spectrometry. Steroid-binding values can be expressed conveniently as the cpm [^{3}H]steroid bound/ immunopellet prepared from 50 μl Sf9 cytosol or 100 μl of L cell cytosol.

Assay of hsp90 and hsp70 Bound to GR

To assay receptor-bound proteins, immunopellets prepared from 100 μl Sf9 cytosol or 300 μl of L cell cytosol and reconstituted with the five-protein system are resolved on 10% SDS-polyacrylamide gels and transferred to Immobilon-P membranes. The membranes are probed with 0.25 μg/ml BuGR2 for GR, 1 μg/ml AC88 monoclonal IgG (StressGen, Victoria, BC) for hsp90, and 1 μg/ml N27F3-4 anti-72/73 kDa hsp monoclonal IgG (StressGen, Victoria, BC) for hsp70.

Note. All three proteins can be blotted on the same transfer. We find it is best to blot for hsp90, develop with peroxidase, and then cut the blot just above and below the hsp90 band for blotting GR and hsp70, respectively. This trimming works fine for mouse or human GR, which have an apparent mass of $\sim$100 kDa; however, some GRs (e.g., rat) migrate at the same position as hsp90 on denaturing gels and separate immunoblots must be developed for GR and hsp90.

Protein Purification

The purification of baculovirus-expressed human hsp90, and recombinant human Hop and p23 expressed in *Escherichia coli* were published by Buchner *et al.*[24,25] in a previous issue of this series. Because we are studying the mechanism of GR-hsp90 heterocomplex assembly, large amounts of hsp90 and hsp70 are required that are active in the assay described above. We perform a three-step purification for both proteins from rabbit reticulocyte lysate. The commercial, untreated reticulocyte lysate preparation used as a source (from Green Hectares, Oregon, WI) contains $\sim$4 μM hsp90 and $\sim$7 μM hsp70.[9] The Toft laboratory purifies human hsp70 (GenBank^{TM}/EBI accession number M11717) and human hsp90β (GenBank^{TM}/EBI accession number M86752) overexpressed in Sf9 cells for use in similar progesterone receptor-hsp90 heterocomplex assembly assays. Their purification procedures[21] contain MonoQ and Superdex 200

[24] J. Buchner, S. Bose, C. Mayr, and U. Jakob, *Methods in Enzymology* **290**, 409 (1998).
[25] J. Buchner, T. Weikl, H. Bugl, F. Pirkl, and S. Bose, *Methods in Enzymology* **290**, 418 (1998).

FPLC steps that we do not find necessary to obtain ~97% pure rabbit hsp70 and hsp90 as assessed by densitometry of SDS-polyacrylamide gels.[7]

Purification of hsp70 and hsp90

Rabbit reticulocyte lysate (25 ml) is adsorbed to a 2.5 × 20 cm column of DEAE-cellulose (DE52) equilibrated with HE buffer, and the column is washed with 150 ml of HE buffer with 20 mM sodium molybdate followed by 150 ml of HE buffer alone. The adsorbed proteins are eluted with a 400 ml gradient of 0–0.5 M KCl. Hsp90, hsp70, and p23 are detected by resolving an aliquot of each fraction by SDS-polyacrylamide gel electrophoresis and Western blotting with the appropriate antibodies. The JJ3 monoclonal IgG[26] (mouse ascites obtained from Dr. David Toft, Department of Biochemistry and Molecular Biology, Mayo Clinic, Rochester, MN) is used at 0.1% to probe for p23. The elution pattern of the three proteins is shown in Dittmar *et al.*[6] Hsp90 elutes at 0.15–0.25 M KCl, and it is completely separated from p23, which is very acidic and elutes at a much higher salt concentration. Most of the hsp70 and the entire Hop elute prior to the hsp90. The peak fractions of hsp90 and hsp70 are pooled and diluted with an equal volume of 10 mM HEPES, pH 7.35, 10 mM K$_2$HPO$_4$, 1 mM EDTA.

The hsp90 or hsp70 pools are loaded on a 2 × 8 cm hydroxylapatite column that is equilibrated with 10 mM HEPES, pH 7.35, 10 mM K$_2$HPO$_4$, 1 mM EDTA. The column is washed with 120 ml of this buffer and bound proteins are eluted with a 300 ml gradient of 0.01–0.4 M K$_2$HPO$_4$. Fractions containing hsp70 and hsp90 are again detected by Western blotting and peak fractions for each are combined. The great majority of hsp70 elutes from hydroxylapatite prior to hsp90, but the hsp90 still contains a small amount of hsp70 that is removed during the next step of ATP-agarose chromatography.

Hsp70 has a high-affinity ATP-binding site and chromatography on ATP-agarose produces the largest enrichment in the purification protocol. The hsp90 pool from the hydroxylapatite column is applied to a 50 ml column of ATP-agarose. Any hsp70 contaminating the hsp90 is bound to the column and the flow-through fraction contains purified hsp70-free hsp90.

To purify hsp70, the hsp70 pool from hydroxylapatite is applied to a 50 ml column of ATP-agarose. Virtually all remaining contaminants are removed by washing with 150 ml of HEK buffer (HE containing

[26] J. L. Johnson, T. G. Beito, C. J. Krco, and D. O. Toft, *Mol. Cell. Biol.* **14**, 1956 (1994).

500 mM KCl), and the hsp70 is eluted with 200 ml of HEK buffer containing 5 mM ATP.

Both the hsp90 and hsp70 preparations are concentrated to 1 ml by Amicon YM10 ultrafiltration, and both preparations are dialyzed overnight against two changes of HKD buffer (10 mM HEPES, pH 7.35, 100 mM KCl, 5 mM dithiothreitol). The preparations are aliquoted into 50 μl fractions, flash-frozen, and stored at $-70°$C.

Note. There are several important comments to make regarding the purification procedures and the products.

1. In purifying proteins from reticulocyte lysate, it is important to wash the DE52 column with buffer containing 20 mM sodium molybdate after the lysate is adsorbed to the column. This step releases all of the reticulocyte hemoglobin into the column wash, greatly facilitating subsequent purification steps.

2. The hsp70 purified from reticulocyte lysate behaves as a dimer (with a small amount of trimer and no monomer) when electrophoresed under nondenaturing conditions.[9] When the Toft laboratory purifies human hsp70 overexpressed in Sf9 cells, they discard the dimer and use only the monomer peak.[21] Insofar as we can determine, the two hsp70 preparations behave similarly in the purified assembly system.

3. The hydroxylapatite column is important for separating hsp70 from hsp40.[7] Although, through conservative selection of hydroxylapatite fractions, we can prepare hsp40-free hsp70,[14] trace amounts of hsp40 may be detectable in our routine hsp70 preparations by Western blotting. This purified hsp70 preparation with trace hsp40 nevertheless responds well to the presence of YDJ-1 in the five-protein assembly system or in stepwise assembly,[20] and it can be used to study the assembly mechanism.

4. Very little separation of hsp70 from Hop occurs on either DE52 or hydroxylapatite chromatography, and trace amounts of Hop that can be detected by Western blotting are present in the final hsp70 preparation eluted from ATP-agarose.[14] Again, this preparation responds well to the presence of added Hop in the five-protein assembly system,[14] and it can be used in studies of the assembly mechanism. Trace amounts of Hop can be eliminated by an additional purification step,[14] but we do not find this to be necessary as a matter of routine.

5. *In purifying the hsp90, it is important to proceed rapidly from one step to another. Longer times of purification results in lower bioactivity of the hsp90 in opening the steroid-binding cleft of the GR as a component of the five-protein system.* Very bioactive purified hsp90 and purified

hsp90 with no bioactivity behave the same on denaturing gel electrophoresis, and the basis for hsp90 inactivation during extended purification is not known. On one occasion, we tested a commercial preparation of purified mammalian hsp90 and found it to be inactive in our five-protein assembly system. It seems that purified hsp90 can be active in passive *in vitro* chaperoning assays, such as ATP-independent inhibition of enzyme aggregation, but inactive in the ATP-dependent GR-hsp90 heterocomplex assembly assay.

6. The percentage of the purified hsp90 preparation that is bioactive in the assembly assay is unknown. We maintain frozen samples of hsp90 that are known to be active and compare each new batch with this standard in the five-protein GR-hsp90 heterocomplex assembly assay. It is perhaps important that almost no monomers of hsp90 are detected on nondenaturing gel electrophoresis of whole reticulocyte lysate, yet the purified hsp90 is roughly an equal mixture of monomer and dimer on nondenaturing gels.[9] Thus, the hsp90 monomers are formed during purification, and these monomers may be inactive in the assembly assay.

7. The purified hsp70 is largely ATP-bound, and the purified hsp90 does not have bound nucleotide (ATP or ADP).[20]

Purification of Hop

Recombinant human Hop is purified from bacterial lysate. The cDNA for the 60 kDa human protein (GenBank[TM]/EBI accession number M86752) cloned by Honoré *et al.*,[27] which is the ortholog of rabbit Hop,[28] was subcloned by Sullivan and Toft (unpublished) into a pET 23C vector (Novagen) using the *Eco*RI and *Not*I sites. This construct was used to transform *E. coli* strain BL21 (DE3), which harbors an integrated T7 polymerase gene. Bacteria expressing Hop were provided by Dr. David Smith (Mayo Clinic, Scottsdale, AZ). Bacteria were grown to an A_{600} of 0.6, induced with 1 mM isopropyl-1-thio-β-D-galactopyranoside for 3 hr at 25°C, and harvested. Bacterial pellets were washed twice in phosphate-buffered saline, resuspended in lysis buffer (10 mM Tris–HCl, pH 7.5, 1 mM EDTA, 10% glycerol), and bacteria ruptured by sonication.

The bacterial sonicate is centrifuged at $8000 \times g$ for 20 min and the supernatant is loaded onto a DE52 column as described above. The column

---

[27] B. Honoré, H. Leffers, P. Madsen, H. H. Rasmussen, J. Vanderkerckhove, and J. E. Celis, *J. Biol. Chem.* **267**, 8485 (1992).

[28] D. F. Smith, W. P. Sullivan, T. N. Marion, K. Zaitsu, B. Madden, D. J. McCormick, and D. O. Toft, *Mol. Cell. Biol.* **13**, 869 (1993).

is washed with 120 ml of HE buffer and eluted with a 400 ml gradient 0–0.5 M KCl. The fractions containing Hop are identified by immunoblotting with 0.1% DS14F5 monoclonal IgG against Hop (provided by Dr. David Smith, Mayo Clinic, Scottsdale, AZ). Hop-containing fractions are pooled, diluted with an equal volume of 10 mM HEPES, pH 7.35, 10 mM K$_2$HPO$_4$, 1 mM EDTA, and loaded on a hydroxylapatite column equilibrated in the same buffer. The column is eluted with a 300 ml gradient of 0.01–0.4 M K$_2$HPO$_4$, and the Hop-containing fractions are identified and pooled. The pooled fractions are concentrated to $\sim$2 ml by Amicon ultrafiltration and dialyzed overnight against two changes of HKD buffer. Samples are then aliquoted, flash-frozen, and stored at $-70°$C. The final preparation is >95% pure as assayed by densitometry of Coomassie blue stained gels, and it is always active in the five-protein assembly assay.

Purification of YDJ-1

Bacteria expressing YDJ-1[29] (GenBank[TM]/EBI accession number X56560) were provided by Dr. Avrom Caplan (Department of Cell Biology, Mount Sinai School of Medicine). YDJ-1 is purified from bacterial lysate by sequential chromatography on DE52 and hydroxylapatite exactly as described above for Hop. Fractions containing YDJ-1 are identified by immunoblotting with 0.5% anti-Hsp40 rabbit polyclonal antibody (StressGen, Victoria, BC). With conservative selection of fractions for pooling at each step, the preparation is $\sim$85% pure as assessed by densitometry of Coomassie blue-stained gels, and it is always active in the five-protein assembly assay. We routinely use this preparation, but $\sim$99% purity may be obtained by adding MonoQ FPLC and Superdex FPLC steps as described by Hernandez et al.[21]

Purification of p23

Bacteria expressing human p23[30] (GenBank[TM]/EBI accession number L24804) were provided by Dr. David Toft (Department of Biochemistry and Molecular Biology, Mayo Clinic, Rochester, MN). The p23 is purified from bacterial lysate by sequential chromatography on DE52 and hydroxylapatite columns. Fractions containing p23 were identified by immunoblotting with 0.1% JJ3 monoclonal IgG[26] (provided by Dr. David Toft). Because p23 is eluted from DE52 at high salt concentrations (0.3–0.35 M), the pooled fractions are diluted by adding 1.5 volumes of 10 mM HEPES, pH

[29] A. J. Caplan, J. Tsai, P. J. Casey, and M. G. Douglas, J. Biol. Chem. **267**, 18890 (1992).
[30] J. L. Johnson and D. O. Toft, J. Biol. Chem. **269**, 24989 (1994).

7.35, 10 mM K$_2$HPO$_4$, 1 mM EDTA prior to adsorption of proteins to hydroxylapatite. Because p23 is so acidic, the major purification occurs upon DE52 chromatography. The final p23 preparation is >90% pure as assessed by densitometry of Coomassie blue-stained gels, and it is always active in the five-protein assembly assay.

Conclusion

The definition of a purified five-protein system that assembles heterocomplexes between nuclear receptors, and hsp90 is enabling advances in our understanding of the mechanism by which the hsp90/hsp70-based chaperone machinery functions. It is likely that the five-protein system described here will regulate the function of a wide variety of proteins. This system, for example, assembles heterocomplexes between Raf-1 and hsp90 (Pratt laboratory, unpublished data), suggesting that it will assemble hsp90 into complexes with more than 50 signaling protein kinases already known to be regulated by this chaperone.[1] The five-protein system has also been used to reconstitute a functional hepadenovirus reverse transcriptase,[31] indicating that the system may be responsible for hsp90 regulation of a variety of other biological processes that are dependent upon hsp90.

Acknowledgment

This work was supported by grant DK31573 from the National Institute of Diabetes and Digestive and Kidney Diseases.

[31] J. Hu, D. Toft, D. Anselmo, and X. Wang, *J. Virol.* **76**, 269 (2002).

[11] Analysis of Receptor Phosphorylation

By BRIAN G. ROWAN, RAMESH NARAYANAN, and NANCY L. WEIGEL

Introduction

The nuclear receptor superfamily is a diverse group of ligand-dependent transcription factors and orphan receptors, whose activities are regulated by reversible phosphorylation. Phosphorylation occurs predominantly on serine and threonine residues in the N-terminal and DNA-binding regions of receptors. A large body of literature has described the role of phosphorylation in regulating a variety of nuclear receptor functions including transcriptional activation–repression, interaction with coregulators, DNA

0076-6879/2003 $35.00

binding, and receptor turnover. Several recent reviews provide complete descriptions of these functions.[1–7]

The study of nuclear receptor phosphorylation necessitates an understanding of cellular signaling pathways and how kinases and phosphatases in these pathways communicate with the nuclear receptor signaling pathways to modulate receptor function. It has become apparent that signal pathway interaction and regulation of nuclear receptor signaling does not in all cases involve known receptor phosphorylation sites,[8–13] suggesting the existence of novel sites in receptors and/or the phosphorylation of receptor-associated coregulators as contributors to crosstalk mechanisms. Furthermore, there is growing recognition that tissue-specific patterns of nuclear receptor and coregulator phosphorylation may underlie the tissue as well as gene-selective actions of nuclear receptors.[14] These latter points emphasize the need for further studies to define the role of phosphorylation in nuclear receptor action.

The following sections describe procedures to analyze nuclear receptor phosphorylation. The majority of the chapter focuses on methods for identifying phosphorylation sites. Many of these procedures are directly applicable to the analysis of other posttranslational modifications such as acetylation and methylation of receptors and other types of proteins.

Types of Phosphorylation Studies

Identification of Novel Phosphorylation Sites In Vivo and In Vitro

For some receptors, there may be no information about phosphorylation (most notably for some of the orphan receptors) and a study to identify

[1] N. L. Weigel, *Biochem. J.* **319**, 657 (1996).

[2] D. Shao and M. A. Lazar, *J. Clin. Invest* **103**, 1617 (1999).

[3] S. Y. Cheng, *Rev. Endocr. Metab. Disord.* **1**, 9 (2000).

[4] K. M. Coleman and C. L. Smith, *Front Biosci.* **6**, D1379 (2001).

[5] A. O. Brinkmann, *Eur. J. Dermatol.* **11**, 301 (2001).

[6] S. Kato, *Breast Cancer* **8**, 3 (2001).

[7] J. E. Bodwell, J. C. Webster, C. M. Jewell, J. A. Cidlowski, J. M. Hu, and A. Munck, *J. Steroid Biochem. Mol. Biol.* **65**, 91 (1998).

[8] H. Lee, F. Jiang, Q. Wang, S. V. Nicosia, J. Yang, B. Su, and W. Bai, *Mol. Endocrinol.* **14**, 1882 (2000).

[9] W. Bai, B. G. Rowan, V. E. Allgood, B. W. O'Malley, and N. L. Weigel, *J. Biol. Chem.* **272**, 10457 (1997).

[10] B. G. Rowan, N. Garrison, N. L. Weigel, and B. W. O'Malley, *Mol. Cell. Biol.* **20**, 8720 (2000).

[11] J. De Mora Font and M. Brown, *Mol. Cell. Biol.* **20**, 5041 (2000).

[12] W. Feng, P. Webb, P. Nguyen, X. Liu, J. Li, M. Karin, and P. J. Kushner, *Mol. Endocrinol.* **15**, 32 (2001).

[13] Y. Zhou, W. Gross, S. H. Hong, and M. L. Privalsky, *Mol. Cell. Biochem.* **220**, 1 (2001).

phosphorylation sites should be initiated.[15–18] Depending upon the method chosen, identifying authentic phosphorylation sites *in vivo* can be costly, time consuming, and/or may require large amounts of purified receptor. Large quantities of [^{32}P]H$_3$PO$_4$ are often required unless sensitive conventional or mass spectrometry sequencing approaches are used (see next section). Once phosphorylation sites have been identified, stoichiometry can be determined using *in situ* double-labeling procedures with [^{32}P]H$_3$PO$_4$ and [^{35}S]methionine.[19] Stoichiometry measurements require much less receptor protein and only low levels of radioisotope.

Phosphorylation sites also may be identified by *in vitro* phosphorylation using purified receptors and kinases (many of which are commercially available).[17] In other cases, antibodies that precipitate kinases in their active form can be used to isolate enzymes for phosphorylation studies.[20] Provided that sources of receptor and kinase are available,[21,22] *in vitro* phosphorylation is less complicated and requires lower levels of radioactivity; however, phosphorylation sites identified *in vitro* must always be confirmed *in vivo* by comparison of the appropriate two-dimensional phosphopeptide maps. The chief disadvantage of using an *in vitro* approach is that kinases of interest are often not known.

Regulation of Previously Identified Phosphorylation Sites and Correlation to Receptor Function

Studies of the effects of hormones, drug treatments, cell cycle, and other conditions on receptor phosphorylation often provide new insights into receptor function. The results obtained can be related to various receptor functions such as transcriptional activation, DNA binding, ligand binding, coregulator interaction etc.[22–27] Site-directed mutagenesis of previously

[14] N. J. McKenna and B. W. O'Malley, *Cell* **108**, 465 (2002).

[15] Z. X. Zhou, J. A. Kemppainen, and E. M. Wilson, *Mol. Endocrinol.* **9**, 605 (1995).

[16] A. Poletti and N. L. Weigel, *Mol. Endocrinol.* **7**, 241 (1993).

[17] Y. Zhang, C. A. Beck, A. Poletti, J. P. Clement, P. Prendergast, T. T. Yip, T. W. Hutchens, D. P. Edwards, and N. L. Weigel, *Mol. Endocrinol.* **11**, 823 (1997).

[18] D. Gioeli, S. B. Ficarro, J. J. Kwiek, D. Aaronson, M. Hancock, A. D. Catling, F. M. White, R. E. Christian, R. E. Settlage, J. Shabanowitz, D. F. Hunt, and M. J. Weber, *J. Biol. Chem.* **277**, 29304 (2002).

[19] D. B. Mendel, J. E. Bodwell, and A. Munck, *J. Biol. Chem.* **262**, 5644 (1987).

[20] M. D. Krstic, I. Rogatsky, K. R. Yamamoto, and M. J. Garabedian, *Mol. Cell. Biol.* **17**, 3947 (1997).

[21] C. Glineur, A. Bailly, and J. Ghysdael, *Oncogene* **4**, 1247 (1989).

[22] S. Kato, H. Endoh, Y. Masuhiro, T. Kitamoto, S. Uchiyama, H. Sasaki, S. Masushige, Y. Gotoh, E. Nishida, H. Kawashima, D. Metzger, and P. Chambon, *Science* **270**, 1491 (1995).

identified phosphorylation sites is useful to assess functional changes in nuclear receptor signaling.[9,22,23,28–31] *In vitro* phosphorylation of receptors with purified kinases can be used to examine several receptor functions including DNA binding.[32–34]

A recently developed approach that is especially amenable to studying known modification sites is the use of phosphorylation site-specific antibodies, that permit detection by Western blotting. The antibodies are particularly useful for detecting phosphorylation of specific sites in response to a variety of signals in proteins with multiple phosphorylation sites. In these instances simple measurement of total labeling is often insufficient to identify cell-signaling pathways responsible for phosphorylation. The use of a site-specific antibody for the hormone-dependent site, Ser^{294}, in the human progesterone receptor revealed that although both the PR-A and PR-B forms of the receptor contain Ser^{294}, only PR-B is phosphorylated on this residue in T47D breast cancer cells.[35]

Site-specific antibodies typically are prepared using synthetic phospho-peptides as antigens. Antibodies that recognize both phosphorylated and nonphosphorylated proteins or only the nonphosphorylated protein are removed using an affinity column comprised of the nonphosphorylated peptide coupled to a resin. The flowthrough will contain antibodies that are dependent upon phosphorylation for recognition. Phosphorylation-specific antibodies have been described for some of the modification sites in the estrogen, progesterone, and glucocorticoid receptors,[35–37] and some of these

[23] G. Bunone, P. A. Briand, R. J. Miksicek, and D. Picard, *EMBO J.* **15**, 2174 (1996).

[24] S. C. Hsu, M. Qi, and D. B. Defranco, *EMBO J.* **11**, 3457 (1992).

[25] L. A. Denner, W. T. Schrader, B. W. O'Malley, and N. L. Weigel, *J. Biol. Chem.* **265**, 16548 (1990).

[26] C. A. Beck, N. L. Weigel, and D. P. Edwards, *Mol. Endocrinol.* **6**, 607 (1992).

[27] J. M. Hu, J. E. Bodwell, and A. Munck, *Mol. Endocrinol.* **11**, 305 (1997).

[28] W. Bai, S. Tullos, and N. L. Weigel, *Mol. Endocrinol.* **8**, 1465 (1994).

[29] W. Bai and N. L. Weigel, *J. Biol. Chem.* **271**, 12801 (1996).

[30] R. A. Campbell, P. Bhat-Nakshatri, N. M. Patel, D. Constantinidou, S. Ali, and H. Nakshatri, *J. Biol. Chem.* **276**, 9817 (2001).

[31] D. Chen, P. E. Pace, R. C. Coombes, and S. Ali, *Mol. Cell. Biol.* **19**, 1002 (1999).

[32] J. C. Hsieh, P. W. Jurutka, M. A. Galligan, C. M. Terpening, C. A. Haussler, D. S. Samuels, Y. Shimizu, N. Shimizu, and M. R. Haussler, *Proc. Natl. Acad. Sci. USA* **88**, 9315 (1991).

[33] D. Katz, M. J. Reginato, and M. A. Lazar, *Mol. Cell. Biol.* **15**, 2341 (1995).

[34] G. Castoria, A. Migliaccio, S. Green, M. Di Domenico, P. Chambon, and F. Auricchio, *Biochemistry* **32**, 1740 (1993).

[35] D. L. Clemm, L. Sherman, V. Boonyaratanakornkit, W. T. Schrader, N. L. Weigel, and D. P. Edwards, *Mol. Endocrinol.* **14**, 52 (2000).

reagents are available commercially. One limitation of this type of antibody is cross-reactivity with other more abundant phosphoproteins. For analysis of proteins produced in transient transfection studies, a mock transfected control frequently is sufficient to demonstrate that the immunoreactive band is receptor specific. Alternatively, the receptor can be immuno-precipitated with a general receptor antibody first followed by Western analysis of the precipitated protein using the phosphorylation site-specific antibody.

Identification of Signaling Pathways and Kinases that Alter Phosphorylation

The amino acid sequence surrounding phosphorylation sites identified *in vivo* often will reveal the kinase and signaling pathway involved. With this information, *in vitro* phosphorylation may be used to determine whether the site is a substrate for a specific kinase. In conjunction with *in vitro* phosphorylation, several other approaches are useful in identifying the signaling pathways that regulate nuclear receptor phosphorylation. Kinase activators and inhibitors may be used to selectively disrupt signaling pathways followed by analysis of receptor phosphorylation *in vivo*.[8,22,23,38] Alternatively, expression of dominant active and/or dominant negative kinases in cells may be used to affect specific pathways.[8] Site-specific antibodies, if available, are particularly useful for monitoring phosphoryla-tion of selected sites.

In Vitro *versus* In Vivo *Phosphorylation*

Caution is needed when comparing *in vitro* to *in vivo* phosphorylation, and when using kinase activators–inhibitors *in vivo*, to avoid drawing false conclusions about signaling pathways and kinases that induce receptor phosphorylation. *In vitro* phosphorylation of proteins may result in low-level phosphorylation at nonphysiological sites. In addition, some authentic phosphorylation sites may be targeted by multiple kinases, both *in vivo* and *in vitro*, indicating that several signaling pathways are involved in phosphorylation of the site. Some kinases, such as glycogen synthase

[36] Z. Wang, J. Frederick, and M. J. Garabedian, *J. Biol. Chem.* **277**, 26573 (2002).

[37] D. Chen, E. Washbrook, N. Sarwar, G. J. Bates, P. E. Pace, V. Thirunuvakkarasu, J. Taylor, R. J. Epstein, F. V. Fuller-Pace, J. M. Egly, R. C. Coombes, and S. Ali, *Oncogene* **21**, 4921 (2002).

[38] Y. Goldberg, C. Glineur, J. C. Gesquiere, A. Ricaurt, J. Sap, B. Vennstrom, and J. Ghysdael, *EMBO J.* **7**, 2425 (1988).

kinase 3 (GSK3), will phosphorylate a consensus site only after an adjacent site has been phosphorylated by a different kinase.[39,40] In this situation, *in vitro* phosphorylation with the authentic *in vivo* kinase would not result in phosphorylation at this site due to the absence of the specifying phosphorylation event.

Although altered activity in the presence of kinase modulators is suggestive of phosphorylation, some of these compounds may have a profound effect on nuclear receptor function without altering receptor phosphorylation. For example, activation of protein kinase A (PKA) with 8-bromo cyclic AMP induced ligand-independent activation of the chicken PR, although there was no change in receptor phosphorylation.[9] Another caveat of the use of kinase activators–inhibitors *in vivo* is that they may alter the activity of a downstream kinase. *In vivo* activation of PKA resulted in an indirect phosphorylation of the coactivator SRC-1 via PKA-dependent activation of Erk-2 kinase that in turn directly phosphorylated SRC-1 at two sites. *In vitro* phosphorylation of SRC-1 with PKA did not result in phosphorylation at these sites.[10] This example also illustrates crosstalk between unrelated kinase pathways.

Methods for Identification of Phosphorylation Sites

In choosing a method for identifying phosphorylation sites, one must consider cost, time involved, equipment availability, technical expertise, and the amount of protein, that can be isolated. Most large research institutions provide core facilities for protein sequencing, but peptide purification is often the responsibility of the investigator. In the last few years, many departments have established mass spectrometry facilities with capabilities for the identification of phosphorylation sites in purified proteins. This section will outline briefly four general approaches that have been used to identify phosphorylation sites in nuclear receptors. The conventional protein sequencing and mass spectrometry techniques typically require more protein than the third and fourth methods. Although the equipment itself is very sensitive (often less than 1 pmol detection limit), much larger amounts of starting material are required due to the substoichiometric nature of most phosphorylations and the inevitable losses of material that occur during enzymatic digestion and peptide purification. Often, it is not feasible to obtain sufficient protein to use these approaches and, in these cases, approaches 3 and 4 described below are particularly valuable.

[39] M. A. Price and D. Kalderon, *Cell* **108**, 823 (2002).
[40] B. Chu, F. Soncin, B. D. Price, M. A. Stevenson, and S. K. Calderwood, *J. Biol. Chem.* **271**, 30847 (1996).

Conventional Isolation and Sequencing of Phosphopeptides[25,41,42]

In this approach, cells or tissue minces are incubated with $[^{32}P]H_3PO_4$, the receptor is purified from extracts, phosphopeptides are generated and then purified using one or more steps. The $[^{32}P]$ label is utilized primarily to monitor purification of the peptides. Gas-phase protein sequencing is used to determine the sequence of the peptide and to identify phosphorylated amino acids.[43] The main advantage of this approach is accuracy. The disadvantages include the need for access to costly equipment and skilled technicians, and the requirement for significant amounts of purified receptor (>pmol range). The same techniques for radiolabeling, isolation, and detection of phosphopeptides can be used to prepare peptides for direct protein sequencing in low-abundance proteins. The added requirements are that the peptides must be highly purified and free of not only other phosphopeptides but also of nonphosphorylated peptides. The addition of a metal affinity column, such as an Fe column, is particularly useful in separating phosphopeptides from unphosphorylated peptides,[44,45] although some nonphosphorylated, negatively charged peptides will also bind to the column. In addition, peptides must be purified in sufficient yield to permit protein sequencing. Individual core facilities can provide information on the amount of protein needed.

Mass Spectrometry[18,44,46]

This method can be used both for the detection of phosphopeptides and their sequencing and identification. In some cases, the investigator will be responsible only for preparing sufficient purified protein. In others, he/she will be expected to prepare and partially purify peptides. A major advantage of mass spectrometry is that the peptides do not have to be highly purified prior to analysis. The majority of the work as well as the analysis of the data is carried out by an expert in mass spectrometry. Thus, the technical aspects of the mass spectrometers and the analyses of the data are not described in

[41] J. E. Bodwell, E. Orti, J. M. Coull, D. J. Pappin, L. I. Smith, and F. Swift, *J. Biol. Chem.* **266**, 7549 (1991).

[42] S. F. Arnold, J. D. Obourn, H. Jaffe, and A. C. Notides, *Mol. Endocrinol.* **8**, 1208 (1994).

[43] P. J. Roach and Y. H. Wang, *Methods Enzymol.* **201**, 200 (1991).

[44] T. A. Knotts, R. S. Orkiszewski, R. G. Cook, D. P. Edwards, and N. L. Weigel, *J. Biol. Chem.* **276**, 8475 (2001).

[45] D. C. Neville, C. R. Rozanas, E. M. Price, D. B. Gruis, A. S. Verkman, and R. R. Townsend, *Protein Sci.* **6**, 2436 (1997).

[46] V. B. Chanta (ed.), "Current Protocols in Protein Science," Chapter 16. John Wiley & Sons, Inc, Hoboken, NY, 2002.

detail here. There are several types of mass spectrometers, and the techniques utilized to detect phosphorylation sites will depend upon the equipment available. Two common machines, electrospray (triple quadrupole, ion trap, quadrupole-TOF) and MALDI-TOF mass spectrometers will be discussed here. Investigators should consult with the director of the core facility prior to preparing samples. Because peptides are stable, it is possible to have samples analyzed commercially or at a distant site if facilities are not available locally.

Mass spectrometers measure the mass–charge ratio (m/z) of a molecule. Peptides must be volatilized in order to be detected by the machine. MALDI-TOF mass spectrometers generally are capable of detecting larger peptides than are electrospray machines. Typically, purified protein will be treated with no more than 5% (w/w) of a proteolytic enzyme, such as trypsin,[44] to produce peptides that can be analyzed. In the case of MALDI-TOF analyses, a second aliquot of protein should be treated with a general phosphatase, such as bacteriophage lambda phosphatase,[46] prior to digestion. A comparison of the m/z ratios of the peptides generated $+/-$ phosphatase treatment will show a loss of a specific peptide with a concomitant appearance of a peptide with a m/z ratio 79 units smaller corresponding to the loss of a phosphate group for each phosphopeptide. For electrospray analyses, the peptides can be analyzed to identify peptides with precursor ions with a characteristic loss of a mass of 79 indicating a phosphopeptide. Note that very large peptides (> 20 amino acids) may not be detected in the electrospray analysis and it may be necessary to use additional proteolytic enzymes to produce smaller peptides. In both cases, the m/z ratio of the peptide is frequently sufficiently unique to identify the peptide containing the phosphorylation site within the known sequence. If the m/z ratio does not yield an unambiguous identification and/or there is more than one candidate phosphorylation site (Ser, Thr, or Tyr), further analysis is necessary. In some cases, the peptide is sufficiently well resolved from other peptides that a quadrupole–ion trap–TOF electrospray tandem mass spectrometer can be used directly to select the phosphopeptide for subsequent fragmentation and analysis of the sequence. More often, the mixture is too complex and partial purification is necessary. Some electrospray mass spectrometers are linked to an HPLC and prefractionation on the HPLC provides a sufficiently purified sample for on-line MS analysis or subsequent analysis of collected fractions. A convenient alternative is to pass the digest through a metal affinity column that specifically binds phosphopeptides.[45] The eluted material is highly enriched for the phosphopeptides and can be used for identification of the phosphorylation within the peptide by mass spectrometry.

Indirect Approach[15,47,48]

Receptor deletion and point mutations can be used to identify phosphorylation sites indirectly. In general, a first step is to express receptor deletion mutants in cells to sublocalize phosphorylation sites. After incubation of cells with $[^{32}P]H_3PO_4$, receptor deletion mutants are purified and phosphorylation detected by SDS PAGE and autoradiography. Once phosphorylation sites are sublocalized, then a second step involving point mutation of candidate phosphorylation sites to alanines is performed. Each mutant is expressed separately in cells and the $[^{32}P]$-labeled receptor visualized by SDS-PAGE and autoradiography. In cases of multiple phosphorylation sites, an additional phosphopeptide mapping step (see section "Approach for use with low protein levels") may be needed to determine whether a specific mutation has eliminated a phosphorylation site. The need for relatively low levels of $[^{32}P]H_3PO_4$ and small amounts of purified receptor represent the greatest benefits of this approach. The obvious disadvantages are those inherent with drawing conclusions based on deletion and point mutations in proteins, including misfolding and nonphysiological subcellular localization of receptors. This approach is useful provided the results are validated with full-length nonmutated receptor. Techniques for labeling, purification, and peptide mapping are described below.

Approach for Use with Low Protein Levels[17,18,49–52]

For most experiments, the limiting parameter of studies aimed at identifying authentic *in vivo* phosphorylation sites is the amount of receptor that can be purified. An approach has been developed to identify *in vivo* phosphorylation sites when purified receptor is limiting (less than a picomole). Similar to the conventional approach, this method uses high levels of $[^{32}P]H_3PO_4$ to label cells or tissue minces, and the substitution of a modified manual Edman degradation procedure for gas-phase sequencing to eliminate the necessity for large amounts of receptor and for highly purified

[47] S. Ali, D. Metzger, J. M. Bornert, and P. Chambon, *EMBO J.* **12**, 1153 (1993).

[48] P. Le Goff, M. M. Montano, D. J. Schodin, and B. S. Katzenellenbogen, *J. Biol. Chem.* **269**, 4458 (1994).

[49] Y. Zhang, C. A. Beck, A. Poletti, D. P. Edwards, and N. L. Weigel, *J. Biol. Chem.* **269**, 31034 (1994).

[50] Y. Zhang, C. A. Beck, A. Poletti, D. P. Edwards, and N. L. Weigel, *Mol. Endocrinol.* **9**, 1029 (1995).

[51] C. A. Beck, Y. Zhang, M. Altmann, N. L. Weigel, and D. P. Edwards, *J. Biol. Chem.* **271**, 19546 (1996).

[52] G. M. Hilliard, R. G. Cook, N. L. Weigel, and J. W. Pike, *Biochemistry* **33**, 4300 (1994).

peptides. Modified manual Edman degradation is combined with phosphoamino acid analysis and secondary proteinase digestion to identify phosphorylation sites. Although this approach is accurate, large and hydrophobic phosphopeptides are not amenable to modified manual Edman degradation.

The following sections describe specific procedures used for identifying phosphorylation sites that requires small amounts of protein. As indicated earlier, some of these techniques are also used in identifying sites when ample protein is available. Several of the procedures are updated and modified from a previous chapter by the authors.[53]

Labeling Proteins for Phosphorylation Analysis

In Vivo *Labeling of Cells and Tissue Minces*

Endogenous nuclear receptors or receptors expressed by transfection can be labeled with [^{32}P]phosphate by incubation of cells or tissue minces with [^{32}P]H$_3$PO$_4$. The number of cells/amount of tissue and the conditions for labeling (quantity of [^{32}P]H$_3$PO$_4$ and the duration of labeling) will depend upon whether the goal of the study is to identify all phosphorylation sites, or simply to detect sites whose phosphorylation is regulated by a specific signal.

For identification of phosphorylation sites, uniform labeling (of at least 8 hr), high levels of [^{32}P]H$_3$PO$_4$ (up to 3 mCi/ml of medium) and sufficient receptor to provide a strongly radiolabeled receptor band are required. The number of cells will depend upon the expression level of the receptor and on the stoichiometry of the phosphorylation. To detect very low stoichiometry phosphorylations, more cells and higher levels of [^{32}P]H$_3$PO$_4$ are needed than for higher stoichiometry sites. More material will also be required for site identification than for simple peptide mapping. Nuclear receptors are ordinarily not expressed at high levels. As many as 5×10^7 cells may be needed to obtain enough labeled protein for phosphorylation site identification.[54] Labeling will be more efficient in cell culture. Some information can be obtained from radiolabeling tissue minces, but these typically cannot be maintained in cell culture for more than a few hours.[25]

If the goal is to examine acute effects on receptor phosphorylation (e.g., effects of growth factors or fast acting drugs), where rapid turnover of phosphorylation may occur, then cells/tissue should be harvested and

[53] B. G. Rowan and N. L. Weigel, *in* "The Nuclear Receptor Superfamily, A Practical Approach" (D. Picard, ed.), Chapter 6. Oxford University Press, Oxford, UK, 1999.
[54] B. G. Rowan, N. L. Weigel, and B. W. O'Malley, *J. Biol. Chem.* **275**, 4475 (2000).

receptor purified, a short time after addition of agents to the medium. If there is no information about whether the agent is fast or slow acting, then a time course of agent addition should be performed with subsequent analysis of phosphopeptides (see next section).

Procedure 1

In Vivo Labeling of Nuclear Receptors with [^{32}P]H$_3$PO$_4$ (See Important Safety Issues Under Comments and Critical Points)

1. Culture cells at 37°C, 5% CO$_2$, 95% O$_2$ in a medium containing 5% dialyzed fetal calf serum (FCS) (Hyclone) that is stripped of endogenous steroid by treatment with dextran-coated charcoal. Incubate cells for 24 hr prior to labeling with [^{32}P]H$_3$PO$_4$.

2. During log phase growth, remove the medium and replace it with serum-free, phosphate-free medium [without sodium pyruvate and sodium phosphate (Gibco/BRL)]. Incubation for 1 hr is sufficient to deplete the culture of excess free phosphate.

3. After 1 hr incubation at 37°C, 5% CO$_2$, 95% O$_2$, remove the medium from step 2 and replace it with additional phosphate-free medium supplemented with 1% dialyzed FCS. Use the minimal amount of medium required to maintain cells at log phase growth.

4. Add [^{32}P]H$_3$PO$_4$ (0.5 mCi/ml, ICN) to the medium. The amount of isotope added will depend on the type of study, but typically will range from 0.1 to 3 mCi/ml. Incubate the cells/tissue at 37°C, 5% CO$_2$, 95% O$_2$ for 1 hr.

5. At this point hormones or reagents under study can be added to the medium. It is important to wait for 1 hr after adding the [^{32}P]H$_3$PO$_4$ to allow time for cells to incorporate [^{32}P]PO$_4$ into cellular phosphate pools. This brief preincubation can magnify changes in regulated sites relative to constitutive sites if incubation times are short. Incubation time with [^{32}P]H$_3$PO$_4$ will depend on the type of study being performed as outlined above and will range from 1 to 15 hr. Uniform labeling of receptors can require as much as an 8 hr incubation of cells/tissue minces with [^{32}P]H$_3$PO$_4$, and should be used for identification of both constitutive and regulated sites. For transient phosphorylations, signaling molecules should be added near the end of the labeling time.

6. Carefully remove the [^{32}P]H$_3$PO$_4$-containing medium from the cells. Avoid aerosolization by slowly pipetting off the medium. Add an equal volume of cold PBS (4°C) to the cells to remove traces of radioactive medium. Remove and discard the PBS wash. Harvest the cells from the culture dishes with a disposable scraper and/or by centrifugation in a small volume of cold PBS.

7. Centrifuge to pellet the cells and to remove the PBS supernatant. The pelleted cells are ready for receptor purification as described below.

Comments and Critical Points

a. The mCi amounts of $[^{32}P]H_3PO_4$ used during *in vivo* labeling may be 1000 times higher than the levels of isotope used in a typical probe labeling for Southern/Northern blots. Extra precautions are needed to ensure adequate protection of the user and coworkers. Avoid aerosolization of $[^{32}P]H_3PO_4$ at every step. Adequate eye protection, beta ray-blocking apron, double gloves, barrier tips, and shielding are needed. The user should inform the radiation safety officer of the amount of isotope used and arrange for appropriate license, extra training, and issuance of additional radiation badges. Prior to labeling, proper disposal procedures should be in place, and defined bench space, centrifuges, tissue culture hoods etc., should be marked and checked before and after each experiment with a Geiger counter.

b. Adding $[^{32}P]H_3PO_4$ to cells past log phase growth results in poor uptake of label and subsequent low incorporation of isotope into proteins.

c. For long-term steady-state labeling, high levels of $[^{32}P]H_3PO_4$ may cause genomic damage to cells.

d. To measure the stoichiometry of protein phosphorylation at specific sites, double labeling of cells with $[^{32}P]H_3PO_4$ and $[^{35}S]$methionine can be performed if the exact stoichiometry is of concern.[55]

In Vitro *Labeling of Nuclear Receptors*

In vitro phosphorylation also may be used to identify phosphorylation sites in nuclear receptors. *In vitro* phosphorylation is resorted to when experimental evidence indicates that activation/inhibition of a specific kinase alters one or more functional properties of a nuclear receptor. Identified sites can then be verified, *in vivo*, by mutation to alanine, expression of the mutated receptor in cells, and comparison of the phosphopeptide maps of *in vitro* and *in vivo* phosphorylated receptors. Sites that exhibit significant stoichiometry *in vitro* (> 10%) are typically good candidates for *in vivo* sites.

[55] E. Orti, L. M. Hu, and A. Munck, *J. Biol. Chem.* **268**, 7779 (1993).

In vitro phosphorylation of receptors is easier to perform, less time consuming, and uses far less radioactive isotope than *in vivo* phosphorylation; however, a source of purified protein is needed as a substrate. Because purified protein is used and because of the higher specific activities inherent with $[\gamma\text{-}^{32}P]$ATP, more $[^{32}P]$phosphate is incorporated into proteins during *in vitro* phosphorylation. Although significant $[^{32}P]$phosphate incorporation is desirable for site identification, very high levels of kinase and substrate may lead to artifactual phosphorylation at sites that are not phosphorylated *in vivo*.

Figure 1 shows an *in vitro* phosphorylation experiment using purified human PR_B and the cyclinA–cdk2 kinase complex. Several controls should be included always during *in vitro* phosphorylation studies. The reaction in lane 1 [(−) kinase, (+) receptor] shows background PR_B phosphorylation, and indicates that a small amount of a copurifying contaminant kinase was present in the purified PR_B preparation. The reaction in lane 3 [(+) kinase, (−) receptor] shows phosphorylation of the cdk2 partner protein, cyclinA. Many kinases autophosphorylate or phosphorylate a partner protein resulting in a phosphorylated protein on the autoradiograph. Not included in Fig. 1 is a set of two other recommended controls using a known substrate (histone H1) of cyclinA–cdk2. One of these samples is incubated with cyclinA–cdk2 and histone H1 to ensure that the kinase is active. The second of these samples includes cyclinA–cdk2, histone H1, and PR_B. If the level of histone H1 phosphorylation is the same for both samples, then the PR_B sample must be free of contaminating phosphatases, kinase

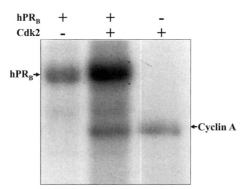

FIG. 1. *In vitro* phosphorylation of human PR_B with cyclinA–cdk2. Human PR_B and cyclinA–cdk2 both purified from baculoviral expression systems were used in *in vitro* phosphorylation reactions containing $[\gamma\text{-}^{32}P]$ATP, nonradioactive ATP and 10X cyclinA–cdk2 kinase reaction buffer as outlined in Procedure 2. Lane 1 is an *in vitro* phosphorylation reaction containing human PR_B alone. In lane 2 both PR_B and cyclinA–cdk2 kinase were included in the reaction. Lane 3 is a reaction with cyclinA–cdk2 kinase alone.

inhibitors, or inhibitory salts. Lane 2 shows a reaction with both PR_B and cyclinA–cdk2 kinase. There is a large increase in the phosphorylation of PR_B indicating that cyclinA–cdk2 phosphorylates the receptor.

Procedure 2

In vitro Phosphorylation of Nuclear Receptors with Purified Kinase

The following protocol uses a final volume of 40 μl per reaction. Before beginning *in vitro* phosphorylation, one should calculate the specific activity of ATP to be used in the assay. The choice of final specific activity will depend upon the amount of receptor used. Typically, a specific activity of ATP of 30,000 dpm/pmol is used.

1. Prepare 40 mM ATP and 10 $\times$ kinase reaction buffer (both can be stored in aliquots at $-70°$C). The composition of the buffer will depend on the kinase. Dilute the 40 mM ATP to 0.4 mM with dH_2O just prior to the experiment.
2. Mix 1–10 μl of 0.4 mM ATP with 1–3 μl of 10 mCi/ml [γ-^{32}P]ATP (DuPont NEN).
3. In a separate tube, mix 4 μl of 10 $\times$ kinase reaction buffer, 0.01–10 pmol of purified receptor, and the purified kinase. Add dH_2O to make a total volume of 32 μl.
4. Mix the solution from step 2 with the solution from step 3 and incubate at 30°C for 1 hr (the time and temperature is dependent on the kinase and should be determined empirically).
5. The reaction is stopped by adding SDS PAGE sample buffer and heating at 100°C for 5 min.
6. Resolve samples on an SDS PAGE gel; cut off the dye front containing the free [^{32}P] ATP. Perform autoradiography on the wet gel and isolate the phosphorylated receptor band as described in Procedure 3. From the Cerenkov counts in the protein (count the acrylamide slice in an empty vial without scintillation fluid), the specific activity of [γ-^{32}P]ATP incorporated into the receptor can be determined.
7. The phosphorylated receptor can be excised and used to identify modification sites.

Comments and Critical Points

a. Many kinases and some nuclear receptors may be obtained commercially in purified form. Other sources include cell tissue extracts, bacterial or baculoviral expression systems, and *in vitro* translation.

b. The amount of kinase used during *in vitro* phosphorylation should
    be titrated to assess maximal [^{32}P] incorporation into the receptor.
c. Receptor or kinase purified on Protein A Sepharose beads by
    immunoprecipitation may be used directly in the *in vitro* phospho-
    rylation reaction if purified components are not available.

Calculations of Specific Activity and % Incorporation for In Vitro
Phosphorylation

In this hypothetical example, the following assumptions are made:

- Amount of receptor protein: 1.5 pmol
- Cerenkov counting of the receptor-containing gel slice: 15,000 cpm
- Efficiency of scintillation counter: 50%
- Only one site in the receptor is phosphorylated by the kinase.

1. Calculate the specific activity of [γ-^{32}P]ATP:
    Moles of ATP: 5 μl of 0.4 mM ATP = 2000 pmol ATP
    Total dpm: 3 μl of 10 μCi/μl [γ-^{32}P]ATP = 30 μCi $\times$ 2.2 $\times$ 10^6
                     dpm/μCi = 6.6 $\times$ 10^7 dpm
    Specific activity of [γ-^{32}P]ATP:

    6.6 $\times$ 10^7 dpm/2000 pmol = 33,000 dpm/pmol ATP

2. Estimate the pmoles of receptor (1.5 pmol). The amount of receptor
    can be estimated by silver staining an SDS PAGE gel containing
    purified receptor and standard marker proteins of known mass or by
    standard protein measurement techniques such as the Bradford assay.
3. Specific activity incorporated in receptor. Following autoradiog-
    raphy, the receptor is excised from the SDS PAGE gel and counted
    by Cerenkov counting (15,000 cpm). Determine the efficiency
    correction factor for the scintillation counter used (correction factor
    of 2 for 50% efficiency) and multiply the Cerenkov counts by the
    correction factor:

    Gel slice: 15,000 cpm $\times$ 2 = 30,000 dpm/1.5 pmol receptor

    = 20,000 dpm/pmol receptor

4. % Incorporation of [^{32}P]PO$_4$ in the receptor:

$$\% \text{ Incorporation} = \frac{\text{specific activity incorporated in receptor}}{\text{specific activity of ATP}}$$

$$= \frac{20,000 \text{ dpm/pmol receptor}}{33,000 \text{ dpm/pmol ATP}} \times 100 = 61\% \text{ or } 0.61 \text{ mol/mol}$$

Comments and Critical Points

a. Common errors are to use too high a specific activity of ATP. This excess results in a strong signal even though the % incorporation into specific phosphorylation sites is low. Other errors include omission of crucial control samples that provide information about potential contaminating kinases, phosphatases, kinase inhibitors, and autophosphorylation.

b. Using the proper control samples as described above and in Fig. 1 will help determine whether a low % incorporation (a few %) is due to contaminating kinase inhibitors or phosphatases, too little kinase in the reaction or endogenous receptor phosphorylation. In the absence of procedural problems, a low % incorporation suggests the receptor is not a substrate for the kinase.

c. Since the mass of $[\gamma\text{-}^{32}P]ATP$ used is small, it does not contribute significantly to the total moles of ATP calculated in step 1.

d. 2.2×10^6 dpm/μCi is a constant.

e. Before determining % incorporation, a two-dimensional phosphopeptide map of the *in vitro* phosphorylated receptor should be prepared to estimate the number of sites phosphorylated by the kinase (see below). Each spot on the two-dimensional map can be counted by Cerenkov counting and the % incorporation for each site determined. One caveat is the possibility of partially digested peptides that could result in one phosphorylation site being represented by more than one spot on the map.

Purification of Nuclear Receptors

With both *in vitro* and *in vivo* phosphorylation experiments, it is important to devise a receptor purification strategy that yields highly purified protein. Even trace levels of proteases, phosphatases or kinase inhibitors may seriously limit recovery of purified, phosphorylated receptor. For cell culture studies in general, immunoprecipitation with high affinity antibodies yields the best results with very little contaminating protein. If high affinity antibodies are not available, then receptor expression vectors can be prepared with epitope tags (2 × FLAG tag on the N-terminus seems to work best) and commercially available antibodies can be used.[18,56] Extraction buffers should contain standard protease inhibitors (1 μl/ml of each of the following from a 1 mg/ml stock in DMSO: leupeptin, antipain, aprotinin, benzamidine–HCl, chymostatin, pepstatin; 2.5 μl/ml of PMSF

[56] M. R. Walters, M. Dutertre, and C. L. Smith, *J. Biol. Chem.* **277**, 1669 (2002).

from a 200 mM stock in dioxane), phosphatase inhibitors (50 mM sodium fluoride, 10 mM sodium vanadate), chelating agents (1 mM EDTA and EGTA), and sufficient detergent and ionic strength to reduce background contamination during immunoprecipitation. Proteolysis is a common problem during immunoprecipitation even when protease inhibitors are included. To circumvent degradation, a denaturing extraction with 8 M urea can be used with immunoprecipitation provided the antibody recognizes the denatured protein.[54]

Following immunoprecipitation, the phosphorylated receptor is electrophoresed using standard SDS-PAGE. The wet gel is marked and exposed to X-ray film for autoradiography and the receptor is excised from the gel using a clean razor blade. Following Cerenkov counting, the gel slice is ready for generation of phosphopeptides.

Preparation of Phosphopeptides

Following excision of the phosphorylated receptor band from the SDS-PAGE gel, the slice is subjected to protease digestion to generate phosphopeptides. Proteases can be incubated directly with the gel slice containing the receptor, or subsequent to receptor electroelution. In terms of ease and time, it is more convenient to directly incubate protease with the gel slice. One can recover $> 70\%$ of the total Cerenkov counts in the eluted phosphopeptides.

Ideally, a proteolytic method should generate small phosphopeptides (< 20 amino acids) in which most of the phosphorylation sites are separated. Trypsin is the protease of choice since it is inexpensive, and cleaves at multiple sites [on the C-terminal side of every arginine (R) and lysine (K) residue]. A list of different enzymes and reagents used to generate phosphopeptides is available.[57]

Procedure 3

Trypsin Digestion of Nuclear Receptor in a Gel Slice
1. Following electrophoresis, the receptor identity should be confirmed by Western blot in a parallel lane.
2. Leave the wet gel on the electrophoresis glass plate, mark the orientation and wrap the gel in plastic wrap. Expose the gel to X-ray film (X-OMAT AR, Kodak) until the phosphorylated protein is visualized.

---

[57] P. van der Geer and T. Hunter, *Electrophoresis* **15**, 544 (1994).

3. Using the X-ray film as a guide, cut out the phosphorylated receptor with a clean razor blade and place the gel slice in a siliconized 1.5 ml microfuge tube.

4. Add 1 ml of 50% methanol and incubate the tube for 1 hr at room temperature with gentle rocking to remove SDS, Tris and glycine from the gel slice.

5. Remove the methanol and add 0.5 ml of HPLC-grade water. Rock for 1 min at room temperature. Repeat one more time and remove the water. This step removes the methanol.

6. Determine the total counts in the gel slice by Cerenkov counting.

7. Add 0.5 ml of freshly prepared 50 mM ammonium bicarbonate (in HPLC grade water) to the gel slice. Add 1–10 μg of trypsin and incubate in a 37°C water bath. Trypsin is prepared at 1 mg/ml in 50 mM ammonium bicarbonate.

8. Add an additional 1–10 μg of trypsin every 4 hr over a 12 hr period.

9. Remove the supernatant to a fresh siliconized microfuge tube. Add 0.5 ml of HPLC water to the gel slice, and combine this wash with the original supernatant.

10. Dry down the eluted tryptic phosphopeptides using a Speed Vac and determine radioactivity by Cerenkov counting.

11. The released phosphopeptides are now ready for phosphoamino acid analysis, separation by reversed phase HPLC, or two-dimensional phosphopeptide mapping (see below).

Comments and Critical Points

a. Use sequencing grade trypsin modified by reductive alkylation (Promega Corp., Madison, WI). The advantage of this modified trypsin over unmodified forms is a marked reduction in autolysis resulting in more complete cleavage of receptors and a reduction in the amount of trypsin peptides that can contribute to the background during HPLC separations.

b. Trypsin cleaves inefficiently at R and K residues that are followed by proline. The possibility of incomplete tryptic digests must be considered when identifying phosphorylation sites. Other protease digestions and/or secondary protease digestions may resolve the problem of incomplete digests.

c. Although the amount of trypsin used should be determined empirically, a balance is needed between too little trypsin (causing incomplete digestion) versus too much trypsin (interferes with subsequent analyses). For proteins in solution, only 5% (mol/mol) of trypsin is required for complete digestion.

Separation of Phosphopeptides

Three general procedures are used to separate phosphopeptides. Reversed phase HPLC separates phosphopeptides based on the strength of interaction with a hydrophobic matrix. Two-dimensional phosphopeptide mapping separates peptides based on charge/mass ratio and hydrophobic character. Alkaline polyacrylamide gels (no SDS) also separate peptides based on charge/mass ratios. Very large phosphopeptides (>40 amino acids) may be separated by tricine gels.[58] The choice of the separation procedure will depend upon the resolution needed, reproducibility, sensitivity required, the total mass of protein, and the availability of equipment. The advantages and disadvantages of several procedures are outlined in Table I.

Procedure 4

Reversed Phase HPLC (Fig. 2)
1. Resuspend the dried phosphopeptides in 50% formic acid in HPLC grade water (100–200 μl), vortex well and centrifuge the sample at 12,000 $\times g$ to remove particulates (Fig. 2).
2. Set up an HPLC system with a C_{18} reversed phase column and equilibrate the system with 0.1% trifluoroacetic acid (TFA)

TABLE I

PROCEDURES FOR SEPARATION OF PHOSPHOPEPTIDES

Separation technique	Advantages	Disadvantages
C_{18} Reversed phase HPLC	rapid; reproducible; can accommodate large protein load [e.g., large amounts of trypsin (up to 1 mg) used for digestion]	lower sensitivity and requires specialized equipment; some phosphopeptides may elute in the same fractions (can be overcome by running fractions on alkaline polyacrylamide gels as a second step)
Alkaline polyacrylamide gels	very sensitive for detection of low levels of [^{32}P]phosphopeptide; highly reproducible; inexpensive	poor resolution of complex mixture of phosphopeptides (best used as a second step for HPLC separation if there are many sites)
Two-dimensional phosphopeptide mapping	very high resolution and sensitivity (most phosphopeptides can be resolved); can accommodate multiple samples	less reproducible than other procedures; there is a limit to the total protein mass that can be spotted onto the plates (<40 μg); requires specialized equipment

[58] H. Schagger and G. Jagow, *Anal. Biochem.* **166**, 368 (1987).

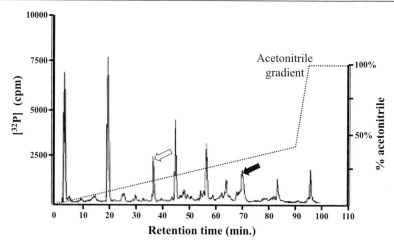

Fig. 2. Reversed phase HPLC profile of tryptic phosphopeptides of human PR_B phosphorylated *in vitro* with cyclinA–cdk2. Purified baculoviral expressed PR_B was incubated with purified cyclinA–cdk2 and [γ-^{32}P]ATP in an *in vitro* phosphorylation reaction. Following separation by SDS-PAGE, the phosphorylated PR_B band was excised from the gel and subjected to extensive trypsin digestion. Tryptic PR_B phosphopeptides were separated by HPLC on a Vydac C18 reversed phase column using an acetonitrile gradient to elute the phosphopeptides from the column. Phosphopeptides were detected with an on-line radio-activity detector. The left axis indicates [^{32}P] CPM eluted and the right axis indicates the % acetonitrile in the elution buffer (dashed line). The open arrow indicates the phosphopeptide peak subjected to modified manual Edman degradation (see Fig. 5) and the closed arrow indicates the phosphopeptide peak subjected to secondary proteinase digestion and alkaline polyacrylamide gel electrophoresis (see Fig. 3).

in HPLC grade water at a flow rate of 1 ml/min. A guard column placed in line before the C_18 column will protect against particulate matter.

3. Inject the sample into the HPLC and immediately begin a 0–100% gradient of acetonitrile containing 0.1% TFA at 1% per minute. Shallower gradients can be used to improve resolution if needed, but this gradient is the best choice for an initial analysis. Phosphopeptide peaks can be detected with an on-line radioactivity detector.

4. Dry down individual 1–2 ml fractions in a Speed Vac and measure radioactivity in the dried fractions by Cerenkov counting. These fractions can be used for manual Edman degradation, secondary protease digestion, or phosphoamino acid analysis. If there is enough mass, the fractions can also be analyzed by direct protein sequencing or mass spectrometry; however, additional purification steps are often required to obtain peptides that are sufficiently purified for

direct sequencing. A metal affinity column is frequently a good choice since it preferentially binds phosphopeptides.[45]

Procedure 5

Alkaline Polyacrylamide Gel Electrophoresis[59]

Figure 3 shows an autoradiograph of an alkaline polyacrylamide gel. The [^{32}P] peak that eluted from the HPLC in the 70 min fraction (Fig. 2, closed arrow) was electrophoresed in lane 1.

Solutions:

- Resolving gel (final concentrations): 40% acrylamide (from a 60% stock), 0.037% bis-acrylamide (from a 2.5% stock), 0.75 M Tris, pH 8.8 (from a 3 M stock), 0.035% TEMED, 0.1% ammonium persulfate (from a 10% stock), deionized water to volume.
- Stacking gel (final concentrations): 3.3% acrylamide (from a 60% stock), 0.16% bis-acrylamide (from a 2.5% stock), 0.125 M Tris, pH 6.8 (from a 2.5 M stock), 0.05% TEMED, 6 M urea, 0.1% ammonium persulfate (from a 10% stock), deionized water to volume.
- Reservoir buffer (no pH): 0.05 M Tris base (from solid), 0.4 M glycine (from solid).
- Sample buffer (final concentrations): 0.125 M Tris, pH 6.8 (from a 2.5 M stock), 6 M urea, 0.01% methylene blue. Sample buffer should be prepared fresh.

1. Prepare resolving and stacking gel in a polyacrylamide gel apparatus with 15 × 18 cm (height × width) glass plates, 0.5 mm spacers and 0.5 mm combs. Add TEMED and ammonium persulfate immediately prior to pouring gel. Cast the resolving gel first about 10 cm in height, overlay with water and let the gel polymerize. Remove the overlay of water and pour the stacking gel on top of the separating gel (about 5 cm in width) and insert the comb.
2. Resuspend the dried phosphopeptides with sample buffer and transfer to individual wells of the gel. Overlay samples with reservoir buffer.
3. Electrophorese at 5–10 mA per gel in reservoir buffer for 10–16 hr or until the methylene blue tracking dye migrates about 2/3 the length of the gel.
4. With the gel still on one of the glass plates, place a piece of Whatman 3MM filter paper on the gel and pull the Whatman paper and the attached gel away from the glass plate.

[59] M. H. P. West, R. S. Wu, and W. M. Bonner, *Electrophoresis* **5**, 133 (1984).

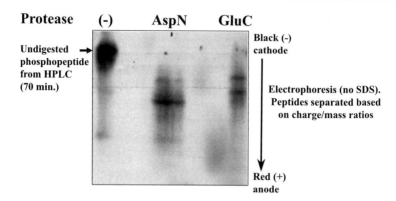

FIG. 3. Alkaline polyacrylamide gel electrophoresis of HPLC fractions from phosphopeptides of human PR$_B$ phosphorylated *in vitro* with cyclinA–cdk2. The control aliquot was resuspended in buffer, another aliquot was resuspended in buffer + endoproteinase Asp-N and the third aliquot was resuspended in buffer + endoproteinase Glu-C. Following incubation for 4 or 8 hr at 37°C, all three aliquots were dried in a Speed Vac, resuspended in sample buffer and electrophoresed by alkaline polyacrylamide gel electrophoresis. The gel was dried and exposed to X-ray film for autoradiography.

5. Dry the gel using a dryer for 1.5 hr with heat. Mark the orientation of the dried gel and expose the gel to X-ray film for autoradiography.

Comments and Critical Points

a. Following autoradiography, individual phosphopeptide bands can be eluted from the gel for subsequent analysis. Excise the phosphopeptide gel slice with a clean razor and place in a 1.5 ml siliconized microfuge tube. Wash the gel slice 2 × 5 min with 0.5 ml 50% methanol in HPLC water to remove electrophoresis buffer and then elute the phosphopeptide in 0.5 ml HPLC water overnight with gentle rocking at room temperature. The eluted phosphopeptides can be dried down in a Speed Vac, subjected to Cerenkov counting and used for gas phase sequencing, manual Edman degradation, secondary protease digestion, or phosphoamino acid analysis.

Procedure 6

Two-Dimensional Phosphopeptide Mapping (Fig. 4)

1. Resuspend the dried phosphopeptides with 0.5 ml of HPLC water, vortex and dry the sample in a Speed Vac. Repeat this step once more to remove ammonium bicarbonate from the tryptic digest (Fig. 4).

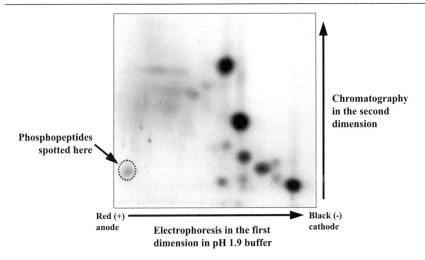

FIG. 4. Two-dimensional phosphopeptide map of phosphopeptides of human PR$_B$ phosphorylated *in vitro* with cyclinA–cdk2. PR$_B$ phosphopeptides were prepared as described in Fig. 2. Phosphopeptides were spotted onto a 20 × 20 cm cellulose plate and electrophoresed in the first dimension at 1000 V for 45 min using the Hunter HTLE 7000 electrophoresis system. The plate was dried and phosphopeptides were separated in the second dimension by chromatography followed by autoradiography to visualize the phosphopeptides.

2. Add 200 μl of pH 1.9 buffer [2.5% formic acid (88% stock), 7.8% glacial acetic acid in dH$_2$O], vortex, and centrifuge at 12,000 × g. Transfer the supernatant to another tube and dry in a Speed Vac.

3. Resuspend the dried sample in 10–15 μl of pH 1.9 buffer and vortex. Apply approximately 1–2 μl of sample as a single spot (about 1 mm diameter) on a 20 × 20 cm × 100 μm thin layer cellulose plates (EM Science). The sample should be spotted towards the bottom, left corner of the plate approximately 3 cm from the bottom edge and 3–6 cm from the left edge. Since most phosphopeptides are positively charged in pH 1.9 buffer, the bottom left of the plate should be placed toward the positive electrode (anode; red) of the electrophoresis unit. During electrophoresis, the phosphopeptides will migrate toward the negative electrode (cathode; black).

4. Cut a double layer square (4 × 4 cm) of Whatman 3MM paper with a 1 cm diameter circle cut out of the middle. Soak in pH 1.9 buffer and center the open circle of the paper over the spotted phosphopeptides. Gently press the edges of the paper so the buffer concentrates the phosphopeptide spot. Subsequent to this, uniformly wet the remaining plate using a larger piece of Whatman 3MM paper soaked in pH 1.9 buffer.

A

hPR_B tryptic phosphopeptide
from HPLC (38 min.)

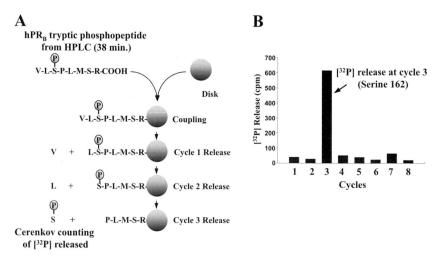

B

Fig. 5. Modified manual Edman degradation of a phosphopeptide of human PR_B phosphorylated *in vitro* with cyclinA–cdk2. The HPLC fraction containing the phosphopeptide peak at 38 min in Fig. 2 (open arrow) was dried in a Speed Vac and subjected to the modified manual Edman degradation to identify the position of the phosphorylated amino acid. (A) The phosphopeptide containing the serine 162 residue was coupled to a disc through the carboxyl terminus and subjected to sequential cycles of Edman degradation. (B) [32P] release data for this phosphopeptide shows a large release of [32P] by Cerenkov counting indicating that the phosphorylated amino acid resides three amino acids from the N-terminus of the phosphopeptide. This information was combined with other data to identify serine 162.

5. Electrophorese the sample in the first dimension at 1000 V for 20–50 min using the HTLE 7000 two-dimensional peptide gel apparatus (CBS Scientific).

6. Allow the plate to dry in a fume hood for 1 hr. Place the plate vertically (phosphopeptide side down) in a chromatography tank (30 cm length × 5 cm width × 30 cm height) containing chromatography buffer (37.5% *N*-butanol, 25% pyridine, 7.5% glacial acetic acid in dH_2O, all v/v) approximately 1.5 cm deep. The phosphopeptides should not be below the level of the chromatography buffer. The tank should be lined with Whatman 3M paper saturated with chromatography buffer to create a humidity chamber prior to inserting the plate. Place the lid on the tank and allow the buffer to rise on the plate until the front is 1–2 cm from the top of the plate (approximately 6–8 hr).

7. Dry the plate (approximately 1 hr), mark the orientation and expose to X-ray film for autoradiography.

Comments and Critical Points

a. For a more detailed description of this procedure and a list of several other electrophoresis and chromatography buffers, see van der Geer and Hunter.[57] Although the buffers listed are appropriate for most phosphopeptides, some phosphopeptides may be resolved more easily with other buffer choices.

b. Streaking of phosphopeptides in the first dimension or failure of the majority of phosphopeptides to leave the origin may indicate that the sample was overloaded, or that the pH of the sample was significantly different from pH 1.9.

c. Performic acid treatment may be used to oxidize phosphopeptides to eliminate oxidation-state isomers that can result in altered migration of phosphopeptides during second dimension chromatography.[57]

d. Phosphopeptide spots can be excised from the plates by scraping the cellulose with a small spatula and collecting the particles in a 1.5 ml siliconized microfuge tube. The phosphopeptides are eluted from the cellulose by adding 100–200 μl of 50% acetonitrile, 0.1% TFA (in HPLC water) and vortexing 30 sec. Following centrifugation for 15 sec to pellet the cellulose, the supernatant is transferred to a fresh tube, dried in a Speed Vac and counted by Cerenkov counting. The dried phosphopeptide can be used for gas phase sequencing, manual Edman degradation, secondary protease digestion, or phosphoamino acid analysis.

Secondary Proteinase Digestion

Secondary proteinase digestion is used to determine whether specific residues are present within the phosphopeptide. The endoproteinase Asp-N cleaves on the N-terminal side of most aspartate residues, and endoproteinase Glu-C cleaves on the C-terminal side of most glutamate residues.[60,61] A comparison of the migration of the Asp-N and/or Glu-C digested phosphopeptide with the undigested phosphopeptide on alkaline polyacrylamide gels will indicate whether the phosphopeptide contains aspartate and/or a glutamate residues, respectively. In Fig. 3, the [^{32}P] peak that eluted from the HPLC column at 38 min (Fig. 2, open arrow) was prepared for Asp-N and Glu-C digestion. Lane 1 shows migration of undigested phosphopeptide. Lanes 2 and 3 show the phosphopeptide

[60] G. R. Drapeau, *J. Biol. Chem.* **255**, 839 (1980).

[61] S. B. Sorensen, T. L. Sorensen, and K. Breddam, *FEBS Lett.* **294**, 195 (1991).

digested with Asp-N and Glu-C, respectively. Both samples show an altered migration of the digested phosphopeptide indicating that the phosphopeptide contains aspartate and glutamate residues.

Procedure 7

Secondary Digestion of Phosphopeptides with Endoproteinases Asp-N and Glu-C (Fig. 3)

1. After elution of individual phosphopeptides from HPLC fractions, two-dimensional phosphopeptide maps or alkaline polyacrylamide gels, resuspend the dried phosphopeptide in 50% acetonitrile, 0.1% TFA (in HPLC water). Divide the solution into three equal volumes and dry in a Speed Vac.
2. Resuspend one sample in 90 μl of 50 mM sodium phosphate buffer, pH 8 and add 5–10 μl of 10 μg/ml Asp-N (Roche) prepared in the same buffer. Incubate for 4 hr at 37°C. Resuspend the second sample in 99 μl of 25 mM ammonium bicarbonate and add 1 μl of 0.5 μg/μl Glu-C (Roche) prepared in the same buffer. Incubate for 8 hr at 37°C. Resuspend the third sample in either sodium phosphate buffer or ammonium bicarbonate buffer and incubate for 4–8 hr at 37°C (control sample).
3. Dry the digests in a Speed Vac.
4. Electrophorese the control, and the Asp-N and Glu-C digests side-by-side on an alkaline polyacrylmide gel. Dry the gel and expose to X-ray film.

Modified Manual Edman Degradation

A modified version of the manual Edman Degradation procedure determines the position of a [^{32}P]-labeled amino acid relative to the N-terminus of the peptide. Unlike standard Edman degradation, the first step of this modified version couples the phosphopeptide to a disk through the carboxyl terminus and then sequential amino acids are cleaved from the N-terminus. A high efficiency acid extraction step recovers the phospho-amino acid released. [^{32}P] Cerenkov counting of the released fraction determines the position of the phosphoamino acid in the peptide (see Fig. 5). Among the advantages of this procedure are low cost, the capacity to analyze many phosphopeptides in one experiment and absence of the need to purify the phosphopeptide from other nonphosphorylated peptides. A disadvantage is that the actual amino acid residue is not identified, although the position of the amino acid within the phosphopeptide is determined. Consequently, further information about the phosphopeptide is needed to identify the phosphorylation site. The combined information of modified manual Edman degradation, phosphoamino acid analysis, and secondary proteinase digestion is usually sufficient to identify the majority of phosphorylation sites in proteins.

Procedure 8

Modified Manual Edman Degradation

1. Resuspend the phosphopeptide in 30 μl of 50% acetonitrile, 0.1% TFA in HPLC water and spot the solution (5 μl at a time) on a Sequelon-AA membrane (kit sold by Millipore contains Sequelon-AA membranes, Mylar sheets, coupling buffer, and carbodiimide). The membrane is placed on a mylar sheet on top of a heating block set at 55°C.
2. Allow the membrane to dry (10–15 min) and remove from the heating block.
3. Add 5 μl carbodiimide to the membrane (10 mg/ml in coupling buffer prepared fresh) and let stand at room temperature for 30 min. This step couples the peptide covalently to the membrane.
4. Place the membrane in a 1.5 ml microfuge tube and wash with 1 ml HPLC water five times (5 min each wash with rocking).
5. Add 0.5 ml of 100% TFA to the disk, invert the tube several times and then remove and discard the TFA. Repeat this step four more times. TFA extracts unbound peptides from the membrane.
6. Wash the disk three times with 1 ml of methanol. Remove the last wash.
7. Add 0.5 ml PITC reagent (methanol–HPLC water–triethylamine–phenylisothiocyanate 7:1:1:1, prepared fresh for each cycle) and place the tube in a heating block (55°C) for 10 min. This step couples phenylisothiocyanate to the N-terminal amino acid sensitizing the peptide bond to cleavage with anhydrous TFA in step 9.
8. Remove the reagent and wash the disk five times with 1 ml methanol.
9. Dry the disk in a Speed Vac for 5 min. Add 0.5 ml TFA and place the tube in a heating block (55°C) for 6 min to cleave the derivatized amino acid.
10. Save the TFA wash and extract the disk with 1 ml of TFA–42.5% phosphoric acid (9:1, v/v).
11. Combine the washes from steps 9 and 10 and determine radioactivity by Cerenkov counting. Subject the disk to counting also.
12. Wash the disk five times with 1 ml of methanol.
13. Start the next cycle at step 7 or store disk at −20°C until ready to begin again.

Comments and Critical Points

a. Minimum starting radioactivity for this procedure should be approximately 500 cpm since typically only about 50% of the purified phosphopeptide will couple to the disk.

b. Phenylisothiocyanate must be flushed with a stream of nitrogen upon opening and closing to prevent oxidation.

c. One of the limitations of this procedure is that some phosphopeptides do not show [^{32}P] release. This failure may be due to the possibility that some of the peptide has linked to the filter through an asp or glu side chain and is released prior to reaching the phosphoamino acid. More commonly, a blocked N-terminus or a phosphoamino acid present toward the carboxy terminus of a large phosphopeptide that cannot be detected due to the loss of efficiency of the degradation through many cycles is the cause.

Phosphoamino Acid Analysis

Phosphoamino acid analysis is used to identify the modified residue (serine, threonine or tyrosine) present in either the intact protein or individual phosphopeptides. An acid hydrolysis step breaks peptide bonds resulting in a pool of amino acids, which then are separated by two-dimensional electrophoresis and visualized by autoradiography. The phosphoamino acids are identified by comparison to the migration of phosphoserine, phosphothreonine or phosphotyrosine standards.

Procedure 9

Phosphoamino Acid Analysis (Van der Geer and Hunter provide a more detailed description of this procedure[57])
Solutions:

- pH 1.9 buffer (2.5% formic acid [88% stock], 7.8% glacial acetic acid in dH$_2$O)
- pH 3.5 buffer (5% glacial acetic acid, 0.5% pyridine in dH$_2$O)

1. Resuspend either purified, [^{32}P]-labeled receptor protein or phosphopeptides (at least 500 cpm) in 100–200 μl of 6 N HCl in a 1.5 ml siliconized microfuge tube. Add 2 μl each of 2 mg/ml phosphoserine, phosphothreonine and phosphotyrosine standards (Sigma Chemicals) to the tube. Secure the cap and incubate the tube at 110°C for 1 hr in a heating block.

2. Dry the sample using a Speed Vac. Resuspend with 15 μl of pH 1.9 buffer (see Procedure 6) and spot the sample onto a 20 × 20 cm × 100 μm thin layer cellulose plate. Each sample should be at least 3 cm

from the edge. Up to four samples can be spotted on a plate with the anticipation that the phosphoamino acids will migrate approximately 5 cm in the first and second dimensions. Since phosphoamino acids are negatively charged even in the low pH 1.9 buffer, they will migrate toward the positive electrode (anode; red) in both dimensions.

3. Concentrate each phosphoamino acid sample spot on the plate as described in step 4 of Protocol 6. Wet the plate uniformly using Whatman 3MM paper soaked in pH 1.9 buffer.

4. Electrophorese the sample in the first dimension for 20 min at 1500 V using the Hunter HTLE 7000 electrophoresis system.

5. Dry the plate completely (at least 30 min at room temperature). Wet the plate uniformly with Whatman 3MM paper soaked in pH 3.5 buffer.

6. Rotate the plate 90° counterclockwise and electrophorese the sample in the second dimension for 15 min at 1300 V using pH 3.5 buffer.

7. Dry the plate completely in a fume hood (at least 1 hr). Use an atomizer (General Glassblowing Inc.) to spray the plate with ninhydrin (0.5 g in 100 ml of acetone; Pierce Chemical Co.). Visualize

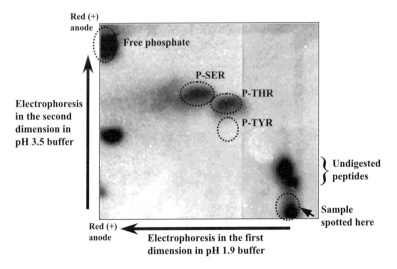

FIG. 6. Phosphoamino acid analysis. Autoradiograph of an undetermined phosphopeptide from the nuclear receptor coactivator, SRC-1. A purified phosphopeptide from *in vivo* phosphorylated SRC-1 was hydrolyzed with 6 M HCl and the resulting amino acids were resolved using the HTLE 7000 two-dimensional electrophoresis system as described in Procedure 9.

the phosphoamino acid standards by heating the plate in an 80°C oven for 1–5 min. The standards will appear purple in color.

8. Use a pen to outline the standards on the back of the plate and to mark the orientation of the plate. Wrap the plate with plastic wrap and expose to X-ray film.

Comments and Critical Points

a. Both free [^{32}P]phosphate and incompletely digested peptides are detected on the phosphoamino acid analysis map (see Fig. 6).

Section III

Analysis of Nuclear Receptor Cofactors and Chromatin Remodeling

[12] Acetylation and Methylation in Nuclear Receptor Gene Activation

By WEI XU, HELEN CHO, and RONALD M. EVANS

Introduction

The human nuclear hormone receptor (NR) superfamily is composed of 48 related genes, whose products include receptors for steroids, fat soluble vitamins, thyroid hormones, xenobiotic (i.e., foreign) compounds, as well as numerous fatty acid and cholesterol derivatives.[1] About half of the family binds no known ligand, and thus function as apparent constitutive activators or repressors. All receptors are presumed to function as transcriptional regulators. Receptors typically contain a DNA-binding domain (DBD) and ligand-binding domain (LBD), and mediate their action by the targeted recruitment of either coactivators or corepressors to the hormone response element (HRE) located in the regulatory regions of target genes. The cofactors are believed to function by controlling the dynamics of activation or repression via local chemical and structural modifications of chromatin.[2,3] Recent studies reveal two typical types of cofactors involved in altering chromatin structure.[4,5] One involves the ATP-dependent remodeling complex (MEDIATOR) that enables the reconfiguration of nucleosomes and increases the accessibility of DNA-binding sites; the others contain enzymatic activities to covalently modify the N-terminal histone tails of nucleosomes. Histone acetyltransferase (HAT), histone deacetylase (HDAC), and emerging members of histone methyltransferases (HMTs), all belong to the second category.[6]

Many of the nuclear receptor coactivators identified including p300/ CBP (CREB-binding protein), PCAF (p300/CBP-associated protein), ACTR, and SRC-1 exhibit HAT activities, and are brought to the promoter through direct and indirect interactions with hormone receptors. Among the coactivators identified, CBP and its homologue p300[7,8] exhibit the most

[1] D. J. Mangelsdorf and R. M. Evans, *Cell* **83**, 841 (1995).
[2] C. K. Glass and M. G. Rosenfeld, *Genes Dev.* **14**, 121 (2000).
[3] B. D. Lemon and L. P. Freedman, *Curr. Opin. Genet. Dev.* **9**, 499 (1999).
[4] N. J. McKenna, J. Xu, Z. Nawaz, *et al.*, *J. Steroid Biochem. Mol. Biol.* **69**, 3 (1999).
[5] M. J. Gamble and L. P. Freedman, *Trends Biochem. Sci.* **27**, 165 (2002).
[6] S. Westin, M. G. Rosenfeld, and C. K. Glass, *Adv. Pharmacol.* **47**, 89 (2000).
[7] J. C. Chrivia, R. P. Kwok, N. Lamb, *et al.*, *Nature* **365**, 855 (1993).
[8] Z. Arany, W. R. Sellers, D. M. Livingston, *et al.*, *Cell* **77**, 799 (1994).

potent HAT activity and appear to be critical in receptor function.[9] Recombinant CBP/p300 acetylates all four core histones in both free and nucleosomal forms. Preferentially, it acetylates Lys 5 of H2A, Lys 12 and 15 of H2B, Lys 14 and 18 of H3, and Lys 5 and 8 of H4.[10] It is not yet known which of these sites specifically become acetylated after hormonal induction.

HAT PCAF (a complex of eight or more proteins) also has been suggested to be part of a coactivator complex mediating transcriptional activation by the nuclear hormone receptors.[11] PCAF directly associates with the DNA-binding domain of nuclear receptors on a distinct interaction surface, such that it may act independently from p300/CBP.[12] Both PCAF and its associated complex can acetylate free and nucleosomal histones. PCAF shows high specificity to Lys 14 of H3 and Lys 8 of H4 in free and nucleosomal histones.[10] However, how the individual components contribute to hormonal regulation has yet not been studied.

Among the numerous NR coactivators identified, the p160 family proteins, including SRC-1, TIF2/GRIP1, and ACTR/AIB1/RAC3/pCIP[13] are well characterized. The p160 proteins form a distinct family of HATs that share several functional domains, including a central receptor interaction domain (RID), an adjacent CBP/p300 interacting domain, a HAT domain, and an amino terminal bHLH/PAS domain. Structural and mutational analyses have indicated that the RID is composed of multiple LXXLL motifs, which mediate direct interaction with agonist-bound receptor LBDs. As there is only one docking site in the LBD for the LXXLL motif, this receptor–cofactor interaction represents a defining step in hormone activation. The three-dimensional structure of this interaction and how ligand helps to create the appropriate interface has now been well described.[14,15] Thus, interaction is stabilized by association with the carboxyl-terminal "AF-2" domain whose position is highly dependent on the presence or absence of ligand.

Once a p160 protein is bound, it is now possible to recruit CBP/p300 as the next step in building the complex. Recent structural and thermodynamic studies by NMR have focused on the ACTR and CBP interaction interface, revealing a novel six helix combined motif of bulky hydrophobic residues that form an extensive and highly stable interaction.[16] Interestingly, NMR

[9] D. Chakravarti, V. J. LaMorte, M. C. Nelson, et al., Nature 383, 99 (1996).
[10] R. L. Schiltz, C. A. Mizzen, A. Vassilev, et al., J. Biol. Chem. 274, 1189 (1999).
[11] A. Vassilev, J. Yamauchi, T. Kotani, et al., Mol. Cell 2, 869 (1998).
[12] J. C. Blanco, S. Minucci, J. Lu, et al., Genes Dev. 12, 1638 (1998).
[13] J. Torchia, C. Glass, and M. G. Rosenfeld, Curr. Opin. Cell Biol. 10, 373 (1998).
[14] B. D. Darimont, R. L. Wagner, J. W. Apriletti, et al., Genes Dev. 12, 3343 (1998).
[15] H. E. Xu, T. B. Stanley, V. G. Montana, et al., Nature 415, 813 (2002).
[16] S. J. Demarest, M. Martinez-Yamout, J. Chung, et al., Nature 415, 549 (2002).

analysis reveals that the two interaction motifs are completely unstructured prior to association and undergo a unique, mutually induced synergistic folding to form the intermolecular complex. Through these interactions, ACTR recruits CBP/p300 to hormone receptors to allow transmission of the hormonal signal to the transcriptional machinery.

Typically, histone acetylation samples are analyzed by gel-electrophoresis, followed by Coomassie staining and fluorography, to identify the [^{3}H]acetate-labeled histones. Because triton–acid–urea (TAU) gels are able to distinguish between mono- or oligoacetylated histones, TAU gels are preferentially used to analyze the specificity of these enzymes. Finally, antibodies against peptides that recognize specific acetylated lysine residues in the H3 and H4 tail are available for Western blot analysis. For convenience, synthetic acetylated peptides are sometimes used for specificity analysis, though they may not be the economically favorable choice. The following TAU gel protocol is typically used to analyze the specificity of histone acetylation. Acetylation reactions with acid-extracted free histones are performed as described.[17]

Interestingly, different from p300, which acetylates mono-, di-, and triacetylated H4 with similar efficiency, PCAF favors nonacetylated H4 as a substrate (Fig. 1). When nucleosomal histones are used as a substrate, H4 becomes less attractive for PCAF (Fig. 1).

Although the p160 family exhibits HAT activity, the specific activity of these enzymes is very low, suggesting that these enzymes may have alternative substrates *in vivo*.

Triton–Acid–Urea Gel Electrophoresis

All acrylamide solutions are freshly prepared, filtered, and degassed prior to polymerization.

1. Resolving gel [bottom gel; 7.5 M urea, 13.5% acrylamide (acrylamide–bis-acrylamide, 150:1), 0.375% Triton X-100, and 5% acetic acid] was poured vertically and polymerized for 1 hr at room temperature (RT).
2. Stacking gel [7.5 M urea, 6% acrylamide (acrylamide–bis-acrylamide, 150:1), 0.37% Triton X-100, and 5% acetic acid] was poured vertically and comb was inserted.
3. After polymerization was completed, gel was assembled into a vertical gel apparatus. Electrophoresis was performed in 5% acetic acid.
4. Each well was carefully washed with running buffer using a syringe to remove excess urea and subsequently loaded with 20 μl of overlay

[17] C. A. Mizzen, J. E. Brownell, R. G. Cook, *et al.*, *Methods Enzymol.* **304**, 675 (1999).

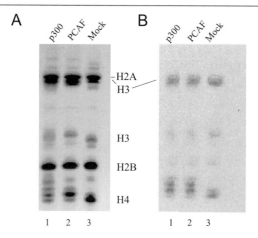

FIG. 1. Histone acetylation specificity of p300 and pCAF. (A) Coomassie stain of histone acetylation reactions of mock (lane 3), PCAF (lane 2), and p300 (lane 1) that were analyzed by TAU gel electrophoresis. Free core histones were incubated with mock, PCAF, or p300 for 1 hr with [³H]acetyl–CoA at 30°C. (B) Autoradiograph of Coomassie-stained gel in panel (A).

solution (2.5 M urea, 5% acetic acid, and 0.02% pyronin Y), and electrophoresed at 20 mA for 18 hr.

5. Old running buffer was discarded and fresh running buffer was poured into the apparatus. Each well was carefully rinsed with running buffer to remove excess urea and subsequently loaded with 20 μl of scavenger solution [2.5 M cysteamine, 2.5 M urea, 5% (v/v) acetic acid, and 0.02% (m/v) pyronin Y], and electrophoresed at 15 mA for 3 hr.

6. Old running buffer was discarded and fresh running buffer was poured into the apparatus. Each well was carefully rinsed and subsequently loaded with 20 μl of protamine sulfate solution [25 mg/ml protamine sulfate, 2.5 M urea, 5% acetic acid (v/v), and 0.02% (m/v) pyronin Y], and electrophoresed at 10 mA for 6 hr.

7. Old running buffer was discarded and fresh running buffer was poured into the apparatus. Each well was carefully rinsed to remove excess urea and subsequently loaded with 20 μl of sample mixed in final 1 × sample solution [2.5 M protamine sulfate, 2.5 M urea, 5% acetic acid (v/v), and 0.002% (m/v) pyronin Y], and electrophoresed at 10 mA for 18 hr.

8. Gel was Coomassie stained [10% (v/v) acetic acid, 40% (v/v) methanol, 0.025% (m/v) Coomassie Brilliant blue R 250], destained (Fig. 1A), and treated with Amplify (Amersham Pharmacia) for fluorography or washed in 5% (v/v) acetic acid–10% (v/v) methanol for 30 min at RT. Autoradiography is shown (Fig. 1B).

In Vivo *Acetylation*

The *in vivo* state of histone acetylation on promoter DNA can be monitored by chromatin immunoprecipitation (ChIP), utilizing specific antibodies against acetylated lysine residues. In MCF-7 cells, 17β-estradiol induces transcription of several target promoters [including pS2, cathepsin D (CTD), c-myc, and EB1], which has been extensively correlated with histone H3 and H4 acetylation in the vicinity of the promoter.[18] Peak acetylation on histone H4 coincides well with maximum occupancy of polymerase on the promoter after 1 hr treatment with 17β-estradiol. Under the same conditions, histones on CD38 promoter are not found hyperacetylated. However, when HL-60 cells are treated with all-*trans* retinoic acid, nucleosomal histones on retinoic acid-targeted CD38 promoter becomes hyperacetylated. The following ChIP protocol is typically used to analyze histone acetylation status on the CD38 promoter in HL-60 cells in response to retinoic acid (Fig. 2).

1. HL-60 cells were grown in RPMI-1640 medium supplemented with 10% charcoal–dextran stripped serum prior to treatment with 10 μM all-*trans* retinoic acid for 2 hr. For each ligand-treated and nontreated cells, typically 10^8 cells were used.

2. Cells were fixed in culture medium directly by adding formaldehyde solution (0.1 M NaCl, 1 mM EDTA, 50 mM HEPES, pH 7.5, 11% HCHO with CH_3OH) to final 1% and incubated at RT for 10 min.

3. Treated cells were harvested by centrifugation at 1000 rpm for 5 min at RT. Then cell pellets were placed on ice.

4. Cell pellets were resuspended in 30 ml ice-cold PBS and spun down at 1500 rpm for 5 min at $4°C$ in 15 ml conical tubes.

5. Cell pellets were washed sequentially in 10 ml of Buffer I (0.25% Triton X-100, 10 mM EDTA, 0.5 mM EGTA, 10 mM HEPES, pH 6.5), and Buffer II (200 mM NaCl, 1 mM EDTA, 0.5 mM EGTA, 10 mM HEPES, pH 6.5) as in step 4.

6. Cell pellets were resuspended in 1 ml of cold lysis buffer (1% SDS, 10 mM EDTA, 50 mM Tris, pH 8.1, 1 mM PMSF, protease inhibitor cocktails) (usually 1 ml per 5×10^7 HL60 cells were used).

7. Cell suspension was sonicated with a microtip (20×0.5 sec pulse at positions 2–3 for 4–6 times with 2 min on ice between each sonication).

8. Sonicated cell suspensions (chromatin solution) were transferred to 1.5 ml eppendorf tubes and centrifuged at 15,000 rpm for 15 min at $6–10°C$.

[18] H. Chen, R. J. Lin, W. Xie, *et al.*, *Cell* **98**, 675 (1999).

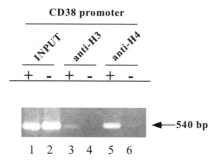

FIG. 2. ChIP of CD38 promoter by anti-acetyl H3 and anti-acetyl H4 antibodies. Lanes 1 and 2 are PCR fragments from DNA isolated from HL60 cells treated (+) or untreated (−) with RA prior to immunoprecipitation. Lanes 3–6 are PCR fragments obtained from cross-linked DNA that coimmunoprecipitated with anti-acetyl H3 (3, 4) or anti-acetyl H4 (5, 6) antibodies as indicated from HL60 cells treated (+) or untreated (−) with RA.

9. The suspension was transferred to a 15-ml conical tube and diluted 10-fold with cold dilution buffer (1% Triton X-100, 2 mM EDTA, 150 mM NaCl, 20 mM Tris, pH 8.1, protease inhibitor cocktails), and placed on ice. One hundred microliter aliquots were taken and saved for input control.

10. The remaining chromatin solution was mixed with antibodies (anti-acetyl H3 or anti-acetyl H4 polyclonal antibodies from Upstate Biotechnology) and incubated at 4°C for 12–16 hr on a nutator.

11. To each immunoprecipitation, 2–5 μg of sonicated salmon sperm DNA was added and mixed briefly followed by addition of 50 μl of protein A-Sepharose 4B beads washed in TE, pH 8.0. The reaction was incubated again on the nutator for 2 hr at 4°C.

12. The beads holding the immune complexes were collected by centrifugation for 2 min at 4°C. The beads were resuspended in 10 ml of TSE (0.1% SDS, 1% Triton X-100, 2 mM EDTA, 20 mM Tris–HCl, pH 8.1) + 150 mM NaCl and mixed on the nutator for another 5 min at 4°C.

13. The beads were collected as in step 12 and washed with 10 ml of TSE + 500 mM NaCl, followed by another wash with Buffer III (0.25 M LiCl, 1% NP-40, 1% deoxycholate, 1 mM EDTA, 10 mM Tris, pH 8.1).

14. Finally, beads were washed twice with 10 ml of TE, pH 8.0.

15. Immune complexes were eluted from the beads with 500 μl of 1% SDS, 0.1 M NaHCO$_3$ at RT for 10 min with agitation.

16. The eluates and saved aliquots (from step 9) were incubated at 65°C to reverse the cross-link for 4–6 hr.

17. The reactions were diluted three-fold with 0.2 M NaCl, and DNA was precipitated with 2.5 volumes of absolute ethanol after incubation at −20°C for O/N.

18. DNA was collected by centrifugation at 14,000 rpm at 4°C for 15 min, washed with 70% ethanol, and vacuum dried.

19. The precipitates were resuspended in 100 μl of TE, pH 8.0, by vortexing the precipitates for 5–10 min.

20. Each reaction was incubated with 100 μl of proteinase K solution (10 mM Tris, pH 7.5, 10 mM EDTA, 20 mM NaCl, 0.5% sarcosyl, and 50 μg/ml proteinase K added before use) for 1.5 hr at 37°C.

21. DNA was extracted once with phenol–chloroform–isoamylalcohol (25:24:1) and once with chloroform–isoamylalcohol (24:1).

22. DNA was precipitated by adding NaCl to final 0.2 M and 2.5 volumes of absolute ethanol, and incubated for 45 min on dry ice.

23. Steps 19–22 were repeated and final DNA was resuspended in 50 μl of TE, pH 8.0.

24. PCR was performed with 10 μl of the DNA sample for 25 cycles and analyzed by agarose gel-electrophoresis. The agarose gel was stained with ethidium bromide for visualization (Fig. 2).

A growing body of evidence suggests that HATs and HMTs not only modify histones but also other cofactors, which encode unique signaling information. One example is ACTR,[19] a p160-family coactivator that becomes acetylated by another HAT CBP/p300. This modification occurs at a key lysine residue in the RID and leads to dissociation of ACTR from estrogen receptor, and subsequently disassembly of the rest of coactivators from estrogen-responsive promoter.[18] The combination of covalent modifications of N-terminal tails of histones, known as the "histone code," appears to play pivotal role in transcriptional regulation.[20] The specific modifications either facilitate or inhibit the binding of chromatin modulators, maintaining chromatin in activated or repressive state. Moreover, multiple histone tails and site modifications can act synergistically or antagonistically. To date, protein arginine methyltransferases (PRMTs) extend the list of enzymatic activities that potentiate transcriptional regulation by nuclear receptors. The coactivator, associated arginine (R) methyltransferase CARM1/PRMT4[21] is part of ACTR coactivator complex;

[19] H. Chen, R. J. Lin, R. L. Schiltz, et al., Cell 90, 569 (1997).
[20] T. Jenuwein and C. D. Allis, Science 293, 1074 (2001).
[21] D. Chen, H. Ma, H. Hong, et al., Science 284, 2174 (1999).

its H3-histone methyltransferase activity is required for transcriptional activation.[22] The other member, PRMT1, facilitates NR-mediated transcriptional activation, possibly through interplay between histone methylation and acetylation.[23] Our recent studies describe that CARM1 methylates CBP/p300 and ACTR, in addition to histone H3, which exerts as a transcriptional switch between CREB and nuclear receptor pathways.[24] The following section summarizes the methods to analyze PRMTs in nuclear receptor action.

Protein Arginine Methyltransferases (PRMTs)

Six members of PRMTs (1–6) in mammalian and one in *Saccharomyces cerevisiae* termed HMT1/RMT1 have been identified.[25–29] PRMTs catalyze the transfer of methyl group from *S*-adenosyl-L-methionine (AdoMet) to the guanidino group of arginines in protein substrates.[30] Based on the symmetry of dimethylarginine products, PRMTs fall into two types: type I forms asymmetric dimethylarginine and type II forms symmetric dimethylarginine (Fig. 3A). All PRMTs except PRMT5 are type I methyltransferases. The most striking feature of PRMTs is the presence of a highly conserved *S*-adenosyl methionine (AdoMet) binding motif (Fig. 3B) and a less-conserved C-terminal domain, which is presumably involved in interaction with arginines in substrate. The predicted CARM1 structure (amino acids 148–458) from SWISS-MODEL, based on two known three-dimensional structures of the core regions of yeast hmt1 and human PRMT3 is shown (Fig. 3C). Many RNA-binding proteins, such as hnRNPs, are targets of arginine methylation, whose function is to regulate RNA processing or transport.[30] It should be noted that methylation does not alter the overall charge on an arginine residue. However, the bulky methyl group is likely to increase the steric hindrance on amino groups, which might lead to disruption of protein–protein interactions. The arginine methylation of the Stat1 by PRMT1[31] and CBP/p300 by CARM1[24] suggest that methylation

[22] D. Chen, S. M. Huang, and M. R. Stallcup, *J. Biol. Chem.* **275**, 40810 (2000).
[23] S. S. Koh, D. Chen, Y. H. Lee, *et al.*, *J. Biol. Chem.* **276**, 1089 (2001).
[24] W. Xu, H. Chen, K. Du, *et al.*, *Science* **294**, 2507 (2001).
[25] A. Frankel and S. Clarke, *J. Biol. Chem.* **275**, 32974 (2000).
[26] A. Frankel, N. Yadav, J. Lee, *et al.*, *J. Biol. Chem.* **277**, 3537 (2002).
[27] J. Rho, S. Choi, Y. R. Seong, *et al.*, *J. Biol. Chem.* **276**, 11393 (2001).
[28] J. Tang, J. D. Gary, S. Clarke, *et al.*, *J. Biol. Chem.* **273**, 16935 (1998).
[29] V. H. Weiss, A. E. McBride, M. A. Soriano, *et al.*, *Nat. Struct. Biol.* **7**, 1165 (2000).
[30] A. E. McBride and P. A. Silver, *Cell* **106**, 5 (2001).
[31] K. A. Mowen, J. Tang, W. Zhu, *et al.*, *Cell* **104**, 731 (2001).

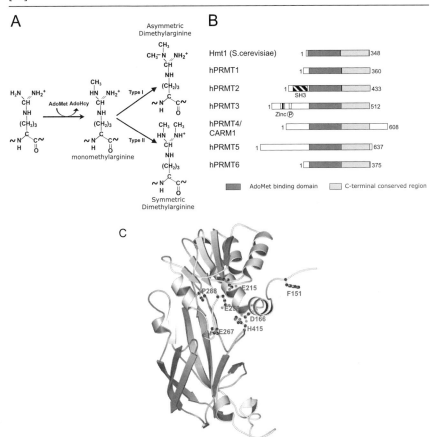

FIG. 3. Structural features of PRMTs. (A) Methylation of arginine residues by PRMTs converts *S*-adenosyl methionine (AdoMet) into *S*-adenosyl homocysteine (AdoHcy). (B) Schematic representation of seven PRMT homologs. The yeast hmt1 and six known human arginine methyltransferases (PRMT1 → 6) share a conserved "core" region of ∼310 amino acids, including AdoMet-binding domain (purple) and a little less-conserved domain (green). (C) The structural model of CARM1 (amino acid 148–458) generated with SWISS-MODEL and the Swiss-Pdb Viewer[32] based on the 3D structures of Hmt1 and PRMT3 (Accession code 1G6Q and 1F3L). CARM1 is folded into two domains connected at proline 288. The N-terminal residues E215, E258, E267 constitute arginine amino methylation catalytic active sites, which are conserved between PRMTs. The two terminal amino groups of the substrate arginine interact with the side chains of E258 and E267. The carboxylate oxygen of D166 forms a hydrogen bond to H415, which is involved in elimination of proton after methyl transfer. The hydrophobic phenyl ring of F151 in N-terminal helix is essential to lock the AdoHcy in position and almost buries the AdoHcy. The movement of N-terminal helix would allow AdoMet into and AdoHcy out of the binding site. (See Color Insert.)

[32] N. Guex and M. C. Peitsch, *Electrophoresis* **18**, 2714 (1997).

may account for switching different signaling pathways by altering specific protein–protein interactions.

Specificity of PRMTs

PRMT1 and CARM1 exhibit different substrate specificity toward core histones *in vitro* (Fig. 4A). PRMT1 methylates histone H4 at R3,[33] while CARM1 methylates H3 at R2, 17, 26[34] with the highest preference of R17. Bacterial-expressed GST-PRMT1 or baculovirus-expressed Flag-tagged CARM1 was used in the typical methylation assay. The 20 μl of reaction mix contained 1.5 μg of core histones purified from HeLa cells, 0.4 μM [methyl-^{3}H] AdoMet (80 Ci/mmol) (Amersham Pharmacia), and 3 μM PRMT1 or CARM1 in 20 mM Tris pH 8.0, 2 mM EDTA, and 1 mM dithiothreitol. The reactions were carried out at 37°C for 1 hr. Methylation was analyzed by 15% SDS-PAGE or by filter-binding assay. For the filter-binding assay, the reaction mix was spotted onto a phosphate cellulose filter paper disc (Whatman, Cat. 3698-023). The paper disc was air-dried for 2 min and washed in the conical tube twice with 20–30 ml of 0.1 M Na$_2$CO$_3$, pH 9.2, by gently swirling for 2 min. Finally, the disc was rinsed once with 10–20 ml of acetone, air-dried before liquid scintillation counting.

Analogous to PRMT1, CARM1 was observed to methylate histones H3 *in vivo*. Since CARM1 specifically methylates R17 in histone H3, an antibody against H3 R17 (Novus Biological Co.) was used to detect the methylation of endogenous histones (Fig. 4B, lane 1). The histones were purified from HeLa cells using the acid extraction method detailed below.

1. Grow 10 × 15 cm plates of HeLa cells, when the cells enter log phase, change to 10 ml fresh media. Add 100 μl of 10 M sodium butyrate (Upstate biotechnology) and 200 μl of ^{3}H-acetic acid (10 mCi/ml, Amersham Pharmacia) to the media to give a final concentration of 10 mM of Na butyrate and 0.2 mCi/ml of ^{3}H-acetic acid.
2. After 4 hr of treatment, harvest cells with rubber policeman.
3. Centrifuge the cells at 700 × g for 10 min.
4. Wash with cold PBS–10 mM Na butyrate once.
5. Lyse in ice-cold lysis buffer 1% NP-40, IB (10 mM Tris–HCl, pH 7.4, 2 mM MgCl$_2$, 3 mM CaCl$_2$, 10 mM Na butyrate, 0.1 mM PMSF).
6. Dounce homogenize.
7. Collect nuclei by centrifugation at 1000 × g for 10 min.

[33] H. Wang, Z. Q. Huang, L. Xia, *et al.*, *Science* **293**, 853 (2001).
[34] B. T. Schurter, S. S. Koh, D. Chen, *et al.*, *Biochemistry* **40**, 5747 (2001).

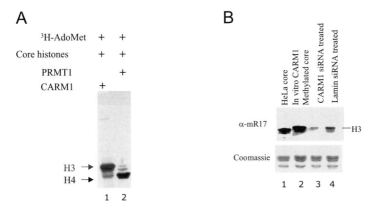

FIG. 4. PRMT1 and CARM1 exhibit different substrate specificity toward core histones. (A) 1.5 μg of core histones were methylated with [methyl-³H] AdoMet either by bacterial-expressed GST-PRMT1 (lane 1) or baculovirus-expressed Flag-tagged CARM1 (lane 2). Methylation was analyzed by 15% SDS-PAGE and autoradiography was taken. (B) Western blotting of histone H3 R17 in HeLa core histones (lane 1), CARM1-methylated core histones (lane 2), CARM1 siRNA-treated HeLa core histones (lane 3), and control lamin siRNA-treated HeLa core histones (lane 4). Commassie staining of histones loaded on each lane is shown in the figure.

8. Wash twice with lysis buffer NIB, then once with NIB + 100 mM NaCl.
9. Wash once with nuclear extract buffer with 400 mM NaCl (containing 10 mM Na butyrate).
10. Wash once with IB buffer, the pellet looks puffy.
11. Resuspend pellet in 1 ml H$_2$O.
12. Add 0.25 ml of 2 N H$_2$SO$_4$.
13. Incubate at 4°C for 1 hr.
14. Centrifuge at 15,000 rpm × 15 min. By then, the pellet looks dry and white.
15. Dialyze with 0.1 N acetic acid twice (1–2 hr at a time), then with H$_2$O overnight.
16. Lyophilize the pellet till dry.
17. Dissolve in appropriate amount of H$_2$O to give concentration of 2 mg/ml.

In PRMT1−/− ES cells,[35] histone H4 R3 methylation is diminished, suggesting that PRMT1 is the predominant methyltransferase that methylates histone H4 on arginine 3 *in vivo*.[33] Due to the unavailability of CARM1−/− cells, our method of choice to deplete endogenous CARM1

[35] M. R. Pawlak, C. A. Scherer, J. Chen, *et al.*, *Mol. Cell. Biol.* **20**, 4859 (2000).

involves the usage of small interfering RNA (siRNA) duplexes that specifically abolish CARM1 expression. RNA interference (RNAi) is the technology to induce sequence-specific posttranscriptional gene silencing by applying double-stranded RNA (dsRNA).[36] Because the synthetic short interfering RNAs (siRNAs), ~22 bp, are the sequence-specific mediator of RNAi and are efficient in shutting down gene expression in a variety of organisms, this method is popular for gene silencing.[37] Following siRNA treatment in HeLa cells, endogeneous histones were acid purified for detection of histone H3 R17 methylation. The siRNA protocol is designed for turning off endogenous genes for one 10-cm tissue culture plate.

1. Dilute 30 μl of 20 μM siRNA (Dharmacon Research Inc., CO) with 500 μl of OPTI-MEM (Invitrogen). The siRNA sequence for CARM1 is AACGGCGAGAUCCAGCGGCAC position 130–150 relative to the first nucleotide of the CARM1 start codon. As a control, lamin A/C duplex CUGGACUUCCAGAAGAACA (Dharmacon Research Inc.) was used.

2. Dilute 30 μl of Oligofectamine into 120 μl of OPTI-MEM, incubate for 8 min at RT.

3. Mix the above two together by pipetting several times, incubate for 25 min at RT.

4. Add 320 μl of OPTI-MEM to make the final volume to be 1 ml.

5. HeLa cells, seeded the day before, have about 50% confluence. When seeded, omit antibiotics in the media.

6. Change the media to 5 ml of DMEM with 10% FBS without antibiotics.

7. Simply add the RNA mixture to the cells dropwise.

8. Incubate at 5% CO_2 incubator for 2 days.

9. Extract histones with acid from cells and detect Me-H3 R17 by Western blotting.

Figure 4B showed that histones extracted from CARM1 siRNA-treated cells (lane 3) had decreased H3 R17 methylation than lamin siRNA-treated cells (lane 4), suggesting that CARM1 is the prominent H3 R17 methylase in cells.

Inter-Correlation Between Acetylation and Methylation of Histones

The effects of histone acetylation on its subsequent methylation or vice versa were usually determined by the *in vitro* methylation or acetylation assay

[36] P. A. Sharp, *Genes Dev.* **15**, 485 (2001).
[37] S. M. Elbashir, W. Lendeckel, and T. Tuschl, *Genes Dev.* **15**, 188 (2001).

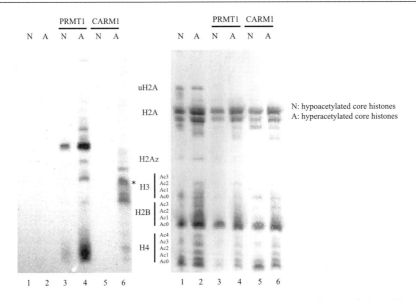

Fig. 5. Intercorrelation between core histones acetylation and methylation. Three micrograms of core histones purified from HeLa cells were methylated with GST-PRMT1 (lanes 3 and 4) or CARM1 (lanes 5 and 6) in the presence of 0.4 μM [methyl-^{3}H] AdoMet. Methylation reactions were loaded on a TAU gel, and autoradiography (left) and Coomassie staining (right) were shown. Hypoacetylated histones (lane 1) and hyperacetylated histones (lane 2) were shown as control. The star (∗) denotes diacetylated histone H3.

on purified histones. Hyperacetylated histones can be purified from sodium butyrate-treated HeLa cells or obtained by acetylation of hypoacetylated histones with specific HAT *in vitro*. Arginine-methylated histones can be obtained by methylation of core histones by specific PRMTs *in vitro*. When hyperacetylated histones from HeLa cells were used, PRMT1 was observed to preferentially methylate hypoacetylated histones rather than hyperacetylated histones (Fig. 5, lane 4).[33] Conversely, CARM1 methylates both hypoacetylated and hyperacetylated histones H3 with the preference for diacetylated H3 (Fig. 5, lane 6). The hypoacetylated histones purified from nontreated HeLa cells probably have possessed some intrinsic modifications that prohibits it from being methylated by CARM1 (Fig. 5, lane 5).

Analysis of Oligomerization of PRMTs

Oligomerization is a common feature among the PRMTs. X-ray crystal structure of yeast Hmt1 indicates that Hmt1 forms a hexamer with approximate 32 symmetry. Importantly, dimerization of Hmt1 is required for its catalytic activity. Another member of PRMTs, PRMT1, also exists as a hexamer in solution. Crystal structure of PRMT3 reveals a dimer interface in the structure, even though the dimer was not seen in solution, probably due to the equilibrium between dimer and monomer. Gel filtration and cross-linking studies are the most commonly used biochemical approaches for elucidation of CARM1 oligomerization status.

Cross-Linking Assay for CARM1

The cross-linking method was adapted from the studies of oligomerization of hmt1.[29] In brief, low concentration of CARM1 (50 μg/ml) in 50 mM Tris, pH 7.8, 1 mM EDTA, 1 mM DTT were cross-linked with 0.025% (v/v) of glutaraldehyde (Sigma) at room temperature for 5 min in a 20 μl reaction. To determine whether the association of oligomers is through hydrophobic or hydrophilic interactions, cross-linking was performed with increasing concentrations of NaCl ranging from 150 mM to 2 M. The reaction was stopped with 100 mM of ammonium acetate. The samples were resolved by 6% SDS-PAGE, and the formation of oligomer was detected by Western blotting with anti-CARM1 antibody (Upstate biotechnology). Different from Hmt1, where the association of dimers to form hexamers is mediated by hydrophilic interactions, the oligomer of CARM1 is not sensitive to salt even at 1 M concentration (Fig. 6A), indicating that CARM1 oligomerization is likely mediated by the hydrophobic interactions.

Gel Filtration Chromatography

Recombinant Flag-tagged CARM1 (2 mg/ml) was loaded onto a Superdex S-200 HR 10/30 column (Pharmacia) and eluted with a buffer containing 20 mM Tris, pH 7.5, 150 mM NaCl, 1 mM EDTA, 1 mM DTT, and 10% glycerol. 0.5 ml fractions were collected and absorption at 280 nm was recorded corresponding to the amount of CARM1 in each fraction. Figure 6B showed that a fraction of CARM1 existed as dimer in solution, even though the majority of it was in the monomer form, suggesting that CARM1 monomer and dimer are in equilibrium in solution. Whether dimerization of CARM1 is required for its enzymatic activity is not yet known since dimerization deficient CARM1 mutants are not available.

In Vitro Chromatin Assembly and Transcription Reactions

Purification of RAR/RXR Heterodimer and p300, ACTR

Flag-tagged RAR, ACTR, and p300 were constructed in pAcSG2 Baculovirus Transfer Vector (PharMingen), RXR was constructed in pAcSG vector without a tag. Recombinant baculovirus were generated using BaculoGold[TM] system (PharMingen). To purify RAR/RXR heterodimer,

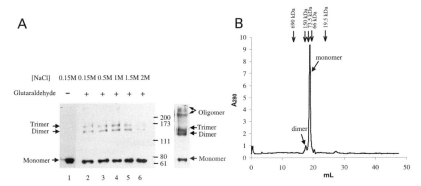

FIG. 6. Oligomerization of CARM1 in solution. (A) CARM1 (50 μg/ml) in 50 mM Tris, pH 7.8, 1 mM EDTA, 1 mM DTT were cross-linked with 0.025% (v/v) of glutaraldehyde (Sigma) at room temperature for 5 min in a 20 μl reaction, with the increasing concentrations of NaCl ranging from 150 mM to 2 M (lanes 1–6). The reaction was stopped with 100 mM of ammonium acetate and resolved by 6% SDS-PAGE. Western blotting with α-CARM1 antibody indicated the formation of monomer, dimer, and trimer (left), longer exposure revealed the presence of oligomers (right). (B) Gel filtration of CARM1 on a Superdex S-200 HR 10/30 column. The elution positions of molecular weight markers (thyroglobulin 670 kDa, alcohol dehydrogenase 150 kDa, chymotrypsin 73.5 kDa, bovine serum albumin 66 kDa, chymotrypsinogen A 19.5 kDa) were indicated.

Sf9 cells were coinfected with recombinant baculovirus of RAR and RXR, and the heterodimer was immunoaffinity purified with anti-Flag-M2 resin (Sigma). After high salt wash (0.4 M NaCl), RAR/RXR dimer was co-eluted with 0.2 mg/ml Flag peptides. Typically, 100 μg of RAR/RXR can be yielded from a 500 ml culture. Ligand binding and DNA-binding properties of the heterodimer were tested with gel-mobility shift assays to confirm that the recombinant proteins were functional. The purification of Flag-tagged ACTR, PCAF, and p300 were similar to the RAR/RXR, with the yield of $\sim$50 μg per 500 ml cell culture.

Chromatin assembly reactions were performed with the S190 extract derived from *Drosophila* embryos.[38] To set up 100 μl chromatin assembly reaction, 1.3 μl core histones ($\sim$0.8 μg) was incubated with 48.7 μl S-190 *Drosophila* embryos extract on ice for 30 min, prior to addition of pHIV-RARE-Luc plasmid ($\sim$1 μg). This plasmid was constructed by inserting five copies of the DR5 RARE from the mouse RARβ2 promoter upstream of human immunodeficiency virus (HIV)-1 promoter in HIV-162 vectors[39] (Fig. 7A). Then, the S-190–histone–DNA mix was combined with 10 μl of 10 × ATP mix (the 100 μl mix includes 60 μl 0.5 M creatine phosphate, 6 μl of 0.5 M ATP, 4.1 μl of 1 M MgCl$_2$, 0.2 μl of 5 mg/ml creatine kinase and

[38] R. T. Kamakaka, M. Bulger, and J. T. Kadonaga, *Genes Dev.* **7**, 1779 (1993).
[39] P. L. Sheridan, C. T. Sheline, K. Cannon, *et al., Genes Dev.* **9**, 2090 (1995).

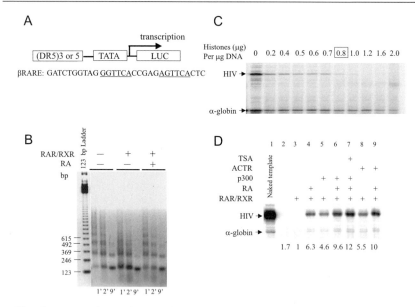

FIG. 7. *In vitro* chromatin assembly and transcription reactions. (A) pHIV-RARE-Luc plasmid was constructed by inserting five copies of the DR5 RARE from the mouse RARβ2 promoter upstream of HIV-1 promoter in HIV-162 vectors. The retinoic acid response of RARE was confirmed in a transient transfection assay. (B) Micrococcal nuclease digestion of chromatin assembled with Sf-9 *Drosophila* embryo extract in the absence of RA, RAR/RXR (lanes 2–4), in the presence of RAR/RXR (lanes 5–7), or RA and RAR/RXR (lanes 8–10). The digestion was stopped at 1, 2, and 9 min. DNA was loaded on 1.5% TAE agarose gel and stained with EtBr. The staph ladder indicated that up to seven oligonucleosomes were formed. (C) The titration of the effect of histone–DNA ratio on transcription with chromatin templates. The amount of HIV-1 transcripts decreased with increasing amount of histones, while the control α-globin transcript remained constant. The optimum amount of core histones is 0.8 μg per microgram of DNA as indicated with box. (D) The effect of coactivator ACTR, p300, and HDAC inhibitor TSA on transcription activation. RA induces six-fold activation on transcription of RAR/RXR (compare lanes 2 and 3). 0.4 μg of ACTR and p300 and 1 μM of TSA stimulate transcription for 12 fold (lane 9).

29.7 μl H_2O). Nuclear receptor RAR/RXR (0.2 μg) and RA (1 μM) were then added along with HeLa dialysis buffer (20 mM HEPES, K^+, pH 7.9, 50 mM KCl, 1 mM DTT, 0.2 mM EDTA, and 10% glycerol) to bring the volume to 100 μl. After 4.5 hr incubation at 27°C, half of the reaction was subjected for structural analysis, while another half of the reaction was used for transcriptional assays as described in the next section.

Chromatin Assembly Structural Analysis

For 50 μl micrococcal nuclease digestion, add 1.5 μl of 0.1 M $CaCl_2$ (final 0.003 M) to 0.5 μg of chromatin. Prior to the addition of micrococcal nuclease (Sigma), set up tubes with 3.75 μl of 5× stop mix [2.5% (w/v) N-lauroylsarcosine, 100 mM EDTA]. Pipet 0.8 μl of micrococcal nuclease to the reaction tubes at the timed intervals, and then remove 15 μl of digested chromatin to the tubes with stop mix at 1, 3, and 9 min. Vortex the mix and set on ice. The sample mix was incubated with 1 μl 10 mg/ml RNase A and 1 μl glycogen carrier at 37°C for 10 min. Then, 2 μl of 10 mg/ml proteinase K and 1 μl of 5% SDS was added, and incubated at 55°C for 10 min. DNA was diluted with TE to bring the final volume to 100 μl and then extracted with phenol–chloroform and chloroform. Finally, DNA was precipitated with final 0.2 M NaCl and EtOH, washed with 70% EtOH, and dried in a speedvac. The samples were resuspended in 8 μl TE and 2 μl of 5× Orange G dye, and loaded on a 1.5% Tris–glycine agarose gel. The gel was run at 200–300 V for 2.5–3 hr in the cold room before staining with EtBr, destaining, and photographing. Micrococcal nuclease digestion in Fig. 7B showed that nucleosomes assembled in the presence of RARα/RXRα and RA had a regular periodicity of 160 bp, indicating that the overall chromatin structure was not affected by RARα/RXRα heterodimer.

Transcription of Chromatin

In vitro transcription reactions were performed with HeLa cell nuclear extracts as described.[40] In the initial experiments, the ratio of histones–DNA was titrated by increasing the amount of histones (0 → 2) μg per microgram of DNA (Fig. 7C). The amount of transcripts decreased as histones–DNA ratio increased, indicating that the assembled chromatin became more and more compact. Typically, core histones–DNA ratio of 0.8 μg/1 μg was used because, at this point, the basal level of transcription on the chromatin template was reduced to the minimal level, which is ideal for detection of transcriptional activation. When coactivators, such as ACTR and p300, were included, approximately 0.4 μg of coactivator proteins were added to the chromatin after the initial assembly reaction was complete, in which case the association of ACTR and p300 with chromatin was promoted by additional 30 min incubation. For a 50 μl reaction containing 500 ng DNA, we combined histone–DNA–coactivator mix with 20 μl of HeLa nuclear extract and incubated on ice. Transcription was initiated by the addition of 80 μl of transcription mix (20 mM HEPES, pH 7.9, 60 mM KCl, 3 mM MgCl$_2$, 0.2 mg/ml BSA, 2 mM DTT, 0.6 mM rNTP, 50 ng α-globin DNA) and incubated at 30°C for 30–45 min. Transcription was stopped by the addition of 250 μl stop mix (20 μg yeast tRNA, 0.1% SDS, 10 mM EDTA, and 20 μg proteinase K) and incubated at 37°C for 15 min. DNA was phenol–chloroform, chloroform extracted, precipitated with EtOH, washed

[40] M. J. Pazin, R. T. Kamakaka, and J. T. Kadonaga, *Science* **266**, 2007 (1994).

with 70% EtOH, dried in a speedvac before subjecting to primer extension analysis. The dried RNA pellet was resuspended with 7 μl of H_2O, 2 μl of 5× annealing buffer (50 mM Tris–HCl, pH 8.0, 5 mM EDTA, 1.25 M KCl) and 1 μl of ^{32}P-labeled primer. Extension reaction was initiated by addition of 23 μl of extension buffer (20 mM Tris–HCl, pH 8.3, 10 mM $MgCl_2$, 5 mM DTT, 0.33 mM each dNTP, and 100 μg/ml actinomycin) containing 10 units of M-MLV RT (Invitrogen) and incubated at 42°C for 1 hr. The reaction was stopped by addition of 300 μl of EtOH (4°C) and mixed by vortexing. After centrifuging in microcentrifuge at the maximum speed for 15 min, the reverse transcribed DNA was washed with 70% ethanol and pellet was dried by speedvac. Samples were resuspended in 8 μl of formaldehyde dye containing NaOH and EDTA (4 μl of 0.1 mM NaOH–1 mM EDTA + 4 μl of formaldehyde dye), boiled for 3 min before loading onto 8% polyacrylamide–8 M urea sequencing gel in 1× TBE. Figure 7D showed that coactivator ACTR and p300 and HDAC inhibitor TSA stimulated RAR/RXR and RA-dependent *in vitro* transcription on chromatin templates, suggesting that acetylation of histones on chromatin at least in part accounts for the transcriptional activation.

Summary

Activation of hormone target genes requires chromatin remodeling and histone modifications. The properties of the two PRMT coactivators, PRMT1 and CARM1, are compared in Table I. One can envision many scenarios in which histone arginine methylation contributes to transcriptional regulation. For example, it could be analogous to histone H3 K4 methylation by Set9,[41] which blocks the HDAC NuRD complex from association and simultaneously impairs Suv39h1-mediated methylation at K9 of H3 (H3-K9). As a result, H3 K4 methylation by Set9 potentiates transcriptional activation. Histone arginine methylation might also promote or antagonize other histone-modifying enzymes. It has been shown that PRMT1-methylated histone H4 becomes a better substrate for p300 and, conversely, the acetylated histones are poor substrates for methylation by PRMT1.[33] As for CARM1, acetylation of multiple lysines within histone H3 facilitates arginine methylation of by CARM1.[24] Since PRMT1 and CARM1 methylate H4 and H3 tails, respectively, and each contributes to activation of the nuclear receptor response, it implicates the "histone code" as the physical template of hormone signaling. However, it remains to be resolved whether p160 family coactivators simultaneously recruit CARM1 and PRMT1 to specific target genes, and the order of the series of modifications on individual histone tails *in vivo*. Time-course studies of

[41] K. Nishioka, S. Chuikov, K. Sarma, *et al.*, *Genes Dev.* **16**, 479 (2002).

TABLE I
COMPARISON OF THE PROPERTIES OF CARM1 AND PRMT1

PRMT1	CARM1
Methylate histone H4 R3 *in vivo* and *in vitro*	Methylate histone H3 R17 *in vivo* and *in vitro*
The most prominent methyltransferase responsible for methylation of H4-R3	The most prominent methyltransferase responsible for methylation of H3-R17
Methylation of histone H4 R3 facilitates subsequent acetylation by p300	p300-acetylated histone H3 is a better substrate for CARM1 methylation
PRMT1 SAM-binding mutants fail to activate transcription by nuclear receptors	CARM1 SAM-binding mutants fail to activate transcription by nuclear receptors
Monomer is 43 kDa; function as 350 kDa oligomer	Monomer is 64 kDa; forms oligomer in solution
PRMT1 knock out is lethal; however, null ES cells are viable	CARM1 knock out mice die at birth*
hnRNPA1, STAT1 > histones	p300, ACTR > core histones > nuleosomes

*Personal communication with Mark Bedford, UT Anderson Cancer Center.

cofactor recruitment by ChIPs will be necessary to decipher the modification patterns. Another useful approach to analyze the function of NR cofactors on target gene transcription is the chromatin-dependent *in vitro* transcription system. As increasing amounts of evidence indicate that one HAT can be acetylated by another HAT, or methylated by HMT, it would not be surprising that transcription factors and their coactivators are bona fide substrates for protein modification.

Acknowledgment

We are grateful to Shen Ye (Syrrx) for preparing CARM1 fragment predicted PDB structure.

[13] Steroid Receptor Coactivator Peptidomimetics

By TIMOTHY R. GEISTLINGER and R. KIPLIN GUY

Introduction

A crucial protein–protein interaction in ligand-dependent nuclear hormone receptor (NR) signaling occurs between steroid receptor coactivators (SRC) and the NR. This interaction is theoretically an excellent point of intervention to better understand NR signaling; however, the development of competitive inhibitors is made difficult by the facts that this interaction is common across the NR superfamily of proteins and that many coactivators bind to the same α-helical $L_1XXL_2L_3$ binding motif in NRs—termed an NR box. The human estrogen receptor α (hERα) and thyroid hormone receptor β (hTRβ), for example, recognize the same coactivator protein *in vivo*, but regulate entirely different gene networks.

Here, we focus on methods useful for designing and synthesizing a series of coactivator peptidomimetics ($a_xXXb_xc_x$) of the $L_1XXL_2L_3$ sequence in order to identify specific inhibitors of particular NR–coactivator interactions. This chapter is organized into four sections: (1) computational design of libraries of potential inhibitors, (2) synthesis of these libraries, (3) screening assays for efficacy and specificity.

Background

Genetic and biochemical studies of NR signaling identified a family of cofactors, the SRCs, which are important in mediating ligand-dependent signaling of many NR (Fig. 1). Subsequent investigations revealed the nuclear receptor interacting domain (NID) of coactivators, which includes multiple interaction motifs known as NR boxes with a consensus sequence of $L_1XXL_2L_3$. Each NR box of the NID can have different affinities for a particular NR–ligand–promoter triad and can bind in cooperative[1] or noncooperative[2–4] manners. While all of the leucines in the NR box are crucial for the interaction,[5] specificity appears to be conferred by sequences immediately flanking the NR boxes,[6–9] rather than the geometry of the leucine side chains. Competitive inhibition of this interaction by fusion proteins containing a minimal NR box antagonizes both estrogen- and thyroid hormone-induced gene transcription.[7] *In vivo*, the estrogen (E$_2$)–estrogen receptor α (E$_2$–hERα) and thyroid hormone (T$_3$)–thyroid

[1] E. M. McInerney, D. W. Rose, S. E. Flynn, S. Westin, T. M. Mullen, A. Krones, J. Inostroza, J. Torchia, R. T. Nolte, N. Assa-Munt, M. V. Milburn, C. K. Glass, and M. G. Rosenfeld, *Genes Dev.* **12**, 3357–3368 (1998).

[2] B. D. Darimont, R. L. Wagner, J. W. Apriletti, M. R. Stallcup, P. J. Kushner, J. D. Baxter, R. J. Fletterick, and K. R. Yamamoto, *Genes Dev.* **12**, 3343–3356 (1998).

[3] R. C. Ribeiro, J. W. Apriletti, R. L. Wagner, W. Feng, P. J. Kushner, S. Nilsson, T. S. Scanlan, B. L. West, R. J. Fletterick, and J. D. Baxter, *J. Steroid Biochem. Mol. Biol.* **65**, 133–141 (1998).

[4] X. F. Ding, C. M. Anderson, H. Ma, H. Hong, R. M. Uht, P. J. Kushner, and M. R. Stallcup, *Mol. Endocrinol.* **12**, 302–313 (1998).

[5] D. M. Heery, E. Kalkhoven, S. Hoare, and M. G. Parker, *Nature* **387**, 733–736 (1997).

[6] C. Chang, J. D. Norris, H. Gron, L. A. Paige, P. T. Hamilton, D. J. Kenan, D. Fowlkes, and D. P. McDonnell, *Mol. Cell. Biol.* **19**, 8226–8239 (1999).

[7] J. P. Northrop, D. Nguyen, S. Piplani, S. E. Olivan, S. T. Kwan, N. F. Go, C. P. Hart, and P. J. Schatz, *Mol. Endocrinol.* **14**, 605–622 (2000).

[8] K. S. Bramlett, Y. Wu, and T. P. Burris, *Mol. Endocrinol.* **15**, 909–922 (2001).

[9] M. Needham, S. Raines, J. McPheat, C. Stacey, J. Ellston, S. Hoare, and M. Parker, *J. Steroid Biochem. Mol. Biol.* **72**, 35–46 (2000).

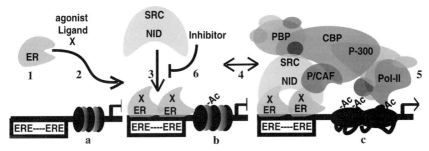

FIG. 1. Schematic model of the dynamic assembly of the transcription–activation complex by agonist-bound ER and function of SRC-binding inhibitors. (1) In the absence of ligand, chromatin is unmodified (a) and transcription at the DNA estrogen response element (ERE) is unaltered. (2) Binding of agonist ligand to the LBD of ER induces a conformational change in ER leading to translocation and homodimerization on the ERE. (3) Liganded ER on the ERE recruits SRCs using the NR box ($L_1XXL_2L_3$) of the NID, resulting in chromatin modification (b). (4) Subsequently, the ER–SRC complex recruits other proteins, such as PBP/DRIP/TRAP, CBP/p300, and RNA Pol-II, to form the activation complex where chromatin is fully modified (c), and (5) transcription of the ERE gene commences. (6) Direct competitive inhibition of SRC binding to NR will block the initial step (3) of activation complex formation and thus prevent transcription. (See Color Insert.)

hormone receptor β (T_3–hTRβ) complexes appear to bind noncooperatively the second NR box of SRC2 (SRC2-2) NH_2-[685]EKHKIL$_1$ERL$_2$L$_3$KDS[697]-COOH.[10] Both, hERα and hTRβ, interact with SRC2-2 peptide through similar surfaces that are highly conserved across the NR superfamily. A shallow hydrophobic groove is formed on the surface of the receptor by residues on the faces of helices 3, 4, 5, and 12 in response to hormone binding. An amphipathic α-helix is formed by the NR box of the SRC with the three leucines on the hydrophobic face of the NR box helix buried in

[10] J. Zhang and M. A. Lazar, *Annu. Rev. Physiol.* **62**, 439–466 (2000).

[11] W. Feng, R. C. Ribeiro, R. L. Wagner, H. Nguyen, J. W. Apriletti, R. J. Fletterick, J. D. Baxter, P. J. Kushner, and B. L. West, *Science* **280**, 1747–1749 (1998).

[12] A. K. Shiau, D. Barstad, P. M. Loria, L. Cheng, P. J. Kushner, D. A. Agard, and G. L. Greene, *Cell* **95**, 927–937 (1998).

[13] H. E. Xu, M. H. Lambert, V. G. Montana, K. D. Plunket, L. B. Moore, J. L. Collins, J. A. Oplinger, S. A. Kliewer, R. T. Gampe, Jr., D. D. McKee, J. T. Moore, and T. M. Willson, *Proc. Natl. Acad. Sci. USA* **98**, 13919–13924 (2001).

[14] R. T. Gampe, Jr., V. G. Montana, M. H. Lambert, G. B. Wisely, M. V. Milburn, and H. E. Xu, *Genes Dev.* **14**, 2229–2241 (2000).

[15] R. T. Gampe, Jr., V. G. Montana, M. H. Lambert, A. B. Miller, R. K. Bledsoe, M. V. Milburn, S. A. Kliewer, T. M. Willson, and H. E. Xu, *Mol. Cell* **5**, 545–555 (2000).

[16] R. C. Ribeiro, J. W. Apriletti, R. L. Wagner, B. L. West, W. Feng, R. Huber, P. J. Kushner, S. Nilsson, T. Scanlan, R. J. Fletterick, F. Schaufele, and J. D. Baxter, *Recent Prog. Horm. Res.* **53**, 351–392 (1998).

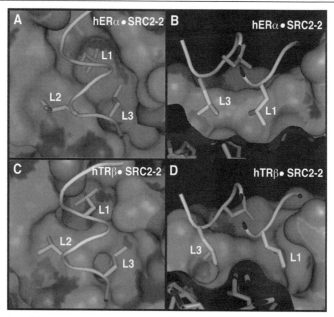

FIG. 2. Panel A: Cocrystal structure of the SRC-binding pocket in E_2–hERα–SRC2-2[12] oriented to show a view from above the pocket. The three leucine side chains of the SRC2-2 ($L_1XXL_2L_3$) motif (yellow wire frame) and labeled L_1 L_2 L_3. The E_2–hERα receptor van der Waals radii surface was generated and color coded according to atom type: red–oxygen, blue–nitrogen, green–carbon and hydrogen, orange–sulfur (*PyMol*[TM]). Panel B: Cross section of the same binding site as in A, approximately through the middle of the L_1 and L_3 subpockets. Panel C: Cocrystal structure of the SRC-binding pocket in T_3–hTRβ–SRC2-2[2] oriented in the same view as that of panel A. Panel D: Cross section of the same binding site in B, oriented as in C. These views demonstrate clear significant differences in steric structure between the hTRβ and hERα SRC-binding pockets. (See Color Insert.)

the groove of the NR during binding (Fig. 2).[2,9,11–16] In many cases, this interaction appears to provide the majority of binding energy for the interaction between NR and coactivator, thus allowing formation of a fully functional transcriptional complex on the response element.

Peptidomimetic analogs of the coactivator NR boxes will inhibit competitively the interaction of the NR and coactivator, often with increased affinity for the NR relative to the native peptide. As peptidomimetics are quite stable to proteolysis owing to their inability to form the extended beta strand conformation normally required by proteases, they can be useful tools for dissecting the function of individual protein interactions. The synthesis and testing of peptidomimetics, in which the leucines of the NR box were replaced with nonnatural amino acids, identified compounds that specifically block the binding of SRC2 to E_2–hERα without affecting the coactivator's interaction with T_3–hTRβ. These results argue that the NR–SRC binding pocket, although evolved to recognize the $L_1XXL_2L_3$ motif *in vivo*, has retained differences in shape and electrostatics between family members that may be exploited by competitive inhibitors mimicking the NR box. The methods for preparation of libraries of potentially selective competitive inhibitors and their evaluation are described in the next section.

Design of Peptidomimetic Libraries

While the differences between NR coactivator interfaces appear small at the conformational level, careful examination of receptor surfaces at the binding site reveals subtle differences. For example, the E_2–hERα–SRC2-2 interaction (Figs. 2A and C) has a ridge adjacent to the L_2 subpocket with a continuous electrostatic potential, whereas the T_3–hTRβ–SRC2-2 interaction (Figs. 2B and D) has a subpocket that is hindered by a more significant plateau. Additionally, a pronounced ridge in hTRβ creates more steric hindrance between the L_1 and L_3 subpockets (Figs. 2C vs D). Similar differences between other receptor pairs can be exploited in the following method.

The design of the library consists of selecting the correct amino acids for inclusion in the peptides. Some 3000 + nonnatural amino acids from the publicly available chemicals directory (ACD) (MDL Information Systems, Inc.) were identified (MDL-ISIS/Draw 2.4) and screened *in silico* by DOCKing to the coactivator-binding surfaces of relevant receptor structures. For example, the SRC2-2 peptide from each of the cocrystal structures, T_3–hTRβ–SRC2-2 and DES–hERα–SRC2-2, was utilized as the scaffold for the presentation of nonnatural amino acid leucine replacements to DOCK into a crystallographically defined negative image of each receptor. The SRC2-2 peptide of individual protein complex crystal structures was edited to remove the distal atoms of the three leucine side chains to Cβ, leaving the rest of the peptide intact. Each of the conformers generated from potential amino acids was then attached to the SRC2-2 structure by anchoring the components according to the Cα and Cβ carbon coordinates of the leucines. The compounds were then evaluated for binding using CombiDOCK in two ways: (1) with each of the leucine positions varied combinatorially to generate 5896^3 possible combinations, and (2) with each position varied separately while holding the other two positions as leucine to generate 5896×3 possible compounds. The SRC2-2 peptide backbone was held rigid to maintain the α-helical structure, while each compound was energy minimized in the receptor pocket to maximize the size–shape complimentarity.

CombiDOCK Method

Selection of diversity reagents was carried out with CombiDOCK[17] (UCSF, Computer Graphics Laboratory)[18] on an Octane workstation (Silicon Graphics Inc.) running IRIX 6.5. Publicly available structures, T_3–hTRβ–SRC2-2 (PDB: 1BSX) and DES–hERα–SRC2-2 (PDB: 3ERD) were utilized for computational evaluation.

1. Preparing the receptor target site by creating a negative image of the binding pocket.
   a. Solvent accessible surface maps were generated for the receptor surface area within the coactivator-binding pocket using a 1.4 Å sphere by the method of Connolly[19] (Sybyl/BASE®[20] and MOLCAD™).
   b. Each receptor coactivator-binding pocket was prepared by creating a negative image of the surface by (*Sphgen*)[17,21–23] using default setting parameters, which generated a sphere file (receptor.sph) for each receptor site.
   c. A contact grid was generated (GRID)[24] with default parameters except for the following modified parameters: (grid spacing 0.3; contact cutoff 4.5; energy cutoff distance 10; bump overlap 0.75). No electrostatics was applied. This routine generated a grid file (receptor.grd) for each receptor site.
2. Generating a library of nonnatural amino acids from the ACD.
   a. The ACD was searched for α-amino acid molecules (AA) with points of diversity at the Cα position (ISIS/Base)™.
   b. AA structures were converted to SMILES strings with unique FCD numbering (UCSelect[25] and *Merlin*).[26] A Daylight fingerprint[27] with fixed length and low density was generated for each

[17] Y. Sun, T. J. Ewing, A. G. Skillman, and I. D. Kuntz, *J. Comput. Aided Mol. Des.* **12**, 597–604 (1998).

[18] http://www.cgl.ucsf.edu/index.html, U.C.G.L.

[19] M. L. Connolly, *Science* **221**, 709–713 (1983).

[20] Tripos Associates, I. 1699 South Hanley Road, Suite 303, St. Louis, Missouri 63144-2913, USA.

[21] I. D. Kuntz, J. M. Blaney, S. J. Oatley, R. Langridge, and T. E. Ferrin, *J. Mol. Biol.* **161**, 269–288 (1982).

[22] R. L. DesJarlais, R. P. Sheridan, G. L. Seibel, J. S. Dixon, I. D. Kuntz, and R. Venkataraghavan, *J. Med. Chem.* **31**, 722–729 (1988).

[23] T. J. Ewing, S. Makino, A. G. Skillman, and I. D. Kuntz, *J. Comput. Aided Mol. Des.* **15**, 411–428 (2001).

[24] E. C. Meng, B. K. Shoichet, and I. D. Kuntz, *J. Comp. Chem.* **13**, 505–524 (1992).

[25] A. G. Skillman, Ph.D. Thesis, UCSF, 1999.

[26] 4.6, D.V. Daylight Chemical Information Systems, Inc., Santa Fe, NM, 1997.

[27] http://www.daylight.com/dayhtml/doc/theory/theory.finger.html.

molecule; selections were based on: hydrophobic side chains, C log P values (a logarithm of octanol–water partitioning), size (MW < 500), and reactive functional groups.

c. AA were clustered hierarchically (*Cluster*)[26] to remove redundant side chains and groups that filled the same conformational space on the basis of the radial overlap between spheres of each molecule. A $n \times n$ distance matrix was calculated for the similarity of the compounds based on nearest neighbors command.[26] The similarity cutoff was changed (final sim.cutoff = 0.78) until a set of < 500 molecules was defined as different, and these were chosen for the library.[28]

d. Each amino acid was then reduced to its side chain, including the alpha carbon (Cα) (*Sybyl*)®.[29] Anchoring numbers were defined as Cα = 3, Cβ = 2, Cγ = 1, and applied to each molecule within the PDB file with an in-house script.[30]

e. All energetically refined conformational rotamers were generated (*Omega*).[31] Input global parameters were: RMS cutoff 0.8 energy window 6.0; maximum conformational output number was evaluated at different levels and a final range of 10–100 per molecule was utilized; maximum population size 200,000; maximum rotors per molecule 17. This routine generated a torlib.txt torsion library.

f. The heavy atoms from the SRC2-2 peptide were then added to the side chain library assembly with anchoring set as: N = 3, Cα = 2, Cβ = 1, for each of the leucine positions (*Sybyl*)®.

g. A final set of molecules was output to a fragment molecule file (fragment.mol2) to be utilized by the CombiDOCK program discussed below.

13. Testing the fit of the leucine replacements to the NR box binding sites. All of the generated peptidomimetics were evaluated *in silico* using CombiDOCK[17] with the following parameters:

a. CombiDOCK scoring parameters—no old grids; bump filter set with no maximum; contact score set; no chemical score; energy scoring and interpolation set; a van der Waals scale range of 1.0–2.0 Å during different evaluations; no electrostatic scale; no energy maximum.

[28] H. Matter, *J. Biolumin. Chemilumin.* **40**, 1219–1229 (1997).

[29] In House Perl Script UCSF.

[30] Anchoring – Script, In House Perl Script UCSF.

[31] Open Eye Software http://www.eyesopen.com/; J. Boström, J. R. Greenwood, and J. Gottfries, Assessing the performance of OMEGA with respect to retrieving bioactive conformations, *J. Mol. Graphics and Modelling* **21**, 449 (2003).

b. CombiDOCK minimization parameters—energy minimization set; probe minimization set; the orientation and position of the peptide was allowed to deviate from the original starting position by a maximum RMSD of 1.5 Å; an energy convergence of 1.0; no torsion minimization; a maximum of 100 iterations; initial translation tolerance of 1.0; initial rotation tolerance of 5.0.

c. CombiDOCK matching parameters—matching of ligand centers; three nodes minimum; six nodes maximum; distance tolerance of 0.2–0.8 were set for different evaluations; distance minimum was 1.0–2.0; 100,000 maximum matches; no match ratio minimum, check degeneracy, critical spheres, or chemical matches were performed.

d. CombiDOCK combinatorial parameters—three total sites (corresponding to a_x, b_x, and c_x, or each leucine); check clashes set; clash dependents were set for each position relative to each other (1 vs 2, 1 vs 3, 2 vs 3); all 5896 ligands were placed at each position in a combinatorial or sequential manner (one position at a time while holding the remaining positions as the original leucines); greedy conformations were varied from 10 to 100; fragments were merged; 20 maximum anchor torsions; precomputed clashes were set to be evaluated; greedy scaffolds were varied from 10 to 500; and one probe was set at each position for initial scaffold configuration; the scaffold was minimized with methyl groups at each position. Side chain spheres that violated van der Waals distances with the SRC2-2 peptide or each other were automatically discarded due to clash violations before docking proceeded. The scaffold was minimized to the receptor site with methyl groups, and then each side chain compound was added and evaluated.

e. General parameters—ligand was scored and oriented in a combinatorial or noncombinatorial fashion; no rule of five; multiple orientations evaluated.

f. Active site mode parameters were: energy window 10.0; chemscore grid energy grid resolution of 0.3; maximum random translation 1.0; maximum random rotation 30.0; number of anchor positions 200; output number of conformers 10.

g. The van der Waals component uses a modification of the Leonard–Jones 6-12 attraction and repulsion components with distance dependence [Eq. (2)].

h. Contact between non-hydrogen atoms that are within 4.5 Å are evaluated and penalized as defined by the user.

4. CombiDOCK reads in the receptor files (receptor.mol, .grd, .sph) and the fragment (.mol) library file. A van der Waals definition file

(vdw.def) is read also in conjunction with grid. Chemical Scoring, which was not employed here, can be used to incorporate empirical theories and constraints, and modifications of the VDW terms.

Final structures were sorted on the basis of lowest energy conformations. The distance dependence van der Waals docking parameter used Eq. (1), where E_{VDW} is the intermolecular interaction energy; C and D are calculated coefficients; e is the well depth of interaction energy; R is the van der Waals radius of the atoms; r is the distance between atoms; and a and b are the van der Waals repulsion and attraction components, respectively.

$$E_{VDW} = Ce(2R/r)^a - De(2R/r)^b. \tag{1}$$

Each top-scoring compound was evaluated visually to ensure that the molecules were minimized into a reasonable conformation in the receptor pocket (Sybyl)[®]. Depending on the parameter settings and scoring constraints, molecules could dock into a conformation that was dramatically different from that of the crystal structure. For example, a large polycyclic monomethyltrityl protecting group on the histidine (MTT) (Fig. 3) was used as a negative control which, when given the appropriate parameter settings, would score well; however, the entire scaffold would then be shifted outside the pocket. The natural leucine side chain served as a positive control. This final subset of molecules was then re-evaluated with a dynamic scaffold (CombiDOCK general parameters, orient ligand—yes) and allowed to deviate in orientation by 3 Å from the starting position by changing the CombiDOCK minimization parameters (max_rmsd, initial translation, and rotation), and were determined to have similar relative scores and final energy-minimized orientations. Often, different results will be obtained depending upon whether or not the scaffold is fixed during the CombiDOCK run. For example, with T$_3$–hTRβ–SRC2-2 and DES–hERα–SRC2-2, the fixed scaffold tended to select residues, such as phenylglycines and phenylglycine analogs (Fig. 3), that were shorter and/or could obtain orientations that allowed the overall helix to bind more closely to the receptor structure. On the other hand, the dynamic scaffold tended to select for residues, such as phenylalanines and related analogues (Fig. 3), that forced the overall helix to bind in an orientation farther from the receptor surface. It is beneficial to include both subsets of side chains in the targeted library.

While this approach is limited by a lack of consideration for dynamics in receptor surface structure, electrostatics, and solvation properties, it was useful in reducing the size of the target library to members that would bind to the receptor in a conformation that is similar to that seen in the crystal structures. These other parameters can be considered with tremendous increase in processor time.

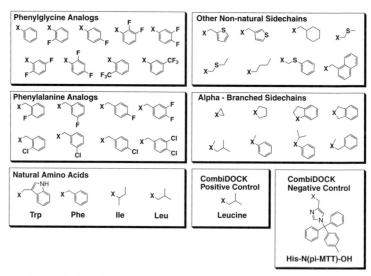

FIG. 3. Targeted diversity elements included in peptidomimetic library as chosen by CombiDOCK. This subset of amino acids was hydrophobic, exhibited good shape complementarity to the leucine-binding pockets, and were structurally diverse. Amino acid side chains shown with the Cα designated as X. Diversity reagents were prepared as Fmoc-protected α-amino acids and incorporated into the peptidomimetic library as individual substitutions at each of the leucine positions. Leucine and His(MTT) were the CombiDOCK controls.

Synthesis of the Peptidomimetic Libraries

Based on the combined results of biochemical and structural studies, indicating the minimal length of an NR box that is physiologically relevant for the interaction of hERα and hTRβ, we chose the 13 amino acid segment of the SRC2-2 peptide, NH_2-^{685}EKHKIL$_1$ERL$_2$L$_3$KDS^{697}-COOH, as a starting point for producing the library. With other NR, a different primary sequence may be more appropriate. Many methods for constraining peptides into a α-helical conformation[32] are available and some have been successfully applied to protein interfaces.[33–39] The starting point is synthesis of a series of lactam-bridged peptidomimetics that have potential to induce

[32] C. Blackburn and S. Kates, *Methods Enzymol.* **289**, 175–198 (1997).
[33] T. R. Geistlinger and R. K. Guy, *J. Am. Chem. Soc.* **123**, 1525–1526 (2001).
[34] M. Uesugi and G. L. Verdine, *Proc. Natl. Acad. Sci. USA* **96**, 14801–14806 (1999).
[35] J. K. Judice, J. Y. Tom, W. Huang, T. Wrin, J. Vennari, C. J. Petropoulos, and R. S. McDowell, *Proc. Natl. Acad. Sci. USA* **94**, 13426–13430 (1997).
[36] J. W. Taylor, N. J. Greenfield, B. Wu, and P. L. Privalov, *J. Mol. Biol.* **291**, 965–976 (1999).
[37] C. Yu and J. W. Taylor, *Bioorg. Med. Chem.* **7**, 161–175 (1999).
[38] T. R. Geistlinger and R. Kiplin Guy, *J. Am. Chem. Soc.* **125** (23), 6852–6853 (2003).
[39] C. Schafmeister, J. Po, and G. Verdine, *J. Am. Chem. Soc.* **122**, 5891–5892 (2000).

an α-helical conformation in some peptide sequences. It is essentially impossible to predict which constraint will yield good functional α-helical character with a given peptide sequence, so synthesis of the entire family is required for each new sequence. Additionally, some constraints may induce a negative interaction with individual receptors, so it is important to validate constraint choices for individual NR boxes by assessing the binding of the trileucine peptidomimetic to the receptor of interest. In targeting T_3–hTRβ, one constraint {c(E^{691}-K^{695}) Ac-^{685}EKHKIL$_1$ERL$_2$L$_3$KDS^{697}-COOH} was effective and gave a 15-fold tighter binding constant. While the same constraint works with E2–hERα, an interaction with the receptor lowers the absolute binding constant for this particular peptidomimetic equivalent to that of the unconstrained peptide. This compound was used as a scaffold for a second-generation library that substituted nonnatural amino acids for each of the leucines and was applied to both receptors.

Safety Note. Many of the chemicals used for peptide synthesis, particularly DIEA, DMF, and TFA, are both volatile and toxic. Make sure to read and understand the appropriate material safety data sheets. Wear appropriate personal protective equipment including goggles, a laboratory coat, appropriate gloves (generally nitrile or viton), and, where appropriate, a respirator capable of blocking fumes. Always work in a fume hood.

Lactam Constrained Peptide Synthesis

A series of nine peptidomimetics was synthesized by solid-phase peptide synthesis, using the Fmoc synthesis strategy with orthogonal protection of the relevant lactam precursor side chains. The macrolactam was formed on bead prior to final deprotection (Fig. 4). These peptidomimetics included variation in the location of the constraining moiety, the length of the constraint, and the orientation of the lactam carbamide group (Fig. 5). Once a particular constraint is proven to be functional, it serves as the starting material for the synthesis of the second-generation library using nonnatural amino acids.

Preparation of Fmoc-Protected Amino Acids

The nonnatural amino acids [NH2-X(1-38)-OH] were Fmoc protected at the primary amine and purified by acid–base extraction and recrystallization.

1. Alpha amino acids (2–3 mmol) were suspended in phosphate buffer (200 mM, pH 10, 300 ml) and kept on ice.
2. Fmoc-succinimidyl ester (NovaBiochem, CA) was prepared freshly in DMF on ice and slowly added at 0, 3, and 8 hr (1 mmol was

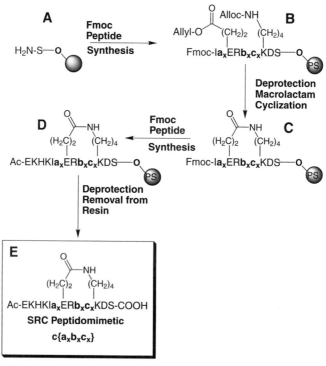

FIG. 4. Solid-phase synthetic scheme for the lactam constrained SRC2-2 peptidomimetic. (A) Wang resin was loaded and Fmoc peptide synthesis proceeded, until the peptide was two amino acids beyond the bridging residues to generate (B). Bridging residues were deprotected and the lactam constraint was formed (C). Fmoc peptide synthesis was continued until the end of the peptide and acetylated (D). Deprotection and removal from the resin under acidic conditions generated the final SRC peptidomimetic (E).

prepared each time), with the reaction being allowed to stir for 24 hr after the final addition. During the reaction, the pH was maintained at 10 by incremental addition of phosphate buffer (500 mM).

3. At the conclusion of the reaction, the mixture was washed twice with hexane, and acidified to pH 2.5 with 1 M HCl. This addition induced a precipitation.

4. The precipitant was extracted into ethyl acetate, washed with acidified brine (three times NaCl sat., pH 2.5), dilute acid (three times HCl, pH 2.5), and dried by rotary evaporation. This step typically gave either a lightly colored solid or oil.

5. The product was washed with hexanes.

6. Purity was measured by TLC on silica plates (Merck EM with 254 indicator) and monitored by UV$_{254\,nm}$ [40:3:2, dichloromethane

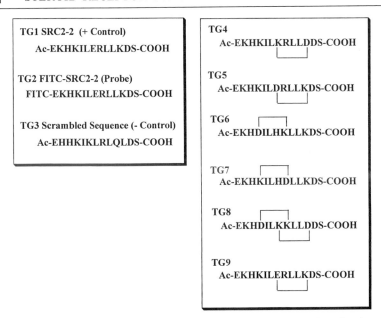

FIG. 5. Library of peptidomimetics designed to evaluate several constraints showing: name, single letter amino acid sequence with cartoon of constraint position, and molecular structure.

(DCM)–methanol (MeOH)–acetic acid (AcOH)—R_f values were 0.0, 0.5, and 0.9 for the free amino acid, the protected amino acid, and the Fmoc group, respectively]. UV exposure identified the Fmoc-protected amino acids, and staining with ninhydrin identified the free amine of the nonprotected amino acid. The resulting products were analyzed by 1D ^{1}H NMR (400 MHz Varian Innova, 1–5 mg/ml, CDCl$_3$ with internal TMS standard) to assess purity. The product can be confirmed by the easily observed downfield shift of the CαH typically from 3.5–3.9 ppm in the unprotected amino acid to 4.5–4.9 ppm in the protected amino acid (ppm relative to TMS). Compounds were determined to be >95% pure.

7. Some of the Cα-branched amino acids (Fig. 3) required further purification by silica gel chromatography (Merck EM silica, 63–230 mesh, 40:3:2, DCM–MeOH–AcOH) to achieve >95% purity.

Preparation of Peptidomimetics

Peptidomimetics were synthesized (Fig. 4) on solid support in Robbins Blocks^{TM} (48 wells/block, 3 ml/well capacity) with a resin loading of

100 mg/well using the Wang™ resin—a 1% cross-linked polystyrene bead functionalized with a TFA labile *p*-amino benzyl alcohol handle. Resins were purchased preloaded with the first amino acid of the sequence with a lower loading capacity (< 0.5 mmol/g) to reduce interpeptide cross-linking during the lactam formation. Alternatively, an underivatized Wang resin can be loaded using 1-(mesitylene-2-sulfonyl)-3-nitro-1,2,4-triazolide (MSNT) (NovaBiochem, CA). Fmoc amino acids were utilized with the standard protecting groups [Fmoc-Lys(Boc)-OH, Fmoc-Glu(o-tButyl)-OH, Fmoc-Asp(o-tButyl)-OH, Fmoc-Arg(Pdf)-OH, Fmoc-His(Trt)-OH, Fmoc-Ile-OH, Fmoc-Leu-OH, Fmoc-Gln(Trt)-OH, Fmoc-Ser(tBut)-OH)], with the exception of the Fmoc-Lys(alloc)-OH, Fmoc-Glu(Allyl)-OH, and Fmoc-Asp(Allyl)-OH used to form the lactam constraint. Natural, alpha amino acids were purchased from NovaBiochem, San Diego, CA (NovaBiochem, CA). Non-natural alpha amino acids (Fig. 3) were purchased from various sources: ABCR GmbH & Co. KG, Karlsruhe, Germany, Peptech Corp., Cambridge, MA, Advanced Chemtech, Louisville, KY, Chem Impex, Wood Dale, IL, Bachem California, Torrance, CA, ACROS Organics, Merris Plains, NJ, Salor, Milwaukee, WI, and protected as the Fmoc amino acid as described earlier.

1. Addition of each amino acid to the growing polypeptide chain followed a cycle of:
   a. Removal of the amino terminal Fmoc protecting group (20% piperidine in DMF, 10 min, room temperature (rt), repeated twice per cycle).
   b. Addition of the next Fmoc amino acid: natural amino acids (3 equivalents), 2-(1*H*-benzotriazole-1-yl)-1,1,3,3-tetramethyl-uronium hexafluorophosphate (HBTU) (3 equivalents), diiso-propylethylamine (DIEA) (3–5 equivalents) in DMF (2 ml), 0.5–2 hr, rt, repeated twice per cycle); nonnatural amino acids—HBTU (1 equivalent), tetramethyluronium hexafluoro-phosphate, benzotriazol-1-yl-oxytripyrrolidiniophosphonium hexafluorophosphate (PyBOP) (1 equivalent), O-97-azabenzo-triazol-1-yl)-1,1,3,3-tetramethyluronium hexafluorophosphate (HATU) (1 equivalent) (Perseptive Biosystems, MA), tetra-methylfluoroformamidinium hexafluorophosphate (TFFH) (1 equivalent), DIEA (5 equivalents) in DMF (2 ml), 2 hr, rt, repeated three times per cycle).
   c. Monitoring each step using the Ninhydrin test.[40]

---

[40] F. Albericio, P. Lloyd-Williams, and E. Giralt, *Methods Enzymol.* **289**, 313 (1997).

2. Once the peptide chain had reached isoleucine[689], lactam formation (cyclization) was performed. Formation of the lactam bridging group followed the sequence of:

  a. Deprotection of allyl and alloc protecting groups using tetrakis(triphenylphosphine)palladium (0) [(Pd(Ph₃P)₄] (0.02 equivalent) [STREM, Newburyport, MA], N,N-dimethyl barbituric acid (3 equivalents), DCM, rt, 2 hr, repeated twice.

  b. Cyclization [PyBOP (1 equivalent), hydroxybenzotriazole (HOBt) (1 equivalent), DIEA (3–5 equivalents); in 1:1 NMP–DMF, shaken, 2 hr, rt, repeated twice].

  c. Both steps were monitored via the ninhydrin test for presence of a free primary amine to evaluate progress. After each deprotection step, a positive blue color was noted. This color was reduced after each round of lactam coupling. The cycle was repeated three times for each of the constrained peptides.

3. Peptide synthesis was continued until the ends of the peptides were reached using the procedure described earlier.

4. The amino termini of the peptides were acetylated with acetic anhydride [Ac₂O (3 equivalents), DIEA (3 equivalents), 30 min, rt].

5. The peptides were cleaved from the resin with concomitant side chain deprotection under acidic conditions with scavengers [95% trifluoroacetic acid (TFA); 1% H₂O, 2% phenol, 2% thioanisol, rt, 3 hr]. (*Warning: Causes severe burns. Toxic by ingestion, inhalation, and through skin contact. Very destructive of mucous membranes.*)

6. For large-scale preparations, compounds were precipitated into diethyl ether, filtered or centrifuged, and decanted, and washed twice with ether. For smaller scale library preparations, the TFA was evaporated in a Genevac HT-4 Series II Evaporating System (Genevac Limited, Ipswich, UK). (*Warning: Diethyl ether may form explosive peroxides upon long standing with exposure to air.*)

7. The peptidomimetics were purified by RP-HPLC (RP-C₁₈, Reliasil, BDX-C18, 5 μm, 50 × 21 mm) preparative column (Column Engineering, Ontario, CA) or RP-C₁₈ Xterra, 50 mm × 19 mm, 5 μm beads, 120A pores (Waters, Wooster, MA), 0.1% TFA in water to 0.1% TFA in 80% acetonitrile, 30 ml/min, 5 min gradient. All compounds eluted at 18–22% ACN, with UV detection at 215 and 280 nm. (Purification was facilitated by the use of a parallel preparative scale BIOTAGE HPLC system.)

8. Compounds were verified by MALDI MS to give satisfactory unit mass [DE-STR-MALDI (Precision Biosystems, Framingham, MA). The matrix was MALDI-Quality alpha-cyano-4-hydroxycinnamic acid in methanol (Hewlett-Packard)]. Purity was then confirmed by

HPLC–ESMS (300 μm × 25 cm Waters Xterra™ C_{18} HPLC column) 100 μl/min, 15–30% MeOH–H$_2$O gradient over 10 min, at 25 and 40°C. Compounds were analyzed by UV$_{215}$, total ion count, and expected mass (m/z) (Finnigan Mat LCQ quadrupole ion trap mass spectrometer).

Synthesis of Labeled Probe FITC–SRC2-2

SRC2-2 peptide was prepared as described earlier except that the amino terminus was left uncapped. A solution of the SRC2-2 peptide (0.8 μmol) in aqueous DMF [100 μl 1:1, 200 mM sodium phosphate (pH 7)–DMF] was treated with Oregon Green 488 succinimidyl ester (OG) (Molecular Probes 0-6147, 3.2 μM in 50 μl DMF) and allowed to stir at rt for 2 hr. The resulting fluorescently labeled peptide 2 (OG–SRC2-2) was purified by RP-HPLC as earlier.

Secondary Structure Evaluation of Peptidomimetics

Solution phase circular dichroism (CD) analysis of the initial constraints will reveal the degree of induction of α-helical character. In the case of the SRC2-2 NR box, one, c(E^{691}-K^{695})Ac-EKHKILERLLKDS-COOH gave helical character. CD spectra were collected on a JASCO spectrapolarimeter (Japan). A solution of peptide (50 μM in 20% acetonitrile in 50 mM Tris–HCl, pH 8.0) was placed in a 0.2 cm path length cuvette and the CD spectra measured from 190 to 250 nm (Fig. 6). The spectrum was normalized using a solution of buffer without peptide.

Assay of the Peptidomimetic Efficacy and Selectivity

While many methods are available for evaluating protein–protein interactions and their competitive inhibition, we chose to utilize fluorescence polarization (FP) competition assays and solid-supported affinity chromatography (AC) competition assays.

The FP assay is an equilibrium homogeneous solution phase assay, suitable for determining affinity and inhibitory constants in systems that have fast off rates, which is a potential concern with these interactions that normally have an affinity constant of nearly 1 μM. The assay utilizes a short peptide probe labeled with a fluorophore that is excited with polarized light. When the probe is not bound, its tumbling rate is rapid relative to the fluorescence lifetime, and the polarization of the emitted light is scrambled. As the probe binds to receptor, the rate of tumbling becomes slower than the

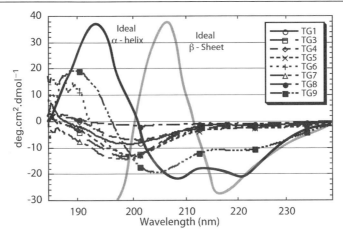

FIG. 6. All peptides were tested by CD. CD spectra of all the linear and constrained macrolactam 13 amino acid peptides were measured on a JASCO™ instrument. An ideal α-helix and β-sheet are illustrated. This figure demonstrates that only **TG9** has any significant α-helix spectrum. Spectra were acquired by scanning solutions of each compound at 50 μM in 20% ACN in 50 mM Tris pH 8.0. Mean residue ellipticity (Θ) reported in deg cm^2/dmol. (B) Temperature dependence of CD spectra of the unconstrained peptide **TG1** and constrained peptide **TG9**. Spectra were acquired by scanning solutions of **TG1** and **TG9** (50 μM in 20% ACN in 50 mM Tris pH 8.0) at various temperatures. Mean residue ellipticity (Θ) reported in deg cm^2/dmol.

fluorescence lifetime, thus increasing the polarization of emitted light (Fig. 7). An effective competitor will release the probe from the protein and reduce the polarization values (Fig. 8). Fortuitously in the coactivator interaction, a complete induced fit of the NR box is observed, so a peptide of 11 or more amino acids spanning the NR box appears to mimic faithfully the affinity of the full-length coactivator protein. We utilize a 13 AA peptide spanning the NR box of interest with an Oregon Green 488 or fluorescein fluorophore attached to the N-terminus. A thorough evaluation of the binding buffer is important at this point, since small changes in buffer composition can have significant effects on the results of the assay.

In Vitro *Assay Conditions*

We studied carefully the effects of pH and salt concentrations on the interaction of TR and ER with SRC2-2, and defined on an optimal buffer (20 mM Tris pH 7.4, 100 mM NaCl, 1 mM DTT, 1 mM EDTA, 0.01% NP-40) (Fig. 7B). Small changes in any of the buffer components have significant effects on the binding assay. Collagen can be included in the

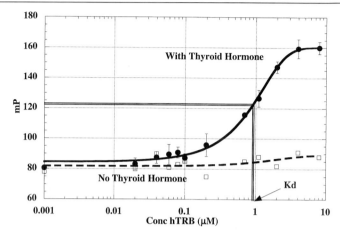

FIG. 7. Fluorescence polarization direct-binding analysis. (●) Direct binding of OG-SRC2-2 (**TG2**) with hTRβ in the presence of thyroid hormone (T_3). The binding curve was saturable at both ends and the K_d of the probe was determined to be 1 μM. (□) No binding of SRC2-2 to hTRβ was seen in the absence of T_3. A final buffer condition was chosen comprising 20 mM Tris pH 7.4, 100 mM NaCl, 1 mM DTT, 1 mM EDTA, 0.01% NP-40, protease inhibitors.

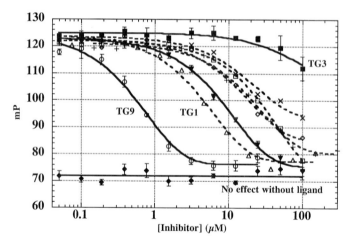

FIG. 8. Fluorescence polarization competition analysis inhibition of SRC2-2 binding to hTRβ by constrained SRC2-2 analogs. [Determined by FP competition experiments using an Oregon Green 488 labeled SRC2-2 peptide (**TG2**).] (◆) No T3-probe binding is dependent upon the presence of T_3 ligand; (▼) unconstrained SRC2-2 peptide (**TG1**) effectively blocks binding of (**TG2**), IC_{50} of 9.6 ± 0.9 μM; (■) control peptide (**TG3**) is unable to block binding of (**TG2**); constrained peptides block binding of (**TG2**): (+) (**TG4**), IC_{50} 21.8 ± 4.0 μM; (△) (**TG5**), IC_{50} 5.8 ± 0.7 μM; (X) (**TG6**), IC_{50} 22.1 ± 2.2 μM; (◇) (**TG7**), IC_{50} 9.1 ± 2.2 μM; (□) (**TG8**), IC_{50} 20.4 ± 8.6 μM; (○) (**TG9**), IC_{50} of 0.62 ± 0.09 μM. [**TG2**] was constant at 10 nM, [hTRβ] 1 μM, [T_3] 10 μM. Binding buffer: 20 mM Tris–HCl pH 8.0, 100 mM NaCl, 10% glycerol, 1 mM DTT, 0.01% NP-40, 1 mM EDTA.

buffer to block nonspecific binding of the probe to the microtiter plate; however, serum albumin should not be used as it often binds nonspecifically to the probe. Because the assay measures a ratio of bound to free probe, it is important that the [protein] $\gg$ [probe] or that K_d Probe $\gg$ [Probe]. In addition, one of the major potential perturbing phenomena in this assay is unexpected quenching or enhancement of fluorescence intensity. We control for this perturbation by simultaneously monitoring overall fluorescence intensity and fluorescence anisotropy for each sample. A G-factor for each probe should be determined, and this value should not change significantly over the course of the assay. Fluorescence intensity also should not change over the conditions of the assay. It is advisable to test the quality of the receptor and probe by performing a binding assay prior to executing competition experiments.

In the case of probe **TG2**, binding to hTRβ and hERα was saturable and reached equilibrium within 10 min, was dependent on the presence of T$_3$ and E$_2$ ligands for hTRβ and hERα, and the observed K_d's were 1 μM and 200 nM, respectively (Fig. 7). Competition for binding to the receptors was exhibited by unconstrained peptide **TG1**, reaching equilibrium within 1 hr, but not by a negative control peptide **TG3** with a sequence-scrambled NR box.

The AC assay is an inhomogeneous nonequilibrium method that depends on a reasonably slow off-rate (minutes or slower) to give an accurate affinity constant or inhibitory constant. It has the distinct advantage over the FP assay of allowing the use of probes that are larger than 25 AA. The AC assay is therefore used as a secondary test for the ability of potential inhibitors to disrupt the interaction between full length (or large domains) of SRCs and full length (or LBD) of NR. While in the example cases of T$_3$–hTRβ–SRC2-2 and E$_2$–hERα–SRC2-2, the binding constants of the SRC2-2 peptide and the SRC2 protein are equivalent; it is important to verify this assumption for each new case.

Equilibrium Binding of the Probe to the Receptor

Binding and competition experiments are typically performed with a FP capable plate reader. We used the LJL Biosystems Analyst (Foster City, CA).

1. Testing the probe stability and G-factor.
   a. The probe (OG-SRC2-2) is serially diluted in binding buffer, for example, 20 mM Tris pH 7.4, 100 mM NaCl, 1 mM DTT, 1 mM EDTA, 0.01% NP-40, and added to a black 384-well microtiter plate.
   b. Samples are read by FP instrument monitoring fluorescence intensity and FP.

   c.   *G*-factor is calculated by Eq. (2) or (3) for each probe, and polarization values are calculated based on Eq. (4) or (5) (see Data Analysis).

2. Purified NR LBD is serially added into the plate with increasing concentration (12 dilutions were made in increments with an estimated K_d of 1 μM as the halfway point).
3. Relative polarization values and the fluorescence intensity are measured for each concentration of LBD.
4. Data are plotted on a semilog scale (Klotz plot) (Fig. 7) and K_d is determined by the concentration of receptor required to bind one-half of the probe, and the polarization is changed to halfway between the baseline and maximum.
5. A control should be performed in the absence of ligand to confirm ligand dependence.

FP Competition Assay

Based on the data from these preliminary studies, competition experiments were carried out using the following conditions: the probe concentration was held constant at 10 nM; the hTRβ LBD concentration constant at 2 μM, and the thyroid hormone (T$_3$) concentration constant at 10 μM. The binding buffer was 20 mM Tris pH 7.4, 100 mM NaCl, 1 mM DTT, 1 mM EDTA, 0.01% NP-40.

FP Competition Procedure

1. Potential competitors are evaluated experimentally by preequilibrating the receptor ($2 \times K_d$) with probe [concentrations of which depend on the limits of linear FP response (1–10 nM)], and hormone ($10 \times K_d$) for 1 hr at room temperature in binding buffer.
2. Ten microliters of the above reaction are added to each well of a black 384-well plate.
3. As an initial assay control, each well is evaluated for fluorescence intensity and FP to ensure no variation across the samples and that the samples are polarized equally relative to free probe.
4. Inhibitors are diluted in a separate 96-well plate for 12 dilutions, usually in 1/2 log intervals.
5. Ten microliters of each diluted inhibitor are transferred to individual wells of the plate. Experiments are conducted in quadruplicate.
6. Samples are incubated and evaluated for competition with the probe at time intervals of 10 min, 1 hr, 8 hr, and 24 hr, to determine equilibrium assay conditions (equilibrium is reached usually by the first timepoint).

Identifying an Effective α-Helical Scaffold and Inhibitor

The ability of the constrained analogs to compete with SRC2-2 peptide for T_3–hTRβ was assessed using FP equilibrium competition assays (Fig. 8). Control experiments indicated that binding of probe **TG2** was dependent upon the presence of T_3 (◆), and that competition for binding to hTRβ was exhibited by unconstrained peptide **TG1**, but not by peptide **TG3**, which has a scrambled NR box (■). While all of the constrained SRC2-2 analogs could successfully compete for binding to hTRβ, most did so with affinities worse than or no better than the unconstrained peptide.

Data Analysis

Calculating a G-Factor

The *G*-factor for each probe on the LJL system should be determined according to Eq. (2), where *S* and *P* are the background-subtracted intensity measurements for the parallel and perpendicular components at a given concentration. P^{true} is a calculated constant based on the literature value for a particular probe. The *G*-factor value should be within the range of 0.8–1.2. The G_{factor} may also be calculated with Eq. (3) for an FP system with two polarizers, one on the excitation and another on the emission

$$G = S/P^*(1 - P^{\text{true}})/(1 + P^{\text{true}}) \qquad (2)$$

$$G_{\text{factor}} = (I_{\text{HV}}/I_{\text{HH}}) \qquad (3)$$

where I_{HV} is the intensity when the excitation polarizer is horizontal and the emission polarizer are vertical, and where I_{HH} is when both polarizers are horizontal. In this case, the *G*-factor can be calculated during the analysis, and the value should not change significantly over the course of the analysis with any one probe.

Actual polarization values on the LJL are determined according to Eq. (4), and anisotropy calculations on a two-polarizer system follow that of Eq. (5)

$$mP = 1000^*(S - G^*P)/(S + G^*P) \qquad (4)$$

$$r = [(I_{\parallel}/I_{\perp}) - 1]/[(I_{\parallel}/I_{\perp}) + 2] \qquad (5)$$
$$I_{\parallel}/I_{\perp} = (I_{\text{VV}}/I_{\text{VH}})^*1/G_{\text{factor}}$$

where *S* and *P* are the background-subtracted data and *G* is the *G*-factor, and where *I* is intensity, and V and H refer to orientation of the positions of the excitation and emission polarizers, respectively.

Calculating IC$_{50}$ Values

The data for each competition are fit via Klotz plots to determine IC$_{50}$ values using nonlinear regression analyses that fit the data to a modification of the model of Heyduk and Lee.[41,42] IC$_{50}$ values can be determined by the following equation (6).

$$y = \frac{\text{bottom} + (\text{top} - \text{bottom})}{1 + 10^{x - \log(\text{IC}_{50})}} \tag{6}$$

GST-SRC2 Pull Down Competition Assay

This is a solid support AC competition assay. It is important to establish whether the competitive ability of peptidomimetics determined in the florescence polarization competition assays is preserved in competition with an SRC NID domain. SRC2 has three NR box motifs in its NID, as well as additional potential protein interaction motifs. Therefore, the ability of a peptidomimetic to compete with intact SRC2 nuclear receptor interaction domain (SRC2 NID), containing all three SRC2 NR boxes, should be tested using a semiquantitative glutathione-*S*-transferase pull down assay. For example, with hTRβ, control experiments indicated that the SRC2 NID bound to hTRβ in the presence of T$_3$ (Fig. 9, lane 2) and failed to bind in the absence of T$_3$ (lane 3). The interaction was blocked by unconstrained peptide 1 at high concentration (lane 4) but not at low concentration (lane 5), and was not blocked by sequence-scrambled peptide **TG3** (lane 6). Constrained peptide **TG9** efficiently blocked the binding of the receptor to the SRC2 NID in a dose-dependent manner (lanes 7–10). Although the assay does not allow for exact determination of IC$_{50}$ values, the relative efficacy of **TG9** to **TG3** was qualitatively in the same range as that observed in the peptide studies. Additionally, the interaction was completely blocked by 10 μM **TG9**, whereas the unconstrained peptide never reached this level of saturation, even with 100 μM concentrations. This study indicates that the constrained peptide **TG9** can block the interaction of receptor and coactivator at the protein level. Assays are carried out using the method of Darimont *et al.*[4]

Conclusion

In this chapter we outline approaches to the design of α-helical coactivator peptidomimetics c{a$_x$b$_x$c$_x$}, and the identification of specific inhibitors that act at a highly conserved interface. These studies have been

[41] T. Heyduk, J. C. Lee, Y. W. Ebright, E. E. Blatter, Y. Zhou, and R. H. Ebright, *Nature* **364**, 548–549 (1993).

[42] T. Heyduk, Y. Ma, H. Tang, and R. H. Ebright, *Methods Enzymol.* **274**, 492–503 (1996).

Lane	1	2	3	4	5	6		7	8	9	10
^{35}S-hTRβ	+	+	+	+	+	+		+	+	+	+
GST-SRC2-2		+	+	+	+	+		+	+	+	+
Ligand T3			+	+	+	+		+	+	+	+
TG-1 (μM)					0.01	100					
TG-3 (μM)							100				
TG-9 (μM)								0.01	0.1	1.0	10.0

^{35}S hTRβ ➤

FIG. 9. Inhibition of SRC2 NID protein binding to hTRβ by TG9. Lane: (1) ^{35}S-hTRβ alone; (2) no T3 hormone, ^{35}S-hTRβ binding is ligand dependent; (3) no competitor maximal binding of ^{35}S-hTRβ to SRC2-NID domain; (4) 0.01 μM **TG1**; (5) 100 μM peptide **TG1** will compete for binding to hTRβ; (6) 100 μM **TG3** showed no competition for binding to hTRβ; (7–10) increasing concentration of **TG9** conformational constraint increases competitive ability for binding to ^{35}S-hTRβ.

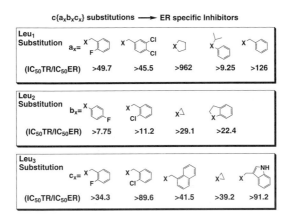

FIG. 10. The peptidomimetic library scaffold (c{a_x,b_x,c_x}) with diversity positions indicated by a_x, b_x, and c_x. Fourteen compounds were specific to hERα due to individual side chain substitutions at $\mathbf{a_x}$, $\mathbf{b_x}$, and $\mathbf{c_x}$ with the ratio of specificity denoted as the IC$_{50}$ of hTRβ/IC$_{50}$ of hERα, and > denoting no inhibition of hTRβ at 100 μM; five at the first leucine position $\mathbf{a_x}$ (A), four at the second leucine position $\mathbf{b_x}$ (B), and five at the third leucine position $\mathbf{c_x}$ (C).

important in identifying the structural, chemical, and energetic details of the coactivator nuclear receptor interface.

At first glance, it appears that the NR–SRC interaction is fairly promiscuous with almost identical structural interfaces between receptors and the NR box motif. Yet, thorough computational evaluation and chemical probing of the binding site reveal that they are not identical.

The testing of the peptidomimetic library identified specific inhibitors (Fig. 10), which take advantage of differences between the leucine-binding pocket of the E$_2$–hERα and that of T$_3$–hTRβ. These results argue that the NR–SRC binding pocket, although evolved to recognize the L$_1$XXL$_2$L$_3$ motif *in vivo*, has retained differences in shape and electrostatics between family members that may be exploited by competitive inhibitors mimicing the NR box. Through the design of such compounds and the disruption of the NR activation complex, we will learn more about NR signaling biology.

Acknowledgment

We thank Dr. Bea Darimont for assistance in the early development of this project, and Dr. Yoko Shibata for recombinant hERα. TRG was funded by the Department of Defense Breast Cancer Research Fund #DAMD17-00-1-0191. We thank the Sidney Kimmel Foundation for Cancer Research, the HHMI Research Resources Program grant #76296-549901, the Academic Senate of the University of California at San Francisco, NIH #DK58080, and the Sandler Foundation for financial support. We dedicate this chapter to the memory of a great mentor and colleague, Dr. Peter Kollman. His thoughts and work will always be with us; he is dearly missed.

[14] Biochemical Isolation and Analysis of a Nuclear Receptor Corepressor Complex

By MATTHEW G. GUENTHER and MITCHELL A. LAZAR

Introduction

A fundamental action of many nuclear hormone receptors is the establishment of active repression in the absence of ligand. While nuclear receptors themselves can exhibit repressive behavior via contacts with the basal transcription machinery,[1,2] the corepressors N-CoR (Nuclear Receptor CoRepressor) and SMRT (Silencing Mediator of Retinoid and Thyroid Receptors) have emerged as the primary mediators of nuclear receptor imposed repression[3-9] (reviewed in Ref. 10). N-CoR and SMRT

[1] A. Baniahmad, I. Ha, D. Reinberg, S. Tsai, M. J. Tsai, and B. W. O'Malley, *Proc. Natl. Acad. Sci. USA* **90**, 8832–8836 (1993).

[2] J. D. Fondell, A. L. Roy, and R. G. Roeder, *Genes Dev.* **7**, 1400–1410 (1993).

[3] A. J. Horlein, A. M. Naar, T. Heinzel, J. Torchia, B. Gloss, R. Kurokawa, A. Ryan, Y. Kamei, M. Soderstrom, C. K. Glass, and M. G. Rosenfeld, *Nature* **377**, 397–404 (1995).

[4] R. Kurokawa, M. Soderstrom, A. Horlein, S. Halachmi, M. Brown, M. G. Rosenfeld, and C. K. Glass, *Nature* **377**, 451–454 (1995).

[5] J. D. Chen and R. M. Evans, *Nature* **377**, 454–457 (1995).

[6] S. Sande and M. L. Privalsky, *Mol. Endocrinol.* **10**, 813–825 (1996).

[7] I. Zamir, H. P. Harding, G. B. Atkins, A. Horlein, C. K. Glass, M. G. Rosenfeld, and M. A. Lazar, *Mol. Cell. Biol.* **16**, 5458–5465 (1996).

[8] P. Ordentlich, M. Downes, W. Xie, A. Genin, N. B. Spinner, and R. M. Evans, *Proc. Natl. Acad. Sci. USA* **96**, 2639–2644 (1999).

[9] E. J. Park, D. J. Schroen, M. Yang, H. Li, L. Li, and J. D. Chen, *Proc. Natl. Acad. Sci. USA* **96**, 3519–3524 (1999).

[10] X. Hu and M. A. Lazar, *Trends Endocrinol. Metab.* **11**, 6–10 (2000).

are highly homologous proteins of approximately 270 kDa that each contain multiple distinct repression domains, a SANT (SW13, ADA2, N-CoR, TFIIIB) domain,[11] and nuclear receptor interaction domains. Deletion of N-CoR in a mouse model system results in disrupted hematopoietic and neural development, thus indicating a nonredundant function between SMRT and N-CoR.[12] While SMRT and N-CoR are likely to have important biological differences related to their unique sequences and expression, the methods described in this article related to SMRT complexes are likely applicable to N-CoR as we, and others, have observed similar results with both corepressors.[13–15]

Reversible acetylation of ε-N-acetyl lysines, present on the amino terminal tails of core histones, has been identified as a major regulator of eukaryotic gene transcription.[16] Gene repression through deacetylation of histone tails is facilitated by histone deacetylases (HDACs), that can be subdivided into three groups: Class I deacetylases, which are homologous to the yeast Rpd3; Class II deacetylases, which are homologous to yeast HDA1; and the Sir2-like deacetylases, which depend on binding of NAD + for activity.[17] Several Class I HDAC-containing complexes have been isolated from mammalian cells. Among these are the NuRD/Mi2,[18–21] Sin3,[22–25] MeCP2,[26–28] and CoREST[29,30] complexes. Each of these

[11] R. Aasland, A. F. Steward, and T. Gibson, *Trends Biochem. Sci.* **21**, 87–88 (1996).

[12] K. Jepsen, O. Harmanson, T. M. Onami, A. S. Gleiberman, V. Lunyak, R. J. McEvilly, R. Kurokawa, V. Kumar, F. Liu, E. Seto, S. M. Hedrick, G. Mandel, C. K. Glass, D. W. Rose, and M. G. Rosenfeld, *Cell* **102**, 753–763 (2000).

[13] M. G. Guenther, W. S. Lane, W. Fischle, E. Verdin, M. A. Lazar, and R. Shiekhattar, *Genes Dev.* **14**, 1048–1057 (2000).

[14] J. Li, J. Wang, Z. Nawaz, J. M. Liu, J. Qin, and J. Wong, *EMBO J.* **19**, 4342–4350 (2000).

[15] J. Zhang, M. Kalkum, B. T. Chait, and R. G. Roeder, *Mol. Cell* **9**, 611–623 (2002).

[16] T. Jenuwein and C. D. Allis, *Science* **293**, 1074–1080 (2001).

[17] C. M. Grozinger and S. L. Schreiber, *Chem. Biol.* **9**, 3–16 (2002).

[18] Y. Xue, J. Wong, G. T. Moreno, M. K. Young, J. Cote, and W. Wang, *Mol. Cell* **2**, 851–861 (1998).

[19] J. K. Tong, C. A. Hassig, G. R. Schnitzler, R. E. Kingston, and S. L. Schreiber, *Nature* **395**, 917–921 (1998).

[20] P. A. Wade, P. L. Jones, D. Vermaak, and A. P. Wolffe, *Curr. Biol.* **8**, 843–846 (1998).

[21] Y. Zhang, G. LeRoy, H. P. Seelig, W. S. Lane, and D. Reinberg, *Cell* **95**, 279–289 (1998).

[22] C. A. Hassig, T. C. Fleischer, A. N. Billin, S. L. Schreiber, and D. E. Ayer, *Cell* **89**, 341–348 (1997).

[23] Y. Zhang, R. Iratni, H. Erdjument-Bromage, P. Tempst, and D. Reinberg, *Cell* **89**, 357–364 (1997).

[24] C. E. Laherty, W.-M. Yang, J.-M. Sun, J. R. Davie, E. Seto, and R. N. Eisenman, *Cell* **89**, 349–356 (1997).

[25] Y. Zhang, Z. W. Sun, R. Iratni, B. H. Erdjument, P. Tempst, M. Hampsey, and D. Reinberg, *Mol. Cell* **1**, 1021–1031 (1998).

complexes contains a unique subset of subunits and exhibits HDAC activity. Early studies of nuclear receptor corepressors suggested that N-CoR and SMRT function in association with Sin3-HDAC1 complexes;[31–33] however, N-CoR and SMRT have not been shown to copurify with cellular Sin3- or HDAC1-containing complexes.[19–22,25,34] This raises the possibility that SMRT/N-CoR may mediate repression via a unique set of associated proteins.

We describe in this article, the initial isolation and characterization of a core SMRT corepressor complex from HeLa cells. Isolation of native SMRT results in copurification of the WD40 repeat protein, TBL1 (Transducin Beta-Like protein-1) and HDAC3.[13] Since many HDACs are either severely impaired or inactive when produced and purified alone, we describe a reconstitution of HDAC3 activity using a rabbit reticulocyte lysate coupled transcription–translation system.[35] This system is used to show that SMRT/N-CoR activates HDAC3 activity without the aid of TBL1, and it can be readily adapted to screen for other potential activators of HDAC activity. Finally, we show how to produce a large-scale preparation of an active SMRT–HDAC3 complex using baculovirus infection of sf9 cells.[35]

Establishing a HDAC Assay

Since many corepressor proteins are increasingly linked to specific deacetylases and deacetylase complexes, it has become necessary to establish HDAC assays to verify this function. These assays may be used to verify the

[26] P. L. Jones, G. J. Veenstra, P. A. Wade, D. Vermaak, S. U. Kass, N. Landsberger, J. Strouboulis, and A. P. Wolffe, *Nat. Genet.* **19**, 187–191 (1998).

[27] X. Nan, H. H. Ng, C. A. Johnson, C. D. Laherty, B. M. Turner, R. N. Eisenman, and A. Bird, *Nature* **393**, 386–389 (1998).

[28] D. E. Ayer, Q. A. Lawrence, and R. N. Eisenman, *Cell* **80**, 767–776 (1995).

[29] G. W. Humphrey, Y. Wang, V. R. Russanova, T. Hirai, J. Qin, Y. Nakatani, and B. H. Howard, *J. Biol. Chem.* **276**, 6817–6824 (2001).

[30] A. You, J. K. Tong, C. M. Grozinger, and S. L. Schreiber, *Proc. Natl. Acad. Sci. USA* **98**, 1454–1458 (2001).

[31] L. Alland, R. Muhle, H. Hou, J. Potes, L. Chin, N. Schreiber-Agus, and R. A. DePinho, *Nature* **387**, 49–55 (1997).

[32] L. Nagy, H.-Y. Kao, D. Chakvarkti, R. J. Lin, C. A. Hassig, D. E. Ayer, S. L. Schreiber, and R. M. Evans, *Cell* **89**, 373–380 (1997).

[33] T. Heinzel, R. M. Lavinsky, T.-M. Mullen, M. Soderstrom, C. D. Laherty, J. Torchia, W.-M. Yuang, G. Brard, S. D. Ngo, J. R. Davie, E. Seto, R. N. Eisenman, D. W. Rose, C. K. Glass, and M. G. Rosenfeld, *Nature* **387**, 43–48 (1997).

[34] E. Y. Huang, J. Zhang, E. A. Miska, M. G. Guenther, T. Kouzarides, and M. A. Lazar, *Genes Dev.* **14**, 45–54 (2000).

[35] M. G. Guenther, O. Barak, and M. A. Lazar, *Mol. Cell. Biol.* **21**, 6091–6101 (2001).

coimmunoprecipitation of HDAC activity from cellular extracts, to serve as a marker to biochemically isolate specific HDAC-containing corepressor complexes, or to verify the *in vitro* reconstitution of previously defined HDAC complexes. Various substrates may be used to measure HDAC activity. These include acetylated core histones that have been prepared by *in vivo* labeling of cultured cells with radiolabeled acetic acid in the presence of HDAC inhibitor, or *in vitro* acetylation with recombinant histone acetyltransferase (HAT). The former yields large amounts of native histones acetylated at random positions, although it should be noted that the latter can produce substrate that may be specifically acetylated on select lysine residues, based on the HAT protein used. Alternatively, N-terminal histone tail peptides that have been acetylated chemically or enzymatically may be used; however, these substrates lack potentially important carboxy terminal sequences that may be important for catalysis or cofactor binding. The HDAC protocol, described in the next section, uses a substrate obtained by isolation of [^{3}H]-acetyl-labeled free histones from HeLa cells.

Preparation of HDAC Substrate

[^{3}H]-Acetyl-lysine labeled histone protein is prepared essentially, as described by Carmen *et al.*,[36] with modifications. Twelve 150 mm plates of HeLa cells are grown to 90% confluency in medium [high glucose Dulbecco's modified Eagle's medium supplemented with 10% FBS and 0.6 mg/ml L-glutamine (all Gibco BRL)]. Cells are grown at 37°C in 5.0% CO_2. Remove medium, and label cells by adding 15 ml fresh growth medium containing 0.1 mCi/ml [^{3}H]-acetic acid (Dupont NEN), 100 μg/ml cycloheximide (Sigma), and 10 mM sodium butyrate (Sigma) to each plate. Incubate at 37°C for 1 hr. Remove labeling medium, and add 15 ml fresh medium containing 10 mM sodium butyrate. Incubate cells at 37°C for an additional 4 hr. ([^{3}H]-Acetic acid may be conserved by scaling back the number of plates to be labeled, or by dividing HeLa cells into three (3) groups of four (4) plates each and consecutively labeling each group by reusing labeling medium. In the latter case, four plates are labeled for 1 hr after which the labeling medium is transferred to the next group of four plates for 1 hr, etc.) Remove growth medium and wash cell monolayers with 10 ml of PBS containing 10 mM sodium butyrate. Remove, wash, and freeze plates at −80°C.

The following day thaw and scrape each plate of cells into 5 ml NIB Buffer [10 mM Tris–HCl pH 7.4, 2 mM MgCl$_2$, 10 mM sodium butyrate, 1 mM PMSF, 1.0% (v/v) NP-40] using a rubber policeman. Collect nuclei by centrifugation at 500 × g. Wash nuclei twice in 40 ml NIB buffer. Wash

[36] A. A. Carmen, S. E. Rundlett, and M. Grunstein, *J. Biol. Chem.* **271**, 15837–15844 (1996).

nuclei once in 40 ml of NIB buffer containing 100 mM NaCl. Wash nuclei once in 40 ml of IB buffer (10 mM Tris–HCl pH 7.4, 2 mM MgCl$_2$, 3 mM CaCl$_2$, 10 mM sodium butyrate, 1 mM PMSF) containing 100 mM NaCl. Extract nuclei twice with IB buffer containing 400 mM NaCl. Extract nuclear pellet twice with 10 pellet volumes of 0.2 M H$_2$SO$_4$ for 90 min on ice, followed by centrifugation at 20,000 × g for 30 min. Pool acid supernatants and dialyze extensively against 100 mM acetic acid at 4°C. Special attention must be made to select a dialysis membrane with an exclusion limit below 10 kDa. Specific activity should be approximately 2.5 × 10^3 cpm/μg. Aliquot 200,000 cpm of radiolabeled histones per 1.5 ml microcentrifuge tube, and add 10 volumes of ice-cold acetone. Precipitate overnight at −20°C. Centrifuge at 20,000 × g for 10 min and remove supernatant. Dry pellet and store at −80°C as 10 × HDAC substrate.

HDAC Assay

1. Resuspend an aliquot of 10× HDAC substrate (200,000 cpm/tube) in 100 μl HDAC buffer (50 mM NaCl, 25 mM Tris–HCl pH 8.0, 10% glycerol, 1.0 mM EDTA).
2. Combine HDAC enzyme source (immunocomplex bound to agarose beads, purified enzyme, or cellular fraction) plus 10 μl substrate (20,000 cpm) and bring to a total volume of 200 μl with HDAC buffer.
3. Rotate at 37°C for 2 hr in tabletop incubator. If the HDAC source is bound to agarose beads, then separate beads from supernatant by centrifugation at 3000 × g and immediately continue to step 4 supernatant. [After the HDAC assay is completed, wash beads twice with HDAC buffer, add 30 μl 1× SDS-loading buffer (50 mM Tris–HCl pH 6.8, 2.0% SDS, 0.3% bromophenol blue, 10% glycerol, 0.25 M β-mercaptoethanol), and subject to Western blot analysis].
4. Stop reaction by addition of 50 μl Stop solution (0.1 M HCl, 0.16 M acetic acid). Vortex.
5. Extract free [^{3}H]-acetate by adding 600 μl ethyl acetate. Vortex mixture, centrifuge at 20,000 × g for 1 min, and remove 550 μl of organic (top) layer taking extreme care not to disturb the interface. Subject organic layer to scintillation counting. A typical HDAC assay is shown in Fig. 1b.

Purification of a Core SMRT Corepressor Complex from HeLa Cells

The isolation of a core SMRT corepressor complex by affinity purification from HeLa cells has identified the TBL1 and HDAC3 proteins

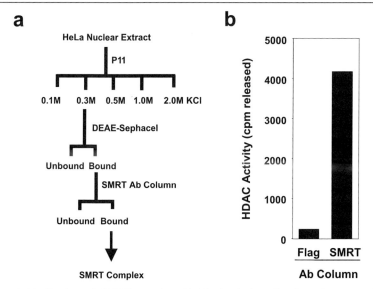

FIG. 1. Purification of SMRT complex. (a) Biochemical purification scheme to obtain SMRT-associated proteins. (b) HDAC assay of SMRT complex as purified in (a). Adapted from Fig. 1 of M. G. Guenther *et al.*, *Genes Dev.* **14**, 1048–1057 (2000).[13]

as associated components.[13] Independent isolation of an N-CoR complex from HeLa cells verified the presence of both TBL1 and HDAC3.[14] Each complex was purified under high salt conditions (500 m*M* KCl), suggesting that the components represent tightly bound core complexes. Subsequent biochemical purification of the N-CoR protein has yielded additional complexes containing the following: (1) GPS2 (G-protein pathway supressor-2),[15] (2) KAP-1 and components of the SWI/SNF complex,[37] and (3) Sin3, and associated components.[37,38] Additional interactions of SMRT/N-CoR with Class II HDACs also have been described by yeast two-hybrid and coimmunoprecipitation studies,[34,39] although only a small fraction of endogenous N-CoR and SMRT were bound to endogenous HDAC4 in 293T cells.[34]

Here, we describe the isolation of a core SMRT corepressor complex from HeLa nuclear extract, using a pool of SMRT specific monoclonal antibodies as an affinity matrix. This complex contains SMRT, TBL1, and HDAC3, but not Sin3 or HDAC1/2. Verification of the importance of this

[37] C. Underhill, M. S. Qutob, S. P. Yee, and J. Torchia, *J. Biol. Chem.* **275**, 40463–40470 (2000).

[38] P. L. Jones, L. M. Sachs, N. Rouse, P. A. Wade, and Y. B. Shi, *J. Biol. Chem.* **276**, 8807–8811 (2001).

[39] H. Y. Kao, M. Downes, P. Ordentlich, and R. M. Evans, *Genes Dev.* **14**, 55–66 (2000).

complex comes from the ability to TBL1 to potentiate nuclear receptor repression,[13] and the necessity of HDAC3 for N-CoR mediated repression.[12,14,40] The scheme for SMRT purification is shown in Fig. 1a. The complex isolated exhibits HDAC activity as shown in Fig. 1b.

Purification of SMRT Corepressor Complex

1. The anti-SMRT antibody pool comprises five SMRT-specific mouse monoclonal antibodies obtained individually as hybridoma supernatants at a concentration of approximately 50 μg/ml. Anti-SMRT affinity matrix is produced by binding 10 ml of each antibody (pooled total volume of 50 ml containing five antibodies) to 0.5 ml protein A agarose (Gibco-BRL). Binding is achieved by high salt coupling followed by diethylpalmidate cross-linking, as described.[41] A control column is produced in parallel by coupling 1.0 mg anti-FLAG antibody (Sigma) to 0.5 ml protein A agarose according to the same protocol. Affinity columns can be produced well in advance of fractionation and may be stored in 0.05% (w/v) sodium azide for up to 6 months.

2. Equilibrate phosphocellulose P11 (Whatman) with 5 column volumes of buffer BC(100) (20 mM Tris–HCl pH 7.9, 10% glycerol, 0.2 mM EDTA, 10 mM β-mercaptoethanol, 0.2 mM PMSF, 100 mM KCl {all "BC" buffer solutions are composed of the above components, except with variable KCl [concentrations are indicated in mM, in parenthesis, e.g., BC(100) = 100 mM KCl]}. Prepare 1.0 ml of solid support per 1.0 mg of total protein to be chromatographed. All manipulations are performed at 4°C. Prepare HeLa nuclear extract as described in Dignam *et al.*[42] Typically, 1.0 g of nuclear extract is produced from 100 liters of HeLa S3 cells at 0.6×10^6 cells/ml. Dialyze nuclear extract against BC(100).

3. Load HeLa nuclear extract onto the equilibrated P11 column and wash with BC(100) until protein ceases to elute from column (as determined by the return of A_{280} to baseline). Perform consecutive step elutions using BC(300), BC(500), BC(1000), and BC(2000). All buffers should be prepared the day of use to maintain PMSF stability. Each elution is complete when the A_{280} returns to baseline. Collect all fractions including the flow through.

[40] E. Jeannin, D. Robyr, and B. Desvergne, *J. Biol. Chem.* **273**, 24239–24248 (1998).
[41] E. Harlow and D. Lane, "Antibodies. A Laboratory Manual." Cold Spring Harbor Laboratory, Cold Spring Harbor, NY, 1988.
[42] J. D. Dignam, R. M. Lebovitz, and R. Roeder, *Nucleic Acids Res.* **11**, 1475–1489 (1983).

4. The presence of SMRT complex is monitored by direct immunoblot analysis. After P11 separation, SMRT is present primarily in the BC(300) fraction, and less so in the BC(500) fraction. Subject each elution of the P11 column to SDS-PAGE followed by anti-SMRT immunoblot to confirm the presence of SMRT in the BC(300) elution. Fractions may be frozen at $-80°C$ between column separations or during immunoblot confirmation.

5. Dialyze the BC(300) eluate against BC(100) buffer until the conductivity is that of a solution of 100 mM KCl, and then subject to anion exchange chromatography on a DEAE-sephacel column (Pharmacia). Equilibrate the DEAE-sephacel column with 5 column volumes of BC(100). Load the SMRT-containing fraction onto the column and wash with BC(100) until the A_{280} returns to baseline. Elute column with BC(350) and collect eluate until A_{280} returns to baseline.

6. Check for presence of SMRT by immunoblot analysis and dialyze the BC(350) eluate against buffer D(150) (150 mM KCl, 20 mM HEPES pH 7.9, 0.25 mM EDTA, 10% glycerol, 0.1% Tween-20, and 0.2 mM PMSF (all "D" buffer solutions are composed of the above components, except that variable KCl concentrations are indicated in parentheses, i.e., Buffer D(300) contains 300 mM KCL total). Incubate this fractionated extract with 5–10 ml of protein A agarose at 4°C for 2 hr to absorb nonspecific protein A binding proteins. Centrifuge at $25,000 \times g$ to remove residual agarose and carefully remove the SMRT-containing supernatant. Incubate SMRT affinity resin produced in step 1 (and control FLAG resin) with the fractionated extract for 12–16 hr at 4°C. Collect the SMRT affinity resin by centrifugation at $3000 \times g$ in a Sorvall RT7 tabletop centrifuge, remove supernatant, and transfer agarose to a gravity flow disposable column (Poly Prep®, BioRad). Wash column with 10 volumes of buffer D(150), 10 volumes of Buffer D(300), and 10 volumes of buffer D(500). Remove 10% of the affinity matrix for HDAC activity assay. Elute the remainder of bound material with 1 volume of 100 mM glycine pH 3.0 at 4°C. Repeat elution five times and neutralize each elution fraction with 1/10 volume of 1 M Tris–HCl pH 7.9. Subject elutions to SDS-PAGE followed by immunoblot analysis and silver staining.

Reconstitution of Enzymatically Active SMRT Corepressor Complex in Rabbit Reticulocyte Lysate

Many HDACs are either inactive or have low enzymatic activity when produced recombinantly. This outcome has been shown for recombinant

HDAC1,[43] HDAC3,[35,44] Class II HDAC4,[35,45] and Sir2.[46] Here, we describe the production of an enzymatically active HDAC3 enzyme in rabbit reticulocyte lysate by coincubation of *in vitro* translated HDAC3 and SMRT. The general scheme for the identification of potential cofactors to activate HDAC3 in rabbit reticulocyte lysate is shown in Fig. 2a. The

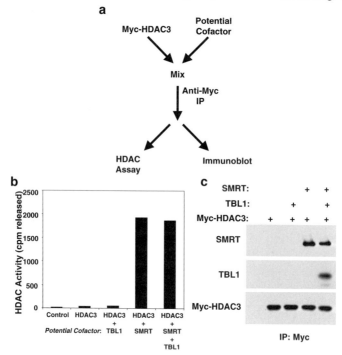

FIG. 2. Formation of an active HDAC using rabbit reticulocyte lysate transcription–translation system. (a) Schematic representation of HDAC reconstitution assay. (b) SMRT core complex components were produced by T7-coupled transcription–translation. Myc-HDAC3 alone or mixed together with Gal-TBL1 and/or FLAG-SMRT were immunoprecipitated with anti-Myc agarose beads to isolate HDAC complexes, and the beads assayed for HDAC activity. "Control" refers to immunoprecipitation of unprogrammed transcription–translation mix. (c) Western blot analysis of complexes formed in (b). Top panel was probed for SMRT (anti-FLAG), middle panel was probed for TBL1 (anti-Gal4), and lower panel was probed for HDAC3 (anti-Myc). Adapted from Fig. 1 of M. G. Guenther *et al.*, *Mol. Cell. Biol.* **21**, 6091–6101 (2001).[35]

[43] Y. Zhang, H. H. Ng, H. Erdjument-Bromage, P. Tempst, A. Bird, and D. Reinberg, *Genes Dev.* **13**, 1924–1935 (1999).

[44] Y. D. Wen, V. Perissi, L. M. Staszewski, W. M. Yang, A. Krones, C. K. Glass, M. G. Rosenfeld, and E. Seto, *Proc. Natl. Acad. Sci. USA* **97**, 7202–7207 (2000).

[45] W. Fischle, F. Dequiedt, M. J. Hendzel, M. G. Guenther, M. A. Lazar, W. Voelter, and E. Verdin, *Mol. Cell* **9**, 45–57 (2002).

[46] S. Imai, C. M. Armstrong, M. Kaeberlein, and L. Guarente, *Nature* **403**, 795–800 (2000).

SMRT-dependent enzymatic activity of HDAC3 (Fig. 2b), and the subsequent purification of the cofactor-containing complex (Fig. 2c) are also demonstrated. Similar results were obtained when N-CoR was used as the activation cofactor.[35] This method is rapid and potentially applicable to other HDACs.

1. All plasmids must contain a T7 RNA polymerase promoter for use with the TNT T7 Quick coupled transcription–translation kit (Promega). Translate 4 μg of pCMX-Myc-HDAC3 and 4 μg pCMX-Gal4-SMRT in separate 100 μl reactions according to the manufacturer's instructions. Briefly, 4 μg plasmid DNA is combined with 80 μl Promega T7 Quick rabbit reticulocyte lysate, 40 μM nonradioactive methionine, 10 μl of 10 × reaction buffer, and water to a final volume of 100 μl. Incubate reactions in a water bath for 2 hr at 30°C.

2. Combine HDAC3 and SMRT reactions, and add 2 volumes (400 μl) of buffer D(300) (300 mM KCl, 20 mM HEPES pH 7.9, 0.25 mM EDTA, 10% glycerol, 0.1% Tween-20) containing freshly added Complete[TM] protease inhibitor (Roche). Add 20 μl anti-FLAG agarose beads (Sigma) and rotate 12–16 hr at 4°C.

3. Collect beads by centrifugation at 3000 × g in an Eppendorf 5415C microcentrifuge. Remove supernatant and wash pelleted beads three times with buffer D(300) using repeated cycles of buffer addition, vortexing at medium speed, and centrifugation at 3000 × g.

4. Subject immobilized HDAC3–SMRT complex to HDAC activity assay as described earlier (HDAC activity assay, Fig. 2b).

5. Verify presence of SMRT and/or TBL1 in HDAC3 immunoprecipitate by immunoblot analysis (Fig. 2c).

Isolation of a Purified SMRT-HDAC3 Complex from Baculovirus

While HDAC complexes may be purified from rabbit reticulocyte lysate, the relative protein amounts are too small for many biochemical assays. To set up potential inhibitor screens or to perform structural and enzymatic studies on HDAC3, large amounts of purified and active enzyme must be available. Detailed analysis of the HDAC3–SMRT complex has determined that a region termed the "SANT domain 1" is important for SMRT to activate HDAC.[35] Here, we describe the production of an active HDAC3 enzyme by coinfection of insect sf9 cells with baculovirus encoding FLAG-HDAC3, and a SANT1 containing fragment of SMRT. Copurification by anti-FLAG affinity chromatography yields a highly purified active enzyme (Fig. 3).

1. Flag-HDAC3 and Myc-SMRT amino acids 1–587 were cloned into pBlueBac4.5, and high-titer recombinant baculoviruses were

produced with the MaxBac 2.0 kit according to the manufacturer's instructions (Invitrogen).

2. Maintain insect sf9 cells in sf-900 II serum-free medium (Gibco BRL) containing $1 \times$ antibiotic–antimycotic solution (GibcoBRL) and 0.1% Pluronic F-68 (Gibco BRL). Sf9 cells are grown at 30°C in 1-liter spinner flasks. Cells should be propagated at 1.0–2.0×10^6 cells/ml, until ready for infection.

3. Allow 500 ml of sf9 cells to grow to a density of 1.0–1.5×10^6 cells/ml. Infect with FLAG-HDAC3 and Myc-SMRT viruses at a multiplicity of infection of 5.0 for each virus. Allow infection to proceed for 48 hr.

4. Pellet cells by centrifugation at $10,000 \times g$ for 10 min. Resuspend cell pellet with 30 ml lysis buffer A (150 mM NaCl, 40 mM Tris–HCl pH 7.6, 10% glycerol, 0.3% NP-40) containing freshly added $1 \times$ protease inhibitor cocktail (Roche, Complete protease inhibitor cocktail). Incubate on ice for 15 min, then centrifuge at $25,000 \times g$ for 20 min. Keep supernatant for affinity purification.

5. Add 1.0 ml anti-FLAG agarose beads (Sigma) to the 30 ml sf9 cell lysate and rotate at 4°C for 12–16 hr. Collect beads by centrifugation at $3000 \times g$ for 1 min in a Sorvall RT7 centrifuge. Remove

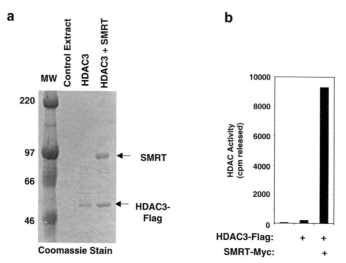

FIG. 3. Purification of an active HDAC3–SMRT complex using baculovirus. (a) Coomassie stain of purified HDAC3 complex. Sf9 cells were mock infected (control) or infected with baculovirus-expressing FLAG-HDAC3, Myc-SMRT amino acids 1–587, or both. Sf9 lysates were purified by anti-FLAG affinity chromatography and resolved by SDS-PAGE. Molecular weight markers (MW) and major polypeptides are shown. (b) HDAC activity of proteins purified in (a). Adapted from Fig. 8 of Ref. 35.

supernatant and resuspend beads in 5 ml lysis buffer A. Load resuspended material onto a 10 ml capacity Poly Prep® gravity filtration column (Biorad) and wash with 10 volumes of lysis buffer A. Wash successively with 10 volumes each of buffer D(300) (300 mM KCl, 20 mM HEPES pH 7.9, 0.25 mM EDTA, 10% glycerol, 0.1% Tween-20), buffer D(500) (500 mM KCl, 20 mM HEPES pH 7.9, 0.25 mM EDTA, 10% glycerol, 0.1% Tween-20), buffer D(700) (700 mM KCl, 20 mM HEPES pH 7.9, 0.25 mM EDTA, 10% glycerol, 0.1% Tween-20), and EB buffer (TBS + 10% glycerol). Do not allow the affinity matrix to dry out. New buffer should be added when prior buffer has nearly reached the top of the solid support.

6. Elute HDAC3–SMRT complex with EBF buffer [TBS, 10% glycerol, 200 μg/ml FLAG peptide (Sigma)]. Perform each elution by adding 1.0 ml EBF buffer to the affinity matrix, sealing each end of the plastic column, and rotating at room temperature for 30 min. Position the column upright, allow matrix to settle, and collect the eluate. Repeat for a total of six elutions and measure the protein content of each fraction by the Bradford assay. Pool eluates with the highest protein content.

7. Dialyze pooled eluate against 2 liters of buffer EB at 4°C for 1 hr. Repeat dialysis with fresh EB buffer for an additional 2 hr and analyze complex by SDS-PAGE, followed by Coomassie blue staining (Fig. 3a) and immunoblot analysis.[35] Assay 10 μl of final complex for HDAC activity (Fig. 3b).

Acknowledgment

This work was supported by grants RO1 DK45586 and RO1 DK43806 to MAL from the National Institute of Health.

[15] Isolation and Functional Characterization of the TRAP/Mediator Complex

By Sohail Malik and Robert G. Roeder

The multiprotein TRAP/Mediator complex has emerged as a central coactivator for transcription by nuclear receptors, as well as by diverse activators (reviewed in Ref. 1). In contrast to many coactivators that possess histone acetyl transferase and other chromatin modifying functions, and are thus necessary for transcription only from chromatin templates, the

[1] S. Malik and R. G. Roeder, *Trends Biochem. Sci.* **25**, 277 (2000).

TRAP/Mediator complex is, in fact, required for activation of naked DNA (nonchromatin) templates *in vitro.*[1-5] With respect to the multiplicity of coactivators involved in the activation of target genes, it has been suggested that (minimally) two steps might be involved.[1,6] The chromatin coactivators may function to facilitate initial penetration of the nucleosome barrier. This breach may then allow entry of TRAP/Mediator, RNA polymerase II (Pol II), and the associated general transcription factors (GTFs) to begin the transcription reaction *per se.* Indeed, efficient TRAP/Mediator-dependent transcriptional activation by a variety of nuclear receptors can be reconstituted *in vitro* with essentially homogeneous preparations of Pol II, the GTFs (TFIIA, TFIIB, TFIID, TFIIE, TFIIF, and TFIIH), and the USA-derived cofactor PC4. This set of proteins, thus defines the *minimum* number of factors required for activated transcription from DNA templates.

Here, we describe methods that we have developed for isolating functionally active preparations of the TRAP/Mediator complex and for testing them in cell-free assay systems.

Isolation of the TRAP/Mediator Complex

The TRAP/Mediator complex was initially isolated from HeLa cells stably expressing FLAG-tagged thyroid hormone receptor (TRα).[2] Propagation of this cell-line in the presence of the cognate ligand (thyroid hormone), but not in its absence, allowed isolation (by subsequent affinity chromatography utilizing the FLAG tag) of a multiprotein complex (TRAP) associated with TRα. Remarkably, the subsequent characterization of a metazoan complex (SMCC), purified on the basis of a resident ortholog of the yeast SRB/Mediator component of the Pol II holoenzyme,[3,4] indicated near structural and functional equivalence between the two complexes, and, at the same time, revealed direct evolutionary links with the yeast SRB/Mediator[7,8] (also reviewed in Ref. 1). In view of these considerations, we now find it more expedient to isolate the human TRAP/Mediator by affinity purification from extracts of HeLa cells that stably express a FLAG-tagged version of an integral TRAP/Mediator subunit.[9]

[2] J. D. Fondell, H. Ge, and R. G. Roeder, *Proc. Natl. Acad. Sci. USA* **93**, 8329 (1996).

[3] W. Gu, S. Malik, M. Ito, C.-X. Yuan, J. D. Fondell, X. Zhang, E. Martinez, J. Qin, and R. G. Roeder, *Mol. Cell* **3**, 97 (1999).

[4] M. Ito, C.-X. Yuan, S. Malik, W. Gu, J. D. Fondell, S. Yamamura, Z.-Y. Fu, X. Zhang, J. Qin, and R. G. Roeder, *Mol. Cell* **3**, 361 (1999).

[5] A. M. Näär, B. D. Lemon, and R. Tjian, *Annu. Rev. Biochem.* **70**, 475 (2001).

[6] R. G. Roeder, *Cold Spring Harb. Symp. Quant. Biol.* **63**, 201 (1998).

[7] T. I. Lee and R. A. Young, *Annu. Rev. Genet.* **34**, 77 (2000).

[8] L. C. Myers and R. D. Kornberg, *Annu. Rev. Biochem.* **69**, 729 (2000).

[9] S. Malik, W. Gu, W. Wu, J. Qin, and R. G. Roeder, *Mol. Cell* **5**, 753 (2000).

Establishment and Propagation of Stable Cell Lines Expressing an Epitope-Tagged TRAP/Mediator Subunit

Whereas the choice of a subunit for epitope-tagging and subsequent affinity purification is largely an empirical matter, the following should serve as general guidelines. First, unless tedious conditional expression systems are to be resorted to, a prerequisite is that the overexpression of the subunit must not be severely toxic for the HeLa cells. Second, the protein (and hence the epitope tag) must be accessible to the affinity antibody (i.e., the protein must not be buried deep within the multiprotein complex). Third, the selected protein must be a bona fide integral subunit (as opposed to a loosely associating albeit relevant protein) that has the potential to bring down relatively homogeneous populations of the cellular complex in biochemically reasonable amounts. Fourth, with an eye toward functional characterization, it is best to avoid subunits (e.g., SRB10/ CDK8 and SRB11/cyclin C) whose overrepresentation in the preparation might confer predominantly negative function.[1,3] Although we have now identified several TRAP/Mediator subunits that meet these criteria, we describe the establishment of a cell-line expressing FLAG-tagged NUT2 (f:NUT2).[9]

Once a positive clone is identified, it is expanded, and adapted to spinner culture. Dignam-type nuclear extracts[10] are prepared and used as the source for obtaining highly purified preparations of TRAP/Mediator.

Materials and Reagents. NUT2 cDNA subcloned into a pIRES-neo (Clontech) expression vector that has been modified to include tandem coding sequences for the FLAG and HA epitope tags, as well as an optimized Kozak sequence;[9,11] HeLa S cells (preferably at an early passage); Lipofectamine (Invitrogen/Life Technologies); G418 (Invitrogen/Life Technologies); M2-agarose (Sigma); FLAG peptide; Standard tissue culture media, equipment, and supplies.

Solutions and Buffers. Phosphate-buffered saline (PBS); high salt lysis buffer: 50 mM Tris–HCl (pH 7.9 at 4°C), 500 mM NaCl, 1% NP-40, 20% glycerol, 0.5 mM PMSF, 5 mM 2-mercaptoethanol; buffer BC: 20 mM Tris–HCl (pH 7.9 at 4°C), 20% glycerol, 0.1 mM EDTA, 0.5 mM PMSF, 2 mM

[10] J. D. Dignam, P. L. Martin, B. S. Shastry, and R. G. Roeder, *Methods Enzymol.* **101**, 582 (1983).

[11] E. Martinez, V. B. Palhan, A. Tjernberg, E. S. Lymar, A. M. Gamper, T. K. Kundu, B. T. Chait, and R. G. Roeder, *Mol. Cell. Biol.* **21**, 6782 (2001).

DTT, or 5 mM 2-mercaptoethanol containing either 100 mM or 300 mM KCl (BC100 or BC300).

General note on buffer composition: Here "BC" followed by a numerical designation indicates the (variable) KCl concentration of buffers that, otherwise, have the same composition. PMSF and reducing agents are added just before use. Supplements (e.g., NP-40, additional protease inhibitors) are specified, as needed. BC-type buffers are used in many of the procedures described herein. Thus, this requirement is not listed at the beginning of each section. Other (non-BC buffers) are explicitly described in the appropriate sections. Note also that although NP-40 is now marketed as IGEPAL CA-630 (Sigma), we have retained the older name here.

Procedure

1. Prepare the expression vector DNA by standard plasmid isolation techniques. We have found that the quality of plasmids obtained by Qiagen maxiprep kits is acceptable. For any new constructs, it is best to authenticate the expression vector by DNA sequencing to ensure that the cDNA insert and the tags are in-frame, and to rule out any PCR-induced artifacts (such as spurious STOP codons). The use of pIRES-type vectors, which contain an internal ribosome entry site, is strongly recommended, since it ensures that G418-resistant clones will also express the cloned polypeptide. Note also that our modified vector contains an additional HA tag.[11] This addition potentially allows purification over an anti-HA resin, and has the added advantage that it allows the screening of NUT2-expressing clones with an anti-HA antibody, which in our hands is more selective than the commonly used FLAG antibody.

2. Several days before the actual transfection, thaw out and plate HeLa S cells in Dulbecco's-Modified Eagle Medium (DMEM) supplemented with 10% fetal calf serum (FCS) and penicillin–streptomycin. For this and following steps, established tissue culture methods are used. Unless otherwise stated, adherent HeLa cells (in DMEM) are grown at 37°C in standard CO_2 incubators.

3. Two days before the transfection, split the cells into eight 100-mm dishes, seeding each with about 3×10^6 cells. Allow the cells to grow to about 50–60% confluence.

4. On the day of the transfection (day 1), set up eight (sterile) tubes with 0.8 ml incomplete DMEM (no serum or antibiotics). Add a different amount of plasmid DNA to each of the tubes (sample amounts: 10, 20, 50, 100, 250, 500 ng). It is important to titrate the

DNA over a wide range because a balance has to be struck between the tendency of the transfected DNA (or its encoded product) to be toxic at higher concentrations, and the need to obtain workable transfection efficiencies. The optimum would be different for each construct. As a positive control for the transfection, it is advisable to include 20 ng of the parent vector; as a negative control, omit DNA. Separately, dilute 400 μl lipofectamine (2 mg/ml) (sufficient for eight reactions at 50 μl per transfection) in 6.4 ml incomplete DMEM. Next combine 850 μl of the diluted lipofectamine with the contents of each of the eight tubes containing DNA in DMEM. Incubate the resulting transfection mix at room temperature for 1 hr.

5. Wash the plates containing the HeLa cells twice with incomplete DMEM to remove any traces of serum, which interferes with lipofectamine-mediated transfection.

6. Cover cells in each plate with 6.4 ml incomplete DMEM and add 1.6 ml of the transfection mix. Swirl the plates to mix.

7. Incubate the cells for 12–14 hr (or overnight).

8. On day 2, add 8 ml DMEM containing 20% FCS (but still no antibiotics) to each plate.

9. After an additional 8 hr of incubation, change the medium to complete DMEM (containing 10% FCS and penicillin–streptomycin).

10. On day 3, cells are finally split into G418-containing medium for selection of resistant clones. As G418-susceptibility of cells appears to be a function of the growth state—at high cell densities G418 is relatively ineffectual—the choice of how much to split is a crucial parameter in the selection process. At the same time, to maintain a high probability of success, seeding the transfected cells at very low densities is also not recommended. Thus, for each original transfection (corresponding to a different amount of input DNA) we routinely trypsinize the cells and, after counting, seed three different amounts (0.8×10^6, 1.0×10^6, and 1.2×10^6) into 150 mm dishes in complete DMEM supplemented with 0.5 mg/ml G418.

11. Incubate the plates for 2–3 weeks, changing the selection medium as needed. By this time, the bulk of the cells should die-off and colonies that are visible to the unaided eye, and which microscopically have a distinct (well-defined) outline, will begin to appear. It is likely that a few colonies will be seen in the negative controls as well; however, these will not be found to be truly G418-resistant upon prolonged culture.

12. Carefully trypsinize individual colonies (as many as 20–24 may be picked, if available) and transfer to 24-well culture plates. Incubate

(in G418-containing medium) until authentic G418-resistant cells are just confluent, whereupon they should be transferred to a larger (12-well) plate. After this stage, any false positives should not survive the selection. Move cells to successively larger dishes, until more than about $40–50 \times 10^6$ cells are available for the analysis of each clone. We normally expand the clones until three confluent 150 mm dishes are available. Cells from two dishes are processed for analysis, while the third serves as a backup from which larger amounts of cells can be propagated, in case a positive is identified.

13. For the analysis of the clones, wash the cells (two near-confluent 150 mm plates for each clone) two times with PBS. Gently scrape-off the cells (in PBS) and collect by centrifugation (1000 rpm for 10 min in a Beckman J-6B or equivalent rotor) into conical 15-ml Falcon tubes, and note the packed cell volume. Resuspend the cell pellets in about 5 packed cell volumes of high salt lysis buffer and allow the cells to lyse on ice for 30 min, while periodically mixing by inverting the tubes.

14. Centrifuge the lysates at 15,000 rpm for 20 min in a Beckman JA-20 or equivalent rotor.

15. Dilute the soluble extracts five-fold with BC300 buffer to reduce the NP-40 concentration to below 0.2% (for affinity purification).

16. For each extract, equilibrate 50–75 μl (settled bed volume) of M2-agarose affinity resin by washing the beads several times in BC300 containing 0.1% NP-40. Add beads to the diluted extract, and incubate for 8–10 hr (at 4°C) with constant rotation.

17. To wash the beads, collect them by gentle centrifugation (1000 rpm for 30 sec in Beckman J-6B or equivalent rotor). For the first wash, resuspend the beads in BC300 containing 0.1% NP-40 and transfer to a fresh tube. Wash the beads with repeated (four times) centrifugation and resuspension in 4 ml BC300 containing 0.1% NP-40. For these subsequent washes, it is not necessary to change tubes each time. Finally, resuspend the beads in 1-ml BC100 containing 0.1% NP-40 and transfer to a microfuge (Eppendorf) tube. Briefly centrifuge the tubes (3000 rpm, 30 sec) in a microfuge swinging bucket rotor. Wash one final time with BC100 containing 0.1% NP-40 to equilibrate the beads in this reduced salt buffer.

18. To elute the bound complex, add 100 μl of 0.5 mg/ml FLAG peptide in BC100 containing 0.1% NP-40 and leave on ice for 45 min. Periodically, mix the contents of the tube by pipetting up and down with a micropipette tip. Because of the small volumes and the tendency of the beads to cling to the walls of the tube, avoid rotation in this case. Collect the eluate by centrifuging in a

microfuge swinging bucket rotor (3000 rpm, 1 min), taking care not to take up any beads with the supernatant. A second centrifugation might be helpful in removing residual beads.

19. Analyze 5–15 μl of the eluates by SDS-PAGE. Perform this analysis in duplicate. One gel can be stained by silver and the other can be probed by immunoblotting, using anti-HA or anti-FLAG antibodies, as well as antibodies directed against selected TRAP/Mediator subunits. Given that the extracts used in this relatively rapid clone analysis are rather crude, it is strongly recommended that appropriate negative controls be included to allow subtraction of background (contaminating) bands. The M2-agarose eluates from extracts of clones resulting from the parent vector (positive control for the transfection step) would be a useful control here. Identify a clone that expresses the FLAG-tagged NUT2 (f:NUT2) polypeptide, and also is enriched in the various TRAP/Mediator subunits relative to the overexpressed NUT2.

20. Expand the positive clone (f:NUT2) to obtain 2×10^8 cells (about 10 near-confluent 150 mm dishes). Trypsinize the cells and resuspend in 200 ml Joklik's medium containing 10% FCS, penicillin–streptomycin, and 0.5 mg/ml G418. Transfer to spinner flasks and incubate at 37°C (no CO_2) with gentle stirring. After an initial period when extensive cell death may be apparent, the cells will begin to adapt to spinner culture conditions and their growth rate will steadily improve to near-normal levels. (However, there may be an occasional clone that fails to thrive under these conditions.) It is advisable to maintain selection (G418) for as long as possible. For larger scale cultures, it is economical to wean the cells to bovine calf serum (BCS) once the cells are well adapted to spinner cultures. This is done in stages. Begin with a 3:1 mixture of FCS:BCS (10% final concentration). Steadily reduce the FCS content until the cells are growing exclusively on 10% BCS.

21. Although large-scale extracts can be made directly from cells grown as earlier, we routinely transfer cells to high-density culture conditions. Cells are grown in DME–phosphate supplemented with 10% BCS and penicillin–streptomycin in large (10 liters nominal capacity) flasks, into which a sterile mixture of air and CO_2 is fed. At this stage, it is usually no longer feasible to include G418 in the medium.

22. For extract preparation, up to 20 liters of cultured cells (growing at more than 1×10^6/ml) are harvested. Nuclear extract preparation is done exactly as described previously.[10] Under favorable conditions, 60–80 ml of nuclear extract (at circa 10 mg/ml protein) is obtained.

It is dialyzed against BC0 until the conductivity corresponds to 100 mM KCl. The extract is quick-frozen in liquid nitrogen and stored at −80°C.

Purification of TRAP/Mediator Complex from Nuclear Extract of the f:NUT2 Cell Line

To obtain highly purified TRAP/Mediator from extracts of f:NUT2-expressing cells, it is helpful to fractionate the extract on a phosphocellulose (P11) ion-exchange column prior to affinity chromatography on M2-agarose. All procedures are carried out at 4°C.

Materials and Reagents. Nuclear extract from f:NUT2-expressing HeLa cells, phosphocellulose (P11, Whatman), M2-agarose, FLAG peptide.

Procedure

1. Carefully (but rapidly) thaw frozen f:NUT2 nuclear extract. To clear the extract, centrifuge it in a Beckman Ti45 rotor at 30,000 rpm for 20 min. Skim off as much as possible of the lipid layer that collects at the top.
2. To the supernatant, slowly add solid ammonium sulfate (ultrapure) to a final concentration of 55%. Once all the salt is dissolved, stir for another 30 min.
3. Collect the ammonium sulfate precipitate by centrifugation (15,000 rpm, 30 min in Beckman JA-20 or equivalent rotor) and resuspend in 50 ml BC0.
4. Dialyze against BC100, with several changes, until the conductivity corresponds to BC100.
5. Pack 50 ml of P11 resin, which has been processed (cycled) according to the manufacturer's instruction, into a column. Equilibrate with at least 10 column volumes of BC100. Check the column effluent to make sure that its pH is the same as that of the input (7.9).
6. Load the clarified nuclear extract supernatant on the P11 column. Use relatively slow flow rates (about 0.5 column volume/hr or 0.4 ml/min).
7. Wash the column extensively with BC100. If the column is monitored by UV, continue to wash at least until the absorbance returns to the base-line.
8. Elute the column with BC300. A relatively sharp elution peak should be observable over 2 column volumes.
9. Once the absorbance returns to base-line, elute the column with BC850. Collect and pool the peak fractions (total just over 1 column volume).

10. Add NP-40 to 0.1% final concentration to the BC850 eluate and dialyze against BC100 containing 0.1% NP-40. Monitor conductivity until it drops to the equivalent of BC300.

11. Equilibrate 400 μl (settled bed volume) of M2-agarose in BC300 containing 0.1% NP-40 and mix with the dialyzed eluate. Rotate for 8–10 hr.

12. Remove the unbound fraction by centrifugation (1000 rpm for 30 sec in a Beckman J-6B rotor). Resuspend the M2-agarose in 5 ml BC300 containing 0.1% NP-40 and transfer to a fresh tube. Centrifuge again. Discard the supernatant and repeat the washing four more times.

13. Wash once in BC100 containing 0.1% NP-40 and transfer to microfuge tubes for elution.

14. To elute the bound f:NUT2 TRAP/Mediator complex, add 400 μl of 0.3 mg/ml FLAG peptide in BC100 containing 0.1% NP-40. Rotate for 45 min and collect the supernatant by centrifugation (3000 rpm, 30 sec) in a microfuge swinging bucket rotor. Alternatively, spin columns, which are somewhat tedious to use but which give slightly higher recoveries, may be used for elutions from small amounts of affinity resins.

15. Repeat the elutions (400 μl of 0.3 mg/ml FLAG peptide in BC100 containing 0.1% NP-40) two more times for durations of 30 min each.

16. Analyze 5–10 μl of each of the eluates for yield and purity by SDS-PAGE (silver staining, Fig. 1), as well as by immunoblotting, before use in functional assays.

Note that this procedure yields almost the entire TRAP/Mediator pool, which may comprise distinct subpopulations, including PC2.[9]

Functional Analysis of TRAP/Mediator Complex

We have successfully used two distinct *in vitro* assay systems to demonstrate that purified TRAP/Mediator complex functions as a coactivator for diverse transcriptional activators, including nuclear receptors.[3,4,9] For our reconstituted assay system (Fig. 1), we obtain TFIIA, TFIIB, TFIIE, TFIIF, and PC4 as recombinant proteins from bacterial expression systems. The Pol II, TFIID, and TFIIH complexes are isolated from corresponding epitope-tagged cell lines. The alternative system comprises unfractionated HeLa cell nuclear extracts that have been selectively depleted of the TRAP/Mediator complex, and has been used for a subset of activators and nuclear receptors.[12,13]

[12] H. J. Baek, S. Malik, J. Qin, and R. G. Roeder, *Mol. Cell. Biol.* **22**, 2842 (2002).
[13] S. Malik, A. W. Wallberg, Y. K. Kang, and R. G. Roeder, *Mol. Cell. Biol.* **22**, 5626 (2002).

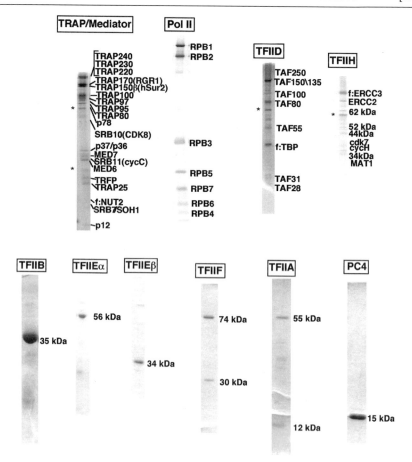

FIG. 1. SDS-PAGE analysis of preparations of TRAP/Mediator, Pol II, TFIID, TFIIH, TFIIB, TFIIE (α and β subunits separately), TFIIF, TFIIA, and PC4. The samples in the top row have been stained with silver, those in the bottom with Coomassie Blue. Asterisks indicate nonspecific bands.

Purification of Pol II and GTFs

Pol II

Materials and Reagents. Nuclear extract from Hela cells (f:RPB9) that stably express the FLAG-tagged RPB9 subunit (C. X. Yuan and R. G. R., unpublished results), DEAE cellulose (DE52, Whatman), M2-agarose, FLAG peptide.

Solutions and Buffers. TGEA (0.1): 20 mM Tris–HCl (pH 7.9), 25% glycerol, 0.1 mM EDTA, 0.5 mM PMSF, 2 mM DTT containing 100 mM

ammonium sulfate; TGEA(0.3): the same buffer containing 300 mM ammonium sulfate.

Procedure

1. Thaw out and clarify 60 ml of f:RPB9 nuclear extract (at 10 mg/ml) as described earlier for f:NUT2 extract.
2. Dialyze against TGEA(0.1) until the conductivity of the extract reaches an equivalent of 0.1 M ammonium sulfate.
3. Pack 65 ml of DE52 resin, which has been processed (cycled) according to the manufacturer's instructions, into a column. Equilibrate with at least 5 column volumes of TGEA(0.1). Check the column effluent to make sure that its pH is the same as that of the input (7.9).
4. Load the clarified nuclear extract supernatant on the column at a flow rate of 0.5 column volume/hr or about 0.5 ml/min).
5. Wash the column extensively with TGEA(0.1) until the absorbance on the UV-monitor/recorder returns to the base-line.
6. Elute the column with TGEA(0.3). Collect and pool the peak fractions (total just under 1 column volume).
7. Dilute eluate two-fold in TGE buffer. Centrifuge (15,000 rpm, for 30 min in Beckman JA-20 rotor) to remove any insoluble material. Add NP-40 to a final concentration of 0.1%.
8. Equilibrate 300 μl (settled bed volume) of M2-agarose in TGEA(0.1) containing 0.1% NP-40 and mix with the diluted eluate. Rotate for 8–10 hr.
9. Remove the unbound fraction by centrifugation. Resuspend the M2-agarose in 5 ml TGEA(0.1) containing 0.1% NP-40 and transfer to a fresh tube. Centrifuge again. Discard the supernatant and repeat the washing four more times with BC300. (Note that because the Pol II is concentrated on the beads by this point, we switch to our standard BC buffers, which have a reduced glycerol content.) Proceed as for the isolation of the TRAP/Mediator complex. Finally, obtain three eluates (300 μl each) and analyze by SDS-PAGE (Fig. 1).

TFIID

Materials and Reagents. Nuclear extract from Hela (f:TBP) cells that stably express FLAG-tagged TBP.[14]

---

[14] C. M. Chiang, H. Ge, Z. Wang, A. Hoffmann, and R. G. Roeder, *EMBO J.* **12**, 2749 (1993).

Procedure

1. Fractionate f:TBP nuclear extract (60 ml at 10 mg/ml) on a 40 ml P11 column essentially as described earlier for f:NUT2, except that clarified extract is directly loaded onto the column (without an additional ammonium sulfate precipitation step) and, in addition to the 0.1 M KCl flow-through and the 0.3 M KCl fractions, the 0.5 M and 0.85 M KCl fractions are separately collected. (See also Ref. 24.)

2. Dialyze the 0.85 M fraction against BC100 until the conductivity drops to the equivalent of 100 mM KCl. Centrifuge to remove any insoluble material.

3. Pack and equilibrate a 20-ml DE52 column with BC100. Load the dialyzed P11 0.85 M fraction containing TFIID onto the column at a flow rate of 0.5 ml/min. Wash with BC100.

4. Next, wash with at least 2 column volumes of BC150 or until the protein peak (containing residual amounts of TRAP/PC2/Mediator) is completely eluted.

5. Elute the bound TFIID with BC300.

6. Add NP-40 to the eluate to a final concentration of 0.1%.

7. Perform M2-agarose affinity chromatography exactly as described earlier for the f:NUT2 TRAP/Mediator complex to obtain finally three FLAG eluates (400 μl each). Check for purity and yield by SDS-PAGE. For quantitation purposes, we also estimate the TBP content of the TFIID preparation by comparison (in immunoblot assays) against known amounts of recombinant TBP.

TFIIH

Materials and Reagents. Nuclear extract from Hela cells (f:ERCC3) that stably express the FLAG-tagged ERCC3 subunit of TFIIH (C. X. Yuan and R. G. R., unpublished results).

Procedure

1. Fractionate f:ERCC3 nuclear extract (60 ml at 10 mg/ml) on a 40-ml P11 column, essentially as described earlier. Because TFIIH appears in both the P11 0.5 M and 0.85 M KCl, the two fractions are pooled before further (affinity) purification. (Alternatively, after washing the P11 column with 0.3 M KCl, elute with 0.85 M KCl.)

2. To the pooled eluates, add NP-40 to a final concentration of 0.1%.

3. Perform M2-agarose affinity chromatography exactly as described earlier for the f:NUT2 TRAP/Mediator complex to obtain finally

three FLAG eluates (400 μl each). Check for purity and yield by SDS-PAGE (Fig. 1).

TFIIB

Materials and Reagents. *Escherichia coli* BL21(DE3)pLysS transformed with pET11d-6His-TFIIB plasmid.[15]

Solutions and Buffers. LB medium (supplemented with 100 μg/ml ampicillin), 1 M isopropyl thio-β-D-galactopyranoside (IPTG), nickel lysis and chromatography buffer [50 mM Tris–HCl (pH 7.9), 0.3 M ammonium sulfate, 10% glycerol, 5 mM 2-mercaptoethanol, 1 mM PMSF, 10 μg/ml each of pepstatin A, leupeptin, and aprotinin].

Procedure

1. Using standard microbiological methods, grow 1 liter of *E. coli* BL21(DE3)pLysS transformed with pET11d-6His-TFIIB plasmid in LB medium supplemented with 100 μg/ml ampicillin to an A_{595} of 0.3.
2. Cool cells to 25°C and induce with 0.5 mM IPTG for 3–4 hr at this temperature.
3. Harvest cells by centrifugation (4000 rpm, 20 min in Beckman J-6B or equivalent rotor).
4. Resuspend cells in 10–20 ml of ice-cold lysis and chromatography buffer.
5. Add NP-40 to the suspension to a final concentration of 0.1% and allow cells to lyse on ice for 10 min.
6. Sonicate the lysate until most of the viscosity is lost.
7. Clear the lysate by centrifugation (19,000 rpm, 30 min in a Beckman JA-20 or equivalent rotor).
8. Add 2 ml of Ni–NTA–agarose resin that has been preequilibrated in lysis and chromatography buffer. Rotate for 2 hr.
9. Pack the slurry into a column.
10. Wash the column with several column volumes of lysis and chromatography buffer.

---

[15] S. Malik, K. Hisatake, H. Sumimoto, M. Horikoshi, and R. G. Roeder, *Proc. Natl. Acad. Sci. USA* **88**, 9553 (1991).

11. Wash the column with 4–5 column volumes of lysis and chromatography buffer containing 20 mM imidazole.
12. Elute the column with 2 column volumes of lysis and chromatography buffer containing 150 mM imidazole. It is advisable to collect small (0.3 ml) fractions, so that only those containing highly concentrated protein (peak) may be pooled, aliquoted, and snap-frozen (Fig. 1). [Note that the final preparation is not in our usual storage buffer (BC100). However, the preparations are extremely concentrated and may be diluted in BC100 buffer for transcription assays.]

TFIIE

The two subunits of TFIIE are expressed and purified separately and combined in equimolar ratios just prior to the transcription assays.

TFIIEα

Materials and Reagents. E. coli BL21(DE3)pLysS transformed with pET11d-FLAG-TFIIEα plasmid.[16,17]

Procedure

1. Grow and induce bacteria expressing FLAG-TFIIEα as described earlier for TFIIB.
2. Resuspend the cell pellet in BC500 buffer containing 5 mM 2-mercaptoethanol, 1 mM PMSF, 10 μg/ml each of Pepstatin A, leupeptin, and aprotinin.
3. Add NP-40 to the suspension to a final concentration of 0.1% and allow cells to lyse on ice for 10 min.
4. Sonicate the lysate until most of the viscosity is lost.
5. Clear the lysate by centrifugation (19,000 rpm, 30 min in a Beckman JA-20 or equivalent rotor).
6. Add 0.4 ml of M2-agarose resin that has been preequilibrated in BC500 buffer. Rotate for 4 hr.
7. Pack the slurry into a column.
8. Wash the column with several column volumes of BC500 containing 0.1% NP-40 and the protease inhibitors.
9. Wash the column with 4–5 column volumes of BC100 containing 0.1% NP-40.

[16] Y. Ohkuma, H. Sumimoto, A. Hoffmann, S. Shimasaki, M. Horikoshi, and R. G. Roeder, *Nature* **354**, 398 (1991).
[17] C. M. Chiang and R. G. Roeder, *Pept. Res.* **6**, 62 (1993).

10. Elute the column with 0.4 mg/ml FLAG peptide in BC100 containing 0.1% NP-40. This step may be done using either a batch-wise method or a spin-column. Repeat the elutions two more times. Aliquot and snap-freeze (Fig. 1).

TFIIEβ

Materials and Reagents. *E. coli* BL21(DE3)pLysS transformed with pET11d-6His-TFIIEβ plasmid.[18]

Procedure. The procedure for expression and purification of TFIIEβ from *E. coli* BL21(DE3)pLysS transformed with pET11d-6His-TFIIEβ plasmid is exactly the same as that described for recombinant TFIIB.

TFIIF

In the procedure described below (adapted from Burton and coworkers[19]), the two subunits of TFIIF, which when individually expressed in *E. coli* are essentially insoluble, are expressed and partially purified separately. The two subunits are then corenatured to reconstitute TFIIF.

Materials and Reagents. *E. coli* BL21(DE3)pLysS transformed with plasmid pET11d-RAP30 and with plasmid pET23-RAP74.[19]

Solutions and Buffers. BA(0.1): 50 mM HEPES (pH 7.9), 100 mM NaCl, 10% glycerol, 0.1% NP-40, 10 mM 2-mercaptoethanol, 1 mM PMSF, and 10 μg/ml each of pepstatin and leupeptin; BA(0.4): same composition as BA(0.1) except that the NaCl concentration is 400 mM; extraction buffer: 50 mM HEPES (pH 7.9), 5% glycerol, 2 mM EDTA, 0.1 mM DTT, 0.05% sodium deoxycholate, 1% Triton X-100; solubilization buffer: 10 mM HEPES (pH 7.9), 6 M guanidine–HCl, 0.2 mM EDTA, 0.2 mM EGTA, 10 mM DTT.

Procedure. Separately grow and induce bacteria (500 ml each) expressing the RAP30 and RAP74 subunits as described earlier for TFIIB. Then proceed as follows with the partial purification of insoluble RAP30 and RAP74 and their eventual reconstitution.

---

[18] H. Sumimoto, Y. Ohkuma, E. Sinn, H. Kato, S. Shimasaki, M. Horikoshi, and R. G. Roeder, *Nature* **354**, 401 (1991).

[19] B. Q. Wang, C. F. Kostrub, A. Finkelstein, and Z. F. Burton, *Protein Expr. Purif.* **4**, 207 (1993).

RAP30

1. Resuspend the pellet in 10 ml BA(0.1).
2. Sonicate the lysate to reduce the viscosity.
3. Centrifuge the sonicated lysate (25,000 rpm, 10 min in a Beckman Ti45 rotor).
4. Add 10 ml of the extraction buffer and resuspend the pellet by homogenization (10 strokes with a B-type pestle).
5. Collect the inclusion bodies by centrifugation (25,000 rpm, 10 min in a Beckman Ti45 rotor).
6. Resuspend the inclusion bodies in 10 ml of solubilization buffer. Rotate 12–16 hr.
7. Centrifuge to clear (25,000 rpm, 15 min in a Beckman Ti45 rotor).
8. Dilute the supernatant by adding 2 volumes of BC300.
9. Dialyze first against BC300 and then against BC150.
10. Remove insoluble material by centrifugation (25,000 rpm, 15 min in a Beckman Ti45 rotor). Save the supernatant containing partially pure, soluble RAP30 for reconstitution.

RAP74

1. Resuspend the induced bacterial pellet in 10 ml of BA(0.4). Sonicate and centrifuge as earlier.
2. To the supernatant add ultrapure urea to a final concentration of 4 M. Add imidazole to a final concentration of 5 mM.
3. Mix with 2 ml Ni–NTA–agarose that has been preequilibrated in BA(0.4) containing 4 M urea. Rotate for 2 hr at room temperature.
4. Pack the slurry into a column. Wash the column with 25 ml of BA(0.4) containing 4 M urea and 5 mM imidazole.
5. Additionally wash the column with 20 ml of BC500 containing 4 M urea and 5 mM imidazole.
6. Elute the column with BC500 containing 4 M urea and 180 mM imidazole.
7. Pool peak fractions and dialyze against BC500.
8. Remove insoluble material by centrifugation (25,000 rpm, 15 min in a Beckman Ti45 rotor). Save the supernatant containing highly pure (but not homogeneous), soluble RAP74 for reconstitution.

Reconstitution of TFIIF

1. Mix equimolar amounts of enriched and solubilized preparations of RAP30 and RAP74 (200 and 500 μg, respectively).

2. Add an equal volume of BC500 containing 8 M urea, so that the final concentration of urea is 4 M.
3. Rotate 2 hr at room temperature.
4. In a series of dialysis steps, gradually remove the urea. Dialyze first against BC500 containing 2 M urea for 2 hr. Then dialyze against BC500 containing 1 M urea for 2 hr. Finally, dialyze against BC500 with several changes over 4–6 hr at 4°C.
5. Remove insoluble material by centrifugation (25,000 rpm, 15 min in a Beckman Ti45 rotor). Aliquot and snap-freeze the supernatant containing highly pure, reconstituted TFIIF (Fig. 1).

Note that although the individual subunit preparations are not quite pure, the final reconstituted (soluble) TFIIF is essentially homogeneous, as the contaminating proteins have a greater tendency to precipitate during the reconstitution. Also, it is helpful (to prevent precipitation) to store the reconstituted TFIIF in a high salt buffer (BC500). Because the preparations are quite concentrated, they can be diluted in BC100 buffer prior to assaying.

TFIIA

Materials and Reagents. *E. coli* BL21(DE3)pLysS transformed with pET11d-6His-TFIIA(p12) plasmid;[20] *E. coli* BL21(DE3)pLysS transformed with pET11d-TFIIA(p55) plasmid.[21]

Solutions and Buffers. TGM(500): 20 mM Tris–HCl (pH 7.9), 10% glycerol, 1 mM MgCl$_2$, 500 mM KCl, 0.5 mM PMSF, 5 mM 2-mercaptoethanol.

Procedure. Natural TFIIA is a trimeric protein, two of whose components (p35 and p19) are derived through proteolytic cleavage of a common 55 kDa precursor.[20,22] Our preparations of recombinant TFIIA comprise p55 and p12. Bacteria (500 ml each) expressing the p55 and p12 subunits are grown separately. The solubilized subunits are purified by chromatography on Ni–NTA–agarose. In the case of p55, this purification relies on an intrinsic run of histidine residues. Finally, the two subunits are corenatured to give reconstituted recombinant TFIIA.

[20] J. DeJong, R. Bernstein, and R. G. Roeder, *Proc. Natl. Acad. Sci. USA* **92**, 3313 (1995).
[21] J. DeJong and R. G. Roeder, *Genes Dev.* **7**, 2220 (1993).
[22] G. Orphanides, T. Lagrange, and D. Reinberg, *Genes Dev.* **10**, 2657 (1996).

p12

1. Resuspend the cell pellet in 10 ml of TGM(500). Add NP-40 to a final concentration of 0.1% and allow cells to lyse on ice.
2. Sonicate and centrifuge to clear (15,000 rpm, 20 min in Beckman JA-20 rotor).
3. Resuspend the pellet, which contains most of the overexpressed (insoluble) p12, in 5 ml of TGM(500) containing 6 M guanidine–HCl and 5 mM imidazole. Homogenize to obtain a uniform suspension (several strokes with a B-type pestle).
4. Centrifuge to remove insoluble material (25,000 rpm, 10 min in a Beckman Ti45 rotor).
5. To the supernatant add 0.5 ml Ni–NTA–agarose that has been preequilibrated with TGM(500) containing 6 M guanidine–HCl and 5 mM imidazole. Rotate for 2 hr at room temperature.
6. Pack the slurry into a column. Wash with 5 ml of TGM(500) containing 6 M guanidine–HCl and 5 mM imidazole.
7. Wash the column with 10 ml TGM(500) containing 6 M guanidine–HCl and 20 mM imidazole.
8. Elute the column with TGM(500) containing 6 M guanidine–HCl and 200 mM imidazole. Pool peak fractions (1 ml or less) and determine protein concentration.

p55

Purify p55 from the insoluble fraction of the corresponding extract exactly as described for p12.

Reconstitution of Recombinant TFIIA

1. Mix equimolar amounts of p55 and p12 (circa 550 and 120 μg, respectively) in TGM(500).
2. Dialyze against BC500 (2 hr) at room temperature. Transfer to 4°C and continue dialysis for 10–12 hr with several changes.
3. Remove insoluble material by centrifugation (25,000 rpm, 15 min in a Beckman Ti45 rotor). Aliquot and snap-freeze the supernatant containing highly pure (albeit partially degraded), reconstituted TFIIA.

PC4

Materials and Reagents. E. coli BL21(DE3)pLysS transformed with pET11a-PC4 plasmid.[23]

---

[23] M. Kretzschmar, K. Kaiser, F. Lottspeich, and M. Meisterernst, *Cell* **78**, 525 (1994).

Procedure

The following procedure is essentially identical to one described earlier.[24]

1. Grow and induce bacteria expressing PC4 as described for TFIIB.
2. Resuspend the cell pellet in BC300 buffer containing 5 mM 2-mercaptoethanol, 1 mM PMSF, 10 μg/ml each of Pepstatin A, leupeptin, and aprotinin.
3. Add NP-40 to the suspension to a final concentration of 0.1% and allow cells to lyse on ice for 10 min.
4. Sonicate the lysate until most of the viscosity is lost.
5. Clear the lysate by centrifugation (19,000 rpm, 30 min in a Beckman JA-20 or equivalent rotor).
6. Load the cleared lysate on a 2-ml heparin-Sepharose column that has been preequilibrated in BC300 buffer.
7. Wash the column with at least 5 column volumes of BC300.
8. Elute recombinant PC4 with BC500.
9. Directly load the eluate onto a 1-ml P11 column that has been preequilibrated with BC500.
10. Wash the column with BC500 (at least 10 ml).
11. Elute PC4 with BC850.
12. Dialyze against BC100 until conductivity corresponds to 100 mM KCl. Aliquot and snap-freeze.

In Vitro *Transcription Assay for TRAP/Mediator Function on DNA Templates*

Using the purified transcription factors from the preceding sections TRAP/Mediator-dependent transcriptional activation can be reconstituted *in vitro*. In addition to these factors, nuclear receptors of interest and DNA templates containing cognate target sites upstream of appropriate core promoters will also be needed. The receptors are best expressed in insect cells via baculovirus vectors, and subsequently purified over affinity (if epitope-tagged) and/or conventional resins. Note that for nuclear receptors, like TRα, which require RXR as a dimerization partner, RXR will also be needed.

In general, two kinds of templates may be used, depending on how the RNA product is to be scored: (1) templates that contain a G-less cassette

[24] H. Ge, E. Martinez, C. M. Chiang, and R. G. Roeder, *Methods Enzymol.* **274**, 57 (1996).

downstream of the core promoter elements and allow detection of synthesized RNA by direct incorporation of labeled nucleotides. (A G-less cassette is a stretch of DNA with no G-residues in the nontranscribed strand, such that its transcription in the absence of GTP results in the formation of a discrete RNA product.) (2) Templates that contain any sequence (including a naturally occurring sequence) downstream of the core promoter elements, but which require an additional primer-extension step for detection of synthesized RNA. In both cases, at least in initial experiments, where the receptor and coactivator dependencies are being explored, it is helpful to have multimerized copies of the receptor DNA-binding element located upstream of relatively weak core promoters. Obviously, if one is examining receptor function in the context of the natural configuration of the promoter elements, this choice is not available.

An additional concern (specifically for nuclear receptors) is the requirement of ligand for full activity *in vitro*. For several receptors (including TR/RXR) that we have studied, the systems described below essentially allow us to by-pass this requirement. This ability may reflect the existence of equilibrium between an inactive and active state of the receptor, and the propensity of the receptor to be driven toward its active state under the conditions of our assays. Alternatively, ligand dependence may only be manifested when additional constraints, such as those effected by chromatin structure and/or by receptor interacting corepressors, are imposed on the assay system. While we have not specifically described it in the methods outlined below, we recommend that, for each given receptor, the ligand requirement for *in vitro* activity be evaluated systematically.

Materials and Reagents. Purified transcription factors (as described here), DNA templates, RNAsin (Promega), standard molecular biology and biochemistry equipment and supplies.

Solutions and Buffers. 10× Assay mix: 0.2 M HEPES–KOH (pH 8.2), 50 mM MgCl$_2$; 10× NTP (low CTP): 5 mM ATP, 5 mM UTP, 50 μM CTP, 1 mM 3′-O-methyl GTP; 10× NTP (low UTP): 5 mM ATP, 5 mM CTP, 50 μM UTP, 1 mM 3′-O-methyl GTP; transcription stop mix: 0.4 M sodium acetate (pH 5.0), 13.3 mM EDTA, 0.33% SDS (sodium dodecyl sulfate), 0.67 mg/ml yeast tRNA; denaturing loading buffer: 50 mM Tris–HCl (pH 7.4), 98% deionized formamide, 0.005% Bromophenol Blue, 0.005% xylene cyanol (or 8 M urea, 0.005% Bromophenol Blue, 0.005% xylene cyanol in Tris–borate–EDTA electrophoresis buffer).

Procedure. The following describes our method for assaying TRAP/Mediator-dependent activity of TRα/RXR from a template containing a G-less cassette.[2] Note that we have described the procedure for one reaction that has been reconstituted from transcription factors whose functionalities have been previously established in a number of independent assays [including EMSA, activity for basal (activator-independent) transcription, etc.]. All the factors also have been previously titrated to yield a system that gives low-background levels in the absence of TRα/RXR and TRAP/Mediator. Thus, while we have listed the amounts of the factors that we typically use, these are to be taken only as guidelines. In practice, one finds that each preparation of a given factor has a different specific activity, and, that the optimum must be established empirically. This fact is especially true for PC4 and TFIIH, whose correct balance is crucial to demonstrating TRAP/Mediator function in the purified system. Note also that rigorous demonstration of activator and TRAP/Mediator dependencies requires, minimally, a set of four reactions: (i) Pol II, GTFs, and PC4; (ii) Pol II, GTFs, PC4, and receptor; (iii) Pol II, GTFs, PC4, and TRAP/Mediator; and (iv) Pol II, GTFs, PC4, TRAP/Mediator, and receptor.

In the following procedure, a 25 μl reaction is set up in two parts, which are then combined. Transcription is allowed to proceed, reactions are stopped, and the product RNA is analyzed by denaturing polyacrylamide gel electrophoresis.

1. For mix A, combine on ice: 2.5 μl of 10× assay mix, 2.5 μl of 10× NTP mix (low CTP), 12.5 mM DTT, 1.25 mg/ml BSA, 5% polyethylene glycol, 50 ng of pTRE$_{3x}$MLΔ53, 50 ng pG5HML (control template); 10 μCi of [α-^{32}P]CTP at 3000 Ci/mmol; 10 units RNasin. Bring to 10 μl with water.

2. For mix B, combine on ice: 5 ng TRα, 20 ng RXRα, 100 ng TRAP/Mediator, 50 ng Pol II, 10 ng TFIIA, 10 ng TFIIB, 50 ng TFIID (containing 5 ng equivalent of TBP), 5 ng TFIIEα, 2.5 ng TFIIEβ, 25 ng TFIIF, 20 ng TFIIH, and 150 ng PC4. Bring to 15 μl with BC100 containing 0.5 mg/ml BSA and 5 mM DTT. (Note that all factor preparations have been previously diluted in BC100 containing 0.5 mg/ml BSA and 5 mM DTT.)

3. Combine mix A and mix B, and incubate at 28°C for 50 min.

4. Add 1 μl of 0.5 mM CTP. Incubate at 28°C for another 20 min.

5. Add 75 μl of the transcription stop mix.

6. Extract once with 100 μl of phenol–chloroform–isoamyl alcohol (25:24:1).

7. To the upper phase add 300 μl of ethanol. Leave on powdered dry ice for 30 min.

8. Collect the RNA precipitate by centrifugation (10 min in a microfuge). Centrifugation at room temperature is recommended to minimize the amount of salt recovered in the nucleic acid pellet.
9. Wash the pellet once with 80% ethanol and once with 100% ethanol. Dry under vacuum.
10. Resuspend in 15 μl denaturing sample loading buffer and resolve transcripts on a 5% polyacrylamide gel containing 8 M urea. Dry the gel and autoradiograph.

Alternative Assay System for TRAP/Mediator

If a nuclear receptor (or other activator) can efficiently activate transcription in unfractionated nuclear extracts, TRAP/Mediator-dependent function may be assayed using extracts that have been immunologically depleted of TRAP/Mediator.[13] In as much as unfractionated extracts more closely recapitulate the intracellular distribution of transcription factors (both positive and negative), this approach has the potential of more accurately reflecting normal (cellular) levels of regulation. An obvious practical advantage is that this assay system is much more convenient to establish; however, unlike the transcription system reconstituted from purified factors, a drawback is that if precise factor requirements (or underlying mechanisms) are to be deduced, the complexity of the extract may complicate the results.

For depleting TRAP/Mediator from HeLa cell nuclear extract, we use antibodies directed against integral subunits. Here we describe a procedure that has worked well for, among others, the anti-NUT2 antibodies. We first antigen-purify the antibodies, covalently cross-link them to protein A-Sepharose, and then use the resulting resin to deplete nuclear extract. Protocols for antigen purification of antibodies and cross-linking to protein A-Sepharose have been adapted, in part, from previously published methods.[25]

Materials and Reagents. Rabbit polyclonal anti-NUT2 antiserum,[9] *E. coli* BL21(DE3)pLysS transformed with pET11d-6His-NUT2 plasmid,[9] CNBr-activated Sepharose 4B (Amersham Pharmacia), Protein A-Sepharose (Amersham Pharmacia); dimethylpimelidate (Sigma), transcriptionally active HeLa cell nuclear extract.[10]

Solutions and Buffers. For bacterial expression and purification of 6His-NUT2: LB medium (supplemented with ampicillin), 1 M IPTG; denaturing

[25] E. Harlow and D. Lane, "Antibodies: A Laboratory Manual." Cold Spring Harbor, NY, Cold Spring Harbor Laboratory, 1988.

lysis and chromatography buffer [8 M urea (ultrapure, from ICN) in 20 mM HEPES (pH 7.9), 500 mM NaCl]. For antigen purification of anti-NUT2 antibodies: 10 mM Tris pH 7.4, 100 mM glycine (pH 2.5), 100 mM triethylamine (pH 11.5). For coupling of purified antibodies to protein A-Sepharose: Tris-buffered saline: 20 mM Tris–HCl (pH 7.4 at 25°C), 150 mM NaCl; 0.2 M sodium borate buffer (pH 9.0); 0.5 M ethanolamine (pH 8.0); 0.1 M acetate buffer (pH 4.0), 0.5 M NaCl; 0.1 M Tris–HCl (pH 8.0), 0.5 M NaCl.

Procedure

1. Grow 1–2 liters of *E. coli* BL21(DE3)pLysS transformed with pET11d-6His-NUT2 plasmid (or enough to yield 2–4 mg of purified recombinant NUT2 protein). Induce with 0.4 mM IPTG for 4 hr. Harvest bacteria by centrifugation (4000 rpm, 20 min in a Beckman J-6B or equivalent rotor). Quick-freeze the cell pellet in liquid nitrogen and store at −80°C until needed.

2. Thaw the pellet on ice and resuspend in up to 20 ml of denaturing lysis and chromatography buffer. Note that HEPES-based buffers are used in this procedure as the amino groups in Tris base will interfere with the coupling of NUT2 to CNBr-activated Sepharose 4B. Rotate (at room temperature) for 90 min until the suspension becomes relatively translucent. If cell lysis is not readily achieved, the cell suspension may be sonicated as needed.

3. Clear the suspension by centrifugation (18,000 rpm in a Beckman JA-20 or equivalent rotor).

4. To the supernatant add 1 ml (bed volume) of Ni–NTA–agarose (Qiagen) resin that has been preequilibrated in denaturing lysis and chromatography buffer. Mix for 90 min by gentle rotation.

5. Pack the slurry into a small column.

6. Once the unbound material has been allowed to drain, wash the resin with at least 10 column volumes of denaturing lysis and chromatography buffer.

7. Elute the bound protein with 100 mM EDTA (in denaturing lysis and chromatography buffer). Again, note that the use of imidazole, which is the conventional eluent, is to be avoided in order to minimize interference with the subsequent coupling reaction. Also note that while we do not carry out any stringent washes (equivalent to the intermediate-strength imidazole washes in standard Ni–NTA–agarose chromatography), the material obtained from this scheme is of sufficient purity for the present purpose (provided the expression levels of the recombinant protein are high). If purity is

a concern, the antigen may be purified by the usual procedure that involves washing with 20 mM imidazole. The purified protein may then be diluted (or dialyzed) and rechromatographed on Ni–NTA–agarose, washed extensively to remove residual imidazole, and eluted with EDTA.

8. Weigh out 0.5 g CNBr-activated Sepharose 4B and allow it to swell in 50 ml of 1 mM HCl. To completely remove preservatives from the dried powder, wash with a total of about 200 ml of 1 mM HCl. This step can be done either by washing on a sintered glass funnel or by repeated centrifugation and resuspension in HCl. Before use, equilibrate the matrix in denaturing lysis and chromatography buffer.

9. Mix the purified recombinant protein and the matrix (1 ml settled-bed volume) in a 15 ml conical tube. Rotate for 2 hr at room temperature. Block the coupling reaction by addition of Tris–HCl (pH 8.0) to 0.1 M and incubate, either at room temperature for 2 hr or at 4°C overnight.

10. For this and subsequent washes, it may be most convenient to pack the matrix in a small column.

11. Gradually reduce the concentration of urea from 8 M to zero. Wash first with 5 ml buffer containing 8 M urea in 0.1 M Tris–HCl (pH 8.0), 0.5 M NaCl. In the next washes, progressively reduce the urea to 4, 2, and 1 M in the same buffer. Finally, wash with 0.1 M Tris–HCl (pH 8.0), 0.5 M NaCl.

12. To remove traces of uncross-linked antigen, wash the antigen-coupled matrix with three cycles of alternating pH. Each cycle comprises an acidic [0.1 M acetate (pH 4), 0.5 M NaCl] and a slightly basic [0.1 M Tris–HCl (pH 8.0), 0.5 M NaCl] wash (5 ml each).

13. Prior to passing serum through the column, additionally wash with elution buffers that will be used for the affinity purification. Wash once with 5 ml of 100 mM glycine (pH 2.5). Wash with 100 mM Tris–HCl (pH 8.8) until the pH of the column effluent is neutral. Wash with 5 ml of 100 mM triethylamine (pH 11.5). Wash with 10 mM Tris–HCl (pH 7.4) until the pH of the column effluent is neutral.

14. Dilute 10 ml (or other appropriate amount, depending on the antibody titer) of crude anti-NUT2 serum with 90 ml of 10 mM Tris–HCl (pH 7.4).

15. Pass the diluted serum over the column at relatively slow rates (10 ml/hr). If a peristaltic pump is used, the flow rates can be controlled easily. If gravity flow is used, and the flow rate is high, passage multiple times over the column.

16. Wash the column with at least 20 column volumes of 10 mM Tris–HCl (pH 7.4).

17. Elute the column with 10 column volumes of 100 mM glycine (pH 2.5), and collect the eluate in tubes containing one-tenth volume of 1 M Tris–HCl (pH 7.4) to neutralize the sample. Check the pH, and adjust with 1 M Tris–HCl (pH 7.4), if necessary. Also adjust the salt to 150 mM NaCl.

18. Wash the column with 100 mM Tris–HCl (pH 8.8) until the pH of the column effluent is neutral.

19. Elute the column with 10 column volumes of 100 mM triethylamine (pH 11.5), and again collect the eluate in tubes containing one-tenth volume of 1 M Tris–HCl (pH 7.4) to neutralize the sample. As for the acid elutions, adjust the pH (if necessary) and salt.

20. Analyze each of the eluates in immunoblotting assays, using the antigen (NUT2) to determine where the antibody elutes. Given the polyclonal nature of the serum, we find that subpopulations of the NUT2 antibodies elute in both (acid and base) fractions. Therefore, for the purposes of the present experiments, we pool the two fractions. Also, estimate the purity and amount of the purified antibody preparations by SDS-PAGE and Coomassie Blue staining.

21. Equilibrate 750 μl of protein A-Sepharose in Tris-buffered saline and mix with 1.5–2 mg of the pooled antigen-purified antibody. Incubate with rotation overnight at 4°C.

22. Collect the protein A-Sepharose by centrifugation (1000 rpm for 30 sec in a Beckman J-6B or equivalent rotor). Wash the beads once with Tris-buffered saline.

23. Wash the beads twice with 10 ml of 0.2 M sodium borate (pH 9.0) by repeated centrifugation and resuspension in the buffer, as earlier.

24. Resuspend in 7.5 ml of 0.2 M sodium borate (pH 9.0) and add about 25 mg solid dimethylpimelidate (to a final concentration of 20 mM).

25. Rotate for 30 min at room temperature.

26. To stop the reaction, collect the beads by centrifugation and wash once in 0.2 M ethanolamine. Resuspend the beads in 0.2 M ethanolamine and incubate for 2 hr at room temperature. Wash with Tris-buffered saline for storage.

27. Coupling efficiency can be monitored by removing the equivalent of 10 μl aliquots of beads, both prior to and subsequent to the addition of dimethylpimelidate. Boil the beads in the presence of SDS-PAGE sample buffer and analyze by Coomassie Blue staining. If the coupling is successful, no immunoglobin heavy chains should be released from the beads.

28. For the depletion reaction, equilibrate the beads in BC200 buffer containing 0.1% NP-40. Adjust the salt concentration of the nuclear extract to be depleted (normally stored at 100 mM KCl) to 200 mM KCl, preferably by dialyzing against BC200 buffer. Alternatively, adjust the salt by adding BC1000 to give the desired final concentration. Centrifuge to clear. For small volumes, this step may be done in a microfuge. Also add NP-40 to a final concentration of 0.1%.

29. Mix 750 μl of the processed nuclear extract with the protein A-Sepharose covalently coupled to the anti-NUT2 antibodies. Incubate overnight at 4°C with very gentle rotation.

30. Centrifuge (3000 rpm, 30 sec) in a microfuge, and remove the supernatant (depleted extract). Centrifuge the supernatant one more time (14,000 rpm, 15 min). Make small aliquots and snap-freeze them in liquid nitrogen.

31. Check for depletion of TRAP/Mediator by immunoblotting using antibodies against several subunits (including NUT2) and by transcription assays. (It is strongly recommended that a mock depletion be performed in parallel as a negative control. Ideally, protein A-Sepharose beads covalently coupled to an irrelevant antibody should be used.)

In Vitro *Transcription Assay for TRAP/Mediator in Depleted Nuclear Extract*

The procedure described below has worked well for the orphan nuclear receptor HNF-4[13] but not yet for TR/RXR, which, in our hands, has not functioned efficiently in unfractionated Hela nuclear extract. Note that, in addition to controls employing mock-depleted extracts, rigorous demonstration of receptor and TRAP/Mediator dependencies requires a minimum set of four reactions, as for the reconstituted system: (i) depleted extract; (ii) depleted extract plus receptor; (iii) depleted extract plus purified TRAP/Mediator; and (iv) depleted extract plus receptor and purified TRAP/Mediator. The actual assay procedure is essentially the same as that described for the reconstituted system (preceding section). The following steps describe a typical 25 μl reaction.

1. For mix A, combine on ice: 2.5 μl of 10 × assay mix, 2.5 μl of 10 × NTP mix (low UTP), 12.5 mM DTT, 1.25 mg/ml BSA, 5% polyethylene glycol, 50 ng of pA$_{4x}$MLΔ53, 50 ng pG5HML (control template); 10 μCi of [α-^{32}P]UTP at 3000 Ci/mmol; 10 units RNasin. Bring to 10 μl with water.

2. For mix B, combine on ice: 25 ng HNF-4, 50 ng purified TRAP/Mediator, and 5 μl depleted Hela cell nuclear extract (or other appropriate volume, depending on the potency of the extract). Make up to 10 μl with BC100 containing 0.5 mg/ml BSA and 5 mM DTT. Add 5 μl BC0 containing 0.5 mg/ml BSA and 5 mM DTT. This adjustment is necessary to maintain the final monovalent salt concentration at 60 mM in the final transcription reaction.

3. Combine mix A and mix B, and incubate at 30°C for 50 min.

4. Add 1 μl of 1 mM UTP. Incubate at 30°C for another 20 min.

5. Add 1 μl of RNase T1 (2 units/μl) and incubate for additional 30 min.

6. Add 75 μl of the transcription stop mix.

7. Extract twice with 100 μl of phenol–chloroform–isoamyl alcohol (25:24:1). (Alternatively, remove protein first by digestion with proteinase K, then extract once with phenol–chloroform–isoamyl alcohol.)

8. To the upper aqueous phase, add 300 μl of ethanol. Leave on powdered dry ice for 30 min.

9. Collect the RNA precipitate by centrifugation (10 min in a microfuge) at room temperature.

10. Wash the pellet once with 80% ethanol and once with 100% ethanol. Dry under vacuum.

11. Resuspend in 15 μl denaturing sample loading buffer and resolve transcripts on a 5% polyacrylamide gel containing 8 M urea. Dry the gel and autoradiograph.

The experimental systems described here have provided important insights into some of the mechanisms by which nuclear receptors activate their target genes. Most significantly, they have helped to establish the dominant role of the TRAP/Mediator coactivator complex in this process, especially as it pertains to transcription from DNA templates. As the emphasis in the transcription field shifts to the study of distinct genes, it is likely that more natural templates, i.e., chromatin templates carrying their physiological arrangements of regulatory elements (e.g., promoters and enhancers) will increasingly be used in *in vitro* experiments. Clearly, the availability of pure TRAP/Mediator, together with the assays utilizing purified factors (to which new cofactors can be added) and with unfractionated extracts (from which cofactors can be selectively removed and resupplied), should provide a foundation for this next phase of investigation.

Acknowledgment

We are indebted to J. Fondell for his pioneering work on the TRAP complex. We thank C. X. Yuan for the epitope-tagged cell lines expressing Pol II and TFIIH subunits; current and past members of our laboratory, especially E. Martinez, T. Oelgeschläger, C. Parada, M. Guermah, and V. Palhan, for protocols, reagents, and ideas; E. Uno and A. Gamper for critical comments on the manuscript; and W. Wu and C. Bhattacharyya for providing expert technical assistance in the establishment and propagation of various stable cell lines.

[16] Study of Nuclear Receptor-Induced Transcription Complex Assembly and Histone Modification by Chromatin Immunoprecipitation Assays

By Han Ma, Yongfeng Shang, David Y. Lee, and
Michael R. Stallcup

Introduction

Nuclear receptor-mediated transcriptional regulation is a complex process involving dynamic assembly of multiprotein complexes and modifications of chromosomal proteins, such as histones, at the target gene promoter. An increasingly used method to study this process in the context of natural chromosome structure is the chromatin immunoprecipitation (ChIP) assay. The ChIP assay employs a combination of formaldehyde-induced *in vivo* cross-linking, immunoprecipitation, and sequence-specific DNA detection methods to observe the specific proteins associated with specific gene promoter regions in tissues or cultured cells.[1,2] A big advantage of using formaldehyde as a fixation reagent is that it readily penetrates living cell membranes and induces cross-linking within minutes, thus freezing the DNA–protein and protein–protein contacts in their native and real-time status. Formaldehyde-induced cross-linking is reversible, so that DNA can be purified from the immunoprecipitated chromatin fragments, and further analyzed by PCR or Southern blotting.[3,4] Thus, ChIP assays allow accurate assessment of the protein complexes associated with a given promoter during transcriptional regulation, spatially and temporally. Proteins of interest that can be examined by the use of specific antibodies include

[1] V. Jackson, *Cell* **15**, 945 (1978).
[2] M. J. Solomon, P. L. Larsen, and A. Varshavsky, *Cell* **53**, 937 (1988).
[3] V. Jackson, *Biochemistry* **29**, 719 (1990).
[4] S. L. Dimitrov, V.Yu. Stefanovsky, L. Karagyozov, D. Angelov, and I. G. Pashev, *Nucleic Acids Res.* **18**, 6393 (1990).

transcription factors that bind directly to the DNA, coactivators, corepressors, RNA polymerase and its associated proteins, and modified histones. Target gene promoters controlled by ligand-regulated nuclear receptors in cultured cells provide a particularly advantageous system for study, since one can examine proteins associated with the promoter before, and at various times after ligand stimulation of the cells.

The procedure of the ChIP assay can be summarized as follows. A nuclear lysate is prepared from formaldehyde cross-linked cells that have been previously treated with or without hormones. Then, relatively short, uniformly sized chromatin fragments are generated by sonication. The sheared chromatin preparations are subjected to immunoprecipitation by antibodies against proteins of interest such as nuclear receptors, cofactor proteins, and modified histones. The DNA–protein cross-linking in the immunoprecipitates is subsequently reversed by heating, and DNA is recovered after the removal of all proteins. Finally, the precipitated DNA is analyzed by semiquantitative PCR. If the protein of interest is associated with a specific genomic region, such as the promoter region of a nuclear receptor target gene, that specific genomic region will be enriched in the immunoprecipitates compared to a reference genomic region. This enrichment is represented as an increase in the PCR amplification signal for the specific genomic region (Fig. 1).

ChIP assays have been applied to study nuclear receptor-mediated cofactor complex assembly and histone modifications at target gene promoters.[5,6] For example, ChIP assays demonstrated that the assembly of transcriptional complexes and the hyperacetylation of histones, H3 and H4, happen dynamically at the promoter region of estrogen receptor target genes upon estrogen stimulation. The kinetics of complex assembly and histone hyperacetylation correlates with the rate of gene transcription. In addition, the time courses for the recruitment of functionally similar coactivators, such as the three histone acetyltransferases CBP, p300, and pCAF are distinct, suggesting different roles for each of these proteins in the transcriptional activation process. Results from ChIP assays also recently helped to illustrate the molecular mechanism underlying the tissue-specificity of selective estrogen receptor modulators.[7] Tamoxifen, which has a partial estrogenic activity in the uterus, induces the recruitment of coactivator complexes to the promoter region of estrogen-responsive c-Myc gene in the endometrial carcinoma cell line, Ishkawa; but it fails to do so for

[5] Y. Shang, X. Hu, J. DiRenzo, M. A. Lazar, and M. Brown, *Cell* **103**, 843 (2000).

[6] H. Chen, R. J. Lin, W. Xie, D. Wilpitz, and R. M. Evans, *Cell* **98**, 675 (1999).

[7] Y. Shang and M. Brown, *Science* **295**, 2465 (2002).

Semi-quantitative PCR analysis
of DNA immunoprecipitated in ChIP assay

Amplification of hormone response element (HRE)
in promoter region of interest

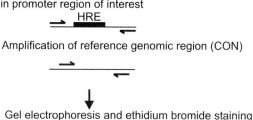

Amplification of reference genomic region (CON)

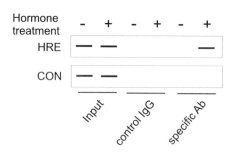

Gel electrophoresis and ethidium bromide staining

FIG. 1. Idealized results of the ChIP assay. To examine the assembly of a transcription complex on the promoter of a gene that is regulated by a ligand-activated nuclear receptor, hormone (i.e., ligand) treated, and untreated cells are fixed in formaldehyde, and ChIP assays are performed as described in the text. Briefly, sheared chromatin is prepared, and an aliquot (*Input* sample) is set aside for later PCR analysis. Antibodies (*specific Ab*) against a nuclear receptor, a coactivator, another component of the transcription machinery, or a modified histone is used to immunoprecipitate chromatin fragments to which the protein in question is bound. As a control, preimmune serum or nonspecific IgG (*control IgG*) is used in a parallel immunoprecipitation, or the primary antibody is omitted. DNA is purified from the immunoprecipitates and input samples, and the presence of the specific ligand-regulated gene promoter is detected by PCR, using primers that flank the promoter or nuclear receptor binding site (e.g., hormone response element or *HRE*) associated with the promoter. As a negative control, a second set of primers is used to amplify another genomic region that is not expected to interact with the nuclear receptor and other components of the transcription machinery (shown in the figure as the reference or control genomic region, *CON*). The PCR products are separated by gel electrophoresis and visualized by ethidium bromide staining. The idealized results show that the HRE region is equally represented in the input chromatin from the ligand-treated (+) and untreated (−) cells; the same is true for the control genomic region. The control IgG fails to immunoprecipitate either the HRE or the control chromatin region. The specific antibody immunoprecipitated chromatin containing the HRE from ligand-treated cells but not from untreated cells; however, the specific antibody did not immunoprecipitate the control region of chromatin. These results indicate that the protein against which the antibodies were directed is specifically recruited to the HRE region in response to hormone treatment of the cells.

the same c-Myc promoter in MCF-7 cells, a carcinoma cell line originated from mammary tissue. Instead in MCF-7 cells, tamoxifen acts as an antagonist. ChIP assays were also used to demonstrate that methylation of arginine 17 of histone H3, mediated by the coactivator CARM1, occurs on target gene promoters as part of the steroid hormone-induced transcriptional activation process.[8,9] The increasing availability of antibodies specific for histone modifications, such as lysine acetylation, lysine methylation, arginine methylation, and serine phosphorylation, makes it possible to investigate the specific circumstances under which each of these histone modifications is involved in nuclear receptor-mediated gene activation or repression. Finally, it will be informative to study whether there are distinct combinations of histone modifications associated with different target genes or physiological signals, as suggested by the "histone code" hypothesis.[10]

Cell Growth, DNA–Protein Cross-Linking, and Preparation of Chromatin Fragments

Materials and Reagents

Appropriate cell culture medium
Charcoal–dextran stripped fetal bovine serum (Gemini)
37% formaldehyde (Fisher Scientific)
1.25 M glycine
1 × phosphate-buffered saline (PBS)
Protease inhibitor cocktail (Roche Molecular Biochemicals)
Hypotonic buffer: 10 mM HEPES–KOH, pH 7.8, 10 mM KCl, 1.5 mM
    MgCl$_2$, 1 × protease inhibitor cocktail
#25 $\frac{1}{2}$ gauge needle and 1 ml syringe
Nuclear lysis buffer: 1% SDS, 50 mM Tris–HCl, pH 8.0, 10 mM EDTA,
    1 × protease inhibitor cocktail

Cell Growth and Hormone Treatment

Equal numbers of cells are seeded in 150-mm diameter dishes 4 days before the planned harvest date and grown in appropriate culture medium. Cells should be switched into medium supplemented with charcoal–dextran

[8] H. Ma, C. T. Baumann, H. Li, B. D. Strahl, R. Rice, M. A. Jelinek, D. W. Aswad, C. D. Allis, G. L. Hager and M. R. Stallcup, *Curr. Biol.* **11**, 1981 (2001).
[9] M. R. Stallcup, *Oncogene* **20**, 3014 (2001).
[10] B. D. Strahl and C. D. Allis, *Nature* **403**, 41 (2000).

stripped serum for at least 48 hr prior to the ligand treatment. The charcoal–dextran treatment removes many small molecules from the serum, including many nuclear receptor ligands. For some nuclear receptors, such as the estrogen receptor, phenol red-free medium should be used to avoid nonspecific stimulation of the receptors by phenol red. At harvest cells should be close to 95% confluency. Depending on cell type, the cell number from each 150-mm dish will be approximately $1.5–2 \times 10^7$. To examine single copy genes by ChIP, chromatin prepared from one near-confluent 150-mm dish is enough for one immunoprecipitation. On the day of harvest, fresh medium supplemented with charcoal–dextran stripped serum and the appropriate ligand is added to the cells for the desired length of time. For glucocorticoid and estrogen receptor regulated target genes, we and others have found that 30–45 min of ligand treatment results in strong recruitment of the nuclear receptors and coactivators to the promoter of the target genes and histone modifications in the promoter region.[5,6,8] Nonetheless, a time course study for ligand treatment is recommended for each specific cell line, ligand, and promoter.

Cross-Linking

After ligand treatment, the cross-linking reagent formaldehyde (37%) is added directly to the medium to a final concentration of 1% and mixed immediately. Formaldehyde cross-linking is carried out at 37°C or at room temperature for 10 min on a shaker platform. The cross-linking reaction is stopped by adding 1.25 M glycine to a final concentration of 0.125 M and incubating for another 5 min at room temperature with shaking. Thereafter, the medium is removed and cells are washed twice with cold PBS solution and harvested in 5 ml of cold PBS supplemented with a protease inhibitor cocktail; all the incubation buffers and lysis buffers hereafter should be supplemented with protease inhibitors. For a time course study in which cells are treated with hormone for various lengths of time, formaldehyde-treated cells from early timepoints can be left in cold PBS at 4°C until all cells are ready to be harvested.

Cell Harvest and Nuclear Isolation

All subsequent steps should be performed on ice or at 4°C whenever possible. Scrape cells off culture dishes with a cell scraper and pool similarly treated cells into one tube (once cells are fixed with formaldehyde, they cannot be detached from cell-culture dishes by trypsin treatment). Cells are then pelleted by centrifugation at $1000 \times g$ for 10 min at 4°C, the supernatant discarded, and the cell pellet resuspended in 1 ml of hypotonic buffer and transferred to a microcentrifuge tube. Cells are allowed to swell on ice in hypotonic buffer for 10 min. The cells are then lysed by forcing the suspension through a #25 $\frac{1}{2}$ gauge needle five times, and nuclei are collected by centrifugation at 13,000 rpm in a microcentrifuge for 30 sec. At this stage the nuclear pellet can be snap-frozen in liquid nitrogen or dry ice–ethanol bath and stored at $-80°C$ until use.

Lysis of Nuclei

The nuclear pellet is resuspended in nuclear lysis buffer, using 100 μl for each 150-mm dish. If the nuclear pellet is very tight, 10 μl of the previous hypotonic buffer is added first and mixed by gentle vortex to loosen the pellet. The nuclei are mixed with the lysis buffer by gentle pipetting and incubated on ice for 10 min before chromatin fragmentation. The concentration of SDS in the lysis buffer affects the efficiency of chromatin fragmentation by sonication. For example, replacing the above specified lysis buffer with buffer containing 1% Triton X-100 and 0.1% SDS decreases sonication efficiency, probably due to incomplete lysis of nuclei.

Chromatin Fragmentation

In order to map precisely the locations of recruited proteins or histone modifications in the promoter regions of nuclear receptor target genes, it is important to generate relatively small, uniformly sized chromatin fragments in the nuclear lysate. Sonication is a fast and efficient method to reduce the size of chromatin fragments. We sonicate 0.5–1.0 ml of chromatin preparation with three 15-sec pulses using a Virtis Versonic 600 sonicator equipped with a tapered microtip at setting 4. After each pulse the sample is cooled on ice for 1 min. Caution should be taken when sonicating samples of small volume (i.e., less than about 0.5 ml) to avoid sample foaming, which will lead to sample loss and a decrease in chromatin shearing efficiency. After sonication, the preparation is centrifuged in a microfuge at 13,000 rpm for 10 min to remove particulate matter. The supernatant may be used immediately for immunoprecipitation or stored at $-80°C$ until use. Our sonication procedure usually generates chromatin fragments with an average size of 300–800 bp, but a minor fraction of the sheared chromatin is larger, up to 2–3 kb. We recommend that sonication strength and time be optimized for each particular sonicator used. To accurately estimate the size of the chromatin fragments generated by sonication, DNA should be purified and checked by electrophoresis after the cross-linking is reversed, as described in the next section.

An alternative approach to reduce chromatin size is to digest cross-linked nuclei with micrococcal nuclease.[11] Digestion of nuclei by micrococcal nuclease gives rise to predominantly mononucleosomes under the conditions described by the authors (0.1 unit/μg DNA at 37°C for 10 min). The resulting mononucleosomes will be released from the nuclei. After insoluble materials are removed by centrifugation, the supernatant containing mononucleosomes is subjected to immunoprecipitation as described in the next section. It is important to note that this method can only be applied to studies of promoter regions for which nucleosomal positioning *in vivo* is well characterized. This permits the design of PCR primers to amplify the genomic sequence within the nucleosome that contains the hormone response elements of interest. A good example of the use of this technique has been in the study of the transcription complex assembly and histone modifications of the mouse mammary tumor virus (MMTV) promoter upon glucocorticoid stimulation.[11] The MMTV promoter is arranged into an array of six well-defined nucleosomes (A to F) *in vivo*, with nucleosome B containing four glucocorticoid response elements.[12,13] This information has permitted the design of primers that amplify within the nucleosome B sequence, so that mononucleosomes generated by micrococcal nuclease digestion can be used for ChIP analysis.

Immunoprecipitation and DNA Recovery from Precipitates

Materials and Reagents

Polyclonal antibodies or monoclonal antibodies to relevant promoter-associated proteins

Preimmune serum or control immunoglobulin gamma (IgG)

IP dilution buffer: 0.01% SDS, 20 mM Tris–HCl, pH 8.0, 1.1% Triton X-100, 167 mM NaCl, 1.2 mM EDTA, 1× protease cocktail

Protein A/G Sepharose beads (Santa Cruz Biotechnology)

Salmon sperm DNA (Gibco)

Wash buffer I: 0.1% SDS, 1% Triton X-100, 2 mM EDTA, 20 mM Tris–HCl, pH 8.0, 150 mM NaCl

Wash buffer II: 0.1% SDS, 1% Triton X-100, 2 mM EDTA, 20 mM Tris–HCl, pH 8.0, 500 mM NaCl

Wash buffer III: 250 mM LiCl, 1% NP-40, 1% sodium deoxycholate, 1 mM EDTA, 10 mM Tris–HCl, pH 8.0

1× TE buffer: 10 mM Tris–HCl, pH 8.0, 1 mM EDTA

---

[11] L. A. Sheldon, M. Becker, and C. L. Smith, *J. Biol. Chem.* **276**, 32423 (2001).

[12] G. Fragoso, S. John, M. S. Roberts, and G. L. Hager, *Genes Dev.* **9**, 1933 (1995).

[13] G. Fragoso and G. L. Hager, *Methods Companion Methods Enzymol.* **11**, 246 (1997).

IP elution buffer: 1% SDS, 100 mM NaHCO$_3$
IP elution buffer for reimmunoprecipitation: 10 mM DTT
IP dilution buffer for reimmunoprecipitation: 1% Triton X-100, 2 mM
  EDTA, 150 mM NaCl, 20 mM Tris–HCl, pH 8.1
5 M NaCl
Phenol–chloroform–isoamyl alcohol (25:24:1)
Chloroform–isoamyl alcohol (24:1)
QIAEX II Gel Extraction kit (Qiagen)
Wheat germ tRNA (Sigma)
20 mg/ml glycogen (Roche Molecular Biochemicals)

Chromatin Quantification

To examine changes in promoter-associated proteins induced by different ligand treatments, it is important to use chromatin samples containing equal amounts of DNA as starting materials for each immunoprecipitation. In practice, 4 μl of each sonicated chromatin sample is mixed with 396 μl of purified H$_2$O and optical density at 260 nm is taken. The volume of chromatin sample used for each immunoprecipitation is adjusted based on the optical density reading.

Immunoprecipitation

Generally, a chromatin sample representing 1.5–2 $\times$ 10^7 cells (about 100 μl in volume) is diluted with 10 volumes of IP dilution buffer and used for each immunoprecipitation. A sample equal to 20% of the chromatin used in each immunoprecipitation is set aside and frozen as input DNA. In addition, for each treatment condition analyzed, one additional sample will be needed for immunoprecipitation with preimmune serum, control IgG, or no antibody.

To reduce nonspecific precipitation of chromatin, the chromatin samples should be precleared with preblocked protein A/G beads (Santa Cruz Biotechnology) before specific antibodies are added. Preblocked protein A/G beads are made by incubating 100 μl (bed volume) of protein A/G beads, 20 μg sheared salmon sperm DNA, and 1 mg/ml BSA in 400 μl IP dilution buffer on a rotating wheel for 1 hr. The beads are centrifuged and resuspended as 50% slurry in IP dilution buffer. Subsequently, 20 μl of the bead slurry is added to each diluted chromatin sample and incubated on a rotating wheel at 4°C for 1 hr. The beads are pelleted by centrifugation. The supernatant is transferred to a clean tube, and the specific antibody is added.

The amount of antibody used in the immunoprecipitation depends on the affinity of the specific antibody, as well as the abundance of the chromatin-associated proteins and protein modifications. We recommend that the amount of each antibody or antiserum used in immunoprecipitation be optimized so that maximal specific signal to noise ratio is achieved. For most purified antibodies, final concentrations of 1–2 μg IgG/ml in the immunoprecipitations yield a sufficient specific signal in the genomic region of interest with low background signal when an irrelevant genomic region is used as a negative control.

After antibody addition, chromatin samples are incubated on a rotating wheel overnight at 4°C. However, for some antibodies and/or proteins of interest, 4–6 hr incubation is sufficient as well as desirable for reducing nonspecific binding. Thus, the optimal length of the incubation time may need to be determined individually. After the incubation, 40 μl of preblocked 50% protein A/G bead slurry is added to each sample and incubated for another 1–2 hr at 4°C. Thereafter, the protein A/G beads with the bound antibody–antigen–DNA complexes are centrifuged at 3000 rpm in a microcentrifuge for 1 min. The pelleted beads are then washed for 5 min at room temperature sequentially with 1 ml of wash buffer I, II, and III. Afterwards, the beads are further washed three times with 1 ml of 1× TE buffer. Since the final step of the ChIP assay is to analyze the precipitated chromatin DNA by PCR amplification, it is very important to reduce the unbound DNA contamination to a minimum by thorough washing of the protein A/G beads. After each wash, the supernatant should be removed as completely as possible without disturbing the pelleted beads.

After the final TE washes, the antibody–antigen–DNA complexes are eluted from the protein A/G beads by incubating the beads twice with 150 μl of IP elution buffer on a rotating wheel for 15 min at room temperature with occasional vortexing. After pelleting the beads by centrifugation at 13,000 rpm in a microcentrifuge for 1 min, the supernatants from the two elutions are pooled.

Optional Secondary Immunoprecipitation

If it is of interest to investigate whether two different proteins (e.g., two different coactivators) are simultaneously present in the same protein complex on the same promoter, a second immunoprecipitation of the eluted primary immunoprecipitates can be performed with an antibody against the second protein. In this case, the complexes from the first immunoprecipitation are eluted from the beads by incubation with 10 mM DTT at 37°C for 30 min. The eluents are diluted 1:50 in IP dilution buffer and immunoprecipitated by an antibody against the second cofactor protein.[5]

Reversal of Cross-Linking and Recovery of DNA

To reverse the formaldehyde-induced protein–DNA cross-linking, an appropriate volume of 5 M NaCl is added to the samples (both input samples and eluted immunoprecipitates) to a final concentration of 300 mM, and the samples are then heated at 65°C for at least 5 hr or overnight. The DNA is purified by extraction with phenol–chloroform, or using QIAEX II Gel Extraction kit (Qiagen) following the manufacturer's instructions. For the phenol–chloroform extraction method, the heated samples are cooled to room temperature and extracted with an equal volume of phenol–chloroform–isoamyl alcohol (25:24:1), once for the immunoprecipitated samples, or twice for the input samples. The aqueous upper layer is transferred to new tubes. For the immunoprecipitated samples, the organic phase is extracted one more time with 100 μl of 1× TE buffer. The pooled aqueous layer is then extracted again once with chloroform–isoamyl alcohol (24:1). Afterwards, 5 μg of tRNA and 20 μg of glycogen are added as carrier to the aqueous phase, and DNA is recovered by ethanol precipitation. DNA pellets from both inputs and immunoprecipitates are air-dried and resuspended in 50 μl of 1× TE buffer.

Analysis of Precipitated DNA by PCR

Materials and Reagents

Primers: 18–24 mers with 50–55% GC content
Taq polymerase (Fisher scientific)
Ultra pure agarose (Invitrogen Life Technologies)
5 mg/ml ethidium bromide
30% (19:1) acrylamide–bisacrylamide solution (Bio-Rad laboratories)
10× Tris–borate–EDTA buffer (TBE): 89 mM Tris–HCl pH 8.3, 89 mM boric acid, 2 mM EDTA

PCR Primer Design

Three issues need to be considered when designing primers for analyses of genomic sequences present in the immunoprecipitates. First, we typically design 24-mer primers with about 50–55% GC content. This length helps to ensure a high level of sequence specificity, and the relatively narrow range of GC content allows similar PCR amplification conditions to be used for all the primer pairs. The second issue is the location of the primers. Ideally, primers for the gene of interest should bracket the hormone response element(s) in the promoter region. If such primers are not available, alternative primers should be designed to amplify the genomic sequence adjacent to the hormone response elements as long as the distance from the amplified region to the hormone response element is within the range of the average chromatin fragment size (300–800 bp). Control primers that amplify a genomic sequence at least 4 or 5 kb upstream from the hormone response elements should be included as a negative control representing a region that should not assemble a transcription factor–coactivator complex in response to promoter stimulation. In addition or alternatively, primers that amplify a sequence from an irrelevant gene or genomic locus that are not responsive to the stimulus under study (e.g., the promoter region of a house keeping gene like β-actin[8]) can also serve as a negative control. Third, it is desirable to design various primer sets such that all the amplified products are of similar length, since this would minimize the variation of PCR amplification efficiency. We usually design primer pairs to produce PCR products of about 300–400 bp in length, which is at the lower end of our average chromatin size range.

PCR Analysis, Gel Electrophoresis, and Data Analysis

PCR is performed in a 25 μl reaction volume with 0.4 μM of each primer, 200 μM of each dNTP, and 1.25 units of *Taq* DNA polymerase (Fisher Scientific). In some cases it may also be necessary to use 5–10% DMSO or other specialized PCR conditions to optimize results for specific primer–template combinations. Our standard PCR conditions include 3 min denaturation followed by 28–30 cycles of amplification with denaturation for 1 min at 95°C, annealing for 30 sec at 55–60°C, elongation for 1 min at 72°C. The reaction is completed with a final elongation period of 5 min at 72°C. Analysis of ligand-induced changes in the proteins associated with a specific promoter site should include PCR amplification with the appropriate primer pairs for the following DNA template samples: (1) input chromatin DNAs, (2) DNAs immunoprecipitated by preimmune serum, control IgG, or no antibody, and (3) DNAs immunoprecipitated by the antibodies against the specific proteins whose presence on the promoter is being investigated. Such analyses are done with the DNA samples prepared from both untreated and hormone-treated cells.

Since ChIP assays are semiquantitative assays analyzing relative enrichment of a specific genomic region in the immunoprecipitates, it is important that both the amount of DNA template used in PCR and the number of PCR cycles are adjusted to ensure that the amplification is in the linear range of the assay. Within the linear range of PCR amplification, the amount of PCR products should be proportional to the amount of DNA template added. We generally use serial dilutions of the input sample DNAs to determine the linear range of the assay. In practice, a series of 2.5-fold increments from 2.5 μl of a 1:1000 dilution to 2 μl of a 1:100 dilution of the input sample DNAs (equivalent to 1/100,000 to 1/12,500 of the chromatin DNA used for the immunoprecipitation) are used as templates to determine the linear range for the PCR amplification. For the immunoprecipitated DNA samples, about 1–3 μl of undiluted DNA is generally used in PCR amplification; however, we again recommend testing serial dilutions of each immunoprecipitated DNA sample to determine the linear range of DNA template amounts.

As stated earlier, the amount of chromatin DNA precipitated is influenced by the affinity of antibodies and the abundance of the targeted chromosome-associated proteins. For example, we found that antibodies against acetylated histone H3 or histone H4 (Upstate Biotechnologies) were more efficient in precipitating the hormone-responsive elements of several steroid hormone-activated gene promoters that we tested, compared with antibodies against receptors and cofactor proteins. This could be explained by a high local concentration of histones, acetylation of multiple lysine residues in one histone molecule, and the high affinity of these antibodies. As a principle of the PCR reaction, the amplification cycles required are in inverse relationship to the amount of DNA template added, i.e., less PCR cycles should be used with more DNA template and vice versa. We typically use 25–30 cycles for the amplification of immunoprecipitated sample DNAs. Again, to validate the results, it is important to demonstrate that the conditions used are such that the amount of PCR products produced are proportional to the amount of DNA template added.

After PCR amplification, 20 μl from each PCR reaction is loaded on 1–3% agarose gels or 6% 0.5× TBE acrylamide gels alongside appropriate DNA size markers, resolved by electrophoresis, and visualized by ethidium bromide staining. Stained gels can then be documented and quantified with appropriate imaging software.

An alternative PCR analysis procedure utilizes ^{32}P-end-labeled primers in the PCR reactions. The PCR products are then electrophoresed on 6% 0.5× TBE acrylamide gels and exposed on a phosphorimager cassette for quantification and image documentation.[14] With the evolution of the more quantitative real-time PCR systems in the past few years, more accurate quantification of changes in chromosome-associated proteins following ChIP assay is now possible.[15–17] The use of real-time PCR should eliminate

[14] R. M. Nissen and K. R. Yamamoto, *Genes Dev.* **14**, 2314 (2000).

[15] M. D. Litt, M. Simpson, M. Gaszner, C. D. Allis, and G. Felsenfeld, *Science* **293**, 2453 (2001).

[16] L. K. Christenson, R. L. Stouffer, and J. F. Strauss, III, *J. Biol. Chem.* **276**, 27392 (2001).

[17] S. K. Chakrabarti, J. C. James, and R. G. Mirmira, *J. Biol. Chem.* **277**, 13286 (2002).

the need to optimize some PCR conditions and demonstrate assay linearity for each input and immunoprecipitated DNA sample.

Acknowledgment

We thank Dr. Myles Brown (Dana-Farber Cancer Institute) for guidance in establishing and application of the ChIP technique, and Mr. Hongwei Li (University of Southern California) for critical comments on the manuscript. Support was provided from United States National Institutes of Health grant number DK55274 to M. R. S., CA57374 to Myles Brown, and training grant number CA09569 to D. Y. L., and United States Department of Defense Breast Cancer Program grant DAMD17-01-1-0222 to Y. S.

Section IV

Identification of Nuclear Receptor Target Genes and Effectors

[17] Serial Analysis of Gene Expression and Gene Trapping to Identify Nuclear Receptor Target Genes

By Hana Koutnikova,* Elisabeth Fayard,* Jürgen Lehmann, and Johan Auwerx

Introduction

Nuclear hormone receptors (NRs) form a family of transcription factors that regulate the expression of target genes in response to binding small, lipophilic ligands, such as hormones, nutrients, as well as endo- and xenobiotic metabolites.[1,2] In addition to these receptors, a large number of proteins has been identified that share many of the structural features of nuclear hormone receptors, but lack identified ligands.[3,4] Despite the recent progress in identifying fatty acids and certain eicosanoids as ligands for the peroxisome proliferators-activated receptors (PPARs), oxysterols as ligands for LXR, bile acids as ligands for FXR, and a diverse set of xenobiotics as ligands for the pregnane X receptor (PXR) and the constitutive androstane receptor (CAR), more than half of the 48 human nuclear receptors remain classified as orphan nuclear receptors. The study of NRs has pointed to the existence of several previously unknown signaling pathways impacting a number of research fields, such as endocrinology, metabolism, development, and oncology. In addition, several synthetic NR ligands have found their way into the clinic for the treatment of disorders as diverse as breast cancer, type 2 diabetes mellitus, and hyperlipidemia.

Often, key insights into the nuclear receptor biology have come from the identification of the underlying transcriptional programs. However, until recently, the identification of NR target genes has not been approached systematically and was often limited to a so-called "candidate gene approach." For instance, PPARα which is activated by a number of fatty acids and eicosanoids, is strongly expressed in the liver. Therefore, it was not too surprising that the expression of a number of genes encoding enzymes involved in liver fatty acid metabolism turned out to be regulated by PPARα. However, this approach is not only cumbersome but also ignores

*Contributed equally.

[1] A. Chawla, J. J. Repa, R. M. Evans, and D. J. Mangelsdorf, *Science* **294**, 1866–1870 (2001).
[2] D. J. Mangelsdorf, C. Thummel, M. Beato, P. Herrlich, G. Schutz, K. Umesono, B. Blumberg, P. Kastner, E. Mark, P. Chambon, and R. M. Evans, *Cell* **83**, 835–839 (1995).
[3] E. Enmark and J. A. Gustafsson, *Mol. Endocrinol.* **10**, 1293–1307 (1996).
[4] D. J. Mangelsdorf and R. M. Evans, *Cell* **83**, 841–850 (1995).

unexpected roles of the receptors, such as the function of PPARα in macrophage maturation.

In view of these limitations, several investigators have begun using systematic genomic strategies to identify additional NR target genes. Microarray-based studies were applied to identify target genes for CAR,[5] progesterone receptor A and B isoforms,[6] PPARα and PPARγ,[7–9] and the retinoic acid receptor alpha-PLZF oncogene.[10] A subtractive hybridization technique was used in an effort to identify genes that are regulated by the PPARγ agonist, rosiglitazone, or retinoic acid,[11,12] and the GeneCalling mRNA profiling technology was applied to identify PPARα and PPARγ regulated genes in liver and adipose,[13,14] to cite some of them.

In this review we will focus on two genomic approaches, which have been used recently in our laboratory to identify additional NR target genes, i.e., serial analysis of gene expression (SAGE) and gene trapping. Since the success of these differential gene expression (DGE) approaches depends to a large extent on the experimental study design, we will first describe the different strategies used to change NR activity.

Two Approaches to Modulate the Transcriptional Activity of Nuclear Receptors

In order to identify NR target genes, gene expression profiles need to be compared between an inactive and an active state of the nuclear receptor of interest. This can be achieved either by using a ligand to change the

[5] A. Ueda, H. K. Hamadeh, H. K. Webb, Y. Yamamoto, T. Sueyoshi, C. A. Afshari, J. M. Lehmann, and M. Negishi, *Mol. Pharmacol.* **61**, 1–6 (2002).

[6] J. K. Richer, B. M. Jacobsen, N. G. Manning, M. G. Abel, D. M. Wolf, and K. B. Horwitz, *J. Biol. Chem.* **277**, 5209–5218 (2002).

[7] K. Yamazaki, J. Kuromitsu, and I. Tanaka, *Biochem. Biophys. Res. Commun.* **290**, 1114–1122 (2002).

[8] M. Cherkaoui-Malki, K. Meyer, W. Q. Cao, N. Latruffe, A. V. Yeldandi, M. S. Rao, C. A. Bradfield, and J. K. Reddy, *Gene Expr.* **9**, 291–304 (2001).

[9] R. A. Gupta, J. A. Brockman, P. Sarraf, T. M. Willson, and R. N. DuBois, *J. Biol. Chem.* **276**, 29681–29687 (2001).

[10] L. Z. He, T. Tolentino, P. Grayson, S. Zhong, R. P. Warrell, Jr., R. A. Rifkind, P. A. Marks, V. M. Richon, and P. P. Pandolfi, *J. Clin. Invest.* **108**, 1321–1330 (2001).

[11] C. M. Steppan, S. T. Bailey, S. Bhat, E. J. Brown, R. R. Banerjee, C. M. Wright, H. R. Patel, R. S. Ahima, and M. A. Lazar, *Nature* **409**, 307–312 (2001).

[12] H. Nakshatri, P. Bouillet, P. Bhat-Nakshatri, and P. Chambon, *Gene* **174**, 79–84 (1996).

[13] B. E. Gould Rothberg, S. S. Sundseth, V. A. DiPippo, P. J. Brown, D. A. Winegar, W. K. Gottschalk, S. G. Shenoy, and J. M. Rothberg, *Funct. Integr. Genomics* **1**, 294–304 (2001).

[14] J. M. Way, W. W. Harrington, K. K. Brown, W. K. Gottschalk, S. S. Sundseth, T. A. Mansfield, R. K. Ramachandran, T. M. Willson, and S. A. Kliewer, *Endocrinology* **142**, 1269–1277 (2001).

transcriptional activity of the receptor or by changing the receptor expression level. The potency, selectivity, and pharmacokinetic properties of the ligand used for the study will have a large influence on the result. The predictive power of such an experiment can be increased by using a pharmacological approach, such as the use of multiple ligands or performing a dose–response curve. Another important consideration is the duration of the ligand treatment. A too short induction period may lead to a change in target gene expression that is too small to be detected reliably, while a longer period may cause late responses leading to the identification of indirect target genes. Of course, such ligand-based approaches are not amenable to the orphan NRs for which ligands have not yet been identified. Instead, techniques such as targeted gene disruption, transient or stable ectopic expression,[15] antisense mRNAs, dominant negative receptors, or the recently developed RNAi technique can be used. Obviously, the chances of success will increase greatly when transgenic or ectopic expression approaches are used in combination with synthetic ligands.

Isolation of Nuclear Receptor Target Genes Using SAGE Technology

Currently, SAGE[16,17] is the only DGE technique available, that allows the cataloging of all expressed genes (reviewed in Refs. 18, 19). The SAGE technology is based on two principles. First, a short nucleotide sequence tag of 10 base pairs is sufficient to identify a transcript, provided that the position of this sequence tag within the transcript is defined. Only one tag is created per transcript, and a 10 bp sequence has a complexity of 1,048,576, providing a large excess over the 40,000 genes thought to be contained within the human genome. Second, the frequency with which a given tag appears is directly proportional to the level of mRNA expression. The main advantage of the SAGE method is that all mRNAs have largely the same probability to become tagged and sequenced. In addition, SAGE is an open approach and thus, gene products that have not yet been described can be identified. However, SAGE is a very expensive technique, which is due in large part to licensing and sequencing costs.

[15] H. R. Kast, C. M. Nguyen, C. J. Sinal, S. A. Jones, B. A. Laffitte, K. Reue, F. J. Gonzalez, T. M. Willson, and P. A. Edwards, *Mol. Endocrinol.* **15**, 1720–1728 (2001).
[16] V. E. Velculescu, L. Zhang, B. Vogelstein, and K. W. Kinzler, *Science* **270**, 484–487 (1995).
[17] V. E. Velculescu, L. Zhang, W. Zhou, J. Vogelstein, M. A. Basrai, D. E. Basset, P. Hieter, B. Vogelstein, and K. W. Kinzler, *Cell* **88**, 243–251 (1997).
[18] V. E. Velculescu, B. Vogelstein, and K. W. Kinzler, *Trends Genet.* **16**, 423–425 (2000).
[19] M. Yamamoto, T. Wakatsuki, A. Hada, and A. Ryo, *J. Immunol. Methods* **250**, 45–66 (2001).

The Underlying Principle of the SAGE Technology

The original SAGE method describes a multistep procedure that is outlined in Fig. 1. In a first reaction, mRNA needs to be reverse transcribed into cDNA using biotinylated oligo(dT) primers that can be tagged at their 3'-end with streptavidin-coated beads. The cDNAs are subsequently digested with a frequently cutting restriction enzyme called the anchoring enzyme (e.g., NlaIII). Since an oligo(dT) primer is used to capture the cDNA, only the NlaIII fragment that is closest to the polyadenylation tail will be tagged. Thus, the position of the tag is defined as adjacent to the anchoring restriction enzyme that is nearest to the canonical polyadenylation signal. The captured cDNA is then ligated to an adapter linker that contains a recognition site for a type II restriction enzyme (e.g., BsmFI). After a second restriction digest with the tagging enzyme has been performed to create the tag, the tags are dimerized by blunt-end ligation, amplified by polymerase chain reaction, ligated into linear concatemers, and finally subcloned into an *Escherichia coli* expression vector to form the SAGE library. This SAGE library is then sequenced to identify the frequency by which the different tags occur.

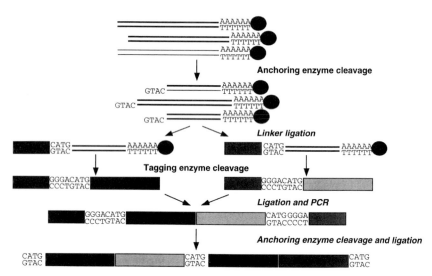

FIG. 1. The principle of SAGE. The mRNA is reverse transcribed into cDNA using biotinylated oligo(dT) primer and tagged with streptavidin beads. The cDNAs are then digested with an anchoring enzyme NlaIII and ligated to an adapter linker that contains a recognition site for the tagging enzyme BsmFI. BsmFI digestion results in a release of tags that are then dimerized by blunt-end ligation and amplified by polymerase chain reaction. The PCR product is digested with NlaIII to release ditags and ligated together into linear concatemers.

The Limitations of the SAGE Technology

Several difficulties and intrinsic problems are inherent to the SAGE methodology. (i) A relatively large amount of total RNA is needed to produce a good SAGE library that represents the 300,000 copies of mRNA, thought to be expressed by the average cell. This limitation, which until recently has prevented the use of tissue biopsies, has been overcome with the development of microSAGE and the SAGE adaptation for downsized extracts, SADE[20,21] (also see detailed protocol for more information). (ii) The restriction enzyme BsmFI does not always release an exactly 14 bp spaced tag, especially when the reaction is carried out at 37°C instead of the recommended 65°C. This leads to tag ambiguity and prevents subsequent assignment to the corresponding gene. (iii) A bias in the representation of SAGE tags can be caused by the specific DNA sequence, which has an influence on the efficiency of the blunt-end ligation process. This problem can be circumvented by a modification of the SAGE procedure[19] that uses A-T cohesive-end ligation instead of the blunt-end ligation. (iv) Short transcripts may not contain a recognition site for the anchoring restriction enzyme and thus, will not be represented in the SAGE library. This limitation can be overcome by constructing independent SAGE libraries using two distinct anchoring restriction enzymes (e.g., NlaIII and Sau3A).

The systematic sequencing of all tags of the SAGE library is needed to calculate the frequency of each tag. Most investigators will perform only a single-pass sequencing to minimize the cost. Thus, the quality of the obtained sequence is of great importance. Indeed, assuming a 1% sequencing error rate, the chance of one sequencing error occurring in a given tag is roughly 10%. This error will cause a significant reduction in the correct tag number and will lead to a shift in the tag distribution. Another important consideration is the number of tags that need to be sequenced. The optimal number depends to a large extent on the change in the expression level that needs to be seen to identify a NR target gene and can be determined using the POWER_SAGE software.[22] Despite all these limitations, the SAGE technology represents a powerful and well-established method to determine the transcriptome of a given cell or tissue.

[20] N. A. Datson, J. van der Perk-de Jong, M. P. van den Berg, E. R. de Kloet, and E. Vreugdenhil, *Nucleic Acids Res.* **27**, 1300–1307 (1999).

[21] B. Virlon, L. Cheval, J. M. Buhler, E. Billon, A. Doucet, and J. M. Elalouf, *Proc. Natl. Acad. Sci. USA* **96**, 15286–15291 (1999).

[22] M. Z. Man, X. Wang, and Y. Wang, *Bioinformatics* **16**, 953–959 (2000).

SAGE Data Analysis

Once the SAGE libraries have been sequenced, a series of bioinformatics analysis need to be performed. For this, two good software packages are available, i.e., SAGE 2000 (http://www.sagenet.org/sage.htm)[16] and USAGE (http://www.cmbi.kun.nl/usage/bin/).[23] Both programs allow the extraction of the individual tag sequences from concatemers, the tag to gene mapping, SAGE library normalization, and the statistical comparison of the tag lists.

The first step of the analysis, the extraction of ditag and tag sequences from each concatemer is straightforward and provides the first information on the total number of tags, the number of individual tags, and their absolute frequencies. Considering that on average 300,000 transcripts are expressed per mammalian cell, this information can then be used to calculate the relative tag abundance, i.e., the number of transcripts per sample. In general, the number of tags or transcripts falls into one of four categories. Transcripts that are present at an abundance of more than 500 copies/cell, between 500 and 50 copies, between 50 and 5 copies, and less than 5 copies per cell. Usually, each of the first three groups represents approximately 20–30% of the total number of tags in the library, while the transcripts present at an abundance lower than 5 copies/cell may represent between 10 and 50% of the total number of tags.[24,25] Indeed, transcripts with an abundance of 5 copies/cell are found at a frequency of one tag per 60,000 tags sequenced. Such a tag, called a singleton, may correspond to an mRNA that is expressed at a very low level or may be due to a sequencing error (SAGE sequencing error estimate is about 10%[26]). Although, a significant portion of these tags will correspond to rare transcripts,[25] these singletons are usually excluded from further analysis.

The second step in the SAGE analysis maps the tag to a specific gene. In general, this is done by searching the GenBank database using the SAGE 2000 program. Alternatively, to avoid false tag to gene assignments due to

[23] A. H. C. van Kampen, B. D. C. van Schaik, E. Pauws, M. C. Michiels, J. M. Ruijter, H. N. Caron, R. Versteeg, S. H. Heisterkamp, J. A. M. Leunissen, F. Baas, and M. van der Mee, *Bioinformatics* **16**, 899–905 (2000).

[24] L. Zhang, W. Zhou, V. E. Velculescu, S. E. Kern, R. H. Hruban, S. R. Hamilton, B. Vogelstein, and K. W. Kinzler, *Science* **276**, 1268–1272 (1997).

[25] D. Sharon, S. Blackshaw, C. L. Cepko, and T. P. Dryja, *Proc. Natl. Acad. Sci. USA* **99**, 315–320 (2002).

[26] V. E. Velculescu, S. L. Madden, L. Zhang, A. E. Lash, J. Yu, C. Rago, A. Lal, C. J. Wang, G. A. Beaudry, K. M. Ciriello, B. P. Cook, M. R. Dufault, A. T. Ferguson, Y. Gao, T. C. He, H. Hermeking, S. K. Hiraldo, P. M. Hwang, M. A. Lopez, H. F. Luderer, B. Mathews, J. M. Petroziello, K. Polyak, L. Zawel, K. W. Kinzler *et al.*, *Nat. Genet.* **23**, 387–388 (1999).

the presence of nonoriented ESTs in the GenBank database, the SAGEmap database (http://www.ncbi.nlm.nih.gov/SAGE/SAGEtag.cgi) can be used.[27] In general, the percentage of tags that match to a GenBank entry varies from 50 to 80%.[24–26] Since the tag to gene mapping procedure is crucial and highly prone to errors, it is recommended to pay special attention to this step. For those tags that do not show a match, the TAGSEARCH program[25] might be useful. This algorithm searches 3′-genomic regions of known genes in an effort to assign the tag to possible alternative splice variants. While, even with the use of the most sophisticated search algorithms, a considerable number of tags will not be assigned to a gene; there will also be a number of instances in which multiple tags will produce a match with one gene. Most likely, this is due to the presence of multiple upstream NlaIII recognition sites and an incomplete restriction digest with the anchoring enzyme. In case the 3′-sequence is identical to several genes, then the derived tag cannot be assigned with confidence. To circumvent these problems, the restriction enzyme MmeI, which generates a 21 bp long tag, can be used for the library construction.[28] Obviously, this approach will lead to a doubling of the sequencing cost. Finally, to assign a tag to a gene with high confidence, the corresponding gene needs to be cloned. For this step, a specific RT-PCR protocol has been developed.[29–31]

Following the gene assignment, a large portion of the most frequently expressed tags is often found to be derived from mitochondrial DNA and most investigators tend to exclude those tags from further analysis.[25] Also, a number of abundantly expressed tags may correspond to ribosomal proteins, ferritins, thymosin β-4, cathepsin B, or vimentin.[32,33] Although those tags may accurately represent the gene expression profile, special consideration is needed, as some of these tags may have been the result of a technical mistake during the SAGE library construction.

[27] H. Caron, B. van Schaik, M. van der Mee, F. Baas, G. Riggins, P. van Sluis, M. C. Hermus, R. van Asperen, K. Boon, P. A. Voute, S. Heisterkamp, A. van Kampen, and R. Versteeg, *Science* **291**, 1289–1292 (2001).

[28] S. Saha, A. B. Sparks, C. Rago, V. Akmaev, C. J. Wang, B. Vogelstein, K. W. Kinzler, and V. E. Velculescu, *Nat. Biotechnol.* **20**, 508–512 (2002).

[29] A. van den Berg, J. van der Leij, and S. Poppema, *Nucleic Acids Res.* **27**, e17 (1999).

[30] J. J. Chen, J. D. Rowley, and S. M. Wang, *Proc. Natl. Acad. Sci. USA* **97**, 349–353 (2000).

[31] J. Chen, S. Lee, G. Zhou, and S. M. Wang, *Genes Chromosomes Cancer* **33**, 252–261 (2002).

[32] N. Tremain, J. Korkko, D. Ibberson, G. C. Kopen, C. DiGirolamo, and D. G. Phinney, *Stem Cells* **19**, 408–418 (2001).

[33] T. Suzuki, S. Hashimoto, N. Toyoda, S. Nagai, N. Yamazaki, H. Y. Dong, J. Sakai, T. Yamashita, T. Nukiwa, and K. Matsushima, *Blood* **96**, 2584–2591 (2000).

SAGE Target Identification and Analysis

After the transcriptome of each individual SAGE library has been established, a comparison of the datasets is performed to identify putative NR target genes. In particular, if the two libraries that are being compared contain different numbers of sequenced tags or if the number of sequenced tags is low, it is important that the statistical analysis is performed on a dataset that has not been previously normalized. Several methods, such as the Monte Carlo simulation, Poisson distribution, or Bayesian analysis, have been applied.[24,34,35] A recent comparison of the different statistical tests revealed that the Chi-square test produces the best power and robustness.[22] For example, if a given gene is expressed at a level of 0.1% of the total number of transcripts and a two-fold change in the expression level is sought with a Chi-square power of $>80\%$, the sequencing of only 25,000 tags is required. However, the majority of genes are expressed in the range of 0.01% of all transcripts. To discover a two-fold change in expression for such a gene, sequencing of even 100,000 tags would produce only a Chi-square power of 44%,[22] while the same number of sequenced tags produces a Chi-square power of $>80\%$ for a three-fold change in gene expression.

Interestingly, the cluster analysis of SAGE data sets derived from normal and diseased tissues and a corresponding cell line revealed that the clusters were formed rather according to the tissue source than the diseased state (http://www.cis.njit.edu/~jason/publications/biokdd2001.htm). Also, two studies involving two different breast cancer cell lines, treated with estrogen or the synthetic selective estrogen receptor modulator tamoxifen, showed a significant larger difference in the gene expression profile between the two cell lines, than between the two treatment groups.[36,37] Therefore, the changes in the gene expression profile between normal and diseased states are quite subtle compared to the difference seen between different tissue types. Special care is required during the preparation of the tissue to ensure a homogenous sample. In case contamination cannot be avoided and SAGE data are not available for the contaminating tissue type, the RNA Abundance Database (RAD), which contains microarray data from a broad range of tissue types and allows a comparison of SAGE

[34] H. Chen, M. Centola, S. F. Altschul, and H. Metzger, *J. Exp. Med.* **188**, 1657–1668 (1998).
[35] S. L. Madden, E. A. Galella, J. Zhu, A. H. Bertelsen, and G. A. Beaudry, *Oncogene* **15**, 1079–1085 (1997).
[36] A. H. Charpentier, A. K. Bednarek, R. L. Daniel, K. A. Hawkins, K. J. Laflin, S. Gaddis, M. C. MacLeod, and C. M. Aldaz, *Cancer Res.* **60**, 5977–5983 (2000).
[37] P. Seth, I. Krop, D. Porter, and K. Polyak, *Oncogene* **21**, 836–843 (2002).

and microarray data, can be used to correct for the contamination.[38] In addition, the comparison with other SAGE libraries may allow the identification of cell-type specific genes.

Despite the use of sophisticated statistical analysis tools, the final verification of the differential expression requires standard molecular biology approaches, such as quantitative RT-PCR, Northern blot hybridization, or microarrays. However, the identification of novel NR target genes does not end here. NRs modulate gene expression by binding to cognate sequences in the promoter regions of target genes. Therefore, the final step in the identification of a novel NR target gene requires the determination of the so-called hormone response element to prove that the regulation is the consequence of a direct transcriptional induction by the NR, in question.

Previous Use of the SAGE Technology in the Nuclear Receptor Field

Two SAGE studies have been performed in estrogen-responsive breast cancer cell lines to identify estrogen receptor (ER) target genes, whose expression is modulated by either estrogen or tamoxifen.[36,37] SAGE was also applied to the search for novel corticosteroid-responsive hippocampal genes.[39] Interestingly, the steroid-responsive genes fell either into the group of mineralocorticoid or into glucocorticoid receptor target genes, while only a few genes were responsive to both receptors.

As mentioned earlier, to unequivocally determine that a gene is a direct target for regulation by a NR, the hormone response element needs to be identified, which requires the isolation of the promoter region of the gene of interest. Whereas SAGE does not distinguish between transcripts whose expression is increased due to transcriptional induction or transcript stabilization, another technique used in our laboratory, called gene trapping, identifies genes whose promoter responds to changes in NR activity.

Isolation of Nuclear Receptor Target Genes Using Gene Trapping

Generalities About Gene Trapping

The gene trapping method is a cell-based approach that uses random insertion of a promoterless reporter gene into the host cell genome. In case

[38] C. Stoeckert, A. Pizarro, E. Manduchi, M. Gibson, B. Brunk, J. Crabtree, J. Schug, S. Shen-Orr, and G. C. Overton, *Bioinformatics* **17**, 300–308 (2001).
[39] N. A. Datson, J. van der Perk, E. R. de Kloet, and E. Vreugdenhil, *Eur. J. Neurosci.* **14**, 675–689 (2001).

the reporter gene has been integrated into a NR target gene, a change in reporter expression is observed in response to NR activation. Subsequently, those cells that show a change in reporter signal are cloned to homogeneity, and the identity of the corresponding targeted gene is determined by sequencing across the fusion boundary. The main advantage of this approach is that it allows the *in vivo* monitoring of the expression of the trapped gene through monitoring the reporter.

Choice of the Targeting Cells

Choosing the right cell line for the gene trap experiments is of great importance and merits some discussion. In general, those cell lines that express the NR of interest and that show a good response for known target genes are well suited, as they should express an appropriate set of nuclear receptor coregulators. Further, cell lines that express a high amount of the NR should be avoided, as those cells will show a very high basal expression of the target gene and only a small change in expression can be expected in response to treatment. Therefore, as a first step, the expression of the nuclear receptor should be analyzed in the candidate cell lines. Embryonic stem (ES) cells often meet the requirements of low or absent expression of the NR, in combination with the possibility to induce known receptor target genes. Since integration of the trapping vector causes a disruption of the endogenous gene in most cases, knockout mice can be directly created starting from such ES cells.

Inducible Over-Expression of the Nuclear Receptor

In the situation when an appropriate ligand is not available, e.g., orphan nuclear receptors, ectopic expression systems can be used to modulate the expression of NR target genes. Several inducible systems are currently available. Noteworthy, are the ecdysone-, the estrogen-, the progesterone-, the chemical inducers of dimerization (CID)-, and the tetracycline-based inducible systems. We refer the interested reader to some recent reviews for more details.[40] In all cases, it is crucial to choose an expression system that is tightly regulated, reversible, and allows easy induction of the NR target genes (for example, tetracycline-based inducible system). Since the integration site can affect the NR expression level, 50–100 cell clones need to be analyzed. In addition, optimal concentration of the inducer needs to be determined to identify clones that

[40] A. D. Ryding, M. G. Sharp, and J. J. Mullins, *J. Endocrinol.* **171**, 1–14 (2001).

give a wide range of inducer-dependent increase in the reporter gene activity. Once the nuclear receptor cDNA is integrated under the control of the inducer-response element, clones with: (i) the lowest background and (ii) the highest/lowest expression level of the NR upon addition/absence of inducer are sought by RT-PCR, Northern blotting, or Western blotting. The response of these selected clones is then confirmed by transfection of a reporter gene, whose expression is controlled by the NR, in question.

Gene Trapping Strategy

The integrating elements of the gene trap vector drive targeting selectivity. There exist several possibilities to do so, but retroviruses appear to be the preferred system, since retroviral integration is not sequence specific and occurs randomly once in every 2×10^4 and 10^5 genes. Furthermore, virus infection could be monitored to restrict the retroviral integration to only once per cell. All gene trapping constructs are based on the use of a reporter, and/or selection marker, to identify the trapping event, and therefore its expression needs to be independent from the retroviral promoter present in the long terminal repeats (LTRs). For this reason, in most trapping constructs, the reporter gene is inserted in the opposite orientation relative to that of viral transcription.

The preferred strategy for gene trapping will enrich for intragenic integration events from the very beginning, to avoid the handling of a large number of cell clones with completely random vector integrations. One commonly used strategy is the *promoter trap*. Promoter trap vectors carry a promoterless reporter gene and a selection marker gene under the control of its own promoter (Fig. 2A). With this strategy, a large number of nonintragenic integration events are still selected, whereas it would be preferable to only select clones that have integrated the trap vector into a gene. An improved strategy comprises a system in which the expression of a promoterless selection marker gene or a promoterless fusion gene, as a reporter/selection marker, would be dependent on the capture of a transcriptionally active promoter in the target cells (Fig. 2B). While this is a very reliable way to focus exclusively on functional genes, genes that are transcriptionally silent in the target cells are not identified by this strategy.

A better way to select for intragenic vector integration is to use a *polyadenylation trap* strategy. In this case, the transcription of the selection marker is driven by a constitutive promoter. The polyadenylation trap vector lacks a polyadenylation (poly A) signal, and hence the transcript is only stabilized as long as the gene trap vector is integrated in the correct

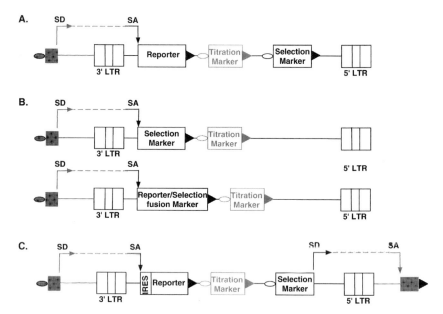

FIG. 2. Different strategies to design the gene trap vector. (A) and (B) Promoter trapping. A promoter trapping event is required for the cells to be selected by reporter expression (A) or by resistance to the selection compound (B). (C) PolyA trapping. A polyA trapping event is required for the cells to become resistant to a selection compound. The promoter activity in this case is monitored through reporter expression. A general feature of all these vectors is the presence of a titration marker cassette to determine the number of retroviruses produced. IRES, internal ribosome entry site; LTR, long terminal repeat; polyadenylation signal, ▶; promoter, ●; SA, splice acceptor; SD, splice donor. The grey color represents components of the endogenous gene.

orientation and captures, using a splice donor signal, the poly A signal of the trapped gene (Fig. 2C). Since poly A trapping occurs independently of the expression levels of the trapped target genes, any gene could be identified at almost equal probability. Inclusion of a promoterless reporter cDNA in the vector, allows the expression level of the trapped gene to be easily monitored in living cells. The combination of a strong splice acceptor and an internal ribosome entry site (IRES) upstream of the reporter gene ensures not only its independent translation, but also that reporter integration is not restricted to endogenous introns, whose splicing yields a correct reading frame. The resulting transcript will be a fusion between the 5′ part of the cellular gene and the reporter gene, whose expression could depend on inherent activity of NR target gene promoter, while the selection marker gene will be transcribed from a constitutive promoter

(Fig. 2C). Successful use of such vectors has been elegantly described by Ishida and Leder.[41]

The Reporter Gene

The reporter system, which serves to monitor the potential NR target gene expression, has to be highly sensitive, easily detectable, and quantitative. The bacterial β-galactosidase, luciferase, or reporter gene– selectable marker gene fusions have all been used. The use of green fluorescent protein (GFP; from the jellyfish *Aequorea victoria*) enables monitoring of the temporal and spatial expression of genes directly in living cells. Unlike other reporters, GFP does not require additional proteins, substrates, or cofactors to emit light. Derived from GFP, enhanced GFP (EGFP) incorporates mutations that allow higher expression in mammalian systems with increased fluorescence intensity. Cells that have effectively integrated the GFP or EGFP reporter downstream from an active promoter can then be efficiently selected by fluorescence-activated cell sorting (FACS).

Production of the Trapping Library

The cells that have randomly integrated a single copy of the gene trap vector, form the trapping library. This library is constructed by retroviral infection, once the gene trap vector is encapsulated using packaging cells. Packaging cells that express an ecotropic envelope, in addition to the retroviral *gag* and *pol* genes, produce ecotropic viruses that infect only murine cells, whereas amphotropic viruses infect also human cells. The gene trap vector needs to be *gag⁻*, *pol⁻*, and *env⁻*, in order for the packaging cells to provide complementation and to avoid retrovirus propagation risk. To avoid multiple integrations, 1 cfu of these retroviruses carrying the gene trap vector is used per 10 cells. The target cells, which have been engineered to inducibly express a specific nuclear receptor, are then selected for expression of the marker, 24–48 hr after infection. As the reporter gene is now under the control of the promoter of the trapped gene, each cell clone will show different degrees of reporter expression depending on the inherent activity of the promoter of the gene trapped. An overview of the production of a cellular gene trapping library, using a poly A trap vector, is shown in Fig. 3.

---

[41] Y. Ishida and P. Leder, *Nucleic Acids Res.* **27**, e35 (1999).

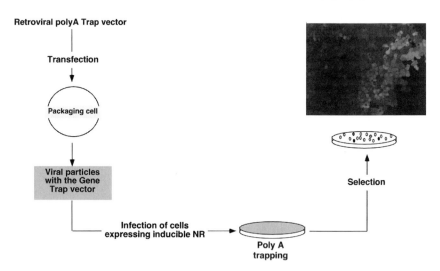

Fig. 3. The construction of the trapping library. Packaging cells are transfected with the retroviral gene trap vector for the production of the corresponding viral particles. The cells chosen for the gene trapping are infected with these retroviruses carrying the gene trap vector and are further selected for the expression of the selectable marker, which only will be produced if spliced to an endogenous polyA signal. Therefore, each cell clone selected will have integrated the gene trap vector into a gene that contains a polyA signal. As the promoterless reporter gene is now under the control of the promoter of the gene whose polyA has been trapped, each cell clone will show different degrees of reporter expression depending on the inherent activity of the promoter of the gene trapped by the polyA trap vector. (See Color Insert.)

Analysis of the Trapping Library

When using EGFP as a reporter, the pool of cell clones can be sorted by FACS into two groups: cells that do and cells that do not express the reporter gene. Cells that do not express the reporter are pooled, treated either with a NR ligand or with an inducer (doxycycline, ecdysone,...) to increase NR expression, and again sorted based on relative reporter levels. The cell clones that become positive for reporter expression upon induction are then isolated and characterized. Cells that express the reporter spontaneously, upon integration of the gene trap vector, will be isolated and treated with NR ligand or inducer. In this case, the level of reporter expression can be compared between the uninduced and induced state by spectrophotometry, and cells which show a significant difference in reporter expression will then be again characterized. An overview of the strategy, used to identify potential target genes from the trapping library of a specific NR, is shown in Fig. 4.

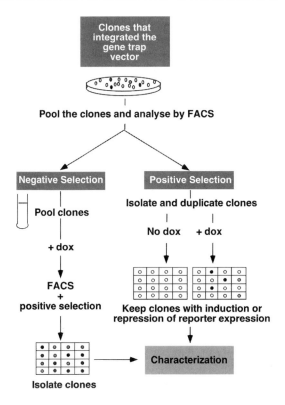

FIG. 4. The analysis of the trapping library. All the cell clones from the trapping library are pooled and analyzed by FACS for the expression of the reporter (e.g., EGFP). The cells are sorted into two groups: (i) the cells that do not express the reporter and (ii) the cells that express the reporter. Both groups of cells are processed as specified in the text. The cells that show a difference in reporter activity upon NR activation are further characterized.

The trapped clones should be characterized by Southern blotting to verify that the integration of the trapping vector occurred into a single locus, and the cDNA fragment corresponding to the trapped gene can subsequently be isolated. In the case of poly A trapping, 5' RACE (rapid amplification of cDNA ends) is performed to isolate the 5' end of the trapped gene. A database search will allow identification of the trapped gene. Similar to SAGE, a direct regulation of the NR target gene has to be proven by additional experiments (e.g., Northern blot, quantitative RT-PCR, or other techniques) and by characterizing the mechanisms of this regulation [e.g., transient transfections, electrophoretic mobility shift assay, and chromatin immunoprecipitation (ChIP)].

Benefits of Gene Trapping

Relative to others methods to identify NR target genes, the gene trapping approach is unique in that the expression of the trapped gene can be monitored *in situ* in living cells. Therefore, the results are not affected by the quality of the extracted RNA, the RNA levels in the cells, kinetics of mRNA induction, mRNA stability, or cell cycle status. The fact that one can monitor changes in gene expression dynamically over a period of time, also allows a better distinction between direct target genes whose expression is affected immediately, and genes whose expression is modulated in an indirect fashion. Moreover, a crucial and powerful feature of this approach resides in the possibility of using ES cells, which are pluripotent, for gene trapping. On one hand, since ES cells can be differentiated into various lineages, the role of NR target genes can be analyzed during differentiation. On the other hand, identified target genes can be further characterized *in vivo*, by directly generating knockout mice from the trapped ES cell. Indeed, since gene trapping is a form of insertional mutagenesis, the resulting mice could in certain cases carry a nonfunctional trapped gene. However, because the recombination event is not specifically targeted, it is not excluded that important functional domains of the target gene may not be disrupted. The direct creation of KO mice from a gene trap library has been successfully reported in a number of cases.[42–47]

Future Perspectives

The number of genomic studies with a specific aim to identify NR target genes will likely grow exponentially in the future. This will enable the identification of previously unknown NR signaling pathways and the creation of new paradigms to study transcriptional regulation of gene expression by NRs. Furthermore, the elucidation of NR function will have a direct clinical impact, since NRs represent important drug discovery targets. To this end, new genomic and proteomic techniques are constantly being developed. For example, the genome-wide location analysis method, that combines a modified ChIP procedure and has been used to study protein–DNA interactions at a small number of specific DNA sites with

[42] E. G. Neilan and G. S. Barsh, *Transgenic Res.* **8**, 451–458 (1999).

[43] M. V. Wiles, F. Vauti, J. Otte, E. M. Fuchtbauer, P. Ruiz, A. Fuchtbauer, H. H. Arnold, H. Lehrach, T. Metz, H. von Melchner, and W. Wurst, *Nat. Genet.* **24**, 13–14 (2000).

[44] W. C. Skarnes, B. A. Auerbach, and A. L. Joyner, *Genes Dev.* **6**, 903–918 (1992).

[45] B. P. Zambrowicz, G. A. Friedrich, E. C. Buxton, S. L. Lilleberg, C. Person, and A. T. Sands, *Nature* **392**, 608–611 (1998).

[46] K. Kitajima and T. Takeuchi, *Biochem. Cell Biol.* **76**, 1029–1037 (1998).

[47] G. Friedrich and P. Soriano, *Genes Dev.* **5**, 1513–1523 (1991).

DNA microarray analysis, is very promising.[48–54] A further interesting development is expected to come from the improvement and wider implementation of proteomics. Combined with genomic strategies, the use of proteomics will enable verification of gene expression profiling, and will subsequently permit characterization of both the quantity and quality (alternative splicing, . . .) of target proteins. Such a comprehensive analysis is already being utilized to chart and characterize complex NR signaling cascades.[55–57] Therefore, we predict that there is a bright future for systematic approaches to identify nuclear receptor targets.

Acknowledgments

The authors acknowledge support from CNRS, INSERM, CHU de Strasbourg, ARC, the European community RTD program (QLG1-CT-1999-00674 and QLRT-2001-00930), Ligue Nationale Contre le Cancer, NIH (1P01 DK 59820-01), and the Human Frontier Science Program (RG0041/1999-M).

[48] B. Ren, H. Cam, Y. Takahashi, T. Volkert, J. Terragni, R. A. Young, and B. D. Dynlacht, *Genes Dev.* **16**, 245–256 (2002).

[49] B. Ren, F. Robert, J. J. Wyrick, O. Aparicio, E. G. Jennings, I. Simon, J. Zeitlinger, J. Schreiber, N. Hannett, E. Kanin, T. L. Volkert, C. J. Wilson, S. P. Bell, and R. A. Young, *Science* **290**, 2306–2309 (2000).

[50] I. Simon, J. Barnett, N. Hannett, C. T. Harbison, N. J. Rinaldi, T. L. Volkert, J. J. Wyrick, J. Zeitlinger, D. K. Gifford, T. S. Jaakkola, and R. A. Young, *Cell* **106**, 697–708 (2001).

[51] A. S. Weinmann, P. S. Yan, M. J. Oberley, T. H. Huang, and P. J. Farnham, *Genes Dev.* **16**, 235–244 (2002).

[52] J. J. Wyrick, J. G. Aparicio, T. Chen, J. D. Barnett, E. G. Jennings, R. A. Young, S. P. Bell, and O. M. Aparicio, *Science* **294**, 2357–2360 (2001).

[53] V. R. Iyer, C. E. Horak, C. S. Scafe, D. Botstein, M. Snyder, and P. O. Brown, *Nature* **409**, 533–538 (2001).

[54] J. D. Lieb, X. Liu, D. Botstein, and P. O. Brown, *Nat. Genet.* **28**, 327–334 (2001).

[55] A. C. Gavin, M. Bosche, R. Krause, P. Grandi, M. Marzioch, A. Bauer, J. Schultz, J. M. Rick, A. M. Michon, C. M. Cruciat, M. Remor, C. Hofert, M. Schelder, M. Brajenovic, H. Ruffner, A. Merino, K. Klein, M. Hudak, D. Dickson, T. Rudi, V. Gnau, A. Bauch, S. Bastuck, B. Huhse, C. Leutwein, M. A. Heurtier, R. R. Copley, A. Edelmann, E. Querfurth, V. Rybin, G. Drewes, M. Raida, T. Bouwmeester, P. Bork, B. Seraphin, B. Kuster, G. Neubauer, and G. Superti-Furga, *Nature* **415**, 141–147 (2002).

[56] A. H. Tong, B. Drees, G. Nardelli, G. D. Bader, B. Brannetti, L. Castagnoli, M. Evangelista, S. Ferracuti, B. Nelson, S. Paoluzi, M. Quondam, A. Zucconi, C. W. Hogue, S. Fields, C. Boone, and G. Cesareni, *Science* **295**, 321–324 (2002).

[57] Y. Ho, A. Gruhler, A. Heilbut, G. D. Bader, L. Moore, S. L. Adams, A. Millar, P. Taylor, K. Bennett, K. Boutilier, L. Yang, C. Wolting, I. Donaldson, S. Schandorff, J. Shewnarane, M. Vo, J. Taggart, M. Goudreault, B. Muskat, C. Alfarano, D. Dewar, Z. Lin, K. Michalickova, A. R. Willems, H. Sassi, P. A. Nielsen, K. J. Rasmussen, J. R. Andersen, L. E. Johansen, L. H. Hansen, H. Jespersen, A. Podtelejnikov, E. Nielsen, J. Crawford, V. Poulsen, B. D. Sorensen, J. Matthiesen, R. C. Hendrickson, F. Gleeson, T. Pawson, M. F. Moran, D. Durocher, M. Mann, C. W. Hogue, D. Figeys, and M. Tyers, *Nature* **415**, 180–183 (2002).

Appendix A. Detailed Technical Protocol of SAGE/SADE

The SAGE protocol described below applies a modification introduced by Virlon *et al.*, known as SADE.[21] The authors replaced the streptavidin–biotin system by a more quantitative approach to purify the mRNA using the oligo(dT)$_{25}$, covalently linked to the magnetic beads. In addition, this protocol integrates a microSAGE procedure, in which PCR-amplified and gel-purified ditags are reamplified by a limited number of cycles.[20] This modification provides sufficient quantity of ditags that do not need to be purified and thus, are digested more efficiently by anchoring enzyme.

Purification of mRNA

1. Tissue/cell specimen of about 100–120 mg is homogenized and immediately lysed in 1000 μl of lysis buffer (Dynabead mRNA direct kit, Dynal, Oslo, Norway). Following lysis, the homogenate is centrifuged at 15,000 × g for 3 min.
2. In the meantime, 120 μl of magnetic beads are washed with 100 μl of lysis buffer, while vortexing gently. The washing buffer is eliminated by the use of a magnet, and 900 μl of clear homogenate is added immediately to the beads, gently vortexed and incubated together for 10 min at RT.
3. The oligo(dT) beads with the bound mRNA are recovered using the magnet for 5 min, and the following washing steps are performed:
   - twice with 800 μl of washing buffer A containing 20 μg/ml of glycogen (Life Technologies, Rockville, USA);
   - three times with 800 μl of washing buffer B containing 20 μg/ml of glycogen;
   - twice with 250 μl of cold first strand buffer (Life Technologies, Rockville, USA) containing 20 μg/ml of glycogen.
4. The beads with the bound mRNA are transferred to a new siliconized eppendorf tube and washed once more with 250 μl of cold first strand buffer containing 20 μg/ml of glycogen.

Synthesis of First and Second Strand cDNA

1. The beads and bound mRNA are resuspended in 10 μl of cold first strand and incubated at 42°C for 2 min.
2. The tube is then placed at 37°C and 10 μl of the following mixture is added:
   5 μl of water
   2.5 μl of first strand buffer
   1.25 μl of 10 mM dNTPs (Life Technologies, Rockville, USA)

2.5 μl of 100 mM DTT (Life Technologies, Rockville, USA)

1.25 μl of SuperScript II RNase H-reverse transcriptase (Life Technologies, Rockville, USA)

3. The reaction mixture is incubated at 37°C for 30 min. Next, an aliquot of 1 μl of SuperScript reverse transcriptase is added and the reaction is continued for an additional 30 min.

4. At the end of the first strand cDNA synthesis, the reaction mixture is stopped on ice and 138 μl of second strand cDNA synthesis mixture, prepared as described below, is added.

101.6 μl of DEPC water

3.3 μl of 10 mM dNTPs

33 μl of second strand cDNA synthesis buffer (Life Technologies, Rockville, USA)

0.5 μl of 20 mg/ml glycogen

4.4 μl of DNA polymerase I $E.$ $coli$ (Life Technologies, Rockville, USA)

1.1 μl of RNase H $E.$ $coli$ (Life Technologies, Rockville, USA)

1.1 μl of DNA ligase $E.$ $coli$ (Life Technologies, Rockville, USA)

5. The reaction mixture is then incubated overnight at 16°C.

Cleavage with Anchoring Enzyme and Linker to cDNA Ligation

1. The eppendorf tube is placed at the magnetic track to eliminate the supernatant containing the first and second strand cDNA synthesis mixture.

2. The beads containing the double-stranded cDNA are then washed:
   - four times with 200 μl of TEN–BSA buffer (10 mM Tris–HCl pH 8, 1 mM EDTA, 1 M NaCl, 1× BSA; New England Biolabs, Beverly, USA); and
   - three times with 200 μl of cold NE–BSA buffer (1× NEB Biolabs buffer 4, 1× BSA) prior to the last wash, the material is transferred into a new siliconized eppendorf tube.

3. The cDNA digestion is then carried out in 98 μl of NE–BSA buffer using 40 U of NlaIII restriction enzyme (New England Biolabs, Beverly, USA) for 3 hr. The mixture is gently vortexed every 30 min.

4. The eppendorf tube is placed in the magnetic track to eliminate the NlaIII restriction mixture and the liberated 5'-cDNA restriction fragments.

5. Then the beads are washed:
   - once with 200 μl of NE–BSA buffer;
   - three times with 200 μl of TEN–BSA buffer.

6. After the last wash, the material is separated into two equal aliquots of 100 μl and each is washed three times with 200 μl of cold TE–BSA buffer (10 mM Tris–HCl pH 8, 1 mM EDTA, 1× BSA).

7. The ligation reaction is carried out in
   28 μl of water
   1 μl of glycogen (1 mg/ml)
   1 μl of each linker at 0.5 pmol/μl
   8 μl of 5× ligation buffer.

8. The ligation reaction mixture is incubated at 45°C for 5 min and cooled on ice. Then 2 μl of T4 DNA ligase (5 U/μl, Life Technologies, Rockville, USA) are added and the reaction is allowed to continue overnight at 16°C.

9. Following the eluate elimination, the beads with material are carefully washed:
   – four times with 200 μl of TEN–BSA buffer
   – three times with 200 μl of NE–BSA buffer

10. Prior to the last two washes, the material is transferred into a new siliconized eppendorf tube.

Release of cDNA Tags Using Tagging Enzyme

1. The washing buffer is eliminated and 100 μl of BsmFI restriction mixture is added as follows:
   87 μl of water
   10 μl of NEB Biolabs buffer 4
   1 μl of 100 × BSA
   2 μl of BsmFI (2 U/μl, New England Biolabs, Beverly, USA).

2. The reaction is incubated at 65°C for 3 hr with gentle vortexing.

3. At the end of restriction digestion, the tube is cooled on ice and supernatant containing the cDNA tags is recovered. The walls of the eppendorf tubes are carefully washed twice with 75 μl of cold TE–BSA buffer and pooled with previous material.

4. The material is extracted with 250 μl of phenol–chloroform in the presence of 60 μg of glycogen. The organic and aqueous phase are separated by centrifugation at 10,000 × g for 10 min at 4°C.

5. The aqueous phase is carefully recovered and precipitated in the presence of 125 μl of 10 M NH$_4$(CH$_3$COO) with 1125 μl of 100% ethanol. The DNA is centrifuged at 15,000 × g for 20 min at 4°C.

6. The DNA pellet is washed twice with 75% ethanol, centrifuged at 15,000 × g for 5 min at 4°C, dried, and resuspended in 20 μl of LoTE (3 mM Tris–HCl pH 7.5, 0.2 mM EDTA).

Blunt-End Filling of Released cDNA Tags

1. The cDNA tags are incubated in the presence of 5 μl of 10× salt mix (200 mM Tris–HCl pH 7.5, 100 mM MgCl$_2$, 250 mM NaCl) at 42°C for 2 min. Next, 25 μl of the following mix is added:
   8.5 μl of water
   5.5 μl of 100 mM DTT
   10 μl of 2 mM dNTPs
   1 μl of T7 DNA polymerase
   and reaction mixture is incubated at 42°C for 10 min.
2. At the end of this reaction, the material corresponding to the linker 1A/1B and 2A/2B are pooled together, while the eppendorf tube being emptied is washed with 150 μl of LoTE containing 20 μg of glycogen.
3. The pooled material is extracted with phenol–chloroform and precipitated as described earlier. The DNA pellet is resuspended in 7 μl of LoTE.

Tags Ligation

1. The ligation of tags to form the ditags occurs in a mixture of
   6 μl of tags in LoTE
   2 μl of 5× ligation buffer
   1 μl of T4 DNA ligase (4 U/μl)
   1 μl of LoTE
2. As a negative control, the reaction without the T4 DNA ligase is carried out using 1 μl of tags in LoTE. Both tubes are incubated overnight at 16°C. Ninety microliters of (LoTE + 1 × BSA) are added to the sample. The negative control is diluted to the final volume of 17 μl.

PCR Amplification of Ditags

1. The PCR reaction mixtures, including the sample and negative control are prepared as follows:
   5 μl of 10× PCR buffer
   3 μl of DMSO
   7.5 μl of 10 mM dNTPs
   1 μl of 10 pmol/μl primer 1
   1 μl of 10 pmol/μl primer 2
   30.5 μl of water
   1 μl of Platinum Taq (5 U/μl, Life Technologies, Rockville, USA)
   1 μl of diluted ligation or negative control product.
2. The PCR reaction is performed at 94°C for 1 min, 20, 23, and 26 cycles at 94°C for 10 sec; 55°C for 20 sec; and 70°C for 20 sec

320 NUCLEAR RECEPTOR TARGET GENES AND EFFECTORS [17]

with final extension at 70°C for 5 min. The sample is diluted with 6× BB-XC and loaded on 12% PAGE gel (Mini-protean 3 cell, BioRad, Hercules, USA), run at 130 V for about 1 hr and stained for 10 min with SYBR Green I (Sigma, St. Louis, USA) diluted 1:5000. No PCR product should be detected in tubes to which negative control was added. A band of 102 bp corresponding to the amplified ditags and, most likely, a weak band of 80 bp corresponding to the linker 1AB–2AB amplification are detected in the sample tube. For further use, the number of PCR cycles corresponding to the product devoid of smear or any other background is selected.

Large-Scale PCR Amplification of Ditags

1. Once the number of cycles was optimized, 14 PCR reactions are carried out in parallel.
2. The obtained PCR products are pooled together, diluted with 6 × BB-XC and loaded on 12% PAGE gel (Protean II xi cell, BioRad, Hercules, USA), run at 90 V overnight and stained for 10 min with SYBR Green I diluted 1:5000.
3. The band corresponding to the 102 bp ditag amplification is cut out and extracted from the gel as follows. The slice of PAGE gel is placed in 0.5 ml eppendorf tube, previously pierced with a 22-gauge needle. This tube is placed in a 2 ml siliconized collector tube and spun together for 5 min at 21°C. Two hundred and fifty microliters of LoTE and 37.5 μl of 10 M NH$_4$(CH$_3$COO) are added to the PAGE matrix, vortexed, and incubated at 65°C for 15 min. Then the material is loaded on SpinX column (Costar, Corning, USA), centrifuged at 15,000 × g for 5 min. The eluate of 300 μl is precipitated in the presence of 2.5 μl of glycogen and 100 μl of 10 M NH$_4$(CH$_3$COO) with 1000 μl of ethanol. Following centrifugation at 15,000 × g for 20 min at 4°C, the DNA pellet is washed twice with 75% ethanol and resuspended in 300 μl of LoTE.
4. This purified product then serves as a DNA template for the large-scale PCR reaction. Prior to this amplification, PCR is performed to optimize the number of cycles required. Both of these PCRs are carried out with appropriate negative controls as follows:

Mix DNA	MixTaq	
5 μl	5 μl	of 10× PCR buffer (Sigma, St. Louis, USA)
5 μl	5 μl	of 1% Triton X-100
8 μl		of 25 mM MgCl$_2$ (Sigma, St. Louis, USA)
5 μl		of 2 mM dNTPs
1 μl		of primer 1 (10 pmol/μl)

Mix DNA	MixTaq	
1 μl		of primer 2 (10 pmol/μl)
	1 μl	of *Taq* DNA polymerase (5 U/μl; Sigma, St. Louis, USA)
24 μl	39 μl	of water
1 μl		of purified product (or no DNA)

The tube of Mix DNA and MixTaq are heated to 80°C for 2 min, and then hot start is performed manually by adding the *Taq* DNA polymerase mixture to the template. The PCR program is as described earlier and the optimal number of PCR cycles corresponding to the appearance of a clear unique band (usually between 8 and 13 cycles) is chosen.

5. Once the number of PCR cycles was optimized, large-scale PCR amplification corresponding to 200 reactions is performed.

Ditag Purification

1. The 200 PCR reactions are pooled together, extracted with phenol–chloroform, precipitated, and washed as described earlier. The DNA pellet is resuspended in 300 μl of LoTE, and reprecipitated again and resuspended in 158 μl of LoTE.

NlaIII Digestion

1. The restriction reaction is carried out with 158 μl of LoTE in the presence of
   20 μl of NEB Biolabs buffer 4
   2 μl of 100× BSA
   2 μl of NlaIII
   for 3 hr at 37°C.

2. The restriction mixture is extracted with phenol–chloroform and precipitated in the presence of 2.5 μl of glycogen and 49.5 μl of 10 *M* $NH_4(CH_3COO)$ with 825 μl of ethanol. The material is allowed to precipitate in a dry ice–ethanol bath for 15 min. Following centrifugation at 15,000 × *g* for 30 min at 4°C, the DNA pellet is washed twice with 75% ethanol and resuspended in 40 μl of TE.

Ditag Purification

1. The 40 μl of purified NlaIII restriction digest are diluted with 6× XC-BB and loaded onto four lanes of 12% PAGE. The band of 24–26 bp corresponding to the ditags is cut out and purified from the gel as described earlier with an exception that the elution step is carried out at 37°C.

2. The 200 μl of DNA eluate is precipitated as described earlier, using the dry ice–ethanol bath for 15 min. Following centrifugation at 15,000 $\times$ g for 30 min at 4°C, the DNA pellet is washed twice with 75% ethanol and resuspended in 7 μl of cold LoTE.

Concatemer Formation

1. The 7 μl of purified tags are ligated for 3 hr in the presence of 2 μl of 5× ligation buffer using 1 μl of T4 DNA ligase at 16°C.
2. At the end of the ligation reaction, the sample is diluted with 6× XC-BB, heated at 65°C for 5 min, deposited onto one lane of 8% PAGE, and run at 130 V for about 1 hr.
3. Following SYBR Green I coloration, a smear of concatemers is observed and two slices corresponding to the low (200–500 bp range) and high (> 500 bp) size concatemers are cut.
4. Concatemer purification from PAGE gel and precipitation is as described earlier. Purified concatemers are resuspended in 6 μl of cold LoTE.

pZerO Vector Preparation, Ligation, and Bacteria Transformation

1. pZerO vector (Invitrogen, Carlsbad, USA) is linearized using the SphI restriction enzyme (New England Biolabs, Beverly, USA) according to the manufacturer's instructions.
2. About 25 ng of linearized vector is ligated 0/N at 16°C with 6 μl of low- and high-size concatemers in a reaction mixture as described earlier.
3. Ligation mixture is precipitated, resuspended in 10 μl of LoTE, and electroporated in DH10B electrocompetent bacteria (Life Technologies, Rockville, USA) forming the SAGE library.

[18] Expression Cloning of Receptor Ligand Transporters

By PAUL A. DAWSON and ANN L. CRADDOCK

Introduction

Transporters for Nuclear Receptor Ligands

In the classic endocrine paradigm, steroid hormones are synthesized and circulated in the body. Upon encountering their target cells, steroid hormones passively diffuse across the plasma membrane and bind their cognate receptors with high affinity. A new paradigm has emerged with the

METHODS IN ENZYMOLOGY, VOL. 364

finding that many orphan receptors function as sensors of dietary and endogenous lipids. Members of this group include receptors for fatty acids (PPARs), oxysterols (LXRs), bile acids (FXR), and xenobiotics (SXR/PXR, CAR).[1] Unlike classic endocrine receptors, the ligands for these orphan receptors are derived from endogenous and exogenous sources. As such, their synthesis is not under strict endocrine control, but instead, the intracellular concentration of the receptor ligand in the target cell may be regulated by the action of uptake and efflux pumps. The best example of this concept is the bile acids.

Bile acids are natural detergents synthesized from cholesterol by the liver. Following their secretion by the hepatocyte, bile acids solubilize dietary lipids and facilitate their absorption in the small intestine. A specific transporter expressed on the enterocytes lining the terminal ileum then efficiently reabsorbs the bile acids, which are sent back to the liver in the portal circulation and cleared by additional transporters expressed on the hepatocyte sinusoidal membrane.[2] In the hepatocyte and ileal enterocyte, bile acids act as ligands for the orphan nuclear receptor FXR[3] to alter expression of a variety of genes involved in lipid metabolism, storage, and transport. The intracellular concentration of the ligand, i.e., bile acids, is dependent as much on the plasma membrane transporters directing their traffic through the cell as on their initial biosynthesis.

Many of the membrane carriers responsible for the trafficking of bile acids through the hepatocyte and ileocyte have been identified in recent years, using expression-cloning techniques. This chapter details methodologies used in expression cloning of transporters with particular emphasis on transporter solutes that function as nuclear receptor ligands.

General Approaches to Expression Cloning

Expression cloning in mammalian cells, or *Xenopus* oocytes, is a powerful method to isolate cDNA clones that encode functional proteins for which no amino acid sequences exist.[4] These methods have been especially useful for the membrane transporters that tend to be very hydrophobic proteins and as such, refractory to conventional protein purification techniques. In theory, definitive identification of a transporter requires protein purification and reconstitution of the activity; however, fulfilling this goal has been extremely difficult for several reasons. First, the hydrophobic nature of membrane proteins precludes their isolation in soluble form

[1] A. Chawla, J. J. Repa, R. M. Evans, and D. J. Mangelsdorf, *Science* **294**, 1866 (2001).
[2] B. L. Shneider, *J. Ped. Gastro. Nutr.* **32**, 407 (2001).
[3] D. W. Russell, *Cell* **97**, 539 (1999).
[4] M. F. Romero, Y. Kanai, H. Gunshin, and M. A. Hediger, *Methods Enzymol.* **296**, 17 (1998).

suitable for amino acid sequencing or immunization. Secondly, the target membranes, such as the hepatic sinusoid or ileal brush border, are complex and harbor many transporter proteins. For example, the intestinal brush border membrane transports glucose, fructose, amino acids, peptides, inorganic anions, fat-soluble vitamins, and other nutrients in addition to bile acids. Copurification of these polypeptides further complicates the isolation, reconstitution, and partial protein sequencing of any single transporter. As a consequence of these difficulties, oligonucleotide probes and antibodies for screening cDNA libraries were not often available. Expression cloning has proven to be a powerful alternative approach. Whereas protein purification relies on differences in physical properties to achieve isolation, expression cloning takes advantage of the high degree of substrate specificity exhibited by different transporters. Expression cloning has been used to isolate a wide variety of transporters, including the hepatic[5] and ileal[6] sodium–bile acid cotransporters. As expression cloning of transporters using *Xenopus* oocytes has recently been reviewed in this series,[4] this chapter will focus on expression cloning of receptor ligand transporters using mammalian cell systems.

Strategy and General Concerns

The ultimate success or failure of expression cloning depends on a number of factors that must be considered before embarking on what may become a quixotic quest. First, an understanding of the underlying biology of the ligand transporter is important. Is there clear evidence for this specific ligand transporter? Passive diffusion of lipophilic ligands may obviate the need for a specific carrier or limit its potential concentrative capacity. Is there evidence for multiple subunits for the ligand transporter? The expression efficiency for a single cDNA in large complex pools of clones is low in transfected cells or microinjected *Xenopus* oocyte. If multiple heterologous subunits are required for transport activity, the low probability of coexpressing all the required subunits in the same cDNA clone pool will preclude most conventional expression-cloning strategies. Second, the assay system used to detect the transport activity must be specific, sensitive, and scalable to screen large numbers of library pools. Since the screen may require analysis of hundreds of library pools, a reproducible and facile assay system is important. A highly sensitive assay will permit larger pools to be screened and reduce the total number of assays required to obtain adequate

---

[5] B. Hagenbuch, B. Stieger, M. Forguet, H. Lubbert, and P. J. Meier, *Proc. Natl. Acad. Sci. USA* **88**, 10629 (1991).
[6] M. H. Wong, P. Oelkers, A. L. Craddock, and P. A. Dawson, *J. Biol. Chem.* **269**, 1340 (1994).

library coverage. Third, the abundance of the target transporter mRNA and its representation in the library are a consideration. Preparation of the highest quality RNA, cDNA, and library is essential before screening can begin.

Library Preparation

Source of Tissue or Cells for mRNA

The starting source of the RNA for construction of the cDNA expression library is an important consideration. A cell line or tissue, where the activity has been extensively characterized, is ideal. Many ligand transporters are expressed in the liver parenchyma, intestinal enterocyte, and kidney cortex. Since the carriers function directly in nutrient lipid homeostasis, it may also be possible to induce the activity of the target transporter by changing the diet or by hormonal treatment. This aspect is especially important for low-abundance carriers that may not otherwise be amenable to expression cloning. Finally, ease of tissue access and cell purity are also important considerations in choosing a starting tissue source. The tissue should be readily accessible and abundant for preparation of the starting RNA.

Isolation of Total RNA

High-quality poly(A)$^+$ RNA is crucial for an expression-cloning project. The RNA should be isolated from the fresh tissue or tissue culture cells. Tissues can be quick-frozen in liquid nitrogen, and then pulverize under liquid nitrogen at a later date for RNA isolation. Tissue culture cells can be solubilized directly in guanidinium isothiocyanate and stored at $-70°C$. Poly(A)$^+$ RNA may be isolated by one of several methods. Most protocols require isolation of total RNA first, followed by poly(A)$^+$ RNA enrichment. The guanidinium isothiocyanate–CsCl centrifugation method of Chirgwin et al.[7] yields RNA of very high quality and purity. The method is not suitable for simultaneously processing large numbers of samples; however, this is not a major concern for construction of an expression library. The very high quality and quantity of the RNA produced by this method still make it a viable alternative. The most common method for RNA isolation, and the method used here, is the single-step technique of Chomczynski[8] in which cells or tissue are lysed in an acidic monophasic solution of guanidinium isthiocyanate and phenol. Chloroform is then added to generate a second (organic) phase containing protein and DNA. The RNA is precipitated from the aqueous phase using isopropanol. The

[7] J. M. Chirgwin, A. E. Przybyla, R. J. MacDonald, and W. J. Rutter, *Biochemistry* **18**, 5294 (1979).
[8] P. Chomczynski, *BioTechniques* **15**, 532 (1993).

yield of total RNA depends on the tissue or cell source, but is generally ~ 5 μg/mg starting tissue or ~ 5 μg/10^6 cells.

Reagents and Procedure

Trizol Reagent (Invitrogen Cat N. 15596-026), chloroform, isopropyl alcohol, 75% ethanol, DEPC-treated water.

1. Weigh-out tissue. Transfer to sterile mortar under liquid nitrogen. Quickly grind tissue using a mortar and pestle under liquid nitrogen.
2. Scrape ground samples into 50 ml polypropylene tube containing 1 ml of Trizol Reagent per 100 mg of tissue. Homogenize using a Polytron homogenizer at maximum speed for ~ 30 sec. Transfer sample to a sterile (baked 3 hr at 120°C) 15-ml glass Corex tube. Incubate homogenate for 5 min to permit complete dissociation of nucleoprotein complexes.
3. Add 0.2 ml of chloroform per 1 ml of Trizol reagent. Shake the tube vigorously for 15 sec and incubate at room temperature for 3 min. Centrifuge samples at $12,000 \times g$ for 15 min at 4°C.
4. Transfer top aqueous phase to a 15-ml glass Corex tube (volume of aqueous phase should be about 60% of the starting Trizol reagent volume).
5. Precipitate the RNA by adding 0.5 ml of isopropanol per 1 ml Trizol reagent added. Incubate at room temperature for 10 min. Centrifuge at $12,000 \times g$ for 10 min at 4°C. Discard supernatant. Wash RNA pellet using 1 ml of 75% ethanol (in DEPC-treated water) per milliliter of Trizol reagent. Centrifuge at $7500 \times g$ for 5 min at 4°C.
6. Briefly dry the RNA pellet for ~ 5 min. Do not dry completely, since doing so will make it more difficult to rehydrate the pellet. Dissolve the RNA pellet in DEPC-treated water. Pass the solution through a pipet tip several times and incubate for 10 min at 60°C to help dissolve the RNA.

Isolation of Poly(A)$^+$ RNA

The mRNA fraction constitute only a small portion of total cellular RNA (about 2–5%) and is heterogeneous in size and length; however, since most mRNA carry tracts of poly(A) at their 3′ termini, mRNA can be readily separated from the bulk of cellular RNA by affinity chromatography on oligo(dT)-linked matrices. The following is a modification of the oligo(dT) cellulose chromatography described by Sambrook and Russell.[9]

[9] J. Sambrook and D. W. Russell, "Molecular Cloning. A Laboratory Manual." Cold Spring Harbor Press, Cold Spring Harbor, New York, 2000.

Note that oligo(dT) cellulose is a high capacity-binding resin. In order to minimize loss, the optimal ratio of total RNA to resin is about 1 mg of total RNA per 0.1 g of oligo(dT) cellulose.

Reagents and Procedure

Oligo(dT) cellulose (Type VII; Pharmacia)
Elution buffer: 10 mM Tris–HCl, 0.05% SDS, 1 mM EDTA (pH 7.5)
2× Binding buffer: 40 mM Tris–HCl, 1 M NaCl, 1% SDS, 2 mM EDTA (pH 7.5)
Washing buffer: 10 mM Tris–HCl, 0.1 M NaCl, 0.1% SDS, 1 mM EDTA (pH 7.5)

1. Suspend 0.5 g of oligo(dT) cellulose in 0.1 N NaOH.
2. Place oligo(dT) cellulose in a DEPC-treated Disposacolumn and wash with 3–5 column volumes of sterile DEPC-treated water.
3. Wash the column with sterile 1× binding buffer until the pH of the eluent is < 8.0.
4. Dissolve the RNA sample in DEPC-treated water. Heat at 65°C for 5 min. Quickly cool to room temperature on ice. Then add an equal volume of 2× binding buffer.
5. Apply RNA to the column and collect the flow-through in a sterile tube. After all the RNA has entered the column, wash with 1 column volume of 1× binding buffer while continuing to collect the flow-through.
6. Heat the flow-through for 3 min at 65°C, quick chill, and reapply to the column. Wash the resin with 5 column volumes of 1× binding buffer followed by 5 column volumes of wash buffer.
7. Elute the poly(A)$^+$ RNA with 3 column volumes of elution buffer. Collect fractions equivalent to about 1/3 of the column volume. Identify RNA-containing fractions by UV spectroscopy. Pool the fractions containing the eluted RNA. The yield of poly(A)$^+$ RNA should be approximately 2–5% of the starting mass of total RNA.

cDNA Synthesis

There is little hope of success with expression cloning, if the cDNA library is not of the highest quality. Ideally, cDNA libraries should be directional to ensure that the cDNA is cloned in the appropriate orientation with regard to the promoter. This orientation reduces the number of clones to be screened; however, many systems for directional cDNA synthesis rely on digestion with the *Not*I enzyme (or in some cases *Eco*RI) to unmask the ligated oligonucleotide linker. As such, this strategy is not suitable for cDNA clones with internal *Not*I sites. One solution to this problem is to use nonpalindromic adapters, such as *Bst*X1 linkers (Invitrogen Librarian cDNA Synthesis Systems), that carry CACA overhangs. This method eliminates concatemerization and precludes the need to cut back linkers with restriction enzymes, such as *Eco*RI or *Not*I. Since the complete coding region is crucial for functional expression, cleavage at internal restriction enzyme sites would eliminate the cDNA from the library and defeat the expression-cloning strategy. The drawback of this strategy is that the inserts are ligated into the expression vector in either orientation and as such, this system doubles the number of cDNA clones to be screened. Another alternative is the directional cDNA synthesis system available from Stratagene (LaJolla, CA). In this system, the first strand synthesis is performed in the presence of 5-methyl-dCTP to produce hemimethylated cDNA. *Eco*RI adapters are then ligated to the cDNA and the product is digested to cleave at an *Xho*I site in the oligo(dT) primer. Internal *Xho*I sites are protected using this strategy, since they are hemimethylated.

cDNA Synthesis Using the Librarian cDNA Synthesis System (Invitrogen)

The reaction is performed essentially as described by the manufacturer. Poly(A)$^+$ RNA (5 μg) is heated to 65°C for 10 min, quick-chilled on ice, and then combined with oligo(dT), placental RNase inhibitor, deoxynucleoside triphosphates (dNTPs), reverse transcriptase buffer, and AMV reverse transcriptase. At this point, 1 μl of $\alpha[^{32}P]$dCTP (3000 Ci/mmol) should be added as a tracer in order to calculate the total yield of cDNA. After incubation at 42°C for 60 min, the reaction is stopped by incubation for 2 min on ice. The RNA is next degraded by RNase H, and the second strand cDNA is synthesized with the addition of dNTPs, *Escherichia coli* DNA polymerase I, and *E. coli* DNA ligase. The double-stranded cDNA is made blunt-ended by addition of bacteriophage T4 DNA polymerase, and then extracted with phenol–chloroform and ethanol precipitated in the presence of ammonium acetate. The pellet is resuspended in sterile water and nonpalindromic *Bst*X1 adapters are ligated to the cDNA. An aliquot of the radiolabeled cDNA can be subject to liquid scintillation counting at this point. The percent conversion of mRNA to cDNA should be greater than 25% before proceeding to the next stage.

cDNA Synthesis Using the SuperScript cDNA System (Invitrogen)

The reaction is performed essentially as described by the manufacturer. Poly(A)$^+$ RNA (2–5 μg), dNTPs, and the oligo(dT) adapter are heated for 10 min, quick-chilled on ice, and then reverse transcribed at 37°C for 1 hr. The RNA is next degraded by RNase H, and the second strand cDNA is synthesized with the addition of dNTPs, *E. coli* DNA polymerase I, and *E. coli* DNA ligase. Finally, a non-self-ligating blunt-end *Sal*I adapter is ligated to the cDNA, and the product is digested with the *Not*I enzyme. The cDNA is then extracted with phenol–chloroform and ethanol precipitated with ammonium acetate.

Size Selection of cDNA

Size selection of the cDNA is of paramount importance after synthesis. The main purpose of this step is to ensure that the library is not composed of small fragments from incomplete reverse transcription reactions. This step can also enrich for the cDNA of interest, thereby reducing the number of clones to be screened. A readily available technique for cDNA fractionation is agarose gel electrophoresis, followed by electroelution of the DNA from gel slices. This method works well if performed using the procedure outlined below, which is a modification of the manufacturer's instruction.

Reagents and Procedure

1× TAE buffer: 40 mM Tris-acetate, 1 mM EDTA (pH 8.0).

1. Prepare a 0.8% agarose gel in 1× TAE (Gibco-BRL midi-gel apparatus) using a 14-well comb. Divide the cDNA into two aliquots and place in two wells of the gel. Pipet appropriate DNA size markers in the flanking lanes. Electrophorese the gel at 5 V/cm until the bromophenol blue reaches the lower end of the gel.
2. Carefully cut out the cDNA lane. Stain the remaining gel with ethidium bromide and use the DNA markers to identify the migration of the appropriate cDNA size range.
3. Cut out the desired cDNA size range. cDNA greater than at least 800 bp should be selected to eliminate small incomplete synthesis products. Divide this piece into 4–5 smaller gel slices.
4. Transfer to GenEluter apparatus or to dialysis tubing. Electroelute the cDNA in 0.25× TAE buffer at 100 V for 2–3 hr.
5. Transfer the contents of the GenEluter or dialysis membrane to a microfuge tube. Add 5 μg of carrier tRNA and phenol–chloroform extract. Ethanol precipitate in the presence of 2 M ammonium acetate in a siliconized microfuge tube.

Plasmid Expression Vectors

The plasmid expression vector should include several of the following features. (1) Restriction endonuclease sites for directional cloning of the cDNA inserts. Alternatively, a unique *Bst*X1 site in the polylinker to allow for ligation of cDNA inserts with nonpalindromic linker/adapters.

(2) A strong promoter, such as the cytomegalovirus, transcription-enhancer unit to drive high levels of expression. (3) An SV40 origin of replication to allow the plasmid to replicate to high copy number in COS cells that express SV40 T antigen.

Ligation into the Plasmid Vector

A small-scale concentration curve should be performed initially to optimize the ligation step. In a 10 μl reaction, add from 5 to 100 ng of linearized expression vector to 1 μl of library cDNA (50 μl total). Add buffer and T4 DNA ligase, and incubate at 15°C overnight. Add 2 μg of tRNA carrier and ethanol precipitate the ligation reaction. Wash the pellet with 70% ethanol to remove any remaining salts. This step is important since residual salts will interfere with the electroporation technique used in the subsequent *E. coli* transformations. Resuspend the pellet in 100 μl of sterile water and use 2 μl for the transformations. Always perform a vector alone control to calculate the background. The optimal cDNA to vector ratio determined at this step will be used in the large-scale ligation reaction. A typical large-scale ligation reaction would include 100 ng of cDNA, 350 ng of linearized vector, ligation buffer, and 20 units of T4 DNA ligase in a 350-μl final volume. Incubate overnight at 15°C. Ethanol precipitate and wash with 70% ethanol to remove salts. Resuspend in 100 μl of sterile water; dispense in 10 μl aliquots and store at −70°C.

Transformation of E. coli

The cDNA library should be transformed into *E. coli* at the highest efficiency possible. Electroporation gives excellent efficiencies of greater than 1×10^{10} transformants per microgram of plasmid DNA. This efficiency is 10- to 100-fold higher above that obtained with chemically competent cells, and will increase the likelihood that low, abundance mRNA species will be represented in the library. Electroporation-competent cells are available commercially (e.g., Electro Max DH10B cells from Invitrogen). A protocol for preparing high-efficiency electroporation competent cells is also included in the following section.[10] With either cell preparation, the electroporation efficiency should be optimized to enhance the clonal representation in the cDNA library. There are few parameters that can be altered to affect electroporation efficiency, among which is the resistance of the electric current that pulses through the cells. The optimal electroporation depends on two electric pulse characteristics, field strength (which is inversely related to cell viability), and pulse length.[10] Using a GenePulser electroporation

[10] W. J. Dower, J. F. Miller, and C. W. Ragsdale, *Nucleic Acids Res.* **16**, 6127 (1988).

device (BioRad Laboratories), the optimal voltage and capacitance were found to be 2500 V and 25 μF, respectively. Resistance was systematically varied and electroporation efficiency, as measured in transformants per microgram of DNA, peaked at 200 Ω and fell dramatically at 1000 Ω. Cell viability also peaks at 200 Ω (pulse length ~ 5 msec), and falls off with increasing resistance and pulse length.

Preparation of Electroporation Competent Cells

Reagents and Procedure

SOC medium: 20 g/liter bacto-tryptone, 5 g/liter bacto-yeast extract, 0.5 g/liter NaCl, 25 mM KCl, 10 mM MgCl$_2$, 0.4% dextrose (Note that rich broths, such as SOC or TB, should be used instead of LB) 10% glycerol (sterile).

1. Inoculate HB101, MC1061/P3, HB10B, or other appropriate *E. coli* strain into 10 ml of SOC media. Grow overnight at 37°C with shaking.
2. Inoculate 1 liter of SOC media with 1/100th volume of the fresh overnight culture.
3. Grow cells at 37°C with shaking (~ 225 rpm) until OD$_{600}$ is 0.5–0.7 (mid-log phase). This growth usually takes 2–3 hr. (Note that stationary phase cultures should not be used to prepare cells.)
4. Chill the flask on ice for 15–30 min. Transfer to 250 ml bottles and pellet cells at $4000 \times g$ for 15 min at 4°C. Discard the supernatant.
5. Resuspend the cell pellets in 1 liter of cold sterile milli-Q water (or equivalent very low resistance water). Pellet the cells at $4000 \times g$ for 15 min at 4°C. Immediately discard the supernatant. (Note that the cell pellets will be very loose at this point. Removing and inverting the bottles immediately after the centrifugation will decrease cell loss.)
6. Repeat the wash step one time.
7. Resuspend the cell pellets in 0.5 liter of cold milli-Q water. Pellet the cells at $4000 \times g$ for 15 min at 4°C.
8. Resuspend the cell pellet in 100 ml of 10% glycerol. Centrifuge as in step 7.
9. Resuspend the cell pellet in a final volume of 2–3 ml of 10% glycerol. This volume dilutes the cells to an OD$_{600}$ of 300 (approximately 6×10^{11} cells/ml).
10. Dispense in 100 μl aliquots. Quick freeze in liquid nitrogen and store at −70°C. Do not thaw and refreeze. After thawing, discard any unused cells.

Transformation of E. coli *by Electroporation*

Reagents and Procedure

LB: 10 g/liter bacto-tryptone, 5 g/liter bacto-yeast extract, 5 g/liter NaCl, 15 g/liter bacto-agar.

Thaw electroporation-competent *E. coli* cells on ice. Use approximately 10 pg of a control plasmid or 1 μl of cDNA library, and mix with 40 μl of electroporation competent cells in a sterile microfuge tube on ice. Carefully transfer the mixture into ice-cold 0.1-cm electroporation cuvettes (BioRad Laboratories, Richmond, CA), and use the following conditions for electroporation: 2500 V, 25 μF, 200 Ω. The cells are then diluted immediately into 0.5 ml of room temperature SOC, transferred to sterile test tubes and incubated in a 37°C shaker at 225 rpm for 1 hr. This step helps the bacteria recover and allows expression of the antibiotic resistance marker encoded by the plasmid. The cells are then plated onto 150-mm LB-agar plates containing 50 μg/ml ampicillin and incubated overnight at 37°C. The exact volume to be plated will depend on the library titer and the pool size required for screening. The titer of the transformation mixture is first determined using an aliquot of the large-scale ligation and making a series of dilutions of the transformation mixture in 100 μl of SOC medium (1:5, 1:10, 1:50, 1:100). Spread these 100 μl dilutions onto 100-mm LB-ampicillin plates and incubate overnight at 37°C. Selection of the library pool size is described in the next section.

Selection of the Library Pool Size and Positive Controls

The total number of clones to be screened can be calculated. For a low-copy number gene, such as a membrane transporter, theoretically 5×10^4 clones must be screened to achieve a 99% probability that the cDNA clone exists in the library.[11] This theoretical estimate comes from the calculation $q = [1 - (n/T)]^N$, where "q" represents the probability that a single cDNA clone is not represented in the library after choosing at random, "N" clones. In this equation, "n" is the number of desired clones present in "T," the total number of different mRNA species in a cell. If we assume that the transporter is represented once in 10,000 mRNAs (based on the approximation that low-abundance hydrophobic membrane proteins comprise less than 0.2% of total membrane proteins), then at least 50,000 clones must be screened for a 99% probability that the transporter will be detected. Other factors involved in determining this value include

---

[11] T. D. Sargent, *Methods Enzymol.* **152**, 423 (1987).

the correct orientation of the cDNA in the expression vector (directional vs nondirectional library) and the size of the full-length cDNA. In expression cloning, only a full-length cDNA in the proper orientation will be detected. Thus, a more realistic target number is between 5×10^5 and 1×10^6 clones. The next step is then to select the library pool size or complexity.

To minimize the amount of work involved in screening, the cDNA library is divided into a series of pools that contain the maximum number of cDNA clones compatible with efficient detection of the transporter activity. Very large pools of clones ($> 100,000$) may not contain sufficient transporter DNA to yield a detectable signal. Alternatively, using very small pool sizes (10–100 clones) would be impractical to screen the estimated 100,000 clones required to obtain a positive pool. In order to determine the minimum acceptable pool size, a positive control should be used. A previously isolated cDNA can be used to determine the assay sensitivity and target pool size. Ideally, a related transporter activity should be used. For the ileal sodium–bile acid transporter cloning, the cloned hepatic sodium–bile acid cotransporter provided an excellent positive control. The liver sodium–bile acid cotransporter cDNA was subcloned into a similar, but distinct expression plasmid from the cDNA library. This expression plasmid used the same CMV promoter as the library, but carried a distinct antibiotic resistance gene to prevent the very real danger of contaminating the expression library. The importance of using a distinct antibiotic resistance gene for the positive control cannot be understated, especially if the control has a similar activity as the transporter cDNA being cloned. Serial dilutions of the positive control can then be assayed to determine the level of sensitivity and maximum pool size. This approach is illustrated in Fig. 1, where the liver sodium–bile acid cotransporter plasmid (CMV-Ntcp) was mixed with increasing amounts of an irrelevant expression plasmid, CMV-βgal (2 μg total mass), and transfected into COS cells. After 72 hr, the cells were assayed for [^{3}H]taurocholate uptake. The COS cells transfected with undiluted CMV-Ntcp-expressed taurocholate uptake activity that was approximately 470-fold higher than that measured in the mock-transfected cells. When the Ntcp-expression plasmid was diluted 10^1, 10^2, 10^3, and 10^4-fold with inert plasmid DNA, the transfected COS cells expressed taurocholate uptake activity that was 283, 55, 6, and 1.6-fold over background. At a $1:10^5$ dilution, bile acid uptake activity was indistinguishable from background; however, the $1:10^4$ dilution was clearly distinguished from the mock-transfected control, suggesting that a DNA pool size of 10^4 clones containing a single bile acid transporter cDNA clone could be identified. Based on this curve, a pool size of approximately 5000 clones in complexity was selected.

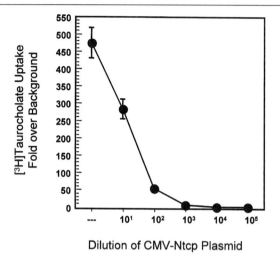

Dilution of CMV-Ntcp Plasmid

FIG. 1. Sensitivity of the bile acid uptake assay. COS cells were transfected with decreasing amounts of CMV-Ntcp expression plasmid. Total plasmid mass was maintained at 2 μg per dish by addition of CMV-βgal DNA. On day 4 posttransfection, [^{3}H]taurocholate uptake was measured. Background values represent COS cells transfected with CMV-βgal alone. The mean ± SEM of triplicate transfections are shown.

Preparation of Library Pools

After electroporation, the transformation is spread onto 150-mm LB plates containing 50 μg/ml ampicillin, and grown for 12–16 hr at 37°C. Colonies are then scraped from the plates, transferred to 50-ml cultures of SOC medium containing ampicillin and chloramphenicol, grown, and the plasmid DNA is purified. The chloramphenicol is added to amplify the plasmid in any slow-growing clones and to prevent the cultures from becoming too dense. Liquid culture is initially avoided, since this may select against poor-growing clones. In addition, the bacterial colonies on the plates can be counted to ensure that the pool size (complexity) has not drifted from the original target number. There are a number of commercially available plasmid purification kits (e.g., Qiagen, Promega) that are rapid, convenient, and yield high-quality plasmid DNA.

Reagents and Procedure

1. Add 25 ml of SOC-ampicillin to a 150-mm LB-ampicillin plate. Gently scrape the colonies from the agar plate using an ethanol-sterilized cell scraper. Transfer to a sterile 250-ml Erlenmeyer flask. Wash the agar plate with another 25 ml of SOC-ampicillin and combine with the first 25 ml.

2. Incubate in a 37°C shaker at 225 rpm. After shaking for 1.5 hr, add chloramphenicol (50 μl of 25 mg/ml in ethanol) to 25 μg/ml final concentration. Continue shaking at 225 rpm for an additional 3.5 hr.

3. Make up duplicate 1-ml glycerol stocks of the library pools. Store at −70°C.

4. Prepare plasmid DNA using the manufacturer's instructions.

5. Resuspend in 50 μl of TE. Measure the absorbance at 260 nm. Typical yields are from 25 to 100 μg of plasmid DNA.

6. To verify the integrity of the plasmid DNA, electrophorese 5 μl on a 0.8% agarose–TBE gel. Store the remaining DNA at −20°C.

Sibling Selection

Once a plasmid pool is identified as having the transport activity of interest, an aliquot of that library pool DNA is diluted 1:100 to 1:1000 in sterile water and transformed into *E. coli*. Typically, the pool complexity is reduced in 10-fold increments, so the exact volume to be plated will depend on the target pool size. Electroporation of the library pool DNA and preparation of the subpool plasmid DNA should be carried out as described earlier. The number of total clones in all the subdivided pools should be approximately twice the number of original clones in the pool. For example, if the original positive pool contained approximately 5000 cDNAs, it should be subdivided into 20 pools of 500 clones each (approximately 10,000 clones) to ensure the positive clone is represented. For pool sizes of 100 cDNAs or less, the colonies should be transferred to gridded filters (10 × 10) using sterile toothpicks. After a filter's grid is full, it is grown for 8–12 hr. Colonies are then picked and pooled from each column and row, and used to inoculate 50-ml cultures of SOC medium containing ampicillin, and the plasmid DNA is purified. After assaying the pools, the clone of interest lies at the intersection of any positive row and column. While labor intensive, this sibling selection procedure is simple and easily tracks the robust enrichment in transporter activity. Alternative methods, such as recovering the plasmid DNA by the Hirt's[12] method from positive-transfected COS cells, have also been used. The method offers no direct advantages for expression cloning of ligand transporters, and for poorly understood reasons, plasmids that have replicated in COS cells often become damaged and cannot be recovered unaltered.

[12] B. Hirt, *J. Mol. Biol.* **26**, 365 (1967).

Functional Screening of the Library

Assay Selection

The expression-cloning strategy requires an extremely sensitive assay for the target activity. Optimally, the chosen expression system should lack the biological activity in question, but possess the ability to express that activity. With expression cloning in *Xenopus* oocytes, the target activity can often be expressed and detected after injection of poly(A)$^+$ RNA. In this manner, the expression system can be validated for the target activity. Also, the sensitivity of the oocyte assay permits larger pools of clones to be screened, allowing for a more rapid identification of initial positive pools. Mammalian cell expression systems lack the ability to test total poly(A)$^+$ RNA or poly(A)$^+$ RNA fractions for the presence of the desired cDNA clone. Thus, positive and negative controls must be used when screening library pools by transfection of mammalian cells. The negative control can be the empty vector or an expression plasmid for an irrelevant protein such as β-galactosidase. As described earlier, the positive control should be a related transporter (in a vector with a different drug resistance marker). These controls must be included with each set of library pools assayed in the screen.

A reliable and robust assay for the target transporter is of crucial importance. For electrogenic transporters, substrate-evoked uptake currents or membrane potential changes can be monitored in *Xenopus* oocytes; however, for transfected mammalian cells, transporters are primarily assayed by monitoring labeled substrate uptake. Variations on this basic method include the use of specific inhibitors, or taking advantage of ion-dependent (such as sodium-dependent) uptake, or coupling the uptake of a radio-labeled precursor to product accumulation. The latter approach is best illustrated by the cloning of a mevalonate transporter.[13]

Cell Recipient

A widely used host cell for expression cloning is COS cells, which are permanent lines of African green monkey kidney cells (CV-1), which stably express the SV40 large T antigen. These cells are hardy, adherent, easily transfected, and will efficiently replicate transfected DNA harboring an SV40 origin of replication. This amplification permits very high levels of transient expression of the library pools. While originally derived from kidney, COS cells appear to express few, if any, of the major receptor ligand

---

[13] C. M. Kim, J. L. Goldstein, and M. S. Brown, *J. Biol. Chem.* **267**, 23113 (1992).

transporters characteristic of a differentiated kidney cell. As such, these cells provide a low transport background for a cDNA library screen. HEK 293 cells (a line of human embryonic kidney cells) also are widely used for expression cloning.[14] Like COS, 293 cells are easy to transfect and provide high levels of transient expression.[15]

Assay Development

As for the pool size selection, a positive control is essential for optimizing the transport assay used to screen the library. Ideally, an already-cloned related transporter cDNA can be used. Parameters to be optimized include the transfection conditions, media used for the uptake assay, the concentration of radioactive ligand, the time course for uptake, and the wash conditions. (1) COS cells are readily transfected by a variety of methods. DEAE-Dextran works well and is inexpensive; however, optimal transfection efficiency requires the use of facilitators such as glycerol and chloroquine. As such, the method is more difficult to optimize. A convenient but more expensive replacement method is the nonliposomal reagent, Fugene 6 (Roche). The basic protocol provided by the manufacturer, yields COS cell transfection efficiencies similar to those of the optimized DEAE-Dextran method. (2) The medium used for the uptake assay should maximize uptake via the target and minimize nonspecific binding to the cells and tissue culture dish. A good starting medium is DMEM plus 0.2% BSA at 37°C. This medium can support the transfected COS cells for incubations over several hours, whereas the BSA will decrease nonspecific binding of the radioactive ligand. (3) The concentration and specific activity of the radioactive ligand is an important consideration. Even under optimal conditions, the transfected library pools will yield only very low levels of transporter expression. As such, the highest possible specific activity radioactive ligand should be used. The concentration of ligand in the assay will depend on the affinity of the transporter and the nonspecific binding by the ligand. In general, transporters have lower affinities for their solutes than receptors (nuclear or cell surface). Using the ligand at a concentration between 0.2 and 0.5 times, the K_m should provide sufficient substrate to yield a strong signal. Higher concentrations may increase the signal, but may be prohibitively expensive. Note that unless the host cell metabolizes the radioactive ligand, the media can be reused after filtering or the radiolabeled ligand can be extracted and reused in a

---

[14] H. Simonsen and H. F. Lodish, *Trends Pharmacol. Sci.* **15**, 437 (1994).
[15] C. M. Gorman, *Curr. Opin. Biotechnol.* **1**, 36 (1990).

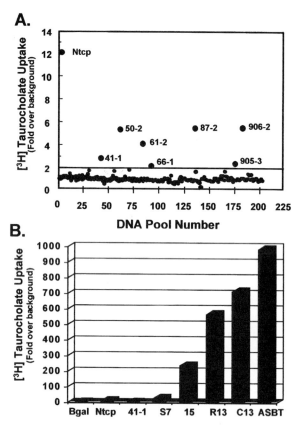

FIG. 2. Expression cloning of the ileal sodium–bile acid cotransporter. (A) For each transfection experiment, 10–24 pools of cDNA plasmids were transiently transfected into duplicate dishes of COS cells. The horizontal axis denotes cDNA plasmid pools of approximately 3000 clones each. The activity is expressed as [3H]taurocholate uptake relative to CMV-βgal (mock) transfected cells. (B) Purification of the ASBT cDNA clone. A single positive ASBT clone was isolated from positive pool 41-1 by sibling selection. Reproduced with permission from Wong.[6]

subsequent screen. (4) The time course of uptake will depend on the transporter and the stability of the radioactive ligand. Longer time courses are suitable for concentrative transporters such as sodium cotransporter. Note that over time the internalized ligand will begin to exit the COS cells, either by diffusion or in some cases via an active transport mechanism. Typical time courses range from 15 to 60 min. (5) Washes are typically done on ice using ice-cold buffers, and should include 0.2% BSA and unlabeled ligand to reduce nonspecific binding of the radiolabeled ligand.

Library Screening

The result of a typical library screen is shown in Fig. 2A. A total of 202 pools representing approximately 6.5×10^5 clones were screened to yield seven pools that exhibited bile acid uptake activity two- to six-fold greater than the CMV-βgal (mock) transfected COS cells. A threshold of two-fold above background was selected based on the narrow standard deviations of the bile acid uptake assay. The pool designated 41-1 was the first positive pool identified in the screen. Figure 2B illustrates the sibling selection results for this pool. In the first round of screening, 18 pools of approximately 500 clones each were prepared and assayed to yield one positive pool (41-1/S7). This pool was subdivided into 48 pools of approximately 50 clones each, and assayed to yield six positive pools. At this point, 200 individual clones from plasmid subpool 41-1/S7/S15 were streaked onto two 10×10 matrices. Plasmid DNA from pools of each row and each column were purified and assayed for [³H]taurocholate uptake activity. The cDNA clone situated at the intersection of the positive row 13 (R13) and column 13 (C13) was isolated and assayed. Transfection of this single clone into COS cells stimulated [³H]taurocholate uptake almost 1000-fold over background.[6]

Acknowledgments

This work was supported in the National Institutes of Health Grants DK47987 and HL49373.

[19] Nuclear Receptor Target Gene Discovery Using High-Throughput Chromatin Immunoprecipitation

By Josée Laganière, Geneviève Deblois, and Vincent Giguère

Introduction

Nuclear receptors are master transcription factors that regulate the development, physiology, and homeostasis of whole organisms through direct control of gene expression in response to diverse ligands and hormonal stimuli.[1] Nuclear receptors regulate the expression of their target genes through association with specific DNA regulatory elements.[2] While a

[1] D. J. Mangelsdorf, C. Thummel, M. Beato, P. Herrlich, G. Schütz, K. Umesono, B. Blumberg, P. Kastner, M. Mark, P. Chambon, and R. M. Evans, *Cell* **83**, 835 (1995).
[2] C. K. Glass, *Endocr. Rev.* **15**, 391 (1994).

significant number of nuclear receptor target genes have been identified to date, it is believed that these genes represent only a small fraction of the regulatory units likely to be under the control of nuclear receptors. Most nuclear receptor target genes identified so far were characterized through "gene oriented" approaches that study the regulation of one candidate gene at a time, and these studies are usually limited to the promoter region. However, in order to understand the complex nuclear receptor-driven transcriptional networks that operate in a living organism, a whole genome approach is required and now feasible. Here, we describe a powerful "nuclear receptor/whole genome-oriented" approach to identify and more accurately study nuclear receptor regulatory networks.

Strategy

We took advantage of the recent advances in chromatin immunoprecipitation (ChIP)[3,4] and associated cloning procedures[5–7] to develop a high-throughput ChIP technique to identify primary nuclear receptor target genes. Nuclear receptors associate with the regulatory elements of their potential target genes leading to transcriptionally active chromatin. Formaldehyde can be used to cross-link the receptors and associated cofactors bound to DNA in living cells. Following isolation and fragmentation, the chromatin is immunoprecipitated using a specific antibody raised against the nuclear receptor of interest. Cross-link reversal and DNA purification is then performed, and the isolated fragments are cloned in a suitable vector for sequencing. Bioinformatic analysis of the fragments obtained is performed using the recently available human genome databases to localize the isolated regulatory elements and thus, identify associated target genes. The strategy is outlined in Fig. 1.

We have adapted a high throughput ChIP technique to clone target sequences bound by the estrogen receptor α (NR3A1, ERα) in the MCF-7 cell line upon estradiol (E$_2$) stimulation. This technique, used for the isolation of regulatory elements bound by the ERα in vivo, can be applied to other sources of materials (cell lines or dissociated primary cells) and other classic or orphan nuclear receptors.

---

[3] V. Orlando, *Trends Biochem. Sci.* **25**, 99 (2000).
[4] Y. Shang, X. Hu, J. DiRenzo, M. A. Lazar, and M. Brown, *Cell* **103**, 843 (2000).
[5] J. Wells and P. J. Farnham, *Methods* **26**, 48 (2002).
[6] A. S. Weinmann, S. M. Bartley, T. Zhang, M. Q. Zhang, and P. J. Farnham, *Mol. Cell. Biol.* **21**, 6820 (2001).
[7] A. S. Weinmann, P. S. Yan, M. J. Oberley, T. H. Huang, and P. J. Farnham, *Genes Dev.* **16**, 235 (2002).

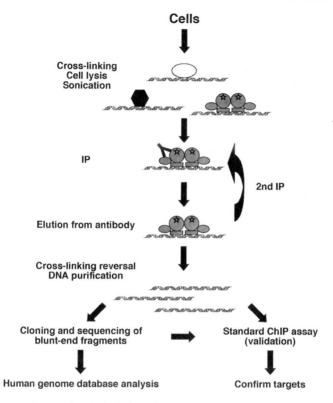

Cells

Cross-linking
Cell lysis
Sonication

IP

2nd IP

Elution from antibody

Cross-linking reversal
DNA purification

Cloning and sequencing of
blunt-end fragments

Standard ChIP assay
(validation)

Human genome database analysis

Confirm targets

Fig. 1. Use of ChIP for the isolation of nuclear receptors DNA regulatory sequences. Following a cross-linking step, the cells are lysed and the chromatin is fragmented by sonication. A specific antibody directed against the nuclear receptor of interest is used in two sequential IP experiments to improve the purity of the complexes. After cross-link reversal and DNA purification, the isolated fragments are repaired, cloned, sequenced, and localized in the human genome database. For new target validation, an independent standard ChIP assay with PCR primers specific for the newly isolated sequence is performed to assess the actual binding efficiency of this sequence to the receptor. Isolated DNA fragments are subcloned into a reporter plasmid and assayed in transient transfection to confirm their regulatory properties. Quantitative RT-PCR is performed in the same cell line used for the ChIP assay, and analyzed for changes in levels of expression of a candidate target gene to confirm the transcriptional regulatory role of the receptor.

Cell Culture and Chromatin Preparation

To study ERα association with regulatory DNA sequences upon E$_2$ stimulation, we used the ER-positive human breast carcinoma cell line MCF-7. Approximately, 1.5×10^7 cells were used for a single ChIP on endogenous ERα. Cells were routinely cultured in phenol-red-free Dulbecco's

Minimal Essential Medium (DMEM) as previously described.[8] Seventy-two hours prior to chromatin extraction, the media containing 10% complete fetal bovine serum (FBS) was replaced by phenol red-free DMEM supplemented with 10% charcoal–dextran treated FBS (steroid deprived). Prior to chromatin extraction, the cells are treated with either 10^{-7} M E_2 (Sigma, St. Louis, MO) or vehicule (ethanol) for 30–45 min.[4]

When studying the association between orphan nuclear receptors and their target regulatory sequences, no addition of ligand is required prior to performing the ChIP, and the cells can be routinely cultured in DMEM supplemented with 10% complete serum. However, in the event that crosstalk in gene regulation between an orphan nuclear receptor and a ligand-inducible nuclear receptor is suspected, the cells should be maintained in a manner not to interfere with specific nuclear receptor signaling, such as in a steroid-deprived serum.

Following E_2 treatment, bound proteins are immediately cross-linked to DNA upon addition of formaldehyde directly into the medium to a final concentration of 1%. After 10 min of incubation at room temperature on a shaking platform, the cross-linking reaction is stopped by addition of glycine to a final concentration of 0.125 M, and incubated for 5 more minutes at room temperature as previously described.[7] The duration of the cross-linking reaction needs to be accurately monitored since extensive exposure of the cells to formaldehyde can lead to a decreased yield in chromatin isolation and to poor immunoprecipitation (IP) efficiency. Following the cross-linking step, the cells are washed twice with ice-cold phosphate-buffered saline (PBS) and scraped in ice-cold PBS.

Chromatin preparation is carried out as follows. The PBS-harvested cross-linked cells are centrifuged at 1500 rpm for 10 min at 4°C, the pellet is resuspended in a suitable volume (200 μl per 1.5×10^7) of lysis buffer [1% SDS, 10 mM EDTA, 50 mM Tris–HCl, pH 8.1, supplemented with complete, mini, EDTA-free protease inhibitor cocktail (Roche, Molecular Biochemicals, Indianapolis, IN)] and incubated on ice for 10 min. The lysates are sonicated five times at setting 10 with the sonicator (model Virsonic 60, Virtis, NY) for 7–8 sec each time. This step needs to be optimized according to the sonicator and the cell line used. At this point, it is important to verify the length of the chromatin fragments resulting from the sonication step. An aliquot of the sonicated lysates is removed, incubated at 65°C for at least 6 hr to reverse the formaldehyde cross-linking,

[8] G. B. Tremblay, A. Tremblay, N. G. Copeland, D. J. Gilbert, N. A. Jenkins, F. Labrie, and V. Giguère, *Mol. Endocrinol.* **11**, 353 (1997).

purified using QIAquick spin PCR purification kit (Qiagen, CA), and analyzed on a 1% agarose gel. For a standard ChIP assay and for cloning ChIP-obtained DNA fragments, the average size of the DNA fragments should be approximately 1000 bp.[7] While this verification step is in progress, the remaining sonicated lysates can be frozen at −80°C. Further sonication steps can be carried out to obtain the desired fragment length. At this point, the sonicated lysates are centrifuged for 10 min, the supernatants collected and diluted 10× in ChIP dilution buffer (0.5% Triton X-100, 2 mM EDTA, 150 mM NaCl, 20 mM HEPES, pH 8) to achieve a final SDS concentration of 0.1%.

Chromatin Immunoprecipitation

Prior to performing the actual ChIP, it is important to remove and freeze an aliquot of the diluted fragmented chromatin corresponding to 10% of the total amount used for one IP. The antibodies (Ab) used to perform ChIP on endogenous ERα in MCF-7 cells were αERα Ab-1 (Neomarker, Fremont, CA) or αERα HC-20 (Santa Cruz Biotechnology, Inc., Santa Cruz, CA).[4] The Abs must have previously been shown to be suitable for IP. The ChIP procedure is essentially carried out as previously described.[4] Immunoclearing of the diluted chromatin is achieved by incubating for 1 hr at 4°C with 40 μl of salmon sperm DNA–protein A agarose (Upstate Biotechnology, Inc., Lake Placid, NY; provided as 50% slurry in 10 mM Tris–HCl, pH 8, 1 mM EDTA, 0.05% sodium azide). IP is performed overnight at 4°C using a specific antibody. Following IP, 40 μl of salmon sperm DNA–protein A agarose is added and the incubation is pursued for 2 more hours. The precipitates are washed sequentially for 10 min each with low-salt wash buffer (0.1% SDS, 1% Triton X-100, 2 mM EDTA, 20 mM Tris–HCl, pH 8.1, 150 mM NaCl), high-salt wash buffer (0.1% SDS, 1% Triton X-100, 2 mM EDTA, 20 mM Tris–HCl, pH 8.1, 500 mM NaCl), and LiCl wash buffer (0.25 mM LiCl, 1% NP-40, 1% deoxycholate, 1 mM EDTA, 10 mM Tris–HCl, pH 8.1). Precipitates are then washed three times with TE buffer and eluted twice for 15–30 min each time on vortex set at 3 with 75 μl of elution buffer (1% SDS, 0.1 M NaHCO$_3$).[7] The pooled eluates and the input are incubated at 65°C for at least 6 hr to reverse the formaldehyde cross-linking. The isolated DNA fragments are then purified according to the QIAquick Spin Kit protocol (Qiagen, CA) and used for quantitative PCR analysis of the ChIP assay. It is advisable to validate that the first ChIP was successful and specific by standard ChIP assay of a known target sequence (refer to section "Validation of the ChIP Experiment (Positive control)" and Fig. 2), and

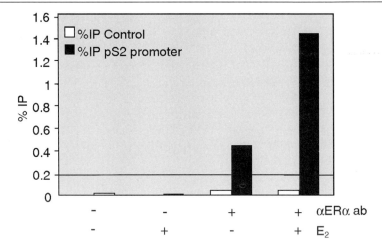

Fig. 2. Standard ChIP assay for ERα in MCF-7 cells on the pS2 promoter. MCF-7 cells were treated with estradiol or vehicle for 45 min and submitted to standard chromatin IP with ERα antibody. The purified DNA was used as a template for quantitative PCR amplification of the pS2 promoter and for a 5-kb upstream region (Control). The percentage of IP (%IP) was determined by comparing with a 2% input sample. Any %IP below 0.2% is considered nonsignificant. The pS2 promoter can be efficiently immunoprecipitated by an ERα antibody in E_2-treated cells, and to a lower extent in nontreated cells.

that the criteria for standard ChIP discussed in this section are achieved before beginning the cloning procedure.

Cloning of Fragments Isolated by ChIP

For cloning the fragments, we suggest that a second purification step be performed; cloning subsequently to a single ChIP may lead to the isolation of a considerable amount of clones containing nonspecific sequences, such as repeat DNA. The first ChIP is carried out as described earlier, with the exception that the eluates are pooled, rediluted $10 \times$ in ChIP dilution buffer, and reimmunoprecipitated using the same Ab for ERα prior to the reversal of the cross-link. The use of a different Ab (if available) for the second IP is a good way to increase the stringency of the technique, further reducing the isolation of nonspecific sequences. The subsequent steps for the second ChIP are carried out as described for the first ChIP.

The ends of the isolated fragments from the double ChIP procedure are then repaired with T4 DNA polymerase (MBI Fermentas Inc., ON) and repurified with the QIAquick Spin Kit (Qiagen, CA). The blunt fragments are then cloned into a suitable vector [such as the Ready-to-go pUC18 *Sma*I/BAP + Ligase kit (Pharmacia, NJ)] for further sequencing analysis.

Following transformation of the ligation reaction and DNA purification, inserts are sequenced using an automated sequencer and appropriate primers.

Criteria for the Analysis of the Isolated Sequences

The cloned sequences are analyzed using BLASTn search on the private CELERA human genome database (http://www.celeradiscoverysystem. com) or the public database (http://www.ncbi.nlm.nih.gov/BLAST/). Several criteria are taken into consideration for the analysis and validation of the sequences obtained.

1. The sequence isolated has to be long enough (> 300 bp) to facilitate following sequence analysis.
2. A perfect match has to be obtained between the isolated sequence and the genome sequence to ensure the localization specified by the search.
3. The BLASTn result has to be unique among the genome, since a sequence leading to multiple blast results could represent a repeat that was isolated in a nonspecific manner.
4. Localization to the closest annotated gene from the isolated sequence is the next criteria to consider. Immediate proximity ($ < \sim 1$ kb) of the isolated sequence to the transcriptional start site of a gene indicates a possible promoter function of the sequence isolated. Sequences situated further 5' or 3' from a gene (we used 50 kb as an arbitrary cutting point) indicates a possible role as an enhancer.
5. The sequence is examined for the presence of consensus half-sites or complete hormone response elements (HRE).[9] However, analysis of the sequences should be extended to sites for other transcription factors, as nuclear receptors can activate the transcription of target genes via interaction with other complexes that associate with DNA (such as Sp1).[10] In addition, as transcription factors are known to bind next to each other in regulatory regions, the presence of other known transcription factor consensus binding sites should also be taken into consideration as it can give insights about the potential regulatory properties of the isolated sequences.

[9] K. Umesono, K. K. Murakami, C. C. Thompson, and R. M. Evans, *Cell* **65**, 1255 (1991).
[10] V. Krishnan, X. Wang, and S. Safe, *J. Biol. Chem.* **269**, 15912 (1994).

Validation and Analysis of the Isolated Sequences

Standard ChIP Assay

As discussed earlier, a standard ChIP assay defined as a PCR on the DNA fragments isolated using a single IP should first be performed to:

1. Validate each ChIP experiment using a known target sequence to which the receptor binds (positive control).
2. Validate the IP of a newly isolated and cloned sequence (standard ChIP).

For convenience and accuracy, we used quantitative PCR for the standard ChIP assay.

Validation of the ChIP Experiment (Positive Control)

Primers are designed to amplify a positive control sequence to which the receptor of interest is known to associate to upon ligand binding. This is done to validate the actual ChIP experiment. Negative control primers situated form 3 to 5 kb upstream or downstream of the region analyzed for nuclear receptor binding are also designed to ensure the specificity of the immunoprecipitated fragment. The PCR product length should be around 200–400 bp as suggested by the manufacturer of the Light Cycler apparatus. For standard ChIP assay, a control IP performed in the absence of Ab (basal control) is carried out for each treatment and used to compare with the actual IP reaction. Samples corresponding to 0.5, 1, and 2% of input are kept for quantification on the Light Cycler to calculate the ChIP efficiency of the target sequence compared to the negative control, for both the IP and the control.

The percentage of sequence immunoprecipitated reflects the affinity of association of the nuclear receptor to the target sequence and can be calculated using the formula:

$$\%\,\text{IP} = 2^{1+n_i-n} \tag{1}$$

where n_i and n are the number of cycles at which the exponential portion of the curves for product formation is most efficient for the 2% input and the sample, respectively. This formula is derived from the assumption that an efficiency of two is achieved for the Light Cycler PCR reactions. In a standard ChIP assay, the sequence has to be immunoprecipitated at higher percentages than the basal control (no Ab); a significant IP percentage (%IP) has to reach at least 0.2%.[6] The %IP of the 5-kb upstream negative control must be insignificant (lower than 0.2% IP) in all the samples. Moreover, the difference in the %IP between the basal control (no Ab) and the sample (with Ab) must be significantly higher in the target sequence compared to the negative control (~ 5 kb upstream). The difference in the IP percentages between the treated and the non-treated purified DNA fragments from the ChIP experiment indicates at which extent the addition of the ligand recruits the receptor to the regulatory element.

Since it had already been shown that the human TFF1 (pS2) gene promoter is occupied by ERα upon E_2 treatment,[11] it was used as a positive control for the validation of our ChIP experiment (see Fig. 2).[4] The pS2 promoter primers were used with an annealing temperature of 62°C on the quantitative Light Cycler, using SYBR Green Light Cycler kit 1 (Roche Molecular Biochemicals, CA) (Fig. 2). To further confirm the specificity of the binding, another set of primers was designed for a negative control sequence situated approximately 4-kb upstream of the studied regulatory sequence. As shown, all negative controls are under the 0.2% IP baseline, and ChIP targets reach higher %IP.

Validation of a ChIP experiment with a known target sequence as positive control is an important step to ensure the accuracy and the specificity of the ChIP technique, before cloning the purified DNA fragments. However, we are aware that it may not always be possible to have access to a known target sequence for some specific nuclear receptors in some particular cell lines. The use of other cell lines with known target sequences for the receptor of interest could also be used (when available) to develop the experimental conditions of the ChIP.

Validation of a Newly Isolated and Cloned Sequence

The standard ChIP assay is similarly used to validate the potential association between the nuclear receptor and the newly cloned sequence. Specific ChIP primers are designed for the cloned sequence of interest, as well as for a negative control sequence (~ 5 kb upstream or downstream), and are used with quantitative Light Cycler to assess the enrichment of this potential regulatory sequence in the ChIP samples. This step allows to discriminate between true and false positives that have been cloned using the double ChIP-cloning technique.

It is interesting to note that one of the clones isolated upon ChIP cloning following a single ER IP was a 389-bp fragment encompassing the pS2 promoter. The isolation of this promoter in the cloning step, thus served as a positive control for our technique.

[11] C. Giamarchi, M. Solanas, C. Chailleux, P. Augereau, F. Vignon, H. Rochefort, and H. Richard-Foy, *Oncogene* **18**, 533 (1999).

Changes in Gene Expression Levels Induced by Nuclear Receptors

Once the association of the receptor with the cloned regulatory sequence has been firmly established using standard ChIP assay, the influence of the nuclear receptor on mRNA expression of a specific target gene close to the sequence of interest needs to be assessed. Quantitative RT-PCR experiment is intended to quantify the levels of specific mRNA corresponding to the expression of this specific gene in the cell line over several time points, following E_2 or other ligand treatment. For example, pS2 RT-PCR primers were designed to perform quantitative RT-PCR on cDNA from MCF-7 cells that were either treated or not with E_2 (10^{-7} M) for a specific period of time. Quantitative RT-PCR using Light Cycler instrumentation and the SYBR Green detection kit 1 was carried out, and the upregulation of the target gene was obvious upon treatment of the cells with E_2, at all the time course tested with respect to the samples obtained from untreated cells (Fig. 3).

This assay should be applied to all the potential target genes, and can also be adapted for orphan nuclear receptors by introducing siRNA (or a vector-based expression of siRNA) directed against the specific orphan nuclear receptor in order to knockdown its expression and analyze changes in the levels of expression of putative target genes.[12]

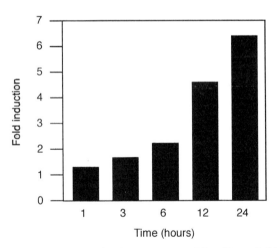

FIG. 3. Quantitative RT-PCR showing the E_2 response of the pS2 gene in MCF-7 cells. Total cDNA from MCF-7 cells treated with E_2 for 1, 3, 6, 12, and 24 hr is used for quantitative RT-PCR of a specific gene, and compared with control cells.

[12] T. R. Brummelkamp, R. Bernards, and R. Agami, *Science* **296**, 550 (2002).

Transient Transfections with Reporter Luciferase Assay

Once it has been established that the cloned sequence is enriched in standard ChIP, and thus associates with the receptor or a complex containing the receptor, it is suggested to test the ability of the cloned sequence to confer nuclear receptor responsiveness to a basal reporter gene. The DNA sequence is subcloned upstream of the TK-Luciferase (TK-Luc) reporter gene (or other suitable basal expression vectors) to be used in transient transfection assays in responsive cells using Fugene 6 reagent (Roche Molecular Biochemicals, CA). The cell line used for transfections should preferentially be the same as the one used as the starting material for the double ChIP cloning reactions. In the case of ligand-induced nuclear receptors, addition of ligand is necessary to observe the transcriptional activity of the receptor on the TK-Luc gene via the sequence of interest. On the other hand, it would be difficult to study the level of transcriptional activation induced by orphan nuclear receptors via its target regulatory sequences, since the absence of ligand for these receptors imply the lack of a basal transcriptional level to compare with. This problem can be overcome by using a different cell line that does not express the orphan receptor to perform the luciferase assay. However, this alternative might as well become problematic, since the cell line used might lack other factors necessary for transcriptional activation by the receptor on the studied regulatory sequence. The use of siRNA as described earlier constitutes a viable alternative.

The pS2 promoter isolated was cloned upstream of a luciferase reporter gene and transfected in MCF-7 cells that were either treated or not with E_2 (10^{-8} M). It was found that the luciferase gene was transcriptionally activated upon E_2 treatment, confirming the regulatory effect of the isolated pS2 promoter on transcriptional activation of the target gene as previously observed.[13]

At this point, it may be suitable to determine which sequence(s) present in the cloned fragment is the site of action of the nuclear receptor. A panoply of well-developed techniques can be used to characterize the regulatory element.[14]

Conclusion

The development of high-throughput ChIP technology, now permits a whole-genome analysis of gene regulation by nuclear receptors. The efficient

---

[13] D. Lu, Y. Kiriyama, K. Y. Lee, and V. Giguère, *Cancer Res.* **61**, 6755 (2001).
[14] M. Shago and V. Giguère, *Mol. Cell. Biol.* **16**, 4337 (1996).

cloning of new regulatory elements harboring high-affinity-binding sites for specific nuclear receptors can now be efficiently achieved, and allows for the identification of new target genes within a particular cell context. These studies will increase our knowledge of gene regulation via their promoters and enhancers. Since these procedures can be adapted to all nuclear receptors and other transcription factors, it should also be possible to discover transcriptional cross-talks between distinct regulatory networks. For this need, ChIP cloning can be used as a basis for the generation of regulatory regions arrays, allowing ChIP-microarray studies to analyze the binding of different nuclear receptors and transcription factors to these regulatory modules in different cell contexts, such as in normal versus cancer cells. In turn, these studies should lead to a better understanding of many diseases and to the development of better drugs for their treatments.

Acknowledgments

This work was supported by the Canadian Institutes for Health Research (CIHR), the National Cancer Institute of Canada, and Genome Québec/Canada. V. G. is a Senior Scientist of the CIHR.

[20] RNA Gel Shift Assays for Analysis of Hormone Control of mRNA Stability

By Robin E. Dodson, Kathryn M. Goolsby, Maria Acena-Nagel, Chengjian Mao, and David J. Shapiro

Introduction

Until recently, nearly all studies of nuclear receptor action focused on the control of gene transcription. It is now clear that steroid/nuclear receptors have diverse regulatory roles that include control of nuclear gene transcription,[1–3] rapid nongenomic actions resulting in the activation of cell signaling pathways[4–6] and, the focus of this chapter, regulation of the

[1] N. J. Mckenna, R. B. Lanz, and B. W. O'Malley, *Endocr. Rev.* **20**, 321 (1999).
[2] M. G. Rosenfeld and C. K. Glass, *J. Biol. Chem.* **276**, 36865 (2001).
[3] J. G. Moggs and G. Orphanides, *EMBO Rep.* **2**, 775 (2001).
[4] E. R. Levin, *Mol. Endocrinol.* **17**, 309 (2003).
[5] E. Falkenstein, H. C. Tillmann, M. Christ, M. Feuring, and M. Wehling, *Pharmacol. Rev.* **52**, 513 (2000).
[6] K. M. Coleman and C. L. Smith, *Front Biosci.* **6**, D1379 (2001).

cytoplasmic stability of specific mRNAs.[7,8] The level of an mRNA in the cytoplasm represents a balance between the rate at which the mRNA precursor is synthesized in the nucleus and the rates of nuclear RNA processing, and export and cytoplasmic mRNA degradation. While it has long been known that steroid hormones regulate mRNA stability,[9,10] until the last few years progress in this area was limited by the dearth of information on the general pathways of eukaryotic mRNA degradation, the lack of suitable cell culture models, and the absence of appropriate tools for analyzing mRNA stability and the protein–nucleic acid interactions that control this process.

Methods for Analysis of Hormone Control of mRNA Stability

Studies of the hormone regulation of mRNA degradation employ an evolving constellation of technologies. These include: RNAi to "knock-down" the level of candidate proteins;[11,12] expression of proteins from stable and transiently transfected cells to produce candidate proteins; the pulsatile intracellular synthesis of wild-type and mutant mRNAs from regulated promoters combined with the use of quantitative RT-PCR to measure their level and decay rates;[13,14] and the analysis of protein–mRNA interactions.[15] Here we focus on the use of RNA gel mobility shift assays to monitor the levels of hormone-regulated mRNA-binding proteins, and to study their interactions with specific mRNAs.[16–18]

Pathways of Eukaryotic mRNA Degradation

The major mechanism by which eukaryotic mRNAs are destroyed is deadenylation-dependent decapping. In this process, the poly(A) tail at the 3'-end of the mRNA is removed. This is followed by removal of the 5'-cap

[7] R. E. Dodson and D. J. Shapiro, *Prog. Nucleic Acids Res. Mol. Biol.* **72**, 129 (2002).
[8] J. Ross, *Microbiol. Rev.* **59**, 423 (1995).
[9] M. L. Brock and D. J. Shapiro, *Cell* **34**, 207 (1983).
[10] R. D. Palmiter and N. H. Carey, *Proc. Natl. Acad. Sci. USA* **71**, 2357 (1974).
[11] G. J. Hannon, *Nature* **418**, 244 (2002).
[12] S. M. Elbashir, J. Harborth, W. Lendeckel, A. Yalcin, K. Weber, and T. Tuschi, *Nature* **411**, 494 (2001).
[13] H. Kanamori and D. J. Shapiro, *Biotechniques* **26**, 1018 (1999).
[14] D. Foster, R. Strong, and W. W. Morgan, *Brain Res. Prot.*, 7137 (2001).
[15] C. W. J. Smith, "RNA:Protein Interactions." Oxford University Press, 1998.
[16] R. E. Dodson and D. J. Shapiro, *Mol. Cell. Biol.* **14**, 3130 (1994).
[17] R. E. Dodson and D. J. Shapiro, *J. Biol. Chem.* **272**, 12249 (1997).
[18] H. Kanamori, R. E. Dodson, and D. J. Shapiro, *Mol. Cell. Biol.* **18**, 3991 (1998).

structure. In yeast most mRNAs are then degraded in a $5' \rightarrow 3'$ direction, while vertebrate mRNAs may be degraded primarily in a $3' \rightarrow 5'$ direction. Passage through the general mRNA degradation pathway determines the intrinsic stability of many mRNAs, whose stability is not regulated. For the many important cell mRNAs whose stability is regulated, regulation of mRNA degradation can be achieved either through activities exerted at the level of the degradation machinery used in the general pathway, or through unique regulatory mechanisms. One alternative pathway for mRNA degradation especially relevant to studies of RNA–protein interactions is endonucleolytic cleavage within the body of the mRNA, followed by exonucleolytic degradation of the remaining mRNA. A major mechanism for regulating endonuclease decay pathways is by regulating the availability of the endonuclease cleavage site by protecting the site through binding of an mRNA-binding protein that blocks access to the cleavage site.[7,19] Hormones regulate the levels of a number of mRNA-binding proteins and thereby modulate mRNA stability.

Estrogen-Mediated Stabilization of Vitellogenin mRNA

The system we have used as a model for steroid hormone regulation of mRNA stability is estrogen regulation of the stability of the egg yolk precursor protein, vitellogenin. In hepatocytes from *Xenopus laevis*, estrogen induces transcription of the vitellogenin genes and stabilizes vitellogenin mRNA against cytoplasmic degradation. In *Xenopus* liver fragment cultures, vitellogenin mRNA exhibits a half-life of 500 hr when estrogen is present, and 16 hr in the absence of estrogen.[9] Simultaneously, estrogen destabilizes albumin mRNA.[20,21] Estrogen stabilization of vitellogenin mRNA requires estrogen receptor,[22] association of ribosomes with the mRNA,[23] and sequences in the mRNA 3'-untranslated region. The data support a model in which estrogen induces the multi-KH domain nucleic acid-binding protein vigilin/SCP160/DDP1, which then binds to sequences in the 3'-UTR that contain cleavage sites recognized by the estrogen-inducible mRNase PMR1 and thereby blocks cleavage of the mRNA.[7] Albumin mRNA also contains

[19] K. S. Cunningham, R. E. Dodson, M. A. Nagel, D. J. Shapiro, and D. R. Schoenberg, *Proc. Natl. Acad. Sci. USA* **97**, 12498 (2001).

[20] D. R. Schoenberg, J. E. Moskaitis, L. H. Smith, and R. L. Pastori, *Mol. Endocrinol.* **3**, 805 (1989).

[21] R. L. Pastori, J. E. Moskaitis, S. W. Buzek, and D. R. Schoenberg, *Mol. Endocrinol.* **5**, 41 (1991).

[22] D. A. Nielsen and D. J. Shapiro, *Mol. Cell. Biol.* **10**, 371 (1990).

[23] J. E. Blume and D. J. Shapiro, *Nucleic Acids Res.* **17**, 9003 (1989).

PMR1 cleavage sites, but it binds vigilin very poorly and therefore is not protected against degradation by the estrogen-inducible PMR-1.[19] RNA gel shift assays played an important role in establishing major features of this model.

Preparation of Vigilin and Polysome Salt Extracts

Purified Recombinant Vigilin

RNA gel shift studies of the interaction of vigilin and other hormone-regulated RNA-binding proteins with nucleic acids, usually employ either the purified recombinant protein or crude salt extracts of polysomes. Purified recombinant vigilin is prepared by expression of FLAG epitope-tagged, full-length, human vigilin in baculovirus-infected SF-9 cells. Despite its size (155 kDa), vigilin is expressed at high levels using the baculovirus system. We recently described extraction and purification of FLAG vigilin by immunoaffinity chromatography.[19]

Preparation of Polysome Salt Extracts from X. laevis

Proteins extracted from polysomes using high salt represent a rich source of vigilin and many other mRNA-binding proteins. Polysomes were isolated from livers of control male *X. laevis*, or from male *Xenopus* injected in the dorsal lymph sac with 0.2 mg/10 g body weight of 17β-estradiol dissolved in redistilled propylene glycol at 10 mg/ml.

1. The livers are excised, weighed, and perfused with Barth X buffer,[24] or with 0.6 × DME/F12 without serum or antibiotics.
2. The tissue is minced, and homogenized in 2 ml/g of homogenization buffer using five strokes of a teflon-glass homogenizer.
3. The homogenate is sedimented by centrifugation for 5 min at 4°C and 700 × g. The supernatant is collected, sedimented in a Beckman SW41 rotor at 2°C for 2 min at 9000 × g, and then for 16 min at 100,000 × g. The supernatant is retained as cytosol. The pellet is resuspended in homogenization buffer and the centrifugation step is repeated.
4. The crude polysome pellet is resuspended in 1.5 volumes of salt extraction buffer and incubated for 15 min on ice. The homogenate is then diluted with the same volume of buffer used for resuspension but lacking KCl, and sedimented by centrifugation at 300,000 × g for

[24] L. G. Barth and L. J. Barth, *J. Embryol. Exp. Morphol.* **7**, 210 (1959).

60 min at 2°C in an SW55 rotor. The supernatant is aliquoted and stored at −70°C.

Materials and Solutions

SW41 and SW55 ultracentrifuge tubes (Beckman)

Teflon-glass homogenizer (A. C. Thomas and Co.)

Perfusion buffer: Barth X buffer,[24] or 0.6 × DME/F12 cell culture medium without serum or antibiotics

Homogenization buffer: 10 mM HEPES, pH 7.6, 0.25 M sucrose, 5 mM MgCl$_2$, 1 mM DTT, 0.1 mM EGTA, 200 U/ml RNasin [Promega], and protease inhibitors (50 μg/ml leupeptin, 1 μg/ml peptstatin, 5 μg/ml aprotinin, 5 μg/ml PMSF, 20 μg/ml benzamidine)

Salt extraction buffer: 10 mM Tris, pH 7.6, 2.5 mM magnesium acetate, 500 mM KCl, 1 mM DTT, 200 units of RNasin/ml 10% glycerol, 0.1 mM EDTA, 0.1 mM EGTA, and protease inhibitors (50 μg/ml leupeptin, 1 μg/ml pepstatin, 5 μg/ml aprotinin, 5 μg/ml PMSF, 20 μg/ml benzamidine)

RNA gel shift assays show that polysome salt extracts from liver cells of estrogen-treated *Xenopus* exhibit increased binding of vigilin to a crucial segment of the vitellogenin mRNA 3′-untranslated region (Fig. 1). Subsequent studies demonstrated that estrogen elicits a similar induction of *Xenopus* vigilin mRNA and vigilin's mRNA-binding activity.[17]

RNA Gel Mobility Shift Assays

The basic principle of DNA and RNA gel mobility shift assays is that polyacrylamide gel electrophoresis is used to separate a labeled nucleic acid probe from the protein–nucleic acid complex. During the electrophoretic separation, the caging effect of the acrylamide gel is thought to stabilize weak RNA–protein interactions and prevent dissociation of the RNA–protein complex. It is also possible that upon dissociation of the RNA–protein complex, the acrylamide matrix prevents the protein and RNA from rapidly diffusing due to random thermal motion, and their extremely high local concentrations induces very rapid reassociation. A fundamental problem in RNA (or DNA) gel shift assays, using relatively crude mixtures of protein, is to identify levels of unlabeled nonspecific competitors sufficient to bind nonsequence-specific RNA-binding proteins without causing much of the sequence-specific RNA-binding protein of interest to bind to the nonspecific competitors. In RNA gel shift assays, especially assays that employ polysome salt extracts, inhibiting RNases that might otherwise

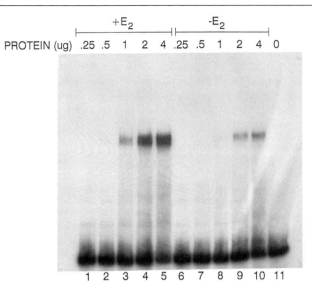

FIG. 1. Estrogen induces binding of vigilin in polysome extracts to the vitellogenin mRNA 3′-untranslated region. An ~120 nucleotide ^{32}P-labeled probe from the 3′ half of the vitellogenin B1 mRNA 3′-untranslated region (lane 11) was incubated with increasing amounts (0.25–4 μg) of polysome salt extract from estrogen-treated ($+E_2$, lanes 1–5) or untreated ($-E_2$, lanes 6–10) *Xenopus* liver. We subsequently identified the 17β-estradiol (E_2) inducible vitellogenin mRNA-binding protein in salt extracts from polysome as vigilin.[17] The autoradiogram is reprinted from Ref. 16 with permission.

degrade the labeled RNA probe is crucial. Because of its sulfated poly-saccharide backbone, heparin is both a reasonably effective nuclease inhibitor, and a useful nonspecific competitor. In our studies with vigilin, we use tRNA as a second nonspecific competitor. Because of its high level of secondary and tertiary structure, tRNA is difficult for many nucleases to cleave, and is therefore useful as an RNase inhibitor and a nonspecific competitor. All of the buffers and glassware used in RNA gel shift assays should be RNase-free. We routinely autoclave solutions for 25 min and prepare labile components with autoclaved water and then sterile filter them.

The protocol presented is the one used to generate the data shown in Fig. 2. General considerations for developing RNA gel shift protocols are presented in the section below on setting up an RNA gel shift assay.

Preparation of the ^{32}P-UTP-Labeled RNA Probe

1. Linearize pB1-15 plasmid by digesting with *Bam*HI.
2. Phenol–chloroform–isoamyl alcohol extract the linearized DNA two times, and ethanol precipitate. Spin down the pellet and rinse it

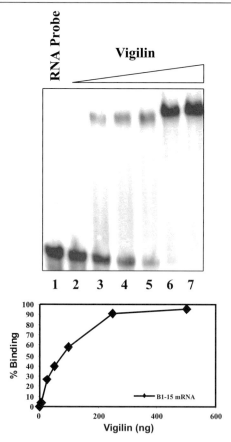

Fig. 2. Binding of purified recombinant human vigilin to the vitellogenin mRNA 3′-untranslated region. Upper panel: 20,000 cpm of vitellogenin B1 mRNA 3′-UTR probe was loaded alone (lane 1) or incubated with increasing concentrations of FLAG epitope-tagged vigilin (lanes 2–7, 5–500 ng, respectively). Lower panel: % binding versus concentration of vigilin. This data was not corrected for the percentage of vigilin that is active and able to bind the probe. % binding was determined by quantitating the volume of the slower migrating RNA species (the RNA–protein complex), and dividing by the total volume of the slower migrating species and the faster migrating species (the free probe).

with 70% ethanol. Resuspend the pellet in water or TE to a final concentration of 0.5 $\mu g/\mu l$.

3. Set up the following reaction at room temperature and add DNA at the end. (If there is spermidine in a buffer, the DNA may precipitate, if it is added at too high concentration or at 4°C):

   5 μl 5× polymerase buffer
   2 μl 100 mM DTT

      1 μl rNTP Mix D
      1 μl RNasin
      30 μCi of ^{32}P-UTP in 3 μl (*see calculation)
      1 μg template in 2 μl volume
      1 μl SP6 RNA polymerase

4. Incubate at 37°C for 1 hr. Add 1 μl of RQ1 DNase. Incubate for 20 min at 37°C.
5. Transfer reaction to a centrisep column to remove free ^{32}P-UTP and use according to the manufacturer's directions (Princeton Separations).
6. Add 1 μl of loading dye. Load the sample onto a 5% denaturing gel, and run in 1 × TBE until the lowest dye is 2/3 of the way down the gel.
7. Remove the top gel plate, wrap the gel and lower plate in saran wrap. In a cassette, align the edge of the X-ray film with the edge of the gel plate. Expose to the X-ray film for approximately 30 sec and develop the film.
8. Wrap film in saran wrap and align under the gel plate, and remove the probe band using an unopened razor blade.
9. To crush the gel slice, use an 18 Ga needle to punch a hole in the bottom of a 0.5 ml conical polypropylene tube, place the gel slice in the 0.5 ml tube, and hang the 0.5 ml tube on the rim of a 1.5 ml conical tube. Centrifuge ~1.5 min at 13,000 rpm in a microcentrifuge. Remove and discard the 0.5 ml tube, and add 0.5 ml of elution buffer to the extruded gel slice in the 1.5 ml tube. Elute the probe overnight at 4°C, or at 37°C, for a few hours. Store the probe at 4°C.

*Calculating the Specific Radioactivity of an RNA Probe.

1. Total ^{32}P-UTP added
      30 μCi of ^{32}P-UTP at 3000 Ci/mmol

Total counts added

$$30 \ \mu\text{Ci} \ (2.2 \times 10^{12} \ \text{dpm/Ci})(0.9 \ \text{cpm/dpm})(10^{-6} \ \text{Ci}/\mu\text{Ci})$$
$$= 5.94 \times 10^{7} \ \text{cpm}$$

Amount added

$$\frac{30 \ \mu\text{Ci}}{(3000 \ \text{Ci/mmol})(10^{6} \ \mu\text{Ci/Ci})} = 10^{-8} \ \text{mmol} = 10 \ \text{pmol}$$

2. Total UTP added
   Cold UTP
   rNTP Mix D has 250 μM UTP

$$\frac{250 \ \mu\text{mol} \ (0.001 \ \text{ml})}{1000 \ \text{ml}} = 0.0025 \ \mu\text{mol} = 250 \ \text{pmol}$$

Total UTP: 250 pmol + 10 pmol = 260 pmol
3. Specific radioactivity of the RNA probe
   Ratio of hot and cold UTP incorporation into probe
   Our probe (B1-15) has 39 U's in a total of 120 nucleotides

$$\frac{5.94 \times 10^7 \ \text{cpm}}{260 \ \text{pmol}} \times \frac{39 \ \text{U's}}{\text{B1-15}} = 8.9 \times 10^6 \ \text{cpm/pmol B1-15}$$

We add 20,000 cpm per reaction:

$$\frac{20,000 \ \text{cpm}}{8.9 \times 10^6 \ \text{cpm/pmol B1-15}} = 2.2 \times 10^{-3} \ \text{pmol B1-15}$$

$$= 2.2 \ \text{fmol B1-15}$$

Materials and Solutions

RNase-free water
rNTP Mix D: 10 mM ATP, 10 mM GTP, 10 mM CTP, 250 mM UTP
   (see sample calculation above to adjust the probe-specific radioactivity with the rNTP mix)
5% Denaturing gel: 14.4 g ultrapure urea, 6 ml 5× TBE, 5 ml 30% 29:1
   acrylamide–bis, 50 μl 30% ammonium persulfate, RNase-free water
   to 30 ml, 20 μl TEMED, 1× TBE
Elution buffer: 66.7 μl of 7.5 M NH$_4$OAc, 10 μl of 1 mM MgSO$_4$, 5 μl of
   20% SDS, 2 μl of 500 mM EDTA, 916 μl of RNase-free water
Plasmid pB1-15
SP6 RNA polymerase
Polymerase buffer
100 mM DTT
RNasin (Promega)
^{32}P-UTP, 3000 Ci/mM
RQ1 DNase (Promega)

Centrisep columns (Princeton Separations)
Loading dye: 90% glycerol, trace bromophenol blue, and xylene cyanol
X-ray film

Gel Shift Assay

1. Just before using the probe: phenol–chloroform extract the probe twice, precipitate with 2 volumes of ethanol by immersing the tube in a dry ice–ethanol bath for 15 min. Pellet the RNA in a microcentrifuge tube, and rinse the pellet with cold 70% ethanol. Resuspend the pellet in water.
2. Assemble following reaction on ice (final volume of 10):
   1× RPI buffer
   tRNA, 1 ng/μl (range in RNA gel shift assays: 0.001–1 μg/μl)
   Heparin 1 ng/μl (range: 0.001–1 μg/μl heparin)
   RNasin 6 U/rxn
   KCl, 60 mM (range: 20–150 mM K$^+$ or Na$^+$)
   RNA probe, 1 μl (20,000 cpm/μl) (Heat to 90°C for ∼2 min just before use, quick cool, and use immediately)
   Protein dilution buffer, 1 μl
3. Incubate for 10 min at 4°C (range: 10 min–2 hr).
4. Add 1 μl of loading dye. Load the samples on the gel and electrophorese at 250 V (about 30 mA) at 4°C, with cold water recirculating, until the xylene cyanol dye is ∼2/3 of the way down the gel (∼3 hr).
5. Remove the gel, and transfer it to filter paper. Dry the gel on a heated gel dryer.
6. Place the dry gel on a phosphor screen and expose for at least 2 hr.
7. Scan the screen using a PhosphorImager and quantitate the shifted bands (ImageQuant software).

Materials and Solutions

RNase-free water
RNasin
10× RPI buffer: 60 mM Tris pH 7.6, 6 mM DTT, 60% glycerol, 10 mM EDTA.
Protein dilution buffer: final concentration, 100 μg/ml BSA, 99.5 mM KCl (for 0.4 ml: BSA, 2 μl of a 20 μg/ml stock; KCl, 398 μl of a 100 mM stock)
RNase-free tRNA: Dilute tRNA (yeast) in water to 10 mg/ml. Add 0.1 volume of 5 M NaCl. Extract four times with phenol–chloroform. Extract once with chloroform–isoamyl alcohol. Precipitate with

2 volumes of ethanol. Resuspend to 10 mg/ml in water. Aliquot and store frozen.

Heparin (Sigma): 10 mg/ml in water. Store frozen.

Purified recombinant FLAG-Vigilin (∼ 1 mg/ml)

RNA probe: 20,000 cpm/μl (2.2 fmol)

BioRad Protean II Gel Box with inner core through which water can be circulated at 4°C. (Recirculating cold water keeps the gel cold, helps stabilize the RNA–protein complex, and improves the sharpness of the bands.)

4°C water cooler

Water circulator

4% Native gel: 6.75 mM Tris pH 7.9, 1 mM EDTA, 3.3 mM NaOAc pH 7.9, 4–8% gel 80:1 acrylamide–bis, 2.5% glycerol, ammonium persulfate, TEMED (after polymerization, cool gel to 4°C and pre-run before use). (Vigilin is a large 155 kDa protein. Assays of smaller RNA-binding proteins may produce better resolution with 6–8% native gels.)

Running buffer: 6.7 mM Tris pH 7.9, 3.3 mM NaOAc pH 7.9, 1 mM EDTA pH 8.0

Gel dryer

Storage phosphor screen

PhosphoImager and ImageQuant software (Molecular Dynamics)

Setting up a New RNA Gel Shift Assay

RNA probes can be from twenty to several hundred nucleotides in length. Because vigilin's preferred RNA-binding sites are > 70 nucleotides long,[18] we use RNA probes 80–150 nucleotides long. For most RNA-binding proteins, shorter probes will be most appropriate. The long-labeled RNA probes used here are not very stable, and should be used within 1–2 days of preparation. Vigilin's preferred RNA-binding sites are relatively unstructured single-stranded sites.[18] For RNA-binding proteins that interact with double-stranded stem and loop structures, tRNA may not be a suitable nonspecific competitor. The precise conditions for RNA gel shifts need to be optimized for each protein. For a high affinity-binding protein, such as vigilin, very high levels of nonspecific competitors (up to 10 μg each of tRNA and heparin) can be used.[16,18] In many systems, generating specific binding with the complex protein mixtures in polysome salt extracts requires high levels of nonspecific competitors.

High salt concentrations and high temperature favor dissociation of RNA–protein complexes. Most RNA gel shift assays use monovalent cation concentrations in the range of 20–150 mM. Mg^{2+} activates some RNases, and we usually leave it out of binding buffers unless the protein requires Mg^{2+} to bind RNA. To minimize RNase activity we generally carry out incubations for the shortest time feasible and at 4°C. The dyes used in loading gel buffers interfere with some RNA–protein interactions and need to be tested on an individual basis. Glycerol alone can be added to load samples.

We usually calculate binding data by quantitating both the shifted and the unshifted band, and determining the percentage of the probe shifted. For protein-excess titrations (as in Fig. 2), binding data may need to be corrected for the percentage of the purified recombinant protein that is able to bind the protein. This can often be determined by titrations using RNA gel shifts in which the labeled RNA, but not the protein, is present in excess.[15]

While techniques, such as surface plasmon resonance, that allow real-time studies of protein–RNA interactions potentially represent a powerful tool for studying hormone-regulated protein–mRNA interactions, there has been little work in this area. RNA gel shift assays remain, perhaps, the most widely used method for analyzing interaction of hormone-regulated nucleic acid-binding proteins with mRNAs.

Acknowledgment

Work from the author's laboratory was supported by NIH grant DK-50080.

[21] Expression Cloning of Ligand Biosynthetic Enzymes

By SHIGEAKI KATO and KEN-ICHI TAKEYAMA

Introduction

Endogenous ligands for steroid and vitamin D nuclear receptors are synthesized from cholesterol by specific converting enzymes expressed in multiple tissues. Many of the enzymes involved in ligand biosynthesis are members of the cytochrome P450 family. Those enzymes that perform the final conversion of ligands into their active forms are thought to play crucial roles in target tissues.

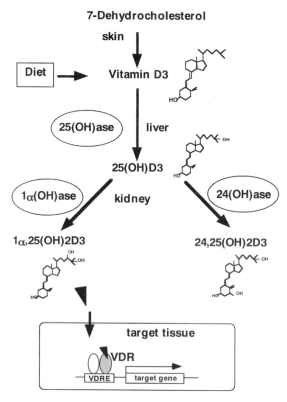

FIG. 1. The schema of vitamin D actions. Vitamin D is metabolized by sequential hydroxylations in the liver and kidney to a family of secosteroids. The two most biologically active forms of vitamin D are $1\alpha,25(OH)_2D_3$ and $24,25(OH)_2D_3$. The binding of $1\alpha,25(OH)_2D_3$ to the nuclear receptor for the hormonally active form of vitamin D (VDR), activates the VDR, with subsequent regulation of physical events, such as calcium homeostasis, cellular differentiation and proliferation.

The active form of vitamin D, $1\alpha,25$-dihydroxycholecalciferol [$1,25(OH)_2D_3$], is a typical receptor ligand with a dedicated biosynthetic pathway (Fig. 1).[1–3] The 1α-hydroxylation of 25-hydroxyvitamin D_3 occurs in the kidney by means of the 25-hydroxyvitamin D_3 1α-hydroxylase (CYP27B1) [$1\alpha(OH)$ase]. The $1\alpha(OH)$ase is the key enzyme in the

[1] A. W. Norman, J. Roth, and L. Orchi, *Endocrinol. Rev.* **3**, 331–366 (1982).
[2] H. F. DeLuca, *Adv. Exp. Med. Biol.* **196**, 361–375 (1986).
[3] M. R. Walters, *Endocrinol. Rev.* **13**, 719–764 (1992).

biosynthesis of $1,25(OH)_2D_3$, which is the most biologically active form of the vitamin D.[4] The $1\alpha(OH)$ase is a cytochrome P450 enzyme that is present in the inner membrane of the mitochondrion.[4-7] High levels of $1\alpha(OH)$ase enzyme activity are expressed in the kidney, and low levels of the enzyme are found in several extrarenal tissues.[5] The expression of $1\alpha(OH)$ase activity is tightly regulated by various factors, such as $1,25(OH)_2D_3$ and calciotropic hormones.[4,6] The latter include parathyroid hormone and calcitonin, which are positive regulators of $1\alpha(OH)$ase expression. Serum calcium and phosphate levels also modulate $1\alpha(OH)$ase activity; however, the regulation by these minerals is indirect. The inhibitory action of $1,25(OH)_2D_3$ on $1\alpha(OH)$ase expression in the kidney is mediated at both the levels of the enzyme and the gene.[6,7] Abnormal vitamin D metabolism occurs in some disease states, including in patients with chronic renal failure.[8] This disregulation is thought to play a role in the development of renal osteodystrophy, a serious complication of renal failure. Moreover, mutations in the $1\alpha(OH)$ase gene that render the enzyme inactive cause vitamin D-dependent rickets type 1.[9,10]

Despite numerous observations on the regulation of the $1\alpha(OH)$ase activity and accumulating clinical interest in the $1\alpha(OH)$ase,[11,12] the cloning of $1\alpha(OH)$ase cDNA was hampered by the enzyme's low abundance in the kidney and its location in the mitochondrial membrane. To overcome these problems and to isolate $1\alpha(OH)$ase cDNAs, we developed an expression-cloning system that exploited a ligand-induced transactivation function of the vitamin D receptor.[13]

[4] S. Kato, J. Yanagisawa, A. Murayama, S. Kitanaka, and K. Takeyama, *Curr. Opin. Nephrol. Hypertension* **7**, 377–383 (1998).

[5] H. L. Henry, *J. Cell Biochem.* **49**, 4–9 (1992).

[6] A. Murayama, K. Takeyama, S. Kitanaka, Y. Kodera, T. Hosoya, and S. Kato, *Biochem. Biophys. Res. Commun.* **249**, 11–16 (1998).

[7] H. Kawashima, S. Torikai, and K. Kurokawa, *Nature* **291**, 327–329 (1981).

[8] K. A. Hruska and S. L. Teitelbaum, *N. Engl. J. Med.* **333**, 166–174 (1995).

[9] S. Kitanaka, K. Takeyama, A. Murayama, T. Sato, K. Okumura, M. Nogami, Y. Hasegawa, H. Niimi, J. Yanagisawa, K. Tanaka, and S. Kato, *N. Engl. J. Med.* **338**, 653–661 (1998).

[10] Fu Gk, D. Lin, M. Y. Zhang, D. D. Bikle, C. H. Shackelton, W. L. Miller, and A. A. Portale, *Mol. Endocrinol.* **11**, 1961–1970 (1997).

[11] O. Dardenne, J. Prud'homme, A. Arabian, F. H. Glorieux, and R. St-Amaud, *Endocrinology* **142**, 3135–3141 (2001).

[12] D. K. Panda, D. Miao, M. L. Tremblay, J. Sirois, R. Farookhi, G. N. Hendy, and D. Goltzman, *Proc. Natl. Acad. Sci. USA* **98**, 7498–7503 (2001).

[13] K. Takeyama, S. Kitanaka, T. Sato, M. Kobori, J. Yanagisawa, and S. Kato, *Science* **277**, 1827–1830 (1997).

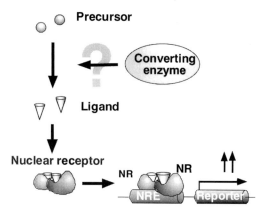

FIG. 2. Converting enzyme produces ligands for nuclear receptor.

Method

Principle

The expression assay is based on the observation that nuclear receptors are activated by their cognate ligands, but not by their precursors or metabolites (Fig. 2). To isolate the $1\alpha(OH)$ase cDNA, we developed a screen that assayed for the conversion of a precursor $[25(OH)D_3]$ into an active form, $[1,25(OH)_2D_3]$, which in turn stimulated transcription mediated by a chimeric vitamin D receptor protein.

Preparation of Mammalian cDNA Expression Library from Kidney of VDR KO Mice

mRNA Isolation

$VDR^{-/-}$ null mutant (VDR KO) mice were generated as described.[14] These mice were used at 7 weeks of age, as $1\alpha(OH)$ase gene expression in the kidney is increased at this time based on the observation that the knockout animals contain 20 times more $1,25(OH)_2D$ in serum, than do wild-type littermates.[14] The kidneys of VDR KO mice (about 1 g wet-weight tissue) were disrupted in 10 ml of 5.5 M GTC solution (Table I), and the DNA sheared by passing the solution through an 18G needle several times. Total

[14] T. Yoshizawa, Y. Handa, Y. Umematsu, S. Takeda, K. Sekine, Y. Yoshihara, T. Kawakami, K. Arioka, H. Sato, Y. Uhciyama, S. Masushige, A. Fukamizu, T. Matsumoto, and S. Kato, *Nature Genet.* **16**, 391–396 (1997).

TABLE I

MEDIA AND REAGENTS FOR EXTRACTION OF TOTAL AND POLY(A)$^+$ RNA

Compound	Volume (final concentration)	Stock solution	Notes
5.5 M GTC solution	10 ml/1 g tissue	5.5 M guanidine thiocyanate, 25 mM sodium citrate/2H$_2$O, 0.5% sodium lauryl sarcosinate, 0.2 M mercaptoethanol* (pH 7.0)	Store at room temperature *Add 0.2 M mercaptoethanol just before extraction
4 M GTC solution	3.6 ml (600 μl/1.5 ml tube)	5.5 M GTC solution 4 ml, dH$_2$O 1.5 ml	Prepare the same day Store at room temperature
CsTFA solution^a	96 ml (16 ml/polyaroma supercentrifuge tube)	Cesium trifluoroacetate ($D = 1.51$) 100 ml, 0.1 M EDTA	Store at 4°C; at room temperature before centrifugation
2 M NaAc	10 μl (0.2 M)	2 M sodium acetate (pH 4.0)	Store at room temperature
TE buffer	600 μl (100 μl/1.5 ml tube)	10 mM Tris–HCl (pH 7.5), 1 mM EDTA	Store at room temperature
2 × Binding buffer	100 μl	20 mM Tris–HCl (pH 7.5), 1 M LiCl, 2 mM EDTA	Store at 4°C
1 × Washing buffer	100 μl	10 mM Tris–HCl (pH 7.5), 0.15 M LiCl, 1 mM EDTA	Store at 4°C
1 × Elution buffer	20 μl	2 mM EDTA	Store at 4°C

^{a}CsTFA is from Pharmacia Biotech (Uppsala, Sweden).
Dynabeads Oligo(dT)25 (5 mg/ml DYNAL#61005-5) and Magnetic Particle Concentrator (DYNAL MPC-E-1 #12004) are used for purification of poly(A)$^+$ RNA.

RNA was extracted by the acid guanidinium thiocyanate–phenol–chloroform method.[15,16] After centrifugation at 3000 rpm for 10 min at 4°C, the supernatant was layered on a 17-ml cushion of CsTFA solution (Table I) in a polyallomer tube, and subjected to ultracentrifugation (25,000 rpm) for 24 hr at 15°C RNA. RNA pellets were suspended in 600 μl of 4 M GTC solution, and total RNA was precipitated by adding 15 μl of 1 M NaAc and 450 μl of 100% ethanol, followed by incubation overnight at -20°C. Precipitates were collected by a 10 min (3000 rpm) centrifugation at 4°C. After resuspension in TE buffer, the $OD_{260/280}$ ratio was measured (1.7–2.0), and the amount of RNA was calculated using the formula, 1 $OD_{260} = 40$ μg/ml.

Polyadenylated RNA [poly(A)$^+$-RNA] was purified by means of oligo(dT) cellulose affinity chromatography (Table I). Total RNA (75 μg in 10 μl H_2O) was denatured at 65°C for 2 min, and 1 mg of Dynabeads that had been prewashed in 2× binding buffer were added. After gentle shaking for several minutes at room temperature, the beads were collected for 30 sec in the Magnetic Particle Concentrator (DYNAL MPC-E-1 #12004). The supernatant was discarded and 200 μl of 1× washing buffer was added to the beads. Beads were collected again in the Magnetic Particle Concentrator. After washing the beads two more times, 10 μl of 1× elution buffer was added, and the mixture was incubated for 2 min at 65°C. The supernatant was then diluted to 100 μl with H_2O and the poly(A)$^+$-RNA concentration was measured by spectroscopy as earlier. Poly(A)$^+$-RNA can be stored at least 6 months at -20°C after precipitation in 10 μl of 2 M NaAc and 250 μl of 100% ethanol.

Preparation of Mammalian cDNA Expression Library

A library was prepared by inserting cDNAs into the pcDNA3 mammalian expression vector,[17] using the protocol accompanying a cDNA synthesis kit from Gibco-BRL[18,19] (Table II). First strand cDNA was synthesized for 1.5 hr at 45°C with a linker primer (oligo)dT containing a *Hin*dIII site, the Superscript II reverse transcriptase[TM], and a methyl–nucleotide mixture as substrates. Next, RNA templates and free RNA were digested by RNase H, and second strand synthesis was

[15] P. Chomczynski and N. Sacchi, *Anal. Biochem.* **162**, 156–159 (1987).
[16] H. Okayama, M. Kawauchi, M. Brownstein, F. Lee, T. Yokota, and K. Arai, *Methods Enzymol.* **154**, 3–28 (1987).
[17] From Invitrogen Corporation (http://www.invitrogen.com).
[18] H. Okayama and P. Berg, *Mol. Cell. Biol.* **2**, 161–170 (1982).
[19] M. Kobori and H. Nojima, *Nucleic Acids Res.* **21**, 2782 (1983).

TABLE II
MEDIA AND REAGENTS* FOR CONSTRUCTION OF cDNA LIBRARY

Compound	Volume (final concentration)	Stock solution	Notes
Superscript II reverse transcriptase	1 μl (20 unit)		Store at −20°C
E. coli RNase H	1.5 μl (2 unit)		Store at −20°C
E. coli DNA polymerase I	10 μl (50 unit)		Store at −20°C
T4 DNA polymerase	3.5 μl (5 unit)		Store at −20°C
T4 DNA ligase	1.5 μl, 1 μl (4 unit)		Store at −20°C
T4 polynucleotide kinase	1 μl (10 unit)		Store at −20°C
Restriction enzyme	3 μl (50–100 unit)	HindIII, SalI	Store at −20°C
0.1 M dithiothreitol (DTT)	2.5 μl, 7.5 μl		Store at −20°C; thaw and keep on ice
Linker primer	1 μl (1.6 μg/μl)	(GA)4CAAGCTTAAG(T)18 in TE buffer	Store at −20°C
RNase inhibitor	0.5 μl (40 U/μl)		Store at −20°C; thaw and keep on ice
dNTP mixture	5 μl	2.5 mM each	Store at −20°C; thaw and keep on ice
γ-ATP	2 μl, 3 μl	10 mM	Store at −20°C; thaw and keep on ice
Adaptor	1 μl	HindIII/SalI d(AGCTTCCCGGG)	Store at −20°C; thaw and dilute
10× First strand buffer	2.5 μl	500 mM Tris–HCl, 750 mM KCl, 30 mM MgCl$_2$	Store at −20°C; thaw and dilute
First strand methyl nucleotide mixture	1.5 μl	10 mM dATP, 10 mM dGTP 10 mM dTTP, 5 mM 5-methyl-dCTP	Store at −20°C; thaw and dilute
10× Second strand buffer	20 μl	188 mM Tris–HCl (pH 8.3), 906 mM KCl, 46 mM MgCl$_2$	Store at −20°C; thaw and dilute
Second strand nucleotide mixture	3 μl	10 mM dATP, 10 mM dGTP 10 mM dTTP, 25 mM dCTP	Store at −20°C; thaw and dilute
10× T4 DNA polymerase buffer	10 μl	500 mM Tris–HCl (pH 8.3), 100 mM MgCl$_2$ 500 mM NaCl, 100 mM DTT	Store at −20°C; thaw and dilute
10× Ligase buffer	2 μl, 3 μl	500 mM Tris–HCl (pH 7.5), 70 mM MgCl$_2$, 10 mM DTT	Store at −20°C; thaw and dilute
10× STE		1 M NaCl, 100 mM Tris–HCl (pH 8.0)	Store at −20°C; thaw and dilute
TE buffer		10 mM EDTA (pH 8.0) 10 mM Tris–HCl (pH 7.5), 1 mM EDTA (pH 8.0)	Store at −20°C; thaw and dilute
LB		Bacto-trypton 10 g, bacto-yeast extract 5 g, NaCl 5 g/H$_2$O 1 liter	Store at room temperature (25°C)

* All reagents are from Gibco-BRL except Linker primer, Adaptor, and LB.

accomplished with *Escherichia coli* DNA polymerase I. The sticky ends of the newly synthesized dsDNA were blunted by bacteriophage T4 DNA polymerase, and then ligated to an adaptor containing *Hin*dIII and *Sal*I sites using T4 DNA ligase. The adaptor was phosphorylated with T4 polynucleotide kinase prior to ligation. The *Hin*dIII site in the linker primer was digested with the cognate restriction enzyme, the digested dsDNAs were fractionated by CROMA SPIN-400 chromatography, and then inserted into the pcDNA3 vector by ligation with the T4 enzyme overnight at 12°C.

Isolation of 1α(OH)ase cDNA by Expression Cloning

Nuclear Receptor Mfcediated Reporter System

This assay is based on the fact that the 25(OH)D$_3$ precursor of vitamin D does not activate the VDR unless hydroxylated by the 1α(OH)ase.[20,21] COS-1 cells were transfected using a lipofection method with the cDNA library made as earlier.[22] At the same time, an expression vector encoding a chimeric protein that included the VDR ligand-binding domain (DEF) fused to the yeast GAL4 DNA-binding domain [GAL4-VDR(DEF)], and a reported plasmid bearing *lac*Z regulated by the GAL4 DNA-binding site (17M2-G-lacZ), were transfected into the COS cells. In addition, because the 1a(OH)ase was known to be a mitochondrial P450, cDNAs encoding two P450 protein cofactors [adrenodoxin (ADX) and adrenodoxin reductase (ADR)] were added[23,24] (Table III). Activation of the VDR by binding of 1,25(OH)$_2$D$_3$ ligand induced *lac*Z expression (Fig. 3), which was detected as β-galactosidase staining.[25] Positively stained cells (Fig. 3B) were harvested by micromanipulation,[25,26] and analyzed by PCR. The amplified products were examined by agarose gel-electrophoresis and cDNA fragments of 2.0–2.5 kbp [predicted size for full-length 1α(OH)ase cDNA] were purified, subcloned into pcDNA3, and screened

[20] M. R. Haussler, G. K. Whitfield, C. A. Haussler, J. C. Hsieh, P. D. Thompson, S. H. Selznick, C. E. Dominguez, and P. W. Jurutka, *J. Bone Miner. Res.* **13**, 325–349 (1998).

[21] S. Kato, *J. Biochem.* **127**, 717–722 (2000).

[22] P. L. Felgner, T. R. Gadek, M. Holm, R. Roman, H. W. Chan, M. Wenz, J. P. Northrop, G. M. Ringold, and M. Danielsen, *Proc. Natl. Acad. Sci. USA* **84**, 7413–7417 (1987).

[23] T. Sakaki, S. Kominami, K. Hayashi, M. Akiyoshi-Shibata, and Y. Yabusaki, *J. Biol. Chem.* **271**, 26209–26213 (1996).

[24] F. J. Dilworth, *et al.*, *J. Biol. Chem.* **270**, 16766–16774 (1995).

[25] M.A Frederick, *et al.*, "Current Protocols in Molecular Biol." Wiley, New York, 1995.

[26] H. S. Tong, *et al.*, *J. Bone Miner. Res.* **9**, 577–584 (1994).

TABLE III

Media and Reagents for X-gal Staining and Amplification of cDNA from Staining Cells

Compound	Volume (final concentration)	Stock solution	Notes
Fixative solution	0.05% (5 ml for 10-cm dish)	25% glutaraldehyde	Store at $-20°C$; thaw and dilute
X-gal solution	5 ml (5 ml for 10-cm dish)	0.1 M K$_3$Fe(CN)$_6$,	Store at 4°C
		0.1 M K$_4$Fe(CN)$_6$/3H$_2$O, 4% X-gal	*Add X-gal just before the assay
		(5-bromo-4-chloro-3-indolyl-β-D-galactoside)*	
10 × PCR buffer	2 μl	100 mM Tris–Cl, pH 8.4, 500 mM KCl,	Store at $-20°C$; keep on ice
		15 mM MgCl$_2$, 0.01% gelatin	
AmpliTaq DNA polymerase	0.2 μl (1 U)	(5 U/μl from Perkin-Elmer Corp.)	
Primers	2 μl (0.5 μM)	5 μM T7 primer	Store at $-20°C$
	2 μl (0.5 μM)	5 μM Sp6 primer	Store at $-20°C$
dNTP	2 μl (250 μM)	dATP, dCTP, dGTP, dTTP (2.5 mM each)	Store at $-20°C$

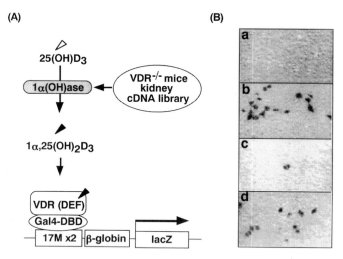

FIG. 3. Isolation of 1α(OH)ase cDNA from the kidney of VDR$^{-/-}$ mice by a novel expression-cloning method (A). (B) β-Galactosidase staining of the transfectants. (a) Nontransfected cells. (b) Effect of active ligand 1α,25(OH)$_2$D$_3$ on cells with expression system, but lacking kidney cDNA library. (c) Detection of 1α(OH)ase-expressing cells transfected with kidney cDNA library. (d) Cells transfected with the cDNA-encoding 1α(OH)ase.

again.[17,27] A total of three independent clones encoding the same enzyme were isolated in this manner from the kidney cDNA library (Fig. 3).

Characterization of $1\alpha(OH)$ase

COS-1 cells were cotransfected with 0.5 μg of GAL4-VDR(DEF), 1 μg of 17M2-G-CAT, 0.5 μg each of the ADX and ADR expression vectors, and 1 μg (+) or 3 μg (++) of the 1α(OH)ase (CYP27B1) expression vector with or without the indicated ligands (Table IV). A representative CAT assay (lower panel) and relative CAT activities (upper panel), corresponding to means ± SEM for three independent transfection experiments[28,29] are shown (Fig. 4A). In Fig. 4B, the metabolic conversion of [^{3}H]25(OH)D$_3$ was analyzed in transfected cells by high pressure

[27] J. Sambrook, E. F. Fritsch, and T. Maniatis "Molecular Cloning," 2nd Ed. Cold Spring Harbor Laboratory Press, 2000.

[28] S. Kato, et al., Science **270**, 1491–1494 (1995).

[29] K. Tayekama, et al., Mol. Cell. Biol. **19**, 1045–1055 (1999).

TABLE IV

MEDIA AND REAGENTS FOR HPLC ASSAY

Compound	Volume (final concentration)	Stock solution	Notes
Chloroform–methanol		Chloroform–methanol (3:1)	Store at room temperature
Normal-phase solvent	15 ml	Hexane–isopropanol–methanol (88:6:6)	Store at room temperature
		Hexane–isopropanol (90:10)	
		Hexane–isopropanol (80:20)	
Reversed-phase solvent	55 ml	Methanol–H$_2$O (80:20)	Store at room temperature
		Methanol–H$_2$O (75:25)	
Vitamin D$_3$ sample		1×10^{-5} M in ethanol	Store at $-20°$C
1α(OH)D$_3$			
25(OH)D$_3$			
24,25(OH)$_2$D$_3$			
1α,25(OH)$_2$D$_3$			
1α,24,25(OH)$_3$D$_3$			
^{3}H-25(OH)D$_3$		10^5 dpm (6.66 tetrabecquerels/mmol) in ethanol	Store at $-20°$C

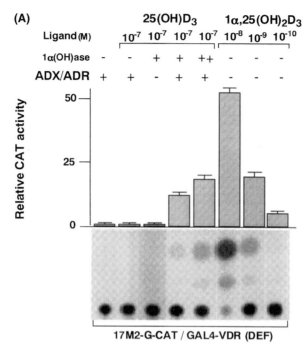

FIG. 4. Conversion of 25(OH)D₃ into an active vitamin D3 serving a VDR ligand by 1α(OH)ase(A), and identification of the converted 25(OH)D₃ by HPLC analysis (B). (A) Analysis of 1α(OH)ase activity by CAT assay. (B) HPLC analysis of 25(OH)D₃ metabolite converted by 1α(OH)ase. Authentic vitamin D derivatives [1α(OH)D₃, 25(OH)D₃, 24,25(OH)₂D₃, 1α,25(OH)₂D₃, and 1α,24,25(OH)₃D₃] (a) for normal phase, and (d) for reverse phase. [³H]25(OH)D₃ was incubated with COS-1 cells transfected with (b, e) or without (c, f) 1α(OH)ase expression vector.

liquid chromatography (HPLC).[30–32] In this experiment, [³H]25(OH)D₃ (10⁵ dpm, 6.66 TBq/mmol; Amersham International) was incubated with COS-1 cells transfected with (b, e) or without (c, f) a 1α(OH)ase expression vector, together with the ADX and ADR plasmids for 6 hr at 37°C. The culture medium were extracted with chloroform–methanol, the organic phase dried, and components were analyzed by normal-phase HPLC (a, b, c) using TSK gel silica 150 column (4.6 × 250 mm, Tosoh Inc.), with hexane–isopropanol–methanol (88:6:6) as the mobile phase at a

[30] M. Burgos-Trinidad, R. Ismail, R. A. Ettinger, J. M. Prahl, and H. F. DeLuca, *J. Biol. Chem.* **267**, 3498–3505 (1992).

[31] M. Warner, et al., *J. Biol. Chem.* **257**, 12995–13000 (1982).

[32] H. L. Henry and A. W. Norman, *J. Biol. Chem.* **249**, 7529–7535 (1974).

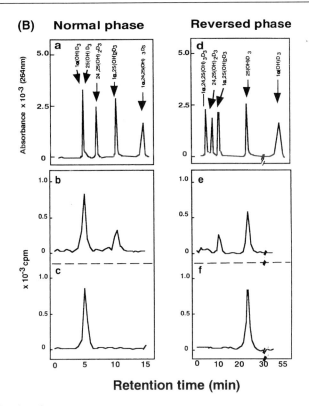

FIG. 4. Continued.

flow rate of 1.0 ml/min (Table IV).[33,34] Eluent fractions were collected and their radioactivity estimated by liquid scintillation counting (Fig. 4B). Authentic vitamin D derivatives [$1\alpha(OH)D_3$, $25(OH)D_3$, $24,25(OH)_2D_3$, $1\alpha,24(OH)_2D_3$, and $1\alpha,24,25(OH)_3D_3$] were chromatographed and their retention times determined by UV absorption at 264 nm [(a) for normal phase, and (d) for reverse phase]. Reverse-phase HPLC (d, e, f) was performed to confirm the presence of [^3H]$1\alpha,25(OH)_2D_3$ using a Cosmosil 5C18-AR packed column (4.6×150 mm, Nakarai Tesque Inc.) with methanol–H_2O (80:20) as a mobile phase and a flow rate of 1.0 ml/min. On both normal- and reverse-phase HPLC performed with other solvent

[33] S. Itoh, et al., Biochem. Biophys. Acta **1264**, 26–28 (1995).
[34] J. Picado-Leonard and W. L. Miller, Mol. Endocrinol. **2**, 1145–1150 (1988).

1   MTQAVKLASRV F HR I HL PLQLDASLGSRGSESVLRSLSDI
mitochondrial target signal

41   PGPSTLSFLAELFCKGGLSRLHELQVHGAARYGPIWSGSF

81   GTLRTVYVADPTLVEQLLRQESHCPERCSFSSWAEHRRRH

121   QRACGLLTADGEEWQRLRSLLAPLLLRPQAAAGYAGTLDN

161   VVRDLVRRLRRQRGRGSGLPGLVLDVAGEFYKFGLESIGA

201   VLLGSRLGCLEAEVPPDTETFIHAVGSVFVSTLLTMAMPN

241   WLHHLIPGPWARLCRDWDQMFAFAQRHVELREGEAAMRNQ

281   GKPEEDMPSGHHLTHFLFREKVSVQSIVGNVTELLLAGVD

321   TVSNTLSWTLYELSRHPDVQTALHSEITAGTRGSCAHPHG

361   TALSQL PLLKAVIKEVLRL YPVVPGNSRVPDRDIRVGNYV
sterol-binding domain

401   IPQDTLVSLCHYATSRDPTQFPDPNSFNPARWLGEGPTPH

441   PFASLPFGFGKRSCIGRRLAELELQMALSQILTHFEVLPE
heme-binding domain

481   PGALPIKPMTRTVLVPERSINLQFVDR*

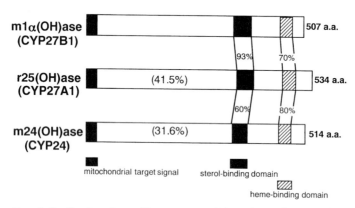

m1α(OH)ase (CYP27B1)   507 a.a.
93%   70%
r25(OH)ase (CYP27A1)   (41.5%)   534 a.a.
60%   80%
m24(OH)ase (CYP24)   (31.6%)   514 a.a.

mitochondrial target signal   sterol-binding domain
heme-binding domain

FIG. 5. Predicted amino acid sequence and homology of mouse vitamin D3 1alpha-hydroxylase [1α(OH)ase]. The mitochondrial target signal is boxed. The sterol-binding domain is underlined. The heme-binding domain is indicated by a dashed line. The 1α(OH)ase protein is homologous to members of the P450 family, particularly to rat vitamin D3 25-hydroxylase [25(OH)ase] (41.5%) and mouse 25(OH)D$_3$ 24-hydroxylase (31.6%).

systems [hexane–isopropanol (90:10), hexane–isopropanol (80:20) for normal phase, methanol–H$_2$O (75:25) for reverse phase], the retention times of the enzymatic product matched those of authentic 1α,25(OH)$_2$D$_3$ (data not shown).

Application

This expression-cloning method, which relies on the conversion of a precursor into the active form of a ligand, is in principle applicable to the isolation of other biosynthetic enzymes. The only requirement for success is that the activity of the ligand produced in the transfected cells must be higher than its precursor in terms of its transactivation function for the nuclear receptor.

Section V

Use of Animal Models to Study Nuclear Receptor Function

[22] Targeted Conditional Somatic Mutagenesis in the Mouse: Temporally-Controlled Knock Out of Retinoid Receptors in Epidermal Keratinocytes

By DANIEL METZGER, ARUP KUMAR INDRA, MEI LI, BENOIT CHAPELLIER, CÉCILE CALLEJA, NORBERT B. GHYSELINCK, and PIERRE CHAMBON

Introduction

The last decade has witnessed an enormous rise in the interest for nuclear receptors (NRs) due to their central role in the coordination of animal development and homeostasis, through their ability to transduce a hormonal signal into modulation of gene activity. The activity of these receptors as regulators of transcriptional responses is controlled by classic hormones and derivatives of vitamin and dietary nutrients, including fatty acids and cholesterol derivatives, with the possible exception of the so-called orphan receptors for which no ligands have been yet identified.[1,2] Nuclear receptors are now seen as integrators of multiple signals and are thought to be involved in the control of most complex processes in metazoans.

In the mammalian skin, the nuclear receptors for vitamin A derivatives (RARs and RXRs), the vitamin D receptor (VDR), the thyroid hormone receptors (TRs), the peroxisome proliferator-activated receptors (PPARs), the liver X receptors (LXRs), the bile acid receptor (FXR), as well as the steroid hormone receptors (ERs, AR, GR), are expressed and may be crucial in its development and homeostasis.[3-10] The skin is composed of the epidermal layer and its appendages (hair follicles), which are separated from the dermal layer by a basement membrane. The epidermis, a

[1] D. J. Mangelsdorf, C. Thummel, M. Beato, P. Herrlich, G. Schutz, K. Umesono, B. Blumberg, P. Kastner, M. Mark, P. Chambon et al., Cell **83**, 835–839 (1995).

[2] V. Laudet and H. Gronemeyer, "The Nuclear Receptor Facts Book." Academic Press, 2002.

[3] G. J. Fisher and J. J. Voorhees, FASEB J. **10**, 1002 (1996).

[4] T. Yoshizawa, Y. Handa, Y. Uematsu, S. Takeda, K. Sekine, Y. Yoshihara, T. Kawakami, K. Arioka, H. Sato, Y. Uchiyama, S. Masushige, A. Fukamizu, T. Matsumoto, and S. Kato, Nature Genet. **16**, 391 (1997).

[5] Y. C. Li, A. E. Pirro, M. Amling, G. Delling, R. Baron, R. Bronson, and M. B. Demay, Proc. Natl. Acad. Sci. USA **94**, 9831 (1997).

[6] Y. C. Li, M. Amling, A. E. Priemel, M. Meuse, J. Baron, R. Delling, and M. B. Demay, Endocrinology **139**, 4391 (1998).

[7] L. G. Kömüves, K. Hanley, Y. Jiang, P. M. Elias, M. L. Williams, and K. R. Feinglod, J. Invest. Dermatol. **111**, 429 (1998).

stratified epithelium made principally of keratinocytes, is a highly dynamic structure.[11] The innermost basal layer that is attached to the basement membrane is a proliferative layer, from which keratinocytes periodically withdraw from the cell cycle and commit to terminally differentiate, while migrating outward into the next layers known as the spinous and granular layers, which together represent the suprabasal layers. Terminally differentiated keratinocytes or squames that reach the skin surface, form the cornified layer or corneum. Squamous keratinocytes are lost daily from the surface of the skin, and are continuously replaced by differentiating cells moving vectorially. Hair follicles that develop through a series of mesenchymal–epithelial interactions during embryogenesis are also dynamic structures. They are mostly composed of keratinocytes, and their outer root sheath (ORS) is contiguous with the epidermal basal layer. Once formed, hair follicles periodically undergo cycles of regression (catagen), rest (telogen) and growth (anagen), through which old hairs are eventually replaced by new ones.[12,13]

Ligand deprivation and pharmacological studies both *in vitro* (keratinocytes in culture) and *in vivo* have indicated that the active retinoid derivatives of vitamin A (retinoic acids, RA) can play crucial regulatory roles in growth, differentiation and maintenance of mammalian epidermis and hair follicles.[3,14] Even though retinoid pharmacological effects have been studied more in skin than in any other tissue, the skin cell-type(s) in which RA regulates gene expression has (have) been elusive until recently.

Retinoids exert pleiotropic effects through two groups of NRs, the Retinoic Acid Receptors (RARα, β and γ) and the Retinoid X Receptors (RXRα, β and γ). RARs bind *all trans-* and *9cis-*RA stereo-isomers, whereas RXRs interact exclusively with *9cis-*RA. RARs, like TRs, VDR, PPARs, LXRs, FXR and several orphan receptors, require heterodimerization with RXRs to regulate target gene expression.[1,2,15–17]

[8] K. Hanley, Y. Jiang, D. Crumrine, N. M. Bass, R. Appel, P. M. Elias, M. L. Williams, and K. R. Feinglod, *J. Clin. Invest.* **100**, 705 (1997).

[9] K. Hanley, D. C. Ng, S. S. He, P. Lau, K. Min, P. M. Elias, D. D. Bikle, D. J. Mangelsdorf, M. L. Williams, and K. R. Feingold, *J. Invest. Dermatol.* **114**, 545 (2000).

[10] K. Hanley, L. G. Kömüves, D. C. Ng, K. Schoonjans, S. S. He, P. Lau, D. D. Bikle, M. L. Williams, P. M. Elias, J. Auwerx, and K. R. Feingold, *J. Biol. Chem.* **275**, 11484 (2000).

[11] E. Fuchs, *Mol. Biol. Cell.* **8**, 189 (1997).

[12] M. H. Hardy, *Trends Genet.* **8**, 55 (1992).

[13] R. Paus and G. Cotsarelis, *New Eng. J. Med.* **341**, 491 (1999).

[14] T. C. Roos, F. K. Jugert, H. F. Merk, and D. R. Bickers, *Pharmacol. Rev.* **50**, 315 (1998).

[15] P. Chambon, *FASEB J.* **10**, 940 (1996).

[16] P. Kastner, M. Mark, and P. Chambon, *Cell* **83**, 859 (1995).

[17] G. M. Morriss-Kay and S. J. Ward, *Int. Rev. Cytol.* **188**, 73 (1999).

RXRα, RXRβ, RARα and RARγ are expressed in epidermis, and RXRα and RARγ are the predominant receptors;[3] however, the precise functions of each receptor in mediating retinoid effects on the epidermis are unknown. In the mouse, the transgenic expression of a dominant-negative (dn) RARα that subverts wild-type RAR functions has suggested that RARs might be involved in keratinocyte differentiation and RA-induced hyperproliferation.[18–20] Furthermore, expression of a dnRXRα in epidermal suprabasal layers indicated that RXRα might also be involved in retinoid-induced cell proliferation in adult mouse skin.[21] As dn receptors may repress the expression of genes even if they are not "normally" repressed by co-repressor-associated, unliganded RXR/RAR heterodimers,[22] and/or interfere through sequestration with a variety of signaling pathways mediated by NRs that heterodimerize with RXRs, the abnormalities exhibited by these transgenic models may not reflect the physiological roles of RA in epidermal homeostasis.

Germline targeted gene disruption through homologous recombination in ES cells has been used extensively to investigate the physiological functions of retinoid receptors in the mouse (reviewed in Refs. 16 and 23). These genetic studies indicate that RARα is apparently dispensable for epidermal homeostasis, whereas RARγ is involved in minor aspects of granular keratinocyte differentiation (our unpublished data). Nevertheless, due to functional redundancy among retinoid receptors (Refs. 16, 24, 25; and references therein), some functions of RARα and/or RARγ in epidermis may have been overlooked. Compound germline disruptions of RARα and RARγ lead to lethality before embryonic day E11.5,[26] thus precluding any analysis of the epidermis. Similarly, it is not possible to investigate the effect of knocking out RXRα on mouse epidermis, as this

[18] S. Imakado, J. R. Bickenbach, D. S. Bundman, J. A. Rothnagel, P. S. Attar, X. J. Wang, V. R. Walczak, S. Wisniewski, J. Pote, and J. S. Gordon, *Genes Dev.* **9**, 317 (1995).

[19] M. Saitou, S. Sugai, T. Tanaka, K. Shimouchi, E. Fuchs, S. Narumiya, and A. Kakizuka, *Nature* **374**, 159 (1995).

[20] J. H. Xiao, X. Feng, W. Di, Z. H. Peng, L. A. Li, P. Chambon, and J. J. Voorhees, *EMBO J.* **18**, 1539 (1999).

[21] X. Feng, Z. H. Peng, W. Di, X. Y. Li, C. Rochette-Egly, P. Chambon, J. J. Voorhees, and J. H. Xiao, *Genes Dev.* **11**, 59 (1997).

[22] J. D. Chen and R. M. Evans, *Nature* **377**, 454 (1995).

[23] M. Mark, N. B. Ghyselinck, O. Wendling, V. Dupé, B. Mascrez, P. Kastner, and P. Chambon, *Proc. Nutr. Soc.* **58**, 609 (1999).

[24] P. Kastner, M. Mark, N. Ghyselinck, W. Krezel, V. Dupé, J. M. Grondona, and P. Chambon, *Development* **124**, 313 (1997).

[25] B. Mascrez, M. Mark, A. Dierich, N. B. Ghyselinck, P. Kastner, and P. Chambon, *Development* **125**, 4691 (1998).

[26] O. Wendling, N. B. Ghyselinck, P. Chambon, and M. Mark, *Development* **128**, 2031 (2001).

mutation leads to lethality at E14.5,[27,28] i.e., at the onset of epidermal morphogenesis.

More generally, germline mutations are inadequate to dissect the functions of NRs in highly pleiotropic signalling pathways. This limitation led us to develop an effective genetic system that allows one to efficiently introduce somatic mutations targeted to a given gene locus, in a selected cell-type, and at a given time of the mouse life. Our strategy is a novel version of the Cre/LoxP system, and is based on cell/tissue-selective expression of chimeric Cre recombinases (Cre-ER[T]s) whose activity is induced by anti-estrogens such as Tamoxifen (Tam), and which are obtained by fusing the Cre recombinase with a mutated ligand-binding domain (LBD) of the human estrogen receptor α (ERα) that binds Tam, but not estrogens.[29] Two such recombinases, Cre-ER[T 30] and Cre-ER[T2,31] harboring G521R and G400V/M543A/L544A mutations in the human ERα LBD, respectively, induce efficient tamoxifen-dependent Cre-mediated recombination in mice (Table IA). Interestingly, the recombinase activity of Cre-ER[T2] is induced with 10-fold lower Tam doses than those required to activate Cre-ER[T].

In this chapter, we illustrate how this conditional mutagenesis approach allowed us to reveal the precise physiological role of nuclear receptors in a given adult tissue. By ablating members of the retinoid receptor family in epidermal keratinocytes, we demonstrated that RXRα plays a key role in the control of hair cycling, most likely through RXRα/VDR heterodimers. Furthermore, our studies revealed that RXRα, probably heterodimerized with other NRs, was implicated in the control of interfollicular keratinocyte proliferation and differentiation, as well as in skin inflammation.[32,33] Importantly, we found that although RAR-mediated signaling pathways were dispensable in epidermis for homeostatic keratinocyte renewal, epidermal hyperplasia induced by topical RA treatment was mediated by

[27] P. Kastner, J. M. Grondona, M. Mark, A. Gansmuller, M. LeMeur, D. Decimo, J. L. Vonesch, P. Dollé, and P. Chambon, *Cell* **78**, 987 (1994).

[28] H. M. Sucov, E. Dyson, C. L. Gumeringer, J. Price, K. R. Chien, and R. M. Evans, *Gene Dev.* **8**, 1007 (1994).

[29] D. Metzger and P. Chambon, *Methods* **24**, 71 (2001).

[30] R. Feil, J. Brocard, B. Mascrez, M. LeMeur, D. Metzger, and P. Chambon, *Proc. Natl. Acad. Sci. USA* **93**, 10887 (1996).

[31] R. Feil, J. Wagner, D. Metzger, and P. Chambon, *Biochem. Biophys. Res. Commun.* **237**, 752 (1997).

[32] M. Li, A. K. Indra, X. Warot, J. Brocard, N. Messaddeq, S. Kato, D. Metzger, and P. Chambon, *Nature* **407**, 633 (2000).

[33] M. Li, H. Chiba, X. Warot, N. Messadeq, C. Gérard, P. Chambon, and D. Metzger, *Development* **128**, 675 (2001).

TABLE I

MOUSE TRANSGENIC LINES EXPRESSING TAMOXIFEN-ACTIVATED CRE-ER RECOMBINASES

Mouse line	Promoter	Tissue/cell selectivity	Reference
(A) Mouse transgenic lines expressing CRE-ER[T 30] and CRE-ER[T2 31]			
CMV-CRE-ER[T]	Cytomegalovirus	Widespread with various efficiencies (skin, kidney, etc. . . .)	30,37
αAT-CRE-ER[T]	α-anti-trypsin	Hepatocytes (∼ 50 % efficiency)	80
K5-CRE-ER[T]	Keratin 5	Basal keratinocytes	35
PrP-CRE-ER[T] (lines 28.4 and 28.6)	Prion protein	Brain (widespread with various efficiencies)	73
PrP-CRE-ER[T] (line 28.8)	Prion protein	Male germ cells	75
Proα2(I)collagen-CRE-ER[T]	Proα2(I)collagen	Fibroblasts	81
Msx2-CRE-ER[T]	Msx2	Ventral ectoderm, apical ectodermal ridge	82
aP2-CRE-ER[T2]	aP2	Adipocytes	71
SM22-CRE-ER[T2]	SM22	Smooth muscle cells	83
K5-CRE-ER[T2]	Keratin 5	Basal keratinocytes	35
K14-CRE-ER[T2]	Keratin 14	Basal keratinocytes	32
Tie2-CRE-ER[T2]	Tie2	Endothelial cells	84
(B) Other mouse transgenic lines expressing Tamoxifen-activated CRE-ER recombinases			
Wnt1-CRE-ER (TM)	Wnt1	Nervous system	65
CAGCRE-ER (TM)	Cytomegalovirus enhancer/chicken β action promoter	Widespread (brain, heart, kidney, etc.)	79
CX-CRE-ER (TM)	Cytomegalovirus enhancer/chicken β action promoter	Widespread (muscle, pancreas, etc.)	85
CreED-30	Ig heavy chain/SV40 early minimal promoter	B cells	64
R26CRE-ER[T]	Rosa 26 KI	Widespread (lung, duodenum, etc.)	76
K14-CreERtam	Keratin 14	Epidermal basal keratinocytes	66
αMHC-MerCreMer	α-MHC	Heart	86
TTR-Cre ind	Transthyretin	Liver	87

RXRα/RARγ heterodimers in suprabasal keratinocytes, which subsequently stimulated basal keratinocyte proliferation through a paracrine signal.[34]

Selective Expression of Cre-ER^{T2} in Basal Keratinocytes: K14-Cre-ER^{T2} Transgenic Mice

To perform temporally controlled conditional somatic mutagenesis in keratinocytes of the epidermis and in the outer root sheath (ORS) of hair follicles, we first generated transgenic mice expressing Cre-ER^{T} under the control of the bovine keratin 5 (K5) promoter.[35] We subsequently generated transgenic mice expressing Cre-ER^{T2} under the control of either the bovine keratin 5 (K5)[35] or the human keratin 14 (K14)[32] promoter. The generation of the K14-Cre-ER^{T2} transgenic mouse line is described in the following paragraph.

The 5.4-kb *Not*I DNA fragment of pK14-Cre-ER^{T2} (Fig. 1A, and Ref. 32) was injected into C57BL/6 × SJL F1 zygotes according to established procedures.[36] The Cre-ER^{T2} transgene was detected in mouse tail DNA of 10 founder mice by PCR, using primers 5'-ATCCGAAAAGAAA ACGTTGA-3' and 5'-ATCCAGGTTACGGATATAGT-3'. PCR amplification was carried out in a buffer containing 10 mM Tris–HCl, pH 8.0, 50 mM KCl, 1.5 mM MgCl$_2$, 0.2 mM dNTPs, 0.25 mM of each primer and 2 U Taq Polymerase, using 1 μg of genomic DNA as template. After 35 cycles (30 sec at 94°C, 30 sec at 55°C), the products were analyzed on ethidium bromide-stained 2.5% agarose gels according to Feil *et al.*[30] (data not shown). Founder mice were bred with C57BL/6 mice to produce transgenic lines.

To analyze the expression pattern of Cre-ER^{T2} in the skin, immunohistochemistry was carried out on tail sections using an anti-Cre antibody. Eight-week-old K14-Cre-ER^{T2} F1 transgenic mice from the 10 lines were intraperitoneally (i.p.)-injected for five consecutive days with tamoxifen (Tam; 1 mg/day) to induce Cre-ER^{T2} nuclear translocation,[37] and tail biopsies were collected the day after the last Tam injection. The Tam stock solution was prepared by suspending 10 mg Tam (free base, ICN) in 100 μl

[34] B. Chapellier, M. Mark, N. Messaddeq, C. Calléja, X. Warot, J. Brocard, C. Gérard, M. Li, D. Metzger, N. B. Ghyselinck, and P. Chambon, *EMBO J.* **21**, 3402 (2002).

[35] A. K. Indra, X. Warot, J. Brocard, J. M. Bornert, J. H. Xiao, P. Chambon, and D. Metzger, *Nucl. Acid Res.* **27**, 4324 (1999).

[36] B. Hogan, F. Costantini, and E. Lacy, "Manipulating the Mouse Embryo: a Laboratory Manual." Cold Spring Harbor Lab, Cold Spring Harbor, NY, 1986.

[37] J. Brocard, X. Warot, O. Wendling, N. Messadeq, J. L. Vonesch, P. Chambon, and D. Metzger, *Proc. Natl. Acad. Sci. USA* **94**, 14559 (1997).

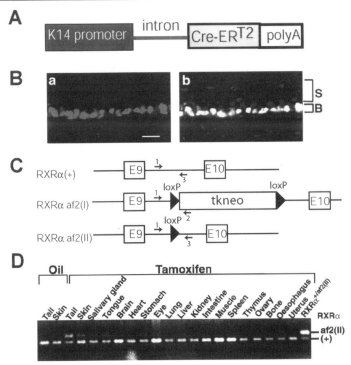

FIG. 1. Characterization of the K14-Cre-ER[T2] transgenic line. (A) Structure of the K14-Cre-ER[T2] transgene. The human K14 promoter, the Cre-ER[T2] coding sequence and the simian virus 40 polyadenylation signal (polyA) are represented by black, grey and open boxes, respectively. The rabbit β-globin intron is depicted by a line. (B) Immunohistochemical pattern of Cre-ER[T2] expression in the tail epidermis of K14-Cre-ER[T2] transgenic mice. The red color (a and b) corresponds to the staining of Cre-ER[T2], and the blue color (b) corresponds to the DAPI staining of the nuclei. The white color of the basal keratinocyte nuclei (b) results from the superimposition of the red color of the anti-Cre signal and the blue color of the DAPI staining. B and S, basal and suprabasal layers, respectively. Scale bar, 25 μm. (C) Genomic structure of the RXRα WT, the RXRα af2(I) target allele and the recombined RXRα af2(II) allele, and PCR strategy to identify the RXRα af2(I) allele and to analyze Cre-mediated excision of the floxed marker. (D) PCR detection of Cre-mediated DNA excision in mice. PCR was performed on DNA isolated from the indicated organs one day after oil- and Tam-treated K14-Cre-ER[T2 (tg/0)]/RXRα[+/af2(I)] mice, as well as from tail of RXRα[+/af2(II)] mice, as indicated. The position of the PCR products amplified from the WT RXRα allele (+) and the RXRα af2(II) allele are shown. (See Color Insert.)

ethanol. Autoclaved sunflower oil was added to obtain a 10 mg/ml emulsion, which was stored at $-20°$C. Tail biopsies were embedded in OCT medium (Sakura Finetek Europe B.V., Zoeterwoude, Netherlands), immediately frozen on dry ice and sectioned with a cryostat (Leica CM 3050). Ten-micrometers-thick longitudinal cryo-sections were mounted on gelatin-coated slides, fixed for 5 min at room temperature in 2% paraformaldehyde, washed in PBT [0.1% Triton X-100 in Phosphate-Buffered Saline (PBS)] and PBS, and incubated in PTB containing 5% normal goat serum (Vector) for 30 min at room temperature. A 1/1000 dilution of the biotin-labeled monoclonal anti-Cre antibody (Cre mAb1; Metzger, D. et al., unpublished results) was applied to the slides for 2 hr at room temperature. After 3 washes in PBT and 3 washes in PBS (10 min each), sections were incubated for 1 hr at room temperature with steptavidine linked to the CY3 fluorochrome (Jackson Immunoresearch, West Grove, PA) at a 1/400 dilution. Slides were washed 3 times 10 min in PBT and 3 times in PBS, and mounting medium for fluorescence (Vectashield, Vector Laboratories, Inc.) containing 0.01% DAPI (4′,6-diamidino-2-phenylindole dihydrochloride—Sigma) was applied. The analysis of pictures taken on a Leica TSD4D confocal microscope revealed that Cre-ER^{T2} expression was restricted to the proliferative keratinocyte basal layer in mice from three K14-Cre-ER^{T2} mouse lines, whereas no significant staining was observed on sections of wild-type mouse skin (Fig. 1B and data not shown). One of these lines apparently expressing Cre-ER^{T2} in most, if not all basal keratinocytes was selected for further studies and was designated as the K14-Cre-ER^{T2} transgenic line. We did not observe any deleterious effects upon Tam i.p. injection (1 mg for five consecutive days), in agreement with reports indicating that short-term tamoxifen treatments have very low acute toxicity and cause no severe abnormalities in mice.[38] Since the translocation of Cre-ER^{T2} from the cytoplasm to the nucleus was as efficient when K14-Cre-ER^{T2} transgenic mice were treated with 5×0.1 mg Tam (data not shown; see also Ref. 35), this lower dose of Tam was routinely used to induce the recombinase activity in K14-Cre-ER^{T2} mice.

The efficiency of excision of a floxed gene segment was analyzed in several organs using a reporter line containing a RXRα allele carrying a floxed tkneo selection marker located into the intron between exon 9 and exon 10 (designated as RXRα$^{af2(I)}$ mice; see Ref. 25, and Fig. 1C). Offspring generated by crossing transgenic K14-Cre-ER^{T2} mice and RXRα$^{af2(I)}$ reporter mice, which were hemizygous for K14-Cre-ER^{T2} transgene (designated as K14-Cre-ER$^{T2(tg/0)}$) and heterozygous for RXRα af2(I) (designated as RXRα$^{af2(I)/+}$) were identified by PCR genotyping of tail biopsies. Primers 1 (5′-CAAGGAGCCTCCTTTCTCTA-3′) and 2 (5′-AAGCGCATGCTCCAGACTGC-3′) were used for the detection of the RXRα$^{af2(I)}$ allele (Fig. 1C). Four week-old K14-Cre-ER$^{T2(tg/0)}$/RXRα$^{af2(I)/+}$ littermates were i.p.-injected once a day with vehicle (oil) or 0.1 mg Tam,

[38] B. J. A. Furr and V. C. Jordan, Pharmac. Ther. 25, 127 (1984).

for five consecutive days. One day after the last injection, mice were killed and genomic DNA was isolated from various organs and analyzed by PCR for excision of the floxed DNA using 1 μg of genomic DNA as template, and primers 1 and 3 (5'-CCTGCTCTACCTGGTGACTT-3') (Fig. 1C). These primers amplify a 156 bp fragment from the RXRα WT allele (+) and a 190 bp fragment from the excised RXRα af2(I) allele that is designated as the RXRα af2(II) allele.[25] After 30 cycles (30 sec at 94°C, 30 sec at 55°C) the products were analyzed on ethidium bromide-stained 2.5% agarose gels. Excision of the floxed marker gene was undetectable in oil-treated control animals, whereas mice injected with Tam reproducibly showed efficient recombination in the tail and skin, as well as in some other epithelia, in which the K14 promoter is also active (for example, tongue, salivary gland, oesophagus, stomach and eye; Ref. 39) (Fig. 1D). These results indicate that Tam treatment of K14-Cre-ER$^{T2(tg/0)}$/RXR$\alpha^{af2(I)/+}$ mice induces efficient excision of the floxed DNA, whereas in the absence of treatment, recombination is undetectable.

The efficiency and kinetics of floxed DNA excision in the epidermis was also analyzed in K14-Cre-ER$^{T2(tg/0)}$/ROSA$^{fl/+}$ double transgenic mice, which are hemizygous for both the K14-Cre-ER^{T2} transgene and the Cre recombinase reporter transgene present in the ROSAR26R transgenic line (Ref. 40; also designated as ROSA$^{fl/+}$), and in which translation of the β-galactosidase encoded in the broadly active Cre reporter transgene occurs only after Cre-mediated DNA excision (Fig. 2A). Eight week-old K14-Cre-ER$^{T2(tg/0)}$/ROSA$^{fl/+}$ double transgenic mice were injected for five consecutive days with tamoxifen (0.1 mg per day). Tail biopsies were taken at various times after Tam injection, embedded in OCT medium, and immediately frozen on dry ice. The analysis of 10 μm-thick longitudinal sections stained with X-gal (5-bromo-4-chloro-3-indolyl β-D-galactoside) according to Takahashi and Coulombe (Ref. 41, and refs. therein), revealed that no X-Gal staining could be detected in the skin of untreated mice (Fig. 2Ba). In contrast, keratinocytes of the epidermis and ORS of hair follicle were stained 5 days after the beginning of Tam injection (Fig. 2Bb). Importantly, 30 and 60 days after Tam treatment, X-Gal staining was observed in all keratinocytes of the epidermis and ORS of hair follicle (Fig. 2Bc and d, respectively). As suprabasal cells are renewed in 7–10 days in the mouse tail epidermis, these results demonstrate that recombination had occurred in most if not all epidermal stem cells. The lack of X-Gal staining in the absence of ligand treatment indicates clearly that the recombinase

[39] X. Wang, S. Zinkel, K. Polonsky, and E. Fuchs, *Proc. Natl. Acad. Sci. USA* **94**, 219 (1997).
[40] P. Soriano, *Nature Genet.* **21**, 70 (1999).
[41] K. Takahashi and P. A. Coulombe, *Proc. Natl. Acad. Sci. USA* **93**, 14776 (1996).

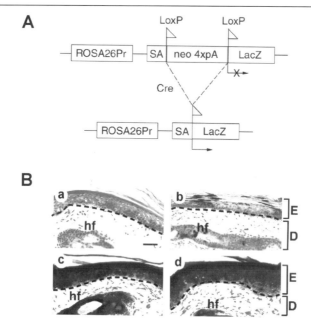

FIG. 2. Characterization of Cre recombinase activity in skin of K14-Cre-ER^{T2} transgenic mice. (A) Structure of the ROSA fl allele before and after Cre-mediated recombination. SA, splice acceptor site; neo 4xpA, neomycin resistance cassette with 4 polyadenylation sites. (B) Kinetics of β-galactosidase expression in tail epidermis of K14-Cre-ER$^{T2(tg/0)}$/ROSA$^{fl/+}$ bigenic mice injected daily with Tam from day 0 to 4. X-Gal-stained tail sections collected just before the first Tam injection (a) and 5 (b), 30 (c) and 60 (d) days after Tam treatment, are presented. D and E, dermis and epidermis, respectively; hf, hair follicle. Scale bar, 16 μm. (See Color Insert.)

activity of Cre-ER^{T2} was fully dependent on Tam binding. The stronger X-Gal staining of suprabasal keratinocytes versus basal keratinocytes (Fig. 2Bc and d) might result from a higher ROSA promoter activity in the former cells, and/or from accumulation of the Lac Z gene product in suprabasal cells over time.

Role of RXRα in Epidermis

Generation of a Floxed RXRα Mouse Line

To carry out the conditional disruption of the RXRα gene in the mouse, we constructed the targeting vector pRXRα^{L2}, which encompasses exons 2 to 4 (E2–E4) and contains a loxP site in the intron located upstream of exon 4, while a tk-neo selection cassette followed by another loxP site is present in the downstream intron 4 (Fig. 3A).[33] Thus, homologous

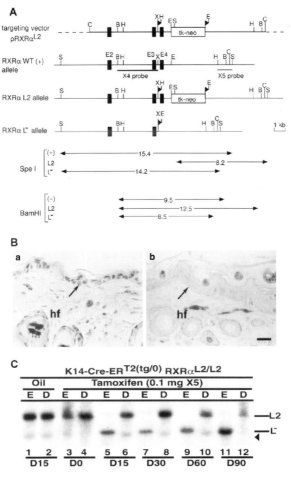

FIG. 3. Temporally controlled RXRα ablation in epidermal keratinocytes. (A) Schematic diagram of the pRXRα^{L2} targeting vector, the RXRα WT (+) genomic locus, the floxed RXRα L2 allele, and the RXRα L⁻ allele obtained after Cre-mediated excision of exon 4. Restriction enzyme sites and the location of X4 and X5 probes are indicated. The numbers in the lower part of the diagram are in kilobases (kb). Abbreviations: B, *Bam*HI; C, *Cla*I; E, *Eco*RI; H, *Hin*dIII; S, *Spe*I; X, *Xba*I. The dashed line corresponds to backbone vector sequences. Arrowheads represent LoxP sites. (B) Immunohistochemical detection of RXRα in skin sections from Tam treated RXRα$^{L2/L2}$ (a) and K14-Cre-ER$^{T2(tg/0)}$/RXRα$^{L2/L2}$ (b) mice. Arrows point to the dermal–epidermal junction. hf, hair follicle. Scale bar, 16 μm. (C) Efficiency of K14-Cre-ER^{T2}-mediated RXRα recombination in adult skin. L2 and L⁻ RXRα alleles were identified by Southern blot on DNA extracted from epidermis "E" or dermis "D" isolated from tail of K14-Cre-ER$^{T2(tg/0)}$/RXRα$^{L2/L2}$ mice before (D0) and 15, 30, 60 and 90 days after the first Tam administration, and 15 days after oil (vehicle) administration, as indicated. The position of the DNA segments corresponding to the RXRα L2 and L⁻ alleles are indicated. The arrowhead points to a nonspecific signal. (See Color Insert.)

recombination (HR) of a wild-type (WT) allele with pRXRα^{L2} should allow a Cre recombinase-mediated excision of exon 4 together with the selection cassette, resulting in deletion of sequences encoding amino acids 149 to 209 that encompass the two zinc finger motifs of the DNA-binding domain.[27,42,43] The 14.3-kb ClaI fragment was purified on a 10–30% sucrose gradient and electroporated into P1 ES cells, as described.[44] After selection with G418, 220 resistant clones were expanded. Genomic DNA was prepared from each clone, restricted with XbaI and analyzed by Southern blotting with probe X5. Nine clones positive for homologous recombination were identified and correct targeting was confirmed by a SpeI digest and probe X4, and a HindIII digest and a neo probe (data not shown). X4 and X5 correspond to 3 kb BamHI–XbaI and 0.5 kb HindIII–SpeI DNA fragments, respectively, isolated from the RXRα genomic clone.[45] Four positive ES clones were injected into C57BL/6 blastocysts, and male chimeras derived from two of them gave germline transmission. Heterozygous and homozygous mice carrying one or two RXRα L2 alleles, respectively, were indistinguishable from WT littermates (data not shown).

Temporally Controlled Selective Disruption of RXRα in Basal Keratinocytes

To selectively disrupt the RXRα gene that is expressed in interfollicular epidermal keratinocytes and hair follicle ORS (see Fig. 3Ba), mice homozygous for the RXRα L2 allele (RXR$\alpha^{L2/L2}$) were bred with hemizygous K14-Cre-ER$^{T2(tg/0)}$ mice to produce K14-Cre-ER$^{T2(tg/0)}$/RXR$\alpha^{L2/L2}$ as well as K14-Cre-ER$^{T2(tg/0)}$/RXR$\alpha^{L2/+}$, K14-Cre-ER$^{T2(0/0)}$/RXR$\alpha^{L2/+}$ and K14-Cre-ER$^{T2(0/0)}$/RXR$\alpha^{L2/L2}$ "control" littermates. Eight-week-old K14-Cre-ER$^{T2(tg/0)}$/RXR$\alpha^{L2/L2}$ mice were i.p.-injected with either vehicle (ethanol) or 0.1 mg Tam per day for five consecutive days. Tail biopsies were taken before and at various times after Tam injection. After treating tail skin with dispase enzyme (Gibco-BRL, 4 mg/ml in PBS at 4°C for 3–4 hr), the epidermis was separated from the dermis.[46] Genotyping of

[42] M. Leid, P. Kastner, R. Lyons, H. Nakshatri, M. Saunders, T. Zacharewski, J. Y. Chen, A. Staub, J. M. Garnier, S. Mader, and P. Chambon, *Cell* **68**, 377 (1992).
[43] D. J. Mangelsdorf, U. Borgmeyer, R. A. Heyman, J. Y. Zhou, E. S. Ong, A. E. Oro, A. Kakizuka, and R. M. Evans, *Genes Dev.* **6**, 329 (1992).
[44] A. Dierich and P. Dollé, *in* "Methods in Development Biology/Toxicology (Symposium Proceedings)," p. 111. Blackwell, Oxford, 1997.
[45] D. Metzger, J. Clifford, H. Chiba, and P. Chambon, *Proc. Natl. Acad. Sci. USA* **92**, 6991 (1995).
[46] K. S. Stenn, R. Link, G. Moellmann, J. Madri, and E. Kuklinska, *J. Invest. Dermatol.* **93**, 287 (1989).

RXRα alleles was performed by Southern blotting on genomic DNA extracted from epidermis and dermis, restricted with *Bam*HI and hybridized with probe X4. Ten days after the last Tam injection (D15) of K14-Cre-ER$^{T2(tg/0)}$/RXRα$^{L2/L2}$ mice, RXRα L2 alleles were found to be fully converted into RXRα L$^-$ alleles in the epidermis, whereas no RXRα L$^-$ alleles could be detected in the dermis of these mice (Fig. 3C), which were therefore designated as RXRα$^{ep-/-}$ mutants. In contrast, no L2 to L$^-$ allele conversion occurred in vehicle-injected animals, or in Tam-treated control animals (Fig. 3C). Moreover, even after 3 months of Tam treatment (D90), only the RXRα L$^-$ allele was detected in K14-Cre-ER$^{T2(tg/0)}$/RXRα$^{L2/L2}$ mouse epidermis, indicating that RXRα was disrupted in most, if not all epidermal stem cells.

Immunohistochemical analysis was performed to verify that the RXRα protein was efficiently ablated in interfollicular epidermis and hair follicles of Tam-treated K14-Cre-ER$^{T2(tg/0)}$/RXRα$^{L2/L2}$ mice. Dorsal skin samples taken 2 months after Tam administration to RXRα$^{L2/L2}$ and K14-Cre-ER$^{T2(tg/0)}$/RXRα$^{L2/L2}$ mice were fixed with periodate–lysine–paraformaldehyde,[47] dehydrated in graded sucrose and embedded into OCT. Ten-micrometer cryosections were blocked in 5% normal goat serum, incubated with biotin-conjugated RXRα monoclonal antibody 4RX3A2,[48] and stained with vectastain ABC kit (Vector) and peroxydase substrate kit DAB (Vector). As expected, the RXRα protein, which was readily detected in interfollicular basal keratinocytes and in the hair follicle of the ORS of control mice (Fig. 3Ba), was not detected in Tam-treated K14-Cre-ER$^{T2(tg/0)}$/RXRα$^{L2/L2}$ mouse keratinocytes (Fig. 3Bb).

Phenotypic Analysis of Mice Lacking RXRα in Epidermal Keratinocytes

Starting 6 weeks after Tam treatment, hair loss (alopecia) was observed in the ventral region of K14-Cre-ER$^{T2(tg/0)}$/RXRα$^{L2/L2}$ mice, but not in oil-treated mice of similar genotype, or in Tam-treated K14-Cre-ER$^{T2(tg/0)}$/RXRα$^{L2/+}$ "control" littermates (identical to WT mice, data not shown). 12–16 weeks after Tam treatment, large regions of ventral skin and smaller regions of dorsal skin were hairless (Fig. 4A and B) and cysts, which enlarged and spread all over the body with time, were visible under the skin surface (Fig. 4C). With increasing age (>20 weeks after Tam treatment), focal lesions with crusts, not caused by fights, appeared on hairless dorsal skin, on chins and behind ears (Fig. 4D).

[47] J. A. Kiernan, "Histological and Histochemical Methods: Theory and Practice," 2nd Ed. Pergamon Press, New York, 1990.

[48] C. Rochette-Egly, Y. Lutz, V. Pfister, S. Heyberger, I. Scheuer, P. Chambon, and M. P. Gaub, *Biochem. Biophys. Res. Commun.* **204**, 525 (1994).

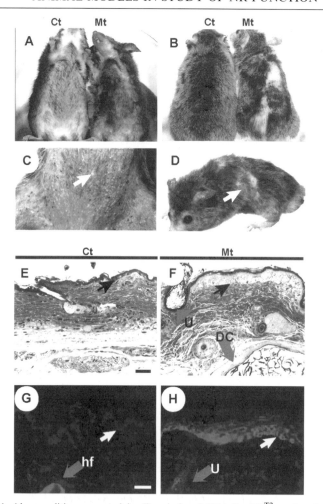

FIG. 4. Abnormalities generated by Tam-induced K14-Cre-ER^{T2}-mediated disruption of RXRα in skin of adult mouse. (A) Ventral view of a female K14-Cre-ER$^{T2(tg/0)}$/RXRα$^{L2/+}$ "control" (ct) mouse (left), and a female K14-Cre-ER$^{T2(tg/0)}$/RXRα$^{L2/L2}$ "mutant" (mt) mouse (right), 16 weeks after the first Tam treatment. (B) Dorsal views of the same animals. (C) Higher magnification of the ventral region of the K14-Cre-ER$^{T2(tg/0)}$/RXRα$^{L2/L2}$ "mutant" mouse, with arrow pointing to one of the cysts. (D) Dorsal view of a female K14-Cre-ER$^{T2(tg/0)}$/ RXRα$^{L2/L2}$ "mutant" mouse, 28 weeks after the first Tam treatment; arrow points to a skin lesion. (E and F) Histological analysis of sections of ventral skin 16 weeks after the first Tam treatment, taken from "control" (E) and "mutant" (F) mice. Scale bar (in E): E and F, 60 μm; (G and H) Keratin 6 (K6) immunohistochemistry on "control" (G) and "mutant" (H) skin sections (16 weeks after the first Tam treatment). The red color corresponds to the staining of the K6 antibody, and the cyan color corresponds to DAPI staining. Scale bar (in G): G and H, 25 μm. Arrows in E–H point to the dermal–epidermal junction. hf, hair follicles; u, utriculi; DC, dermal cysts. (See Color Insert.)

To further analyze the phenotype of these mice, skin biopsies of sex-matched animals were taken from ventral hairless regions of K14-Cre-ER$^{T2(tg/0)}$/RXRα$^{L2/L2}$ and control mice, 16 weeks after Tam treatment. After fixation in glutaraldehyde (2.5% in 0.1 M cacodylate buffer pH 7.2) overnight at 4°C, and post-fixation with 1% osmium tetroxide in cacodylate buffer for 1 h at 4°C, tissue samples were dehydrated with graded concentrations of alcohol and embedded in Epon 812. Histological analysis of 2 μm semi-thin sections, stained with toluidine blue, showed hair follicle degeneration, resulting in utriculi and dermal cysts[49,50] (Fig. 4, compare panels E and F). Interfollicular epidermis was hyperplastic with increased incorporation of BrdU and expression of the Ki67 proliferation marker (Fig. 4F, and data not shown). Dermal cellularity was increased and capillaries were dilated (compare E and F, Fig. 4) underneath the thickened epidermis, reflecting an inflammatory reaction (see also Refs. 32 and 33). Keratin 6 (K6) expression, which "normally" occurs in hair follicle ORS but not in interfollicular epidermis,[50] was analyzed by immunohistochemistry. To this end, after fixation in 2% paraformaldehyde, 10 μm skin frozen sections from control and mutant mice were blocked in 5% Normal Goat Serum (Vector), and incubated with rabbit polyclonal anti-MK6 (Babco). After washing in PBS/0.1% Tween 20, sections were incubated with CY3-conjugated donkey anti-rabbit IgG antibody (Jackson ImmunoResearch), and mounted with Vectashield medium (Vector Laboratories) containing DAPI (Boehringer Mannheim).[37] K6 was aberrantly expressed throughout all layers of hyperproliferative interfollicular epidermis of mutant mice (Fig. 4H), demonstrating that terminal differentiation is altered in epidermal keratinocytes lacking RXRα.

Role of RARs in Epidermis

Temporally Controlled Ablation of both RARα and RARγ in Adult Mouse Epidermis does not Impair Homeostasic Keratinocyte Self-renewal

As RARβ was not detected in epidermis (Ref. 51, and see below) and epidermis of RARβ-null mice is apparently normal,[52] the cell-autonomous requirement of the RAR-mediated signaling pathway in epidermis

[49] J. P. Sundberg and L. E. King, *Dermatol. Clin.* **14**, 619 (1996).

[50] R. M. Porter, J. Reichelt, D. P. Lunny, T. M. Magin, and B. Lane, *J. Invest. Dermatol.* **110**, 951 (1998).

[51] G. H. Fisher, H. S. Talwar, J. H. Xiao, S. C. Datta, A. P. Reddy, M. P. Gaub, C. Rochette-Egly, P. Chambon, and J. J. Voorhees, *J. Biol. Chem.* **269**, 20629 (1994).

[52] N. B. Ghyselinck, V. Dupé, A. Dierich, N. Messaddeq, J. M. Garnier, C. Rochette-Egly, P. Chambon, and M. Mark, *Int. J. Dev. Biol.* **41**, 425 (1997).

homeostasis can be studied in mutant mice lacking RARα and RARγ in keratinocytes. To this end, an RARα targeting vector encompassing exons 3–9 (E3–E9, encoding amino acid residues 1–153 in RARα2, Ref. 53) and containing a floxed neomycin resistance gene upstream of exon 8 and a loxP site downstream of exon 8 was established.[54] After homologous recombination in embryonic stem (ES) cells, clones containing the loxP-flanked L3 RARα allele were identified (data not shown). After electroporation of 5×10^6 cells originated from one of these clones with 10 μg of pSG-Cre expression vector,[55] ES cell subclones in which the floxed marker gene was selectively excised (RARα L2 allele) were identified (Fig. 5A; see also Ref. 54). Chimeric males derived from one of these subclones transmitted the RARα L2 allele through their germ line, yielding the floxed RARα$^{L2/L2}$ mouse line (Fig. 5B). Heterozygous and homozygous mice carrying one or two RARα L2 alleles, respectively, were indistinguishable from WT littermates and expressed normal amounts of RARα (data not shown).

To generate compound mutants, RARα$^{L2/L2}$ mice were bred with RARγ$^{+/-}$[56] and K5-Cre-ER$^{T(tg/0)}$ transgenic mice that express the tamoxifen-inducible Cre-ER^T recombinase specifically in basal keratinocytes,[35] to generate RARα$^{L2/L2}$, RARα$^{L2/L2}$/RARγ$^{-/-}$ and K5-Cre-ER$^{T(tg/0)}$/RARα$^{L2/L2}$/RARγ$^{-/-}$ animals. Fourteen-week-old mice of the various genotypes were subjected to a first tamoxifen treatment (5 days, 1 mg/day), were treated again 2, 4 and 6 weeks later for 5 days, and RARα gene disruption was quantified by Southern blotting.[54] Under such conditions, excision at the RARα L2 locus in Tam-treated mice harbouring the K5-Cre-ER^T transgene was efficient, permanent and selective, as (i) RARα L2 alleles were converted into L$^-$ alleles in tail epidermis, (ii) no RARα L2 allele could be recovered in tail epidermis even 12 months after Tam treatment, and (iii) no excision could be detected in the dermis (thus yielding RARα$^{ep-/-}$ mice; Fig. 5B). To check whether RARβ was aberrantly expressed in epidermis upon RARα and/or RARγ ablation, RARβ RNA levels were determined in skin samples taken 6 and 12 months after Tam administration to RARα$^{L2/L2}$ (control), RARγ$^{-/-}$ and K5-Cre-ER$^{T(tg/0)}$/RARα$^{L2/L2}$/RARγ$^{-/-}$ animals. RNA was prepared using Trizol

[53] P. Leroy, A. Krust, A. Zelent, C. Mendelsohn, J. M. Garnier, P. Kastner, A. Dierich, and P. Chambon, *EMBO J.* **10**, 59 (1991).

[54] B. Chapellier, M. Mark, J. M. Garnier, M. LeMeur, P. Chambon, and N. B. Ghyselinck, *Genesis* **32**, 87 (2002).

[55] J. Clifford, H. Chiba, D. Sobieszczuk, D. Metzger, and P. Chambon, *EMBO J.* **15**, 4142 (1996).

[56] D. Lohnes, P. Kastner, A. Dierich, M. Mark, M. LeMeur, and P. Chambon, *Cell* **73**, 643 (1993).

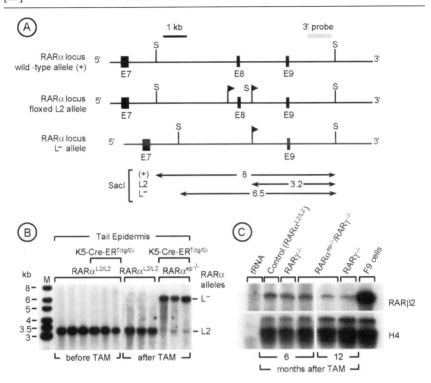

FIG. 5. Conditional mutagenesis of RARα in epidermis. (A) Schematic representation of wild-type RARα locus (+), floxed L2 and excised (L⁻) alleles. Black boxes stand for exons 7 to 9 (E7–9). Restriction sites and location of the 3′ probe are indicated. Sizes of restriction fragments are in kilobases (kb). S, SacI. Arrowhead flags represent loxP sites. (B) Efficiency of Cre-ER^T-mediated RARα gene disruption in K5-Cre-ER^(T(tg/0))/RARα^(L2/L2) mice. Fourteen-week-old mice (genotypes as indicated) received Tam (5 days, 1 mg/day), and were treated again 2, 4 and 6 weeks later. RARα L2 and L⁻ alleles were identified by Southern blotting on tail epidermis genomic DNA restricted by SacI and hybridized with 3′ probe, isolated before and 6 months after Tam injection. M: DNA ladder. (C) RARβ expression in skin samples. RNase protection assay was performed on total RNA (20 μg) from control (RARα^(L2/L2)) and mutant mice (RARγ^(−/−) and RARα^(ep−/−)/RARγ^(−/−)), 6 and 12 months after Tam administration. A tRNA sample and total RNA from RA-treated F9 teratocarcinoma cells were used as controls. A histone H4 protection assay was included for quantitation of the RNA samples.

reagent (Gibco-BRL) and analyzed by RNase protection assay as described.[56] RARβ1/3 isoforms were not detected, whereas similar levels of RARβ2 isoform were present in control and RARα^(ep−/−)/RARγ^(−/−) mice (Fig. 5C). As RARβ is selectively expressed in dermis (Ref. 57; our unpublished data), these data show that it is neither up-regulated in dermis

[57] C. P. F. Redfern and C. Todd, *J. Cell Sci.* **102**, 113 (1992).

nor ectopically expressed in epidermis of $RAR\alpha^{ep-/-}/RAR\gamma^{-/-}$ mice, which therefore exhibit a "panRAR-null" epidermis (panRAR$^{ep-/-}$ mice).

Histology of skin biopsies did not reveal any alteration in adult panRAR$^{ep-/-}$ mice (data not shown, and see next section). Moreover, basal keratinocyte proliferation analyzed by BrdU labeling was similar in control and panRAR$^{ep-/-}$ animals, 12 months after Tam treatment, and there was also no alteration of keratin 5, 6, 10 and 13 expression (data not shown; see Ref. 34). Thus, homeostatic keratinocyte proliferation is maintained for more than a year in absence of $RAR\alpha$, β and γ in epidermis, indicating that RA signaling is not required for self-renewal of epidermal keratinocytes in the adult mouse.

Role of RARs in RA-induced Epidermal Hyperproliferation

Generation of Mice Lacking either RXRα and RARα or RXRα and RARγ in Epidermal Keratinocytes Reveals that the RA Hyperproliferative Signal is Transduced by RXRα/RARγ Heterodimers

Topical RA treatment causes epidermal thickening due to basal cell hyperproliferation.[3] Accordingly, suprabasal layers were thickened when ~ 6 cm^2 shaved dorsal skin of 8–12 week-old female WT mice was topically treated for four consecutive days with 0.4 ml of acetone containing 40 nmoles RA, due to marked increases in numbers of spinous and granular keratinocytes (Fig. 6A and B). Keratinocyte proliferation was evaluated by immunodetection of the proliferation marker Ki67.[58] To this end, frozen sections (10 μm thick) from unfixed skin samples taken the day after the last RA application were postfixed in 2% paraformaldehyde (PFA) in sodium phosphate buffer (PBS), for 10 min at room temperature (RT). The antigen was unmasked by high temperature treatment in 10 mM sodium citrate buffer pH 6.0 and the slides were incubated for 2 hr at RT with an antibody against Ki67 diluted 1/500 in PBS (according to Novocastra's protocol). CY3-conjugated anti-rabbit IgG (Jackson Immunoresearch), diluted 1/400 in PBS was used as secondary antibody. Counterstaining was performed with 0.01% DAPI (4',6-diamidino-2-phenylindole dihydrochloride, Boehringer Mannheim) in Vectashield mounting medium (Vector).

The number of basal keratinocytes expressing the proliferation marker Ki67 was ~ 10-fold higher in RA- than in vehicle-treated epidermis ($56 \pm 5\%$ versus $6 \pm 2\%$, Fig. 6C and D; see also Ref. 34). As males were

---

[58] C. Schlüter, M. Duchrow, C. Wohlenberg, M. H. G. Becker, G. Key, H.-D. Flad, and F. Gerdes, *J. Cell. Biol.* **123**, 513 (1993).

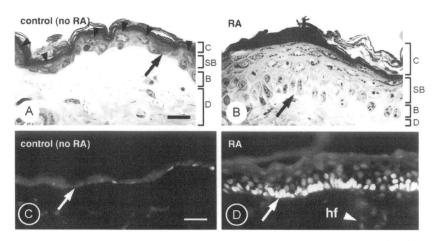

FIG. 6. Histology and proliferative response of wild-type skin upon topical retinoid treatment. Dorsal skin was topically treated for four consecutive days with RA (see text) (B and D) or acetone vehicle (A and C, control) and analyzed 24 hr after the last RA application. (A and B) Histology of control and RA-treated epidermis. Semi-thin sections (2 μm-thick) were stained with toluidine blue. Arrowheads point to the spinous and granular keratinocytes in control epidermis. (C and D) Immunohistochemical detection of the proliferation marker Ki67 (white color) on skin sections counterstained with DAPI (blue color). Arrows point to the dermal–epidermal junction. RA, retinoic acid. B, basal layer; C, cornified layer; D, dermis; hf, hair follicle; SB, suprabasal layers (spinous and granular keratinocytes). Scale bar (in A): 15 μm in A and B; (in C) 25 μm in C and D. (See Color Insert.)

less responsive to RA-induced epidermal hyperproliferation than females (data not shown), female mice were analyzed in the experiments described below.

To determine whether RXRα heterodimerized with either RARα or RARγ transduces RA signaling, RXR$\alpha^{L2/L2}$ mice were bred with either RAR$\alpha^{+/-59}$ or RAR$\gamma^{+/-}$ mice to generate RXR$\alpha^{L2/L2}$/RAR$\alpha^{+/-}$ or RXR$\alpha^{L2/L2}$/RAR$\gamma^{+/-}$ mice. These animals were further mated with K5-Cre-ER$^{T(tg/0)}$ transgenic mice (see above) to generate RXR$\alpha^{L2/L2}$, RAR$\alpha^{-/-}$, RAR$\gamma^{-/-}$, K5-Cre-ER$^{T(tg/0)}$/RXR$\alpha^{L2/L2}$, K5-Cre-ER$^{T(tg/0)}$/RXR$\alpha^{L2/L2}$/RAR$\alpha^{-/-}$ and K5-Cre-ER$^{T(tg/0)}$/RXR$\alpha^{L2/L2}$/RAR$\gamma^{-/-}$ animals, which were treated with Tam (1 mg/day from day 1 to day 4) at 14 weeks of age (Fig. 7A). Under these conditions, RXRα expression was efficiently and selectively abolished in epidermis of RXR$\alpha^{L2/L2}$ mice bearing the K5-Cre-ER^T transgene, which were therefore designated as RXR$\alpha^{ep-/-}$ mutant mice (data not shown; see also Fig. 9A). The epidermis of

[59] T. Lufkin, D. Lohnes, M. Mark, A. Dierich, P. Gorry, M. P. Gaub, M. LeMeur, and P. Chambon, *Proc. Natl. Acad. Sci. USA* **90**, 7225 (1993).

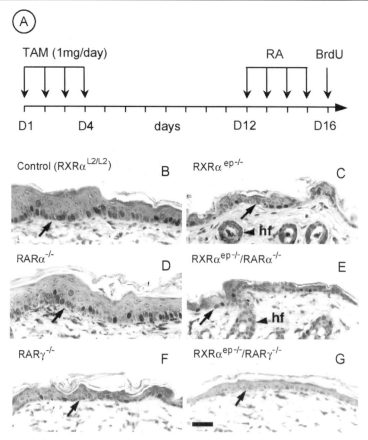

FIG. 7. RA-induced proliferative response in the skin of retinoid receptor mutants. (A) Scheme of the experimental protocol. I.P. injection of Tam (1 mg) was done from day 1 (D1) to day (D4). 0.4 ml of 40 nmoles RA in acetone was topically applied on ∼9 cm² shaved back skin once a day from D12 to D15. BrdU (50 mg/kg) was injected on D16, 2 hr before skin sampling. (B to G) Representative skin sections from mice of the indicated genotype, labelled with BrdU (brown color). Arrows point to the dermal–epidermal junction. hf, hair follicles. Scale bar (in G): 50 μm in B to G. (See Color Insert.)

RA-untreated RXRα$^{ep-/-}$ mice did not exhibit interfollicular hyperplasia shortly after the Tam treatment (data not shown), in agreement with results described above, and showing that hyperproliferation is a late event occurring 10–12 weeks after RXRα ablation (see also Ref. 32).

RA was topically applied as above on ∼6 cm² shaved dorsal skin of mice of the various genotypes from day 12 to 15, and a BrdU injection (50 mg/kg) was given on day 16, 2 hr before skin sampling. In skin sections from "control" mice (RXRα$^{L2/L2}$), about half (54 ± 6%) of the basal keratinocytes were BrdU-positive after RA treatment (Fig. 7B; in contrast, $5 \pm 2\%$ of the basal keratinocytes were BrdU-positive in vehicle-treated mice). RA-induced epidermal thickening and cell hyperproliferation in $RAR\alpha^{-/-}$ mice was similar to that of controls (Fig. 7D; BrdU-positive cells, $50 \pm 4\%$), while it was much reduced in $RAR\gamma^{-/-}$ mice (Fig. 7F; BrdU-positive cells, $12 \pm 3\%$) and in $RXR\alpha^{ep-/-}$ mice (Fig. 7C; BrdU-positive cells, $12 \pm 4\%$). Ablation of RXRα in epidermis of $RAR\alpha^{-/-}$ mice (i.e., $RXR\alpha^{ep-/-}/RAR\alpha^{-/-}$ mice) did not further decrease RA-induced proliferation (Fig. 7E; BrdU-positive cells, $13 \pm 4\%$). In contrast, RA treatment had almost no effect in $RXR\alpha^{ep-/-}/RAR\gamma^{-/-}$ mice (Fig. 7G; BrdU-positive cells, $7 \pm 2\%$). Thus, RXRα/RARγ, but not RXRα/RARα heterodimers, appear to be required within the epidermis to mediate RA-induced hyperplasia.

Selective Ablation of RARγ in Suprabasal Keratinocytes Abrogates RA-induced Epidermal Hyperplasia

To determine in which epidermal cell layer (i.e., basal or suprabasal) RA exerts its primary effect, we ablated RARγ in suprabasal layers. A targeting vector encompassing exons 8–13 (E8–E13, encoding amino-acid residues 63–392; Ref. 60) and containing a loxP site upstream of exon 8, as well as a floxed neomycin resistance gene downstream of exon 8 was constructed (see Ref. 61). After homologous recombination in ES cells, clones containing a floxed RARγ L3 allele were identified by Southern blotting (Fig. 8A). Chimeric males derived from one of these ES clones transmitted the RARγ L3 allele through their germline, yielding the floxed $RAR\gamma^{L3/L3}$ line (Fig. 8C). $RAR\gamma^{L3/L3}$ mice were crossed with CMV-Cre-$ER^{T(tg/0)}$ hemizygous transgenic mice, in which Tam administration can induce Cre-mediated recombination in suprabasal layers but not in the basal layer,[37] to produce "control" (CMV-Cre-$ER^{T(tg/0)}/RAR\gamma^{+/L3}$) and experimental (CMV-Cre-$ER^{T(tg/0)}/RAR\gamma^{L3/L3}$) animals. At 14 weeks of age, all mice were treated with Tam (1 mg at day 1, 2 and 3, and then every second day for 10 days; Fig. 8B), and RARγ gene disruption was quantified by Southern blotting on D9. No RARγ L⁻ allele was found before Tam treatment (Fig. 8C, left panel). As expected from the selective expression of the CMV-Cre-ER^T transgene in suprabasal (sb), but not in basal keratinocytes,[37] this treatment resulted only in a partial conversion of RARγ L3 alleles into L2Neo (in which the neo gene was not removed, see Fig. 8A) and L⁻ alleles (Fig. 8C, middle panel). Importantly, no RARγ L⁻

[60] A. Zelent, A. Krust, M. Petkovich, P. Kastner, and P. Chambon, *Nature* **339**, 714 (1989).
[61] B. Chapellier, M. Mark, J. M. Garnier, A. Dierich, M. LeMeur, P. Chambon, and N. B. Ghyselinck, *Genesis* **32**, 95 (2002).

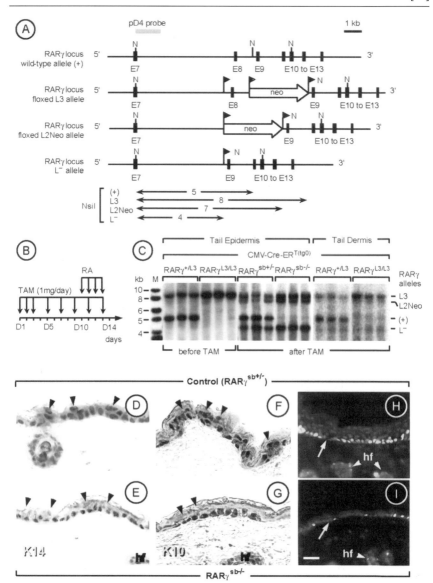

FIG. 8. Conditional mutagenesis of RARγ and RA-induced proliferative response in mice lacking RARγ in suprabasal layers (RARγ$^{sb-/-}$ mice). (A) Schematic drawing of RARγ wild-type (+), L3, partially excised L2Neo and fully excised L$^-$ alleles. Sizes of *Nsi*I fragments obtained for each allele are in kilobases (kb). Black boxes, exons 7 to 13 (E7–13); neo, neomycin resistance gene; N, *Nsi*I. Arrowhead flags represent loxP sites. (B) Experimental protocol. (C) Efficiency of RARγ gene disruption in mice bearing the CMV-Cre-ER^T transgene. RARγ wild-type (+), L3, L2Neo and L$^-$ alleles were analyzed on tail epidermis and dermis genomic

and L2Neo alleles could be detected several weeks after the end of Tam treatment (data not shown), proving that no disruption of the RARγ gene had occurred in basal keratinocytes (see Fig. 8E and G). Thus, Tam administration to CMV-Cre-ER$^{T(tg/0)}$/RARγ$^{L3/L3}$ mice effectively resulted in selective disruption of the RARγ gene in epidermal suprabasal layers, yielding RARγ$^{sb-/-}$ mice.

To verify that the RARγ protein was selectively and efficiently ablated in suprabasal keratinocytes, RARγ was analyzed immunohistochemically on skin samples collected at day 9. To this end, 10 μm thick frozen sections collected onto SuperFrost Plus slides (Menzel Gläser; Germany) were fixed in Zamboni's fluid (2% paraformaldehyde, 0.21% picric acid in 0.15 M sodium phosphate buffer (PBS), pH 7.3) for 10 min at room temperature (RT), and rinsed in PBS-0.05% Tween 20 (PBT). To block nonspecific binding, slides were incubated for 30 min at RT in PBT containing 10% heat-inactivated normal goat serum, and overnight at 4°C with the purified rabbit polyclonal antibody RP453[62] diluted 1/1000 in PBT plus goat serum. Biotinylated anti-mouse IgG (Vector), diluted 1/400 in PBS was used as secondary antibody and revealed using an ABC system (Vector, USA) according to the manufacturer's instructions, and images were acquired with a CCD camera. To identify basal and suprabasal keratinocytes, the sections were probed with antibodies for keratin 14 and keratin 10, respectively, after removing the anti-RARγ antibodies. To this end, the slides were washed in water, incubated for 1 hr in 0.1 M glycine–HCl (pH 3), rinsed in PBS, fixed again for 10 min in Zamboni's fluid, washed in PBS and stored overnight at 4°C. Nonspecific binding of the anti-keratin antibodies was prevented by incubating frozen sections with 5% heat-inactivated goat serum diluted in PBT for 30 min. Sections were then incubated with either Keratin 10 or 14 commercial antibodies (Covance, BabCo) at 1/500 and

---

DNA before and after Tam administration. Note that the minute amount of RARγ excised L$^-$ allele present in dermis most probably originates from contaminating epidermal keratinocytes.[32] M: DNA ladder. (D to G) Immunohistological detection of RARγ (brown color), K14 (D and E; yellow false color) and K10 (F and G; yellow false color) on back skin sections taken at D9 from control (D and F; RARγ$^{sb+/-}$) and mutant (E and G; RARγ$^{sb-/-}$) mice. Arrowheads in D and F point to suprabasal keratinocytes that express RARγ, whereas arrowheads in E and G point to suprabasal keratinocytes that do not express RARγ. (H and I) Immunohistological detection of the Ki67 proliferation marker on skin sections taken at D14 from control (H) and mutant (I) animals (white color, Ki67 signal; blue color, DAPI nuclear staining). Arrows point to the dermal–epidermal junction. hf, hair follicles. Scale bars (in I): 50 μm. (See Color Insert.)

---

[62] C. Rochette-Egly, Y. Lutz, M. Saunders, I. Scheuer, M. P. Gaub, and P. Chambon, *J. Cell. Biol.* **115**, 535 (1991).

1/1000 dilutions, respectively, as described.[32,33] CY3-conjugated anti-rabbit IgG (Jackson Immunoresearch), diluted 1/400 in PBS was used as secondary antibody. Digital images were generated using a CCD camera and assembled with those obtained with anti-RARγ antibodies, using Photoshop Adobe software.

RARγ protein was detected easily throughout the epidermal layers and in hair follicles of "control" mice, whereas as expected K14 and K10 keratins were found in basal and suprabasal layers, respectively (Fig. 8D and F). RARγ was detected in RAR$\gamma^{sb-/-}$ mice in basal and hair follicle keratinocytes, but not in suprabasal keratinocytes (Fig. 8E and G), thus demonstrating that RARγ disruption was indeed restricted to suprabasal cells.

RA was then applied topically on shaved RAR$\gamma^{sb-/-}$ back skin for 4 days (D10–D13) and keratinocyte proliferation was evaluated on D14, using sections post-fixed in 2% paraformaldehyde and an antibody against the proliferation marker K67. In RA-treated "control" mice, the epidermis was thickened markedly and about half ($56 \pm 5\%$) of basal keratinocytes expressed Ki67 (Fig. 8H), instead of $6 \pm 2\%$ in vehicle-treated mice (Fig. 6C). In contrast, epidermal proliferation was much lower in RA-treated RAR$\gamma^{sb-/-}$ mice (Fig. 8I, $13 \pm 4\%$), a situation similar to that observed in RAR$\gamma^{-/-}$ mice (data not shown, $14 \pm 3\%$; see also Fig. 7F). As expected, RA-induced epidermal hyperproliferation in RAR$\gamma^{sb-/-}$ mice was restored to "control" levels several weeks after the end of Tam treatment (data not shown), indicating that the RARγ gene was not disrupted in the basal cell layer.

Together, these results indicate that RARγ expressed in suprabasal keratinocytes is required to transduce the retinoid signal that triggers proliferation of basal keratinocytes.

Selective Disruption of the RXRα Gene in Suprabasal Keratinocytes does not Result in RXRα Protein Depletion

To determine whether RXRα could be the heterodimeric partner of RARγ in suprabasal keratinocytes, RXR$\alpha^{L2/L2}$ mice were crossed with hemizygous CMV-Cre-ER$^{T(tg/0)}$ transgenic mice, to produce "control" (CMV-Cre-ER$^{T(tg/0)}$/RXR$\alpha^{+/L2}$) and "experimental" (CMV-Cre-ER$^{T(tg/0)}$/RXR$\alpha^{L2/L2}$) mice. Before Tam administration, as expected, no RXRα L$^-$ allele was found in epidermis (Fig. 9A, left panel), whereas a similar Tam treatment as in Fig. 8B induced a partial conversion of RXRα L2 alleles into L$^-$ alleles in "experimental" mice (Fig. 9A, right panel). The remaining L2 alleles originated from basal keratinocytes, as this species was absent in DNA from an RXR$\alpha^{ep-/-}$ mouse skin sample. Importantly, as in the case for the RAR$\gamma^{L3/L3}$ gene (see previous paragraph), no RXRα

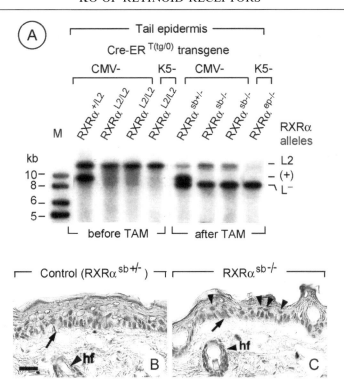

FIG. 9. Conditional mutagenesis of RXRα in epidermis suprabasal layers, and RA-response in RXRα$^{sb-/-}$ mice. (A) Efficiency of RXRα gene disruption in mice bearing the CMV-Cre-ER^T transgene as compared to the K5-Cre-ER^T transgene (as indicated). The experimental protocol is the same as in Fig. 8B. RXRα wild-type (+), L2 and L$^-$ alleles were analyzed on tail epidermis genomic DNA before and after Tam administration. M: DNA ladder. (B and C) Immunohistochemical detection of RXRα on skin sections from control (B, RXRα$^{sb+/-}$) and experimental (C, RXRα$^{sb-/-}$) mice, after RA treatment. Arrowheads indicate nuclei of suprabasal keratinocytes expressing RXRα. Arrows point to the dermal–epidermal junction. hf, hair follicles. Scale bar (in B): 50 μm. (See Color Insert.)

L$^-$ allele could be detected several weeks after the end of Tam treatment (data not shown), indicating that the RXRα gene was not disrupted in the basal cell layer.

Thus, Tam administration successfully induced disruption of the floxed RXRα gene in suprabasal, but not basal keratinocytes of mice bearing the CMV-Cre-ER^T transgene, yielding RXRα$^{sb-/-}$ mice. Nevertheless, at D14 RXRα was detected readily by immunohistochemistry throughout epidermis, including in suprabasal keratinocytes (arrowheads in Fig. 9C) of both RA-treated "control" and RXRα$^{sb-/-}$ mice (compare Fig. 9B with 9C). Thus, the RXRα protein was still present in epidermis suprabasal cells of experimental mice, even though Tam administration induced efficient Cre-mediated disruption of the RXRα L2 alleles in these cells. This result indicates that there is little turnover of RXRα mRNA and/or protein upon differentiation of suprabasal cells from basal cells. Accordingly, the RA-induced proliferation was similar in Tam-treated "control" and "experimental" (RXR$\alpha^{sb-/-}$) mice (Figs. 9B and C).

Conclusion and Future Directions

We demonstrate here the effectiveness of Cre-ER^T recombinases for generating *in vivo* temporally controlled targeted somatic mutations in adult mouse keratinocytes. As the recombinase activity of Cre-ER^{T2} is induced by tamoxifen doses 10-fold lower than those required to activate Cre-ER^T or Cre-ER fusion proteins described by others (Refs. 63–66; and Table I), Cre-ER^{T2} should be used preferentially to generate spatio-temporally controlled targeted somatic mutations in the mouse.

The selective ablation of RXRα, RARα and RARγ in epidermal keratinocytes of adult mice revealed that RXRα and RARγ play several important functions that could otherwise not be unveiled through germline mutagenesis, as both RXRα-null and compound RARα/RARγ-null fetuses die before birth. The data described here establish that (i) selective disruption of RXRα in epidermal keratinocytes of the mouse results in an alopecia, a hyperproliferation of interfollicular epidermis, an abnormal differentiation of keratinocytes and a skin inflammatory reaction, (ii) as this alopecia is similar to that generated by disruption of the vitamin D3 receptor (VDR) in mice[4–6,67] and by mutations of this receptor in humans,[68] it is highly probable that RXRα/VDR heterodimers play a crucial function in the control of the hair cycle,[32,33] (iii) none of the other skin abnormalities generated by RXRα ablation in epidermal keratinocytes are exhibited by mice in which RARα and/or RARγ are disrupted in these cells. As it has been shown that RARβ is not expressed in the epidermis,[34] it follows that the complete absence of RARs in keratinocytes of adult mice does not

[63] Y. Zhang, C. Riesterer, A. M. Ayrall, F. Sablitzky, T. D. Littlewood, and M. Reth, *Nucl. Acid. Res.* **24**, 543 (1996).

[64] F. Schwenk, R. Kühn, P. O. Angrand, K. Rajewsky, and A. F. Stewart, *Nucl. Acid. Res.* **26**, 1427 (1998).

[65] P. S. Danielian, D. Muccino, D. H. Rowitch, S. K. Michael, and A. P. McMahon, *Curr. Biol.* **8**, 1323 (1998).

[66] V. Vasioukhin, L. Degenstein, B. Wise, and E. Fuchs, *Proc. Natl. Acad. Sci. USA* **96**, 8551 (1999).

[67] Y. Sakai and M. B. Demay, *Endocrinology* **141**, 2043 (2000).

[68] P. J. Malloy, W. Pike, and D. Feldman, *Endocrine Rev.* **20**, 156 (1999).

alter their homeostatic self-renewal. Further keratinocyte-specific targeted somatic mutagenesis in adult mice will be necessary to reveal which other signaling pathway(s), involving NRs known to heterodimerize with RXRs and to be present in epidermis and hair follicles [e.g., LXRs[9] and PPARs (Ref. 69; and refs. therein)], are implicated in the generation of the skin abnormalities that cannot be ascribed to the lack of RXRα/VDR heterodimers, (iv) even though RARs do not appear to be required for the control of homeostatic epidermal keratinocyte proliferation, the hyperproliferative effect of topical RA administration is mediated through RXRα/RARγ heterodimers located in suprabasal cells, thus revealing that the RA-induced hyperproliferation of basal keratinocytes involves a paracrine signal(s) induced by RA in suprabasal cells. This paracrine system may operate under conditions of stress, such as wound healing, as it is known that vitamin A deficiency causes delayed wound healing, whereas pretreatment with either vitamin A or RA improves epidermal regeneration.[70]

The final most important question is whether the present spatio-temporally controlled somatic mutagenesis procedure is applicable to genes in any cell-type for which a promoter(s) has been shown to be selectively active. The answer is most probably affirmative as spatio-temporally controlled somatic mutagenesis of RXRα and other floxed genes has been efficiently achieved in our laboratory for several cell-type/tissue, notably for adipocytes,[71] hepatocytes,[72] skeletal muscle (M. Schuler, D. Metzger and P. Chambon, unpublished data), prostatic epithelial cells (M. Jiang, D. Metzger and P. Chambon, unpublished data), nervous system (Ref. 73, and data not shown), and testis germ cells,[74,75] while an increasing number of Cre-ER[T] or Cre-ER[T2] cell specifically expressing transgenic lines have been or are being generated by others (Table I, and unpublished results). These lines, combined with the increasing number of established lines harboring floxed nuclear receptor genes (Table II), should help elucidate the *in vivo* functions of nuclear receptors, as well as the contribution of nuclear receptor signaling to pathophysiology.

[69] L. Michalik, B. Desvergne, N. S. Tan, S. Basu-Modak, P. Escher, J. Rieusset, J. M. Peters, G. Kaya, F. J. Gonzalez, J. Zakany, D. Metzger, P. Chambon, D. Duboule, and W. Wahli, *J. Cell Biol.* **154**, 799 (2001).

[70] T, K. Hunt, *J. Am. Acad. Dermatol.* **15**, 817 (1986).

[71] T. Imai, M. Jiang, P. Chambon, and D. Metzger, *Proc. Natl. Acad. Sci. USA* **98**, 224 (2001).

[72] T. Imai, M. Jiang, P. Kastner, P. Chambon, and D. Metzger, *Proc. Natl. Acad. Sci. USA* **98**, 4581 (2001).

[73] P. Weber, D. Metzger, and P. Chambon, *Eur. J. Neurosci.* **14**, 1777 (2001).

[74] P. Weber, F. Cammas, C. Gérard, D. Metzger, P. Chambon, R. Losson, and M. Mark, *Development* **129**, 2329 (2002).

[75] P. Weber, M. Schuler, C. Gerard, M. Mark, D. Metzger, and P. Chambon, *Biol. Reprod.* **68**, 553 (2003).

TABLE II

Floxed nuclear receptors	Reference
RARα	54
RARβ	88
RARγ	61
RXRα	33, 89
HNF4	90, 91
ERα	92
PPARβ	93
PPARγ	94, 95
GR	96
SF1	97
LXRα	98
LXRβ	98
FXR	99

There is, however, an obvious limitation to the use of somatic mutagenesis to determine the function of gene products in a given cell type, whenever the gene mRNA and/or the protein products have a long half-life as compared to that of the cell. This limitation is well illustrated by the lack of RXRα protein depletion in suprabasal keratinocytes, even though the RXRα gene was effectively mutated.[34] The other limitations that we have encountered occasionally are related either to the selective relative resistance or, on the contrary, to the selective high sensitivity of some floxed genes to tamoxifen-induced Cre-mediated excision, leading in the former case to inefficient gene mutagenesis and in the second case to tamoxifen-independent constitutive mutagenesis. These variations in recombination efficiencies amongst floxed genes, which can occur for different genes in a given cell type or for the same gene in different cell types (Refs. 76,77; our unpublished results), are likely to reflect variations in the accessibility of loxP sites in different chromatin environments. Nevertheless, and in spite of these limitations, our spatio-temporally controlled targeted somatic mutagenesis approach clearly is paving the way to the study of the

[76] M. Vooijs, J. Jonkers, and A. Berns, *EMBO Reports* **21**, 292 (2001).

[77] M. Holzenberger, C. Lenzner, P. Leneuve, R. Zaoui, G. Hamard, S. Vaulont, and Y. L. Bouc, *Nucl. Acid Res.* **28**, E92 (2000).

[78] L. Vallier, J. Mancip, S. Markossian, A. Lukaszewicz, C. Dehay, D. Metzger, P. Chambon, J. Samarut, and P. Savatier, *Proc. Natl. Acad. Sci. USA* **98**, 2467 (2001).

[79] S. Hayashi and A. P. McMahon, *Develop. Biol.* **244**, 305 (2002).

[80] T. Imai, P. Chambon, and D. Metzger, *Genesis* **26**, 147 (2000).

[81] B. Zheng, Z. Zhang, C. M. Black, B. de Crombrugghe, and C. P. Denton, *Am. J. Pathol.* **160**, 1609 (2002).

physiological and pathophysiological functions of a gene in any tissue at any time throughout the life of the animal, including during development in utero (our unpublished data; Refs. 65, 78, 79).

Acknowledgments

We are grateful to P. Soriano for the generous gift of the RosaR26R mice, and C. Rochette-Egly for antibodies. We thank J.M. Bornert, J. Brocard, S. Bronner, N. Chartoire, H. Chiba, C. Gérard, R. Feil, B. Ferret, G. Kimmich, R. Lorentz, M. Mark, B. Mascrez, N. Messaddeq, J.L. Vonesch, X. Warot and O. Wendling for their enthusiastic participation to the studies presented in this review, as well as E. Metzger and A. Van Es and the animal facility staff for animal care, the secretarial staff for typing and the illustration staff for preparing the figures. This work was supported by funds from the Centre National de la Recherche Scientifique, the Institut National de la Santé et de la Recherche Médicale, the Collège de France, the Hôpital

[82] R. A. Kimmel, D. H. Turnbull, V. Blanquet, W. Wurst, C. A. Loomis, and A. L. Joyner, *Gene Develop.* **14**, 1377 (2000).

[83] S. Külbandner, S. Brummer, D. Metzger, P. Chambon, F. Hoffmann, and R. Feil, *Genesis* **26**, 147 (2000).

[84] A. Forde, R. Constien, H. J. Gröne, G. Hämmerling, and B. Arnold, *Genesis* **33**, 191 (2002).

[85] C. Guo, W. Yang, and C. G. Lobe, *Genesis* **32**, 8 (2002).

[86] D. S. Sohal, M. Nghiem, M. A. Crackower, S. A. Witt, T. R. Kimball, K. M. Tymitz, J. M. Penninger, and J. D. Molkentin, *Circ. Res.* **89**, 20 (2001).

[87] M. Tannour-Louet, A. Porteu, S. Vaulont, A. Kahn, and M. Vasseur-Cognet, *Hepatology* **35**, 1072 (2002).

[88] B. Chapellier, M. Mark, J. Bastien, A. Dierich, M. Le Meur, P. Chambon, and N. B. Ghyselinck, *Genesis* **32**, 91 (2002).

[89] J. Chen, S. T. Kubalak, and K. R. Chien, *Development* **125**, 1943 (1998).

[90] G. P. Hayhurst, Y. H. Lee, G. Lambert, J. M. Ward, and F. J. Gonzalez, *Mol. Cell. Biol.* **21**, 1393 (2001).

[91] F. Parviz, J. Li, K. H. Kaestner, and S. A. Duncan, *Genesis* **32**, 130 (2002).

[92] S. Dupont, A. Krust, A. Gansmuller, A. Dierich, P. Chambon, and M. Mark, *Development* **127**, 4277 (2000).

[93] Y. Barak, D. Liao, W. He, E. S. Ong, M. C. Nelson, J. M. Olefsky, R. Boland, and R. M. Evans, *Proc. Natl. Acad. Sci. USA* **99**, 303 (2002).

[94] T. E. Akiyama, S. Sakai, G. Lambert, C. J. Nicol, K. Matsusue, S. Pimprale, Y. H. Lee, M. Ricote, C. K. Glass, H. B. Brewer, and F. J. Gonzalez, *Mol. Cell. Biol.* **22**, 2607 (2002).

[95] J. R. Jones, K. D. Shelton, Y. Guan, M. D. Breyer, and M. A. Magnuson, *Genesis* **32**, 134 (2002).

[96] F. Tronche, C. Kellendonk, O. Kretz, P. Gass, K. Anlag, P. C. Orban, R. Bock, R. Klein, and G. Schütz, *Nat. Genet.* **23**, 99 (1999).

[97] L. Zhao, M. Bakke, Y. Krimkevich, L. J. Cushman, A. F. Parlow, S. A. Camper, and K. L. Parker, *Development* **128**, 147 (2001).

[98] S. Alberti, G. Schuster, P. Parini, D. Feltkamp, U. Diczfaluzy, M. Rudling, B. Angelin, I. Björkem, S. Pettersson, and J. A. Gustafsson, *J. Clin. Invest.* **107**, 565 (2001).

[99] C. J. Sinal, M. Tohkin, M. Miyata, J. M. Ward, G. Lambert, and F. Gonzalez, *Cell* **102**, 731 (2000).

Universitaire de Strasbourg, the Association pour la Recherche sur le Cancer, the Fondation pour la Recherche Médicale, the Human Frontier Science Program, the Ministère de l'Éducation Nationale de la Recherche et de la Technologie and the EEC (CT 97-3220). M.L. was supported by fellowships from the Association pour la Recherche sur le Cancer and the Fondation pour la Recherche Médicale, A.K.I. by a fellowship from the Université Louis Pasteur (Strasbourg), and B.C. by fellowships from the Ministère de l'Education Nationale, de la Recherche et de la Technologie, the Association pour la Recherche sur le Cancer and the Fondation pour la Recherche Médicale.

[23] Analysis of Small Molecule Metabolism in Zebrafish

By Shiu-Ying Ho, Michael Pack, and Steven A. Farber

Introduction

Lipids play essential roles in all living cells, not only as key components of membranes and energy sources, but also as important mediators of cellular signaling. At the organismal level, individual differences in lipid metabolism can impact on a variety of human diseases. This observation has fueled the search for genes that regulate lipid metabolism with the goal to determine how lipids and their metabolic products interact with essential signaling pathways. Although studies have shown how lipids are absorbed, transported, deposited, and mobilized, there remains much to be learned at the genetic and molecular levels.

Since the pioneering work of George Streisinger and his colleagues,[1] the zebrafish, *Danio rerio*, has become an ideal model system for studying vertebrate development.[2] Advantages of the zebrafish include a short generation time, external fertilization, optically clear embryos, and large numbers of offspring produced from a single female (often > 300 larvae); however, what has made the zebrafish even more powerful as an experimental organism is the ability to perform genetic studies by mutagenizing the entire genome and screening for particular phenotypes.[2,3] These large-scale screens have relied primarily on inducing large deletions and translocations by gamma ray radiation,[4] point mutations by soaking founder fish in a variety of mutagens, most frequently ethylnitrosurea (ENU),[2] and retroviral insertions.[5] Because the zebrafish and mammals

[1] G. Streisinger, C. Walker, N. Dower, D. Knauber, and F. Singer, *Nature* **291**, 293–296 (1981).
[2] P. Haffter *et al.*, *Development* **123**, 1–36 (1996).
[3] W. Driever *et al.*, *Development* **123**, 37–46 (1996).
[4] S. Fisher, S. L. Amacher, and M. E. Halpern, *Development* **124**, 1301–1311 (1997).
[5] W. Chen, S. Burgess, G. Golling, A. Amsterdam, and N. Hopkins, *J. Virol.* **76**, 2192–2198 (2002).

shared a common ancestor some 350 million years ago, almost every gene in zebrafish can be mapped to a mammalian orthologue, which usually contains greater than 70% similarity at the protein level.[6] A number of laboratories have utilized zebrafish as a model to study human diseases.[7-9] In this article, we review new techniques and reagents developed to make zebrafish more amenable to studying lipid signaling and metabolism.

Zebrafish Genetics and Vertebrate Physiology

The vertebrate genome contains a predicted 35,000 to 50,000 genes, of which many have no known function. Genetic studies are one approach to identifying genes that direct vertebrate development. Since the completion of large-scale chemical mutagenesis screens in 1997, the phenotypic and molecular characterization of many mutations have been reported.[2,3] Analyses of mutations affecting early developmental processes, such as the specification of the embryonic axis and germ layers, have been particularly rewarding.[10] Related work with mutations that affect organogenesis led to the recognition that zebrafish can also be an important model system for biomedical research.[11] Given the many aspects of organ physiology conserved during vertebrate evolution, the importance of genetic screens that can assay organ function by visualizing biochemical reactions and the transport of molecules *in vivo* in this optically transparent organism is apparent.

We identified nine recessive lethal mutations that perturb the development of digestive organs.[12] Although these mutations were revealed using morphological criteria, their phenotypic analyses suggested that the affected genes regulated developmental processes that also were relevant to digestive physiology. Through the analysis of these and other zebrafish mutants, we gained an understanding of the limitations inherit to genetic screens that are based solely on morphological criteria. First, not all organs are suitably visualized in zebrafish larvae, and mutations that perturb the development of these are often overlooked. Second, since it is difficult to visualize specific cell populations within many larval organs, mutations that affect the

---

[6] J. H. Postlethwait *et al.*, *Nat. Genet.* **18**, 345–349 (1998).
[7] B. A. Barut and L. I. Zon, *Physiol. Genomics* **2**, 49–51 (2000).
[8] A. C. Ward and G. J. Lieschke, *Front. Biosci.* **7**, d827–d833 (2002).
[9] J. F. Amatruda, J. L. Shepard, H. M. Stern, and L. I. Zon, *Cancer Cell* **1**, 229–231 (2002).
[10] B. Feldman *et al.*, *Nature* **395**, 181–185 (1998).
[11] C. Thisse and L. I. Zon, *Science* **295**, 457–462 (2002).
[12] M. Pack *et al.*, *Development* **123**, 321–328 (1996).

development or function of these cells can be missed. Third, despite the transparency of the zebrafish larva, the function of few organs can be effectively assayed by visual inspection. For these reasons, morphology-based screens are best suited for the identification of genes that regulate specification and patterning of embryonic structures. We believe that screens designed to assay physiological processes directly will be the most effective way to study particular biochemical pathways.

In addition to genetic screens, the development of a knockdown technology in zebrafish has enabled genome-wide, sequence-based, reverse genetic screens in this model vertebrate.[13,14] This technology is now widely used in the zebrafish community, and numerous studies have successfully phenocopied mutants by targeting particular genes.[13] Methodologically, a small amount of a morpholino-modified oligonucleotide (1–3 nl, 1 ng/nl) that is complementary to the 5′ UTR of a zebrafish gene of interest is injected into the yolk of a one to eight cell embryo. Morpholino-modified antisense oligonucleotides effectively inhibit translation of target genes, and they are more specific than unmodified oligonucleotides. These reagents offer the potential to explore the molecular mechanisms of organ development and physiology. Specifically, modified oligonucleotides can be directed against any gene to determine its effect on lipid metabolism in live zebrafish larvae.

Zebrafish as a Model to Study Human Physiology and Disease

Several zebrafish mutants share a number of characteristics reported in humans carrying mutations in orthologous genes.[8,15–18] For example, a mutation in the *urod* gene results in impaired heme synthesis and hepatoerythropoietic porphyria in zebrafish and humans.[16] The *Sih* mutation affects *tnnt2*, a gene encoding the thin-filament contractile protein, cardiac troponin T. Mutations in the human gene are responsible for 15% of cases of familial hypertrophic cardiomyopathy, which is a leading cause of sudden death in young athletes.[19] The zebrafish *sau* encodes the erythroid-specific isoform of δ-aminolevulinate synthase that is required for the first step in heme biosynthesis. Mutations in this gene cause congenital

[13] A. Nasevicius and S. C. Ekker, *Nat. Genet.* **26**, 216–220 (2000).

[14] J. Heasman, *Dev. Biol.* **243**, 209–214 (2002).

[15] A. Brownlie *et al.*, *Nat. Genet.* **20**, 244–250 (1998).

[16] H. Wang, Q. Long, S. D. Marty, S. Sassa, and S. Lin, *Nat. Genet.* **20**, 239–243 (1998).

[17] J. F. Amatruda and L. I. Zon, *Dev. Biol.* **216**, 1–15 (1999).

[18] G. Vogel, *Science* **288**, 1160–1161 (2000).

[19] A. J. Sehnert, A. Huq, B. M. Weinstein, C. Walker, M. Fishman, and D. Y. Stainier, *Nat. Genet.* **31**, 106–110 (2002).

sideroblastic anaemia and represent the first animal model of this human disease.[15] The zebrafish *ferroportin1* gene is responsible for the hypochromic anemia of the zebrafish mutant *weissherbst*, and is required for the transport of iron from maternally derived yolk stores to the circulation. Human Ferroportin1 is found at the basal surface of placental syncytiotrophoblasts, suggesting that it also transports iron from mother to embryo. Therefore, *weissherbst* is proposed to be a model of mammalian disorders of iron deficiency or overload.[20] In addition, the visual system of zebrafish is similar to other vertebrates and is now accepted as a useful model for vision science.[21]

The genes important for zebrafish lipid metabolism are highly conserved. For example, cytoplasmic phospholipase A_2 (PLA_2),[22,23] and apolipoprotein E and A-I orthologues have been identified in zebrafish, and these are expressed during embryonic development.[24] A fatty acid desaturase was isolated from zebrafish that had both $\Delta5$ and $\Delta6$ fatty acid desaturase activities.[25]

Lipid Metabolism in Fish

Lipid metabolism in teleoste fish can be classified into four major components; absorption, transport, storage, and mobilization. Numerous studies have shown that lipid transport and mobilization are similar in fish and mammals, although the mechanisms of lipid absorption and storage differ slightly.[26] Fish transport both non-esterified fatty acids (FA) and triacylglycerol (TG)-enriched chylomicrons to the liver through the circulation, and then subsequently to peripheral tissues.[26] In fish plasma, lipoprotein profiles are comparable to those of mammal, with apolipoproteins A- and B-like molecules.[27] Lipolysis in fish is generally similar to that in mammals, and is achieved by various lipases, including lipoprotein lipase, TG lipase, hepatic lipase and lysosomal lipase.[26,28] In addition, hormonal regulation of lipid metabolism in fish is similar to that of mammals.[28]

[20] A. Donovan *et al.*, *Nature* **403**, 776–781 (2000).

[21] J. Bilotta and S. Saszik, *Int. J. Dev. Neurosci.* **19**, 621–629 (2001).

[22] S. A. Farber, E. S. Olson, J. D. Clark, and M. E. Halpern, *J. Biol. Chem.* **274**, 19338–19346 (1999).

[23] J. D. Clark *et al.*, *Cell* **65**, 1043–1051 (1991).

[24] P. J. Babin, C. Thisse, M. Durliat, M. Andre, M. A. Akimenko, and B. Thisse, *Proc. Natl. Acad. Sci. USA* **94**, 8622–8627 (1997).

[25] N. Hastings *et al.*, *Proc. Natl. Acad. Sci. USA.* **98**, 14304–14309 (2001).

[26] M. A. Sheridan, *Comp. Biochem. Physiol., B: Comp. Biochem.* **90**, 679–690 (1988).

[27] P. J. Babin and J. M. Vernier, *J. Lipid. Res.* **30**, 467–489 (1989).

[28] M. A. Sheridan, *Comp. Biochem. Physiol., B Comp. Biochem.* **107**, 495–508 (1994).

The major difference in lipid absorption between teleostes and mammals is that fish contain both a fast FA and a slow TG delivery system.[29,30] The fast FA delivery system can absorb and deliver nonesterified FA (either free or bound to carrier protein) directly to the peripheral tissues via the blood.[30] The slow TG delivery system re-esterifies FA to TG in the intestine and then transports the TG-rich particles to the periphery in a manner similar to that of mammals.[29] With respect to lipid storage, fish deposit lipids not only in mesenteric adipose tissue, but also in muscle and liver. Most of the stored lipids are TG and polyunsaturated FAs but various lipid classes, such as glyceryl ether analogues and alkoxydiacylglycerol have been documented in some fish species.[28]

Optical Biosensors to Visualize Lipid Metabolism in Live Larvae

To assay directly enzymatic function and lipid metabolism in live larvae, we developed a family of fluorescent lipid reporters whose spectral characteristics change upon processing by lipid modifying enzymes.[22,31] We focused on PLA_2 activity in the developing embryo because of its importance in generating lipid-signaling molecules, host defense, lipid absorption and cancer.[32,33] For example, PLA_2 cleavage of PED6 [N-((6-(2,4-dinitro-phenyl)amino)hexanoyl)-1-palmitoyl-2-BODIPY-FL-pentanoyl-*sn*-glycerol-3-phosphoethanolamine] substrate (Fig. 1A), results in the release of a fluorescent BODIPY acyl chain that reveals organ-specific PLA_2 activity at the subcellular level.[31] Larvae, at 5 days post-fertilization (dpf), a stage when digestive organs are fully functional, exhibit a labeled intestine, gallbladder and hepatic duct following PED6 ingestion (Fig. 1B). Time course studies following PED6 incubation show that fluorescent PED6 metabolites appear in the liver first, followed by the gallbladder.[34] Intestinal fluorescence is seen last, after gallbladder secretion of bile containing the PED6 derived fluorescent BODIPY acyl chain (Fig. 2).

To explain these findings we hypothesized that following ingestion, quenched PED6 is cleaved by PLA_2 in the intestine, and the unquenched fluorescent PED6 metabolites are rapidly transported to the liver.

[29] M. F. Sire, C. Lutton, and J. M. Vernier, *J. Lipid. Res.* **22**, 81–94 (1981).

[30] M. A. Sheridan, W. V. Allen, and T. H. Kerstetter, *Comp. Biochem. Physiol. B.* **80**, 671–676 (1985).

[31] H. S. Hendrickson, E. K. Hendrickson, I. D. Johnson, and S. A. Farber, *Anal. Biochem.* **276**, 27–35 (1999).

[32] M. MacPhee, K. P. Chepenik, R. A. Liddell, K. K. Nelson, L. D. Siracusa, and A. M. Buchberg, *Cell* **81**, 957–966 (1995).

[33] E. A. Dennis, *Trends Biochem. Sci.* **22**, 1–2 (1997).

[34] S. A. Farber *et al.*, *Science* **292**, 1385–1388 (2001).

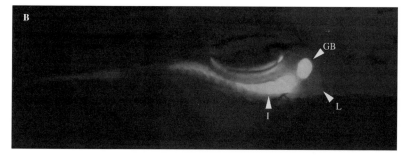

FIG. 1. PED6 as a biosensor to visualize lipid metabolism in live zebrafish larva. (A) The structure of PED6. The emission of BODIPY-labeled acyl chain at *sn*-2 position is quenched by dinitrophenyl group at *sn*-3 position when this molecule is intact. Upon PLA$_2$ cleavage at *sn*-2 position, BODIPY-labeled acyl chain emits green fluorescence. (B) 5 dpf zebrafish larva labeled with PED6 (0.3 μg/ml) for 6 hr. Arrow shows liver (L), gall bladder (GB) and intestine (I). (See Color Insert.)

Thereafter, fluorescent metabolites of the BODIPY acyl chain are secreted into newly formed bile that is then stored in the gallbladder. Following contraction of the organ, fluorescent bile enters the intestine where it is readily visualized. To test this hypothesis further, another fluorescent lipid reporter, BODIPY FR-PC (Fig. 3A), was used to detect PLA$_2$ activity.[34] This fluorophore utilizes fluorescence resonance energy transfer (FRET) to emit different spectra. When excited at a wavelength of 505 nm, the intact substrate emits orange light 568 nm. Upon PLA$_2$ cleavage, the same excitation now results in a green emission 515 nm. As expected, we observed only green fluorescence in the gallbladder and liver. Orange fluorescence was observed only within the intestinal epithelium (Fig. 3B) after 1 hr of BODIPY FR-PC labeling at a concentration of 0.1 μg/ml.[34] These studies supported our hypothesis that lipid digestion and absorption systems in zebrafish were similar to those of mammals.

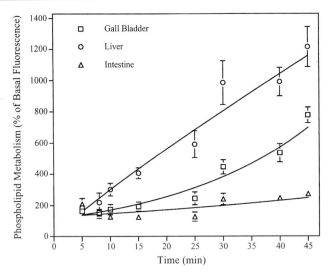

FIG. 2. Time course of PED6 labeling. 5 dpf larvae were immersed in the embryo media containing PED6. At different time points, larvae were anesthetized with 0.16% tricine, and the fluorescence in the digestive organ was digitally quantified by analyzing fluorescent images. The fluorescence of PED6 appeared in the liver first, then the gall bladder, followed by the intestine.

Because these reporter lipids provided a rapid readout of metabolism and digestive organ morphology in living zebrafish larvae, we used them to perform a physiological genetic screen *in vivo*. In a pilot ENU mutagenesis, a recessive lethal mutant was identified, *fat-free*, whose digestive system appeared morphologically normal yet failed to accumulate fluorescently labeled lipid in the gallbladder following PED6 ingestion. Phenotypic analysis of this mutant indicated that swallowing and PLA$_2$ activity were normal.[34] Cholesterol absorption in *fat-free* was also assayed in live larvae using NBD-cholesterol (22-[N-(7-nitronbenz-2-oxa-1,3-diazol-4-yl) amino]-23,24-bisnor-5-cholen-3-ol) (Fig. 4). Here too, *fat-free* mutant larvae exhibited significantly reduced fluorescence in the digestive organs following immersion in medium containing the fluorescent cholesterol analogue.[34] In contrast, *fat-free* fish placed in medium containing 0.5 μM of BODIPY FL-C5, a short chain fatty acid analogue, had nearly normal fluorescence in the digestive tract. We hypothesized that these lipids were absorbed differently in *fat-free* because short chain FA are less hydrophobic and are more soluble in aqueous solution, and therefore emulsifiers (such as bile acids) are not required for their absorption. In contrast, the absorption of PED6 and NBD-cholesterol required biliary emulsification. We therefore proposed that the *fat-free* mutation might

A

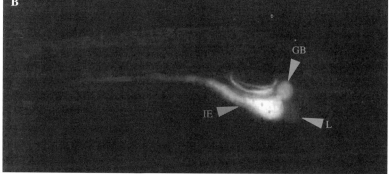

B

FIG. 3. BODIPY FR-PC label in zebrafish larvae. (A) The structure of BODIPY FR-PC. When the molecule is intact in the cell, excitation at 505 nm results in orange (568 nm) emission due to fluorescence resonance energy transfer (FRET) between the two BODIPY labeled moieties. Upon PLA_2 cleavage at *sn*-2 position, BODIPY moiety at *sn*-1 position showed green (515 nm) emission when excited at 505 nm. (B) BODIPY FR-PC (5 μg/ml) labeled 5 dpf zebrafish larva. The liver (L) and gall bladder (GB) showed green fluorescence (green arrow), indicating the accumulation of cleaved products. Uncleaved orange BODIPY FR-PC (orange arrow) is observed only in the intestinal epithelium (IE). (See Color Insert.)

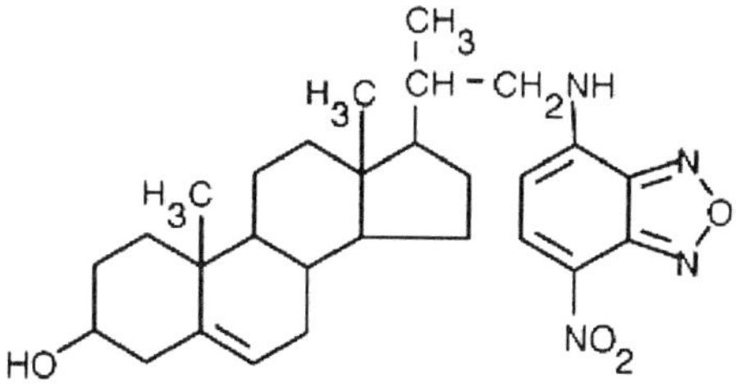

FIG. 4. Structure of NBD-Cholesterol. NBD-Cholesterol in which the fluorophore replaces the terminal segment of cholesterol's alkyl tail, and it labels the 5 dpf larval zebrafish.

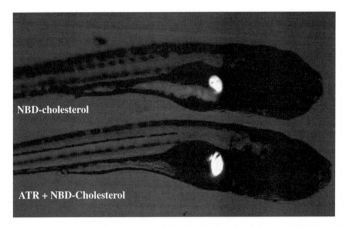

FIG. 5. Atorvastatin (ATR) interferes with NBD-cholesterol labeling. Wild-type larvae had reduced fluorescence in the intestinal lumen, while gallbladder fluorescence was preserved after 1.5 hr labelling of NBD-cholesterol (2 μg/ml) solubilized with tilapia's bile. (See Color Insert.)

attenuate biliary synthesis or secretion because the absorption of short chain FA was nearly normal.

Previously, we demonstrated that atorvastatin (trade name Lipitor), a 3-hydroxy-3-methylglutaryl-CoA (HMG-CoA) reductase inhibitor, had a potent effect on PED6 processing in larval zebrafish similar to that observed in *fat-free* larvae;[34] however, the effect of Lipitor on NBD-cholesterol processing in wild-type larvae was slightly different from that observed in *fat-free* larvae. The experiment was performed using dried NBD-cholesterol re-solubilized with extracted bile from the tilapia fish (*Orechromics mossambicus*) to a final concentration of 2 μg/ml in the medium. After adding 2 μg/ml Lipitor to this medium, wild-type zebrafish larvae were introduced. After 1.5 hr at 28°C, we found that the larvae had markedly reduced fluorescence in the intestinal lumen, while gallbladder fluorescence was preserved (Fig. 5). Because the addition of exogenous fish bile reversed the blocking effect of Lipitor, we proposed that Lipitor inhibited the synthesis of cholesterol-derived biliary emulsifiers that were required for lipid absorption. The fluorescence seen in the gallbladder of Lipitor-treated larvae could be due to NBD-cholesterol absorbed in the presence of pre-existing bile. In contrast, we did not observe much fluorescence in the gallbladder of the *fat-free* mutant, indicating these fluorescent substrates were poorly absorbed. This observation further supported the idea that *fat-free* had impaired biliary synthesis or secretion. Numerous studies show that bile acid metabolism is regulated by nuclear receptors and dietary lipids,[35–37] and the *fat-free* mutant may provide a potential animal model to study the roles of these proteins. Positional cloning of the *fat-free* gene is underway in this laboratory.

Radioactive Lipid Precursors to Study Lipomics in Single Larva

To study genes that regulate lipid metabolism, we combined genetic (ENU mutagenesis screen) approaches with classic isotopic labeling techniques to study lipid profiles of mutant larvae. Single larva were immersed in medium containing radioactive oleic acid (specific activity 250 μCi/mmol) and antibiotics for 20 hr. After washing the larva three times with medium, the embryos were homogenized using a sonic dismembrator (Fisher Scientific, Pittsburg, PA). Lipid extraction was done by chloroform:methanol:water (2:1:1 by volume). The extracted lipids were spotted on 20 × 20 cm thin layer chromatography (TLC) plates (silica gel G, Whatman, Hillsboro, OR), and a solvent system of chloroform: ethanol:triethylamine:water (30:34:30:8) was used to resolve polar lipids. Nonpolar lipids were resolved to half height of the plate in a solvent system of chloroform:benzene:ethanol:acetic acid (40:50:2:0.2, by volume), allowed to dry in a hood, and then developed up the entire height of the plate in a solvent of hexane:ethyl ether (96:4 by volume). Radioactive lipids were detected by scanning (Bioscan 200). When labeling with ^{14}C oleic acid, we found that the major labeled lipid fraction is phosphatidylcholine (PC) at 5 dpf, and remains as FA at 1 dpf (Fig. 6). When labeling with [^{14}C] acetate, the results showed that the major labeled lipid fraction is phosphatidyl-ethamine (PE) (Fig. 7) at all developmental stages examined and that cholesterol was not detectable. In other experiments, we found that *fat-free* larvae exhibited significantly reduced PC synthesis ($P < 0.05$) after 3 hr of [^{14}C] oleic acid labeling (Fig. 8C).

Screening Strategies

Studies have shown that zebrafish can be used to screen for compounds that affect angiogenesis[38] embryogenesis,[39] and for compounds with

[35] M. Makishima *et al.*, *Science* **296**, 1313–1316 (2002).
[36] A. Chawla, J. J. Repa, R. M. Evans, and D. J. Mangelsdorf, *Science* **294**, 1866–1870 (2001).
[37] B. Goodwin and S. A. Kliewer, *Am. J. Physiol. Gastrointest Liver Physiol.* **282**, G926–G931 (2002).
[38] J. Chan, P. E. Bayliss, J. M. Wood, and T. M. Roberts, *Cancer Cell.* **1**, 257–267 (2002).
[39] R. T. Peterson, B. A. Link, J. E. Dowling, and S. L. Schreiber, *Proc. Natl. Acad. Sci. USA* **97**, 12965–12969 (2000).

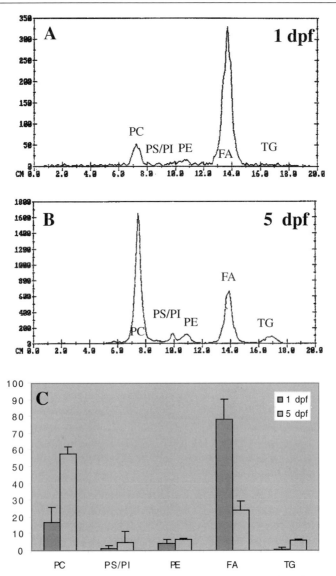

FIG. 6. Lipomics during development. Larvae (1 and 5 dpf) were incubated with radioactive oleic acid for 20 hr, followed by lipid extraction and TLC. The solvent chloroform:ethanol: triethylamine:water (30:34:30:8) was used to develop the TLC plate. The radioactivities were then scanned. The major metabolites derived from oleic acid (FA) are phosphatidylcholine (PC), phosphatidylserine (PS), phosphatidylinositol (PI), phosphatidylethanolamine (PE), and triacylglycerol (TG). (A) One representative lipomics of 1 dpf larva. (B) One representative lipomics of 5 dpf larva. (C) Comparison lipomics between 1 and 5 dpf larvae ($n=9$, mean ± SD). (See Color Insert.)

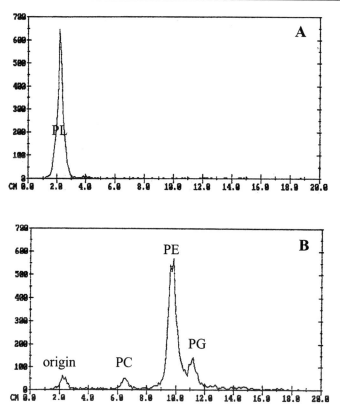

FIG. 7. One dpf larva labeled with radioactive sodium acetate. Larva (1 dpf) was immersed in radioactive sodium acetate for 24 hr, followed by lipid extraction and TLC separation, then radioactivity scanning. (A) Nonpolar TLC plate was developed to the half height of the plate in the solvent system choloroform:benzene:ethanol:acetic acid (40:50:2:0.2 v/v), allowed to dry under the hood, then developed up the entire height of the plate in the solvent of hexane:ethyl ether (94:6 v/v). Phospholipids (PL) are the only detectable metabolite. (B) Polar TLC plate was developed using solvent chloroform:ethanol:triethylamine:water (30:34:30:8 v/v) for the entire plate. Phosphatidylethamine (PE) is the major metabolite of sodium acetate, and trace amount of phosphatidylcholine (PC) and phosphatidylglycerol (PG).

anti-cancer activity.[9] Our studies demonstrate that zebrafish larvae also can be used to identify mutants with perturbed lipid metabolism using ENU mutagenized fish coupled with classic isotope labeling. Reverse genetic approaches can be used as well to screen for genes that mediate lipid metabolism as the fluorescent lipid analogues (such as PED6 and NBD-cholesterol) provide an easy readout for high throughput screening. Zebrafish larvae can be arrayed in multi-well plates that contain different chemical compounds and fluorescent lipid reporters to facilitate rapid screening.

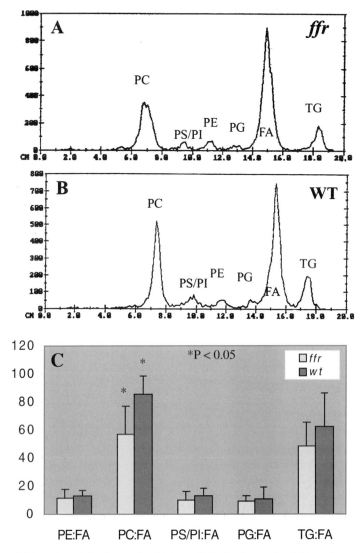

FIG. 8. Lipomics of *fat-free* and wild-type. Both *fat-free* and wild-type larvae (4 dpf) were incubated with radioactive oleic acid for 20 hr, followed by lipid extraction and TLC. The solvent chloroform:ethanol:triethylamine:water (30:34:30:8 v/v) was used to develop the TLC plate. The radioactivities were then scanned. The major metabolites derived from oleic acid (FA) are phosphatidylcholine (PC), phosphatidylserine (PS), phosphatidylinositol (PI), phosphatidylethanolamine (PE), phosphatidylglycerol (PG), and triacylglycerol (TG). (A) One representative lipomics of *fat-free* larva. (B) One representative lipomics of wild-type larva. (C) Comparison lipomics between *fat-free* and wild-type larva. *Fat-free* has significantly decreased PC production when expressed as metabolite/FA ($n = 9$, mean $\pm$ SD). (See Color Insert.)

Zebrafish as a Model System to Study Prostanoid Metabolism

Eicosanoids are lipid signaling molecules that regulate a wide range of cellular and physiological processes in vertebrates. Perhaps the largest and best studied class are the prostanoids (prostaglandins and thromboxanes), whose synthesis from the endoperoxides PGG2 and PGH2 is dependent upon the actions of cyclooxygenases (COXs), the enzymes targeted by aspirin and other nonsteroidal anti-inflammatory drugs (NSAIDs).[40] Humans and other mammals synthesize prostanoids via two COX isozymes that are encoded by separate genes.[41] The COX-2 isoform is induced in various pathological conditions, whereas COX-1 is expressed constitutively in a wide range of cell types and therefore is predicted to play a role in homeostasis. The finding that the two COX isoforms regulate prostanoid synthesis in different biological contexts has led to the development of COX inhibitors with selective isozyme specificities. Numerous clinical trials have shown that COX inhibition has important therapeutic effects in inflammatory states, cardiovascular diseases and cancer.[42] Continued characterization of the synthesis, regulation and activity of prostanoids and other eicosanoids is an active area of research.

Prostanoid Synthesis and Signaling

Cyclooxygenases and prostanoid formation have been extensively studied in mammalian systems. Prostaglandins (PGs) and thromboxanes (TXs) are by a family of synthases that convert PGH2 to bioactive PGs and TXs. PGH2 itself is derived from COX-mediated oxygenation and peroxidation of arachidonic acid (AA), a fatty acid component of membrane phospholipids (PL). AA liberation from membrane PLs is generally accepted as the rate-limiting step in prostanoid biosynthesis. AA bio-availability appears to be regulated by three distinct PLA_2 gene families.[43] One model suggests that *calcium independent* PLA_2 regulates AA release during membrane PL recycling *cytosolic* PLA_2 is activated in response to physiological or pathological stimuli, whereas *secretory* PLA_2 sustains levels of cytosolic AA in response to continued stimulation.[44]

Eicosanoids appear to function as autocoid signaling molecules. PGs and TXs are secreted by cells, which, in turn, are auto-stimulated through

[40] C. N. Serhan and E. Oliw, *J. Clin. Invest.* **107**, 1481–1489 (2001).

[41] W. L. Smith and R. Langenbach, *J. Clin. Invest.* **107**, 1491–1495 (2001).

[42] C. Patrono, P. Patrignani, and L. A. Garcia Rodriguez, *J. Clin. Invest.* **108**, 7–13 (2001).

[43] F. A. Fitzpatrick and R. Soberman, *J. Clin. Invest.* **107**, 1347–1351 (2001).

[44] J. Balsinde and E. Dennis, *J. Biol. Chem.* **271**, 6758–6765 (1996).

the activation of specific PG and TX receptors on the cell surface. So far, at least eight such receptors have been identified and splice variants of three receptors are known.[45] Tissue specificity of prostanoid bioactivity is achieved through the selective expression of the PG/TX synthases and the prostanoid receptors; however, the fidelity of receptor specificity is by no means absolute, as individual PGs can activate more than one PG receptor. It has also been shown that in some instances, PGs produced in one cell type can function transcellularly to activate PG production in neighboring cells.[40] Furthermore, prostanoids also appear to activate PPAR-type nuclear hormone receptors in some cell types, including adipocytes,[46] monocytes[47] and macrophages.[48]

Functional Analyses of Vertebrate COX Proteins

Given the complexity of this system, it is not surprising that aspirin and other NSAIDs, which inhibit prostanoid synthesis in many tissues, have side effects that affect their use in clinical practice. These side effects, particularly gastrointestinal bleeding, were the principal motivating factors in the development of selective COX inhibitors. Biochemical and structural analyses have defined regions common to both COX proteins that are required for the conversion of AA to PGH2. In mammals, the two COX isoforms share high sequence homology, although important differences in their tertiary structure have been identified.[43] Such differences account for the selectivity of COX-2 inhibitors as well as the recent observation that the acetylation of COX-2 by aspirin, a nonselective COX inhibitor, does not completely inactivate this isoform, as occurs with COX-1.[49] Residual acetylated COX-2 activity is postulated to lead to the production of eicosanoids with novel actions that may play an important role in aspirin's anti-inflammatory effects.

Molecular analysis has also defined regulatory regions of both COX genes, although far more is known about this aspect of COX-2 than COX-1. This may be expected given that COX-1 is constitutively expressed at high levels in most cell types. Targeted disruption of both COX isoforms has also been accomplished. Whereas COX-1 mutant mice are viable,[50] COX-2 mutants are infertile and commonly develop progressive renal disease that

[45] S. Narumiya and G. A. FitzGerald, *J. Clin. Invest.* **108**, 25–30 (2001).
[46] S. A. Kliewer, J. M. Lenhard, T. M. Willson, I. Patel, D. C. Morris, and J. M. Lehmann, *Cell* **83**, 813–819 (1995).
[47] C. Jiang, A. T. Ting, and B. Seed, *Nature* **391**, 82–86 (1998).
[48] M. Ricote, A. C. Li, T. M. Willson, C. J. Kelly, and C. K. Glass, *Nature* **391**, 79–82 (1998).
[49] J. Claria and C. N. Serhan, *Proc. Natl. Acad. Sci. USA* **92**, 9475–9479 (1995).
[50] R. Langenbach *et al.*, *Cell* **83**, 483–492 (1995).

affects longevity.[51] These aspects of the COX-2 mutant phenotype present certain limitations to the analysis of COX function in various physiological processes.

Orthologues of the mammalian COX proteins have been identified in nonmammalian vertebrates such as the trout and zebrafish.[52] Our interest in using the zebrafish model system to study COX pharmacology and biology arose from the potential to apply large-scale genetic analysis and gene targeting studies to questions relevant to prostanoid biology.

Molecular, Biochemical, Pharmacological and Functional Analyses of Zebrafish COXs

Studies of zebrafish eicosanoid biology were designed to address several simple questions. First, while it was apparent that zebrafish expressed cDNAs with sequence homology to mammalian COXs, it was necessary to show that these genes were functional orthologues of the mammalian COXs. Second, it was important to learn whether the putative zebrafish COX orthologues share isoform specific properties with their mammalian counterparts. Third, it was essential to know whether established physiological functions of the COX isoforms are conserved evolutionarily. Answers to these questions would likely determine the suitability of the zebrafish model system for analyses clinically relevant aspects of prostanoid biology.

To address these issues, full-length cDNAs encoding the putative zebrafish COX-1 and COX-2 isozymes were obtained and their presence in various cell types was assayed using reverse transcriptase-PCR.[53] These data revealed widespread expression of both isoforms in adult tissues and at different embryonic stages. Expression of COX-2, and to a lesser degree COX-1, was prominent in the vasculature of developing larvae suggesting a role for COX activity during zebrafish blood vessel development. This finding is of interest because COX-2 is believed to play an important role in angiogenesis associated with human cancers.[54] Sequence analysis of the zebrafish COXs revealed a high degree of conservation with their mammalian orthologues. Particularly noteworthy were the conservation of amino acids crucial for catalysis, aspirin acetylation, heme coordination,

[51] S. G. Morham *et al., Cell* **83**, 473–482 (1995).
[52] L.D. Roberts SB, Goetz FW., *Mol Cell Endocrinol.* **160**, 89–97 (2000).
[53] T. Grosser, S. Yusuff, E. Cheskis, M. A. Pack, and G. A. FitzGerald, *Proc. Natl. Acad. Sci. USA* **99**, 8418–8423 (2002).
[54] M. J. Thun, *Gastroenterology Clin. North America* **25**, 333–348 (1996).

and the presence of multiple N-glycosylation sites. A comparison of the amino acid residues that determine the volume of the arachidonate-binding channel for each enzyme gave surprising results. Whereas the mammalian COXs differ in the identity of three important residues, only one of these substitutions (Ile-434-Val) is present among the zebrafish COX isoforms. The volume of this channel in the different COX isoforms is believed to underlie the pharmacological specificity of COX inhibitors. These differences raised the question of whether the pharmacological specificity of COX inhibition achievable in mammals was feasible in zebrafish.

Biochemical analysis of the zebrafish COXs was first addressed in a heterologous system.[53] Full-length cDNAs for each isoform were transfected into COS cells lacking endogenous COX activity. Following stimulation with AA, PG synthesis was measured using mass spectrometry. Introduction of either of zebrafish COX genes led to the production of PGE2, whereas there was minimal PGE2 production in COS cells transfected with vector alone. It was also shown that adult zebrafish produce PGE2, PGI2 and TXB2. Most importantly, both nonselective and selective COX inhibitors (indomethacin and NS-398, respectively) perturbed prostanoid synthesis in a dose-dependent manner in transfected COS cells and live fish. Furthermore, the dose required of each drug to achieve 50% inhibition of PGE2 production (IC50) was similar to that required for the mammalian COX enzymes. Finally, it was shown that the pharmacological specificity of two COX inhibitors is maintained in zebrafish. The nonselective inhibitor indomethacin effectively reduced PGE2 production from both COX isoforms, whereas the selective inhibitor NS-398 affected PGE2 production only from COX-2.

Functional assessment of zebrafish COX isoforms was performed in adult fish and developing embryos. Pre-treatment of adult fish with the nonselective inhibitor idomethacin perturbed aggregation of thrombocytes *ex vivo*, but the COX-2 selective inhibitor NS-398 did not. These findings strongly suggest that thrombocytes of zebrafish, like mammalian platelets, only express COX-1. This finding is important because the specificity of COX expression is, in large measure, responsible for the cardioprotective effects of aspirin, which blocks the aggregation of human platelets.[42] Knocking down of COX isoform translation using antisense morpholino oligonucleotides allowed assessment of the developmental role of zebrafish COXs. As seen in mammals, loss of COX-2 protein had no discernable effect on embryonic development;[51] however, oligonucleotides to zebrafish COX-1 caused a significant delay in epiboly, a developmental process dependent upon cell proliferation and migration. In contrast, COX-1 mutant mice develop normally.[50] The apparently discordant embryonic phenotypes produced by inhibition of teleost versus mammalian COX-1 may result from the fact that anti-sense oligonucleotides are predicted to inhibit translation of both maternal and zygotic COX transcripts in zebrafish, whereas gene targeting in mammals perturbs only zygotic gene expression.

Future Directions

Given the high degree of structural and functional conservation between zebrafish and humans COX genes, further analysis of the regulatory mechanisms that control prostanoid production and bioactivity seem feasible using this model system. The relative ease of high throughput genetic analyses in zebrafish is particularly attractive to questions of gene regulation. For example, mutagenesis strategies could be devised using immunohistochemical techniques or COX-GFP transgenic fish to identify mutants that perturb COX RNA or protein expression or stabilization. Such mutants could lead to the identification of novel COX-1 regulators, which to date have largely eluded detection. Additionally, COX-2 mutants (as well as the COX-1 mutants), which would be predicted to be fully viable, could be used to generate compound mutants using fish that carry established mutations affecting other tissues. These fish, in turn could be assayed for a variety of biochemical or physiological affects. Given that many zebrafish mutants are predicted to arise from the perturbation of novel vertebrate genes, this strategy could reveal novel protein–protein interactions essential to COX function. Mutations that affect COX expression, activity or stability may also identify motifs in either isoform that are relevant to the pharmacological inhibition of the enzyme.

Other mutagenesis strategies are feasible using zebrafish. High throughput assays of prostanoid production using mass spectrometry are one example. A physiological mutagenesis screen of this design would not only identify mutations that perturb COX activity directly, but also mutations that perturb the function of genes predicted to couple COXs to PLA2s or PG synthases. This strategy could identify both upstream and downstream regulators of prostanoid biosynthesis as well. Zebrafish can also be used to assay the function of known genes predicted to regulate prostanoid synthesis or activity. Anti-sense morpholino modified oligonucleotides micro-injected into fertilized zebrafish eggs provide a method to assay gene function rapidly, which is not feasible in other vertebrate model organisms. New techniques for directly identifying specific gene mutations from mutagenized sperm also offer the promise of generating libraries of mutant alleles that can be assayed in live fish generated through *in vitro* fertilization (Draper BW and Moens CB, personal communication). This methodology, commonly referred to as

"TILLING,"[55] offers a chance to perform a comprehensive analysis of genes regulating prostanoid synthesis and activity.

Summary

Recent work has shown that it is possible to assay phospholipid metabolism and prostanoid synthesis in zebrafish.[34,53] These preliminary studies suggest that important questions of lipid biology are amenable to large-scale high-throughput analyses in this model system. Lipid metabolism can now be added to the growing list of vertebrate developmental and physiological processes that can be assayed in zebrafish. The potential to identify new genes, or novel functions of known genes that regulate dietary lipid metabolism, or the generation of lipid signaling molecules, may lead to the development of treatment strategies for common human diseases.

[55] C. M. McCallum, L. Comai, E. A. Greene, and S. Henikoff, *Plant. Physiol.* **123**, 439–442 (2000).

[24] Analysis of Nuclear Receptor Function in the Mouse Auditory System

By Matthew W. Kelley, Pamela J. Lanford, Iwan Jones, Lori Amma, Lily Ng, and Douglas Forrest

Introduction

Hearing is a prominent example of a sensory function that is regulated by nuclear receptors. Thyroid hormone receptors are essential for the development of hearing[1] and other receptors, including those for retinoids and estrogen have been detected in the inner ear.[2,3] Although the requirement for a nuclear receptor gene may be suggested in some human inherited diseases with an incidence of hearing loss,[4] the investigation of the underlying mechanisms requires a suitable model species. The mouse

[1] D. Forrest, T. A. Reh, and A. Rüsch, *Curr. Opin. Neurobiol.* **12**, 49 (2002).
[2] R. Romand, V. Sapin, and P. Dolle, *J. Comp. Neurol.* **393**, 298 (1998).
[3] A. E. Stenberg, H. Wang, L. Sahlin, and M. Hultcrantz, *Hear Res.* **136**, 29 (1999).
[4] F. Brucker-Davis, M. C. Skarulis, A. Pikus, D. Ishizawar, M.-A. Mastroianni, M. Koby, and B. D. Weintraub, *J. Clin. Endocrinol. Metab.* **81**, 2768 (1996).

provides a valuable model because of the availability of spontaneous mutants and because it is amenable to genetic manipulation.

Selected methods are presented here for studying the mouse auditory system, with a focus on the cochlea, which contains the sensory hair cells. The cochlea within the inner ear houses a microcosm of cellular and physiological machinery that mediates auditory transduction.[5] The hair cells transduce the mechanical sound stimulus into neural impulses that are relayed through the auditory nerve to the auditory centers of the brain. Despite its small size and relative inaccessibility within the temporal bones of the skull, the cochlea can be subjected to anatomical, physiological and molecular analyses.

Analysis of Auditory Function: the Auditory-Evoked Brainstem Response (ABR)

The ABR is a form of evoked potential and it provides a sensitive measure of the overall function of the auditory system. The ABR is a useful first screen for auditory defects in mice and it can be determined rapidly in a noninvasive, nonterminal procedure. Related systems are used for testing hearing in humans. The ABR represents the electrical potentials evoked by a sound stimulus at several levels in the auditory system. A computer averages the evoked responses from multiple sweeps allowing an ABR waveform to be detected above the spontaneous background of the electroencephalogram (EEG). A characteristic waveform in mice is detected as a series of four or five peaks.[6,7] The first peak is thought to represent the cochlear/auditory nerve response and the subsequent major peaks represent responses in central brainstem nuclei. The sensory epithelium along the basal-to-apical axis of the cochlea responds to a range of high-to-low frequencies. The test frequencies used to measure the response should reflect this range and the ability of mice to hear at relatively high frequencies. In mice, the ABR is often used to assess the degree of deafness, detected as an increase in the sound pressure level (SPL), the threshold, required to evoke a response. Other informative features of the ABR include the waveform pattern and the latencies of the waveform peaks, which may suggest an origin of the defect between the ear and the brainstem.

[5] P. Dallos, A. Popper, and R. Fay, "The Cochlea." Springer, New York, 1996.
[6] Q. Y. Zheng, K. R. Johnson, and L. C. Erway, *Hearing Res.* **130**, 94 (1999).
[7] K. Parham, X.-M. Sun, and D. Kim *in* "Handbook of Mouse Auditory Research" (J. Willott, ed.), p. 37. CRC Press, Boca Raton, FL, 2001.

ABR Testing Set-up

A suitable apparatus, for example, is the SmartEP system manufactured by Intelligent Hearing Systems (Miami, FL). The device should be located in a sound-proof, or at least a sound-protected, room that minimizes noise sources such as those from air-conditioning vents, fluorescent tube lighting, machinery, and people entering and leaving the room.

The mouse is anesthetized using Avertin (tribromoethanol) to keep it still for the short duration of the test (about 15 min). A stock solution of Avertin is made by dissolving 1 g of tribromoethanol (Sigma) in 50 ml PBS. The stock solution tube is wrapped in aluminum foil and may be stored at room temperature for 1–2 weeks. Avertin is injected intraperitonally at a dose of 2–5 mg/10 g body weight. The lower dose range is used for younger, weanling mice. The mouse resuscitates fairly soon after testing and can be placed on a 37°C warm pad for a few minutes until it is active and can be returned to a cage.

For recording an ABR, platinum needle electrodes are placed sub-dermally on the mouse, which is placed lying on its belly (Fig. 1). The active electrode is placed at the crown of the head between the ears, the reference electrode under the left ear and the ground electrode under the right ear (Fig. 1B). The high-frequency transducers are placed loosely into the ears of the mouse with the tube placed just at the opening of the ear canal, taking care not to insert it too far.

The stimulus frequency, intensity and rate of presentation are controlled by adjusting the settings on the computer. The apparatus should be calibrated by the manufacturer. The test ear switch on the transmitter may be set to left (assuming similar hearing from either ear). Binaural stimulation through both ears and detection with one channel is suitable for most purposes in mice, as no consistent left–right ear asymmetry in response has been noted.[6] The apparatus generates multiple stimuli (up to 512) during the test and computes an average response that is amplified, filtered and displayed on the computer screen. A suitable rate of presentation of stimuli may be between 19 and 25/s. Pure tone bursts can be presented with a rise–fall time of 1.5 ms. The artifact rejection level for the EEG signal is set at 31.5%. The response is amplified 100,000 times and is filtered between 100 Hz (high pass) and 3000 Hz (low pass).

Testing Procedure

The frequencies tested should span the sensitive hearing range of mice and may include a click (a broadband of tones from 2 to 20 kHz), and pure

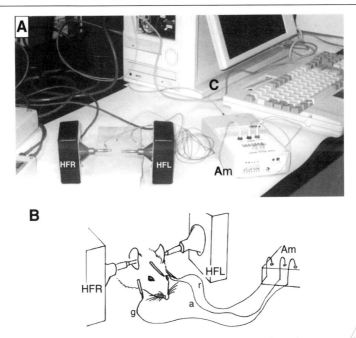

FIG. 1. ABR test apparatus. (A) The computer (C) controls the intensity, presentation rate and frequency of the test stimulus. The left and right high frequency transducers (HFL, HFR, respectively) are used to introduce the sound stimulus into the ears. The response is detected using subdermally placed electrodes and the signals passed through the pre-amplifier (Am) to the computer which summates the average response and displays a waveform on the screen. (B) Diagram of electrode placements: a, active, g, ground and r, reference (set for testing responses at the left ear). HFR, HFL, high frequency transducers.

tones of 8, 16 and 32 kHz.[6,8] The response to a given stimulus is initially tested at 90 dB SPL. The SPL intensity is then decreased in 10 dB steps to arrive at the minimal SPL (the threshold) at which a specific waveform can be discerned visually on the computer screen. Figure 2 shows an example of a normal mouse with a threshold at 43 dB SPL for a click stimulus. When approaching the threshold, smaller steps of 5 dB and then 3 dB may be used. The ABR is recorded on a normalized (relative) scale, which enhances the detection sensitivity. The characteristic four or five peaks of the mouse ABR are seen within the first 5–6 ms of stimulation. In humans, the ABR waveform has five well-characterized peaks, whereas in mice, the initial peaks are the most prominent and peaks 4 and 5 are less distinct.[6] After testing responses to a click stimulus, the procedure is repeated for the 8 kHz tone, and so on for the other test frequencies. A study of

[8] L. C. Erway, J. F. Willott, J. R. Archer, and D. E. Harrison, *Hearing Res.* **65**, 125 (1993).

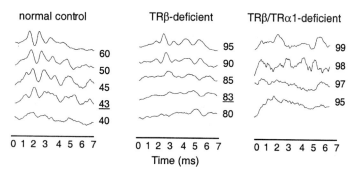

FIG. 2. Examples of ABR waveforms detected for a click stimulus in a normal hearing control mouse and in mice lacking thyroid hormone receptor β (TRβ-deficient) or lacking all known thyroid hormone receptors (TRβ/TRα1-deficient). The stimulus SPL is indicated to the right of each trace (in dB SPL). Threshold values are underlined. The waveforms are shown on a normalized scale for enhanced detection sensitivity. Specific responses are detected for the normal and for the TRβ-deficient mice, although responses in the mutant mouse are evoked only at an elevated threshold. No specific waveform is detected in the TRβ/TRα1-deficient mouse; rather only increased noise is recorded in the trace at the highest stimulus used (99 dB SPL).

many mouse strains at several months of age, indicates average normal thresholds for click, 8, 16 and 32 kHz stimuli at around 38, 30, 15 and 40 dB SPL, respectively.[6] At younger ages in some strains, thresholds may be lower (15–25 dB) for high frequency stimuli (32 kHz).[9]

Data Recording

For a given stimulus, waveforms for each SPL used are recorded and may be ordered to show a series of traces above and below threshold. After detection on a normalized scale, the ABR results may be converted to a fixed voltage scale (e.g., 3 or 4 μV) for a clearer summary of the results that minimizes noise in the trace.[9] In addition, the average threshold $\pm$ S.D. for groups can be plotted.

The absence of an ABR when using the highest possible SPL (e.g., ≥ 100 dB SPL) indicates profound deafness. The data of Fig. 2 show an example of mice lacking all known thyroid hormone receptors (TRβ and TRα1) that have no specific response to a click, even when using a stimulus SPL of 99 dB.[9] The traces of Fig. 2 also show examples of a moderately severe defect in mice lacking only TRβ, which exhibit an ABR

[9] A. Rüsch, L. Ng, R. Goodyear, D. Oliver, I. Lisoukov, B. Vennström, G. Richardson, M. W. Kelley, and D. Forrest, *J. Neurosci.* **21**, 9792 (2001).

but one with a significantly elevated threshold (83 dB SPL shown). This pattern suggests that the initial cochlear response and subsequent brainstem responses are severely diminished in the single knockout mouse although these auditory signals are not absent entirely. Other types of ABR abnormality may include the detection of a first peak, but loss of subsequent peaks, which may suggest defects in postcochlear, central brain functions.[10] Another pattern is exemplified by mutants with myelination defects in the auditory nerve and brain, which may display ABR waveforms with delayed peak latencies.[11,12]

Age and Background Strain Considerations

Auditory function in rodents is considered to begin around postnatal day 13 (P13);[13] however, testing the ABR in mice before weaning (before P21) is difficult because of their small size. The ABR can be tested more readily in post-weaning mice when placement of the transducers is easier and when the ABR waveforms are more robust. Mice can be tested as weanlings and later as adults (7–8 weeks of age) to establish the time-course of any auditory defect. A failure of development would cause consistent defects at young and older ages. This outcome would be distinguishable from a progressive loss of hearing after normal hearing had first developed.

The background strain is an important consideration in transgenic or targeted mutant mice because many inbred strains carry hearing loss genes that may confound the analysis.[6] For example, a B6CBAF1/J hybrid background may be useful for transgenic mice because the parental C57BL/6J and CBA/J strains retain reasonable hearing up until several months of age. For gene targeting, the choice of embryonic stem (ES) cell line is important since several of the 129 substrains used to derive ES cells, including 129/J, 129/ReJ and 129/SvJ, exhibit hearing loss.[6,14] In contrast, 129/Sv *steel* substrains have relatively normal thresholds. Targeted mutations are often transmitted on a mixed 129 × C57BL6/J background, such that with successive generations, the segregation of 129 substrain-derived hearing loss loci may result in deafness even in mice that are wild type for the targeted gene of interest. Many strains also have late-onset

[10] N. J. Parkinson, C. L. Olsson, J. L. Hallows, J. McKee-Johnson, B. P. Keogh, K. Noben-Trauth, S. G. Kujawa, and B. L. Tempel, *Nat. Genet.* **29**, 61 (2001).

[11] S. N. Shah and A. Salamy, *Neuroscience* **5**, 2321 (1980).

[12] T. Fujiyoshi, L. Hood, and T. Yoo, *Ann. Otol. Rhinol. Laryngol.* **103**, 449 (1994).

[13] R. Pujol, M. Lavigne-Rebillard, and M. Lenoir *in* "Development of the Auditory System" (E. Rubel, A. Popper, and R. Fay, eds.), p. 146. Springer, New York, 1997.

[14] E. Simpson, C. C. Linder, E. E. Sargent, M. T. Davisson, L. E. Mobraaten, and J. J. Sharp, *Nat. Genet.* **16**, 19 (1997).

(age-related) hearing loss, often first manifested at higher frequencies,[8] which may complicate analyses in older mice.

These problems may be bypassed by backcrossing the mutation onto a strain with reasonably good hearing, such as C57BL/6J. Five generations of backcrossing will reduce the 129 substrain component to only ~ 1.5% of the genome, while 10 or more generations are considered to yield a C57BL/6J congenic strain. In practice, even a few backcrosses should obviate most of the problems associated with endogenous hearing loss genes.

Cochlear Histology

The overall function of the cochlea is highly dependent on its biomechanical structure. Early studies by von Bekesy demonstrated that the tuning ability of the cochlea is in part structurally determined and is independent of active physiological processes.[15] Thus, a histological analysis of cochlear anatomy can be very informative. Tissue sections can be obtained from either frozen or paraffin embedded samples, but to obtain high quality sections a harder medium such as glycol methacrylate is required. After sectioning on a microtome, individual cell types and cellular structure can be identified by staining. A variety of stains are available, but we find that thionin is simple and fast, and offers good histological resolution.

Tissue should be fixed in a solution of 3% glutaraldehyde (glut)/2% paraformaldehyde (PFA) in 0.1 M phosphate buffered saline (PBS). Glutaraldehyde provides good preservation of fine structures while paraformaldehyde penetrates into tissues rapidly; therefore cellular structures are rapidly stabilized by paraformaldehyde and then more permanently fixed by glutaraldehyde.

Materials

    A.   3% Glutaraldehyde/2% PFA Fixative (100 ml)
        84 ml Water
        10 ml 10× PBS
        2 g PFA (Sigma)
        6 ml 50% EM grade glutaraldehyde (Sigma)
            Heat water on a stir plate to 55°C (do not exceed 60°C). Add PFA, mix until completely dissolved. Add 10× PBS and 50% EM

---

[15] W. S. Rhode, *J. Acoust. Soc. Am.* **67**, 1696 (1980).

grade glut. Confirm that pH is 7.4; if not adjust accordingly with HCl or NaOH. Cool to 4°C.

B. Decalcifying solution
    2% (w/v) EDTA in 1× PBS.
C. Methacrylate embedding medium
    Embedding medium is made according to the manufacturer's instructions (see below).
D. Thionin stain stock solution
    Dissolve 500 mg thionin (Sigma) in 49 ml water and 1 ml glacial acetic acid.

Cochlear Dissection

The easiest access to the cochlea is gained from the dorsal or upper surface of the base of the skull. Skin and dorsal cranial bones should be removed to expose the brain. Next, the brain is removed to expose the dorsal surface of the base of the skull. This dissection reveals the bilateral temporal bones containing the bony labyrinths, each consisting of the cochlea and vestibular parts (semicircular canals, sacculus and utriculus) of the inner ear (Fig. 3). The bony labyrinth can be removed using forceps to cut around the structure. In embryos, or pups younger than P4 or P5, the temporal bone and bony labyrinth consist of relatively soft cartilage. As development proceeds, the labyrinth and temporal bone become more ossified and begin to fuse to one another. It is possible to isolate the bony labyrinth from an adult, but care must be taken to separate the labyrinth from the rest of the temporal bone.

Following isolation, the bony labyrinth should be turned over to expose the lower (ventral) surface of the cochlea (Fig. 3B, C). This orientation places the sensory structures of the cochlea distal to the area of dissection, decreasing the chance that they will be damaged during subsequent steps. Using fine forceps, small holes should be opened in the cartilaginous capsule (or bony capsule in older samples) to expose the spiral cochlea. In adults, holes can be gently poked in the oval and round windows located at the base of the bony capsule, taking care not to damage the spiral. It may assist handling if the vestibular portions of the labyrinth are kept attached to the cochlea.

Fixation and Decalcification

Depending on the age of the animal from which the cochlea is dissected, the tissue can be fixed at any of several steps in the dissection procedure described above and the dissection then completed at a later time. For

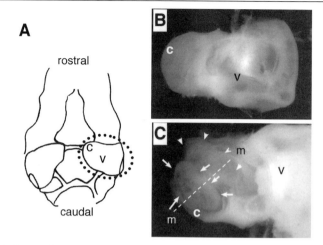

Fig. 3. Isolation of a neonatal mouse cochlea. (A) The temporal bones can be observed on each side of the upper surface of the base of the skull (shown after prior removal of the brain). The right-side temporal bone is circled with a dotted line. The portions that enclose the cochlea (c) and vestibular (v) components of the labyrinth are indicated. The entire temporal bone is removed by picking around the structure with forceps. (B) A magnified photograph of a temporal bone in the same orientation as it would be *in situ* in the skull shown in panel A. (C) The entire structure shown in panel B is flipped over, such that the lower side is now uppermost to make access to the cochlea easier. The cartilaginous surrounding capsule has been picked away carefully to reveal the spiral structure of the cochlea. The basal (arrowheads) and apical (arrows) turns of the cochlear duct marked; m–m. the central axis of the modiolus in the cochlear spiral.

embryos younger than embryonic day 16 (E16), it is only necessary to remove the brain prior to fixation of the skull base, while for embryos between E16 and E18, the bony labyrinth should be isolated prior to fixation. For adults, the bony labyrinth should be isolated and windows opened in the capsule prior to fixation. Quick dissection is particularly important for adults, as cochlear tissues are very sensitive to low oxygen.[16,17] In our experience, whole animal perfusion is unnecessary and does not improve the fixation. If the dissection cannot be completed within a few minutes, the removal of the temporal bones from the skull and subsequent steps may be performed in fixative. After dissection, cochleae should be immersed in fixative and gently rocked at 4°C for at least 2 hr.

For animals older than P7, decalcification of the temporal bone is required after fixation and prior to sectioning. Decalcify by immersing the

[16] T. Konishi, *Acta Otolaryngol.* **87**, 506 (1979).
[17] A. L. Nuttall and M. Lawrence, *Arch. Otolaryngol.* **105**, 574 (1979).

temporal bones in 2% (w/v) EDTA in 1× PBS at 4°C with gentle rocking for 1 to 4 weeks; the longer times are used for older mice. The EDTA solution can be exchanged several times during this period. At the end of the procedure, bone should be soft and cartilaginous in nature.

Embedding, Sectioning and Staining

Glycol methacrylate is a relatively soft plastic and allows sections to be cut with a steel rather than glass knife, using a regular microtome at thicknesses of 1–5 μm. Methacrylate-based embedding mediums include JB-4 and Immunobed (PolySciences, Warrington, Pennsylvania). To embed, samples should first be dehydrated through a graded series of ethanols (a mimimum of 15 min each in 30%, 50%, 70%, 90%, then twice in 100% ethanol), then embedded and polymerized in the embedding medium following the manufacturer's guidelines.

After polymerization, the tissue should be oriented to yield the desired plane of section (typically mid-modiolar) and the sample block is then shaped into a pyramid with the mounting face of the block forming the base of the pyramid. In addition, the cutting face should be made trapezoidal in shape such that the knife strikes the wider part of the trapezoid first. Cut sections are picked off the knife with a pair of forceps and placed on a slide in a drop of water. As the water dries, the section flattens onto the slide.

Sections are stained with thionin. The stock solution can be used directly (staining time 2–5 s) or diluted to allow extended staining times. If sections are stained too darkly, they may be destained with ethanol. After staining, slides are cover-slipped with a permanent mounting medium such as Permount (Fisher Scientific). Thionin is a nissl stain that stains epithelial cell nuclei dark blue and epithelial cytoplasm a lighter blue (Fig. 4). It is important to obtain sections representing all of the basal to apical turns of the cochlea since the different frequency sensitivities along the length of the cochlea have structural and cellular correlates.

Analysis of Gene Expression

Although increasing numbers of genes involved in cochlear function are being identified, the analysis of the regulation of these genes by nuclear receptors or other factors is still at an elementary stage. The limited amounts of tissue available in the small cochlea present some constraints on these analyses; however, the determination of gene expression patterns is a very important part of such studies because of the specialized nature of

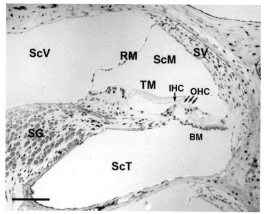

Fig. 4. Cross-section of an apical turn of the cochlea of a normal adult mouse prepared from a methacrylate plastic-embedded sample (3 μm section thickness). Each turn of the cochlear spiral is comprised of three chambers: the scala vestibuli (ScV) and scala media (ScM) separated by Ressner's membrane (RM), and the scala tympani (ScT). The floor of the scala media contains the organ of Corti, including the single row of inner (IHC) and triple row of outer (OHC) hair cells, which reside on the basilar membrane (BM). The lateral wall of the scala media contains the stria vascularis (SV). The noncellular tectorial membrane (TM) overlies the hair cells. Neutrites from the spiral ganglion (SG) extend into the organ of Corti to innervate both inner and outer hair cells. Scale bar = 100 μm.

the cochlear cell types. This analysis can be accomplished by standard methods of *in situ* mRNA hybridization with some modifications. Given the intricate three-dimensional structure of the cochlea, the analysis of both wholemounts and sections of the cochlea is useful (see examples in Fig. 5).

Tissue sections or wholemounts can be analyzed by *in situ* mRNA hybridization using nonradioactive, digoxigenin-UTP (DIG)-labeled riboprobes.[18,19] The presence of DIG in the riboprobe allows detection of probe/target mRNA hybrids using an antibody-coupled enzyme (alkaline phosphatase) in a colorigenic reaction. Antisense and control sense riboprobes are prepared using standard methods[20,21] that are not reiterated in detail here. Rather, we focus on the steps that are useful in the application of *in situ* mRNA hybridization to the cochlea.

[18] P. Lanford, Y. Lan, R. Jiang, C. Lindsell, G. Weinmaster, T. Gridley, and M. W. Kelley, *Nat. Genet.* **21**, 289 (1999).

[19] A. Campos-Barros, L. L. Amma, J. S. Faris, R. Shailam, M. W. Kelley, and D. Forrest, *Proc. Natl. Acad. Sci. USA* **97**, 1287 (2000).

[20] D. Wilkinson and M. Nieto, *Meth. Enzymol.* **225**, 361 (1993).

[21] F. Ausubel, R. Brent, R. E. Kingston, D. D. Moore, J. G. Seidman, J. A. Smith, and K. Struhl, "Current Protocols in Molecular Biology." John Wiley & Sons, Inc., 1997.

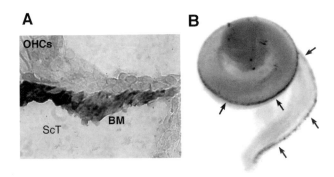

FIG. 5. *In situ* hybridization analysis of gene expression in the cochlea. (A) An example of gene expression detected in a cochlear section. A novel gene was isolated and found to be expressed in the cells below the basilar membrane of the cochlea in a P8 pup (L. Amma, J. Faris, DF, unpublished results). (B) Wholemount analysis used to detect expression of *Brn3.1*, in the sensory epithelium (arrows) of the cochlea in P0 mouse. Magnification: 100 × (A), 4 × (B).

In situ mRNA Hybribization of Cochlear Sections

Materials

Solutions are made up in autoclaved, distilled water unless otherwise noted.

A. NTE buffer
  500 mM NaCl
  10 mM Tris–HCl (pH 7.5)
  1 mM EDTA
B. Hybridization/Prehybridization buffer
  40% (v/v) formamide
  10% (w/v) dextran sulfate
  1× Denhardt's solution
  4× SSC
  10 mM dithiothreitol
  1 mg/ml yeast tRNA
    Before use, boil salmon sperm DNA and add to a final concentration of 100 μg/ml.
C. 0.1 M TEA (Triethanolamine), pH 8.0
    Stored wrapped in foil at room temperature. At the time of use, add 0.25% acetic anhydride (v/v) to the 0.1 M TEA.
D. TN Buffer, pH 7.5
  100 mM Tris–HCl,
  150 mM NaCl

E. TNTS Buffer, pH 7.5
    100 mM Tris–HCl
    150 mM NaCl
    0.1% (v/v) Triton X 100
    2.0% (v/v) sheep serum

F. TNM Buffer, pH 9.5
    100 mM Tris–HCl
    100 mM NaCl
    50 mM MgCl$_2$

G. Maleic acid buffer, pH 7.5
    100 mM Maleic Acid
    150 mM NaCl

H. Blocking Solution
    10% (w/v) blocking powder (from DIG Nucleic Acid Detection Kit, Roche Diagnostics) in maleic acid buffer. Heat in microwave to dissolve, then autoclave and aliquot into microfuge tubes. Store at −20°C.

I. Color Solution made in TNM buffer
    2% (v/v) NBT-BCIP stock (from DIG Nucleic Acid Detection Kit, Roche Diagnostics)
    2 mM levamisole

Dissection, Fixation and Sectioning

Cochleae should be fixed and decalcified as described above in Cochlear Histology, except that the fixative does not contain glutaraldehyde; instead 4% PFA in 1× PBS is used. All instruments and solutions should be RNase free, using usual precautions for RNA analyses. After fixation/decalcification, cochleae should be placed in 30% (w/v) sucrose/1× PBS overnight. Cochleae are then cryoprotected in OCT medium (Tissue-Tek) and sectioned on a cryostat at a thickness of 8–10 μm. For the best presentation of cochlear structures, cochleae should be sectioned in the plane of the modiolus (see Fig. 3C and Fig. 4). Store sections at −20°C.

Post-fixation

Slides should be air dried for 10 min prior to post-fixation. It is then convenient to lay slides flat and gently apply solutions using a disposable pipette during the steps below, as some tissue may detach when processed vertically in jars.

1. Fix for 5 min in 4% PFA in 1× PBS.
2. Rinse for 5 min in 3× PBS.

3. Rinse twice for 5 min each in 1× PBS.
4. Dehydrate through a graded series of ethanols: 30%, 60%, 80%, 95% and 100% ethanol (2 min for each step).
5. Store slides at −20°C (for up to several weeks).

Plasmid Digestion and Riboprobe Preparation

A plasmid with initiation sites for T7, T3 or SP6 RNA polymerases is used. The plasmid is linearized such that runoff transcription will yield probes of about 200–350 bases for hybridization of sections on slides. For hybridization of cochlear wholemounts (see below), longer probes are preferred (700–1200 bases). It may be convenient, in advance, to sub-clone a short fragment of the cDNA probe. Alternatively, longer cDNA inserts can be transcribed and then the RNA products partially hydrolyzed using an equal volume 50 mM NaHCO$_3$/120 mM Na$_2$CO$_3$ to give a probe population of the desired length.[20,21]

1. Restriction digest a stock of plasmid DNA (5–10 μg) to completion.
2. Extract linearized DNA with an equal volume of phenol/chloroform.
3. Precipitate the aqueous phase with 1/10 volume of 3 M sodium acetate, pH 5.2, and 2.5 volumes of 100% ethanol at −20°C overnight or at −80°C for 2 hr.
4. Isolate the precipitate via microcentrifugation for 30 min at 4°C.
5. Rinse the pellet with 70% ethanol.
6. Dry pellet and resuspend in 10 μl water.
7. Perform runoff transcription on 1 μg of DNA in a 20 μl volume following instructions in the DIG RNA Labeling Kit (Roche Diagnostics). The reaction is then treated with DNase I (20 units at 37°C for 30 min) to digest plasmid DNA and the riboprobe products purified by phenol/chloroform extraction, ethanol precipitation and washing in 70% ethanol.
8. To confirm successful transcription and removal of plasmid, 1 μl aliquots of the reaction both before and after DNase treatment are tested in parallel by electrophoresis on a 0.8% TBE gel.

Color Spot Test

The incorporation of DIG-labeled nucleotides into the riboprobe may be confirmed as follows:

1. Spot 100 ng of riboprobe on a small (∼ 1 cm) square of nitrocellulose.
2. Fix with a UV cross-linker or by baking at 80°C for 2 hr.
3. Using a 12-well tissue culture plate, wash filter twice covering the riboprobe squares for 5 min with 1 ml of TN Buffer.

4. Decant TN buffer and block for 15 min with 1 ml of TNTS.
5. Decant TNTS buffer and add antibody solutions at several test dilutions (e.g., 1:500, 1:1000, 1:2500), made in TNTS buffer. Incubate for 1 hr.
6. Wash twice for 8 min each in TN buffer.
7. Incubate for 10 min in TNM buffer 2.
8. Incubate in the NBT-BCIP color solution. A color product should appear within 15 min.

Pretreatment and Hybridization of Sections on Slides

1. Wash with 1× PBS for 5 min.
2. Incubate in 1 μg/ml proteinase K in 1× PBS at room temperature for 5 min.
3. Fix for 5 min in 4% PFA in 1 × PBS.
4. Rinse 3 times in 1× PBS for 5 min each.
5. Acetylate sections with 0.1 M TEA/0.25% acetic anhydride for 10 min.
6. Incubate in 1% Triton X-100 in 1× PBS at room temperature for 30 min.
7. Wash 3 times with 1× PBS for 5 min each.
8. Prehybridize at room temperature for 2 hr (add 100 μg/ml boiled salmon sperm DNA).
9. Heat probe at 80°C for 5 min, then add to hybridization mix at 200 ng/ml.
10. Cover the tissue sample with hybridization mix, and lay a coverslip on it. If signals prove to be weak, higher concentrations of probe may be used (e.g., 400 ng/ml).
11. Hybridize overnight at 55–62°C in a humid chamber. These temperatures provide only a guide and may be suitable for probes of 250–350 bases in length and 50–58% GC content. The temperature may be varied according to the predicted T_m of the probe/target RNA hybrids.

Post-hybridization Washes

Wash sense and antisense samples in separate coplin jars (with screw-cap lids) placed in a shaking water bath.

1. Float off cover slips in 2× SSC in a coplin jar.
2. Wash in 50% formamide/2× SSC at 65°C for 15 min.
3. Wash in 50% formamide/2× SSC at 65°C for 30 min.
4. Wash twice with 25% formamide/1× SSC/0.5× PBS at 65°C for 30 min each.

5. Wash in 1× PBS at room temperature for 5 min.
6. Treat with RNase A in NTE Buffer (20 μg/ml final concentration) at 37°C for 30 min.
7. Wash in 1× PBS at room temperature for 5 min.
8. Block sections with 10% sheep serum, 1% blocking reagent in maleic acid buffer at room temperature for 1 hr.
9. Incubate sections with antibody (at the lowest concentration that gave a signal in the test color reaction) in 1% sheep serum, 1% blocking reagent in maleic acid buffer for 2 hr.
10. Wash 3 times in maleic acid buffer at room temperature for 15 min.
11. Rinse in TNM buffer for 5 min.
12. Overlay color detection solution on slides in a humid chamber. Incubate in the dark.
13. Incubate for a few hours to overnight. Signals should appear purplish under the microscope.

In situ mRNA Hybridization of Cochlear Wholemounts

The wholemount *in situ* mRNA hybridization method is modified from Henrique et al.[22]

Materials

Solutions are made up in autoclaved, distilled water.

A. PTW
    1× PBS in 0.1% Tween-20
B. Hybridization Mix:
    50% Formamide
    5× SSC, pH 5.0
    50 μg/ml Yeast tRNA
    1% SDS
    50 μg/ml heparin
C. Detergent mix (adjust to pH 7.0)
    1% (v/v) IGEPAL ((Octylphenoxy)polyethoxyethanol, Sigma; equivalent to Nonidet P-40)
    1% (w/v) SDS
    0.5% (w/v) sodium deoxycholate
    50 mM Tris–HCl, pH 8.0

[22] D. Henrique, J. Adam, A. Myat, A. Chitnis, J. Lewis, and D. Ish-Horowicz, *Nature* **375**, 787 (1995).

1 mM EDTA
150 mM NaCl
D. Solution X, pH 6.2–6.6
    50% (v/v) Formamide
    2× SSC
    1% (w/v) SDS
E. NTMT:
    0.1 M NaCl
    0.1 M Tris–HCl, pH 9.5
    0.05 M MgCl$_2$
    0.1% (v/v) Tween-20

Tissue Processing

Cochleae are dissected and fixed as described above for hybridization of sections, except that during dissection the scala vestibuli and Reissner's membrane (Fig. 4) should be opened to allow adequate access by the probe. The extent of this opening depends on the specific tissues of interest. If for example, the internal organ of Corti is of interest then the scala vestibuli and Reissner's membrane should be opened widely; however, if more superficial tissues, such as the scala vestibuli or Reissner's membrane are of interest, then less intrusive, periodic openings can be made along the length of the spiral, leaving the tissue as intact as possible.

Dehydration/Rehydration of Tissue Samples

1. Wash 2× for 5 min in PTW
2. Wash 1× for 5 min in 50% methanol in PTW
3. Wash 2× for 5 min in 100% methanol (tissues can be stored at this step at −20°C for a few weeks)
4. Wash 1× for 5 min with 75% methanol in PTW
5. Wash 1× for 5 min with 50% methanol in PTW
6. Wash 2× for 5 min in PTW

Pre-treatments/Deproteination

1. Treat with 10 μg/μl proteinase K in PTW for 6 min
2. Remove proteinase K
3. Carefully rinse with Detergent Mix (2 × 5 min)
4. Post-fix in 0.2% glutaraldehyde/4% PFA for 20 min
5. Rinse with PTW (5 × 5 min)

Hybridization

Transfer tissue to a 1.5 ml tube with screw-on cap with an O-ring.

1. Rinse (10 min) in a 1:1 mix of PTW to hybridization mix (500 μl of each)
2. Rinse (10 min) in 1 ml hybridization mix
3. Incubate tubes horizontally at 70°C on shaking platform in 1 ml fresh hybridization mix for 2 hr. A heating block may be used lying on its side. The shaking speed should be sufficient to keep the solution and tissue moving.
4. Replace with 1 ml of fresh, pre-warmed hybridization mix and 1 μg DIG-labeled probe.
5. Incubate horizontally at 70°C on shaker overnight.

Post-hybridization Washes

1. Quickly rinse in Solution X
2. Wash 4 × for 30 min at 65°C in Solution X
3. Wash 10 min at 65°C with a 1:1 mixture of hybridization mix: PTW
4. Rinse (3 × 10 min) with PTW/levamisole
5. Incubate 3 hr in PTW/Levamisole with 10% heat-inactivated sheep serum on a shaker at room temperature.
6. Incubate overnight at 4°C on shaker in PTW, 10% sheep serum, and 1/2000 dilution of anti-DIG antibody.
7. Next day, rinse 3 × for 1 hr in PTW/levamisole at 4°C

Histochemistry

1. Wash 2 × for 10 min in 5 ml NTMT
2. Incubate at room temperature in the dark while shaking in 2.0 ml NTMT and 60 μl NBT-BCIP mix. Change the solution every 2 hr until color develops. Incubation may require 1 hr-overnight. Signal is detected as a purple color.
3. Remove solution when color is developed
4. Wash for 10 min with 5 ml NTMT
5. Wash for 10 min with PBS. Samples can also be stored in PBS at 4°C.
6. Hybridized wholemount samples can be analyzed directly under a microscope. It is also possible to make cryosections of the stained wholemount which may assist in identifying the cell types that are stained (see below).

Bleaching (Optional)

Bleaching may be used to minimize background color.

1. Cochleae should be at room temperature.
2. Replace the PBS with 70% or 90% ethanol
3. Watch closely under microscope for background to start fading (about 3 min).
4. When desired color is obtained, quickly remove ethanol and replace with PBS

Tissue Preparation for Cryostat Sectioning

1. Gradually replace PBS with sucrose using stepwise changes (20 min per step on a shaking platform) through the following mixes (w/v) of sucrose in $1\times$ PBS: 5%, 10%, 15%, 20%, 25% and finally 30% sucrose.
2. Gradually add OCT embedding medium to the tissue by stepwise replacement of the solution with 2:1, 1:1 then 1:2 (v/v) mixes of 30% sucrose in $1\times$ PBS to OCT.
3. Transfer tissue to 100% OCT, orient and freeze on a mixture of ethanol and dry ice.
4. Wrap in foil and store at $-20°C$ until sectioned.

Detection of β-Galactosidase Marker Gene Expression in Cochlea

Transgenic reporter genes fused to a β-galactosidase marker may be used for monitoring expression of cloned receptor or target genes *in vivo* using essentially standard methods. Although to date these analyses have been relatively little used in the cochlea compared to other tissues, they are likely to provide an additional, useful experimental approach.

Materials

A. Blue I solution (pH 6.9)
    34.74 g PIPES (disodium salt)
    0.41 g $MgCl_2 \cdot 6H_2O$
    0.76 g EGTA
    Make up to 1 liter with distilled water and filter sterilize. Store at room temperature.
B. Blue II solution (pH 7.3)
    100 ml $10\times$ PBS
    1.625 g Potassium ferricyanide
    2.1 g Potassium ferrocyanide $\cdot 3H_2O$
    0.4 g $MgCl_2 \cdot 6H_2O$

200 μl Nonidet P-40
0.1 g sodium deoxycholate
1 mg/ml 5'-Bromo-4-Chloro-3-indoyl-β-D-galactopyranoside (X-gal; Roche Molecular Biochemicals)
Make up to 1 liter with distilled water and filter sterilize. Store at room temperature.

Dissection

1. Cochleae are dissected as described above for Cochlear Histology and fixed for 2 hr on ice with 2% (w/v) paraformaldehyde in 1× PBS. A small hole should be made in the cochlear capsule to permit fixative to permeate into the sample.
2. Rinse cochleae in three changes of 1× PBS.
3. Transfer cochleae into 30% (w/v) sucrose in 1× PBS at 4°C (2 hr overnight).

β-Galactosidase Staining

1. Rinse cochleae for 5 min in 2 ml of Blue I Solution at room temperature.
2. Remove Blue I Solution and add 2 ml of Blue II Solution. Incubate at 37°C overnight for detection of reporter genes that are expressed at low abundance. Shorter times may be used for more abundantly expressed genes.
3. Rinse cochleae in three changes of 1× PBS.
4. Wash stained cochlea in 70% ethanol for 20 min on a rocking platform at room temperature to remove any crystalline deposits.
5. Rinse stained cochlea in three changes of 1× PBS.
6. Store stained cochlea at 4°C in 30% (w/v) sucrose in 1× PBS.
7. Samples can be analyzed under a microscope, as described above for *in situ* mRNA hybridization of wholemount samples.

Cochlear Explant Cultures

In vitro preparations provide a powerful technique to examine the direct effects of pharmacological agents, including ligands for nuclear receptors, on specific cell types.[23] Although single-cell dissociation of either the embryonic or adult cochlea leads to the death of most cochlear cell types, including sensory hair cells, organotypic explant cultures of both the embryonic or early postnatal cochlea have been shown to survive and develop over

---

[23] M. W. Kelley, X.-M. Xu, M. A. Wagner, M. E. Warchol, and J. T. Corwin, *Development* **119**, 1041 (1993).

extended periods of time. Sobkowicz and colleagues have demonstrated that postnatal cochleae can be maintained *in vitro* for up to 4 weeks and that both cytologic differentiation and afferent neuronal innervation proceed normally during this period,[24,25] Similarly, Kelley *et al.* demonstrated that hair cells and other features of the organ of Corti develop in explant cultures isolated from embryos.[23,26] In older mice, however, the ossification of the cochlea, the thinning of the epithelia within the membranous labyrinth and the differentiated nature of the epithelium make maintenance of the explants more difficult. Thus, successful explants have not been maintained for more than a few days when made from mice over one postnatal week of age.

Materials

   A.   HBSS, pH 7.4 (for 1 liter)
       100 ml 10× HBSS (Hank's Balanced Saline Solution; Invitrogen 14065-056)
          5 ml of 1 *M* HEPES (Sigma)
          895 ml sterile, distilled water
   B.   Culture medium, pH 7.2 to 7.4.
       DMEM (Invitrogen 12430-054)
          10% fetal bovine serum (Invitrogen)
          1× N2 supplement (from 100× stock; Invitrogen)
          1× Antibiotics (from 100× stock of 1500 units/ml penicillin G [Sigma], 9 mg/ml Fungizone [Invitrogen])
   C.   Matrigel substrate (BD Biosciences).
      1.   Dilute 300 μl of Matrigel in 8 ml of ice-cold, sterile HBSS.
      2.   Cover the culture surface with approximately 200 μl of the diluted Matrigel.
      3.   Incubate in a 37°C incubator for 40 min.
      4.   Remove diluted Matrigel and replace with HBSS. Coated surfaces can be prepared 24–48 hr prior to use. Store at room temperature (at 4°C, Matrigel will resolubilize).

Cochlear Dissection

     The initial dissection is as described for histological preparations except that all steps are performed with sterile instruments in sterile, ice-cold HEPES/HBSS either in a laminar flow hood or clean bench. Following isolation of the bony labyrinth and removal of the cochlear capsule, the cochlear spiral is exposed. This spiral is comprised of a duct that, in mice, by

[24] H. M. Sobkowicz, B. Bereman, and J. E. Rose, *J. Neurocytol.* **4**, 543 (1975).
[25] J. E. Rose, H. M. Sobkowicz, and B. Bereman, *J. Neurocytol.* **6**, 49 (1977).
[26] M. W. Kelley, D. R. Talreja, and J. T. Corwin, *J. Neurosci.* **15**, 3013 (1995).

P0 has reached its mature length of approximately 1.5 turns. When viewed from the ventral side, the roof of the duct comprises the developing Reissner's membrane and scala vestibuli, while the lateral wall of the duct will develop as the stria vascularis, and the floor of the duct will develop as the organ of Corti, the inner sulcus and the outer sulcus (see Fig. 4). To expose the organ of Corti, Reissner's membrane is carefully removed by breaking the medial edge of the roof of the duct along the medial side of the length of the cochlear spiral. Next, the roof of the duct can be deflected laterally and separated from the floor in the region of the outer sulcus.

Once the organ of Corti has been exposed, the spiral should be cut into non-overlapping pieces of less than one complete turn. Each piece can then be placed with the hair cell side of the organ of Corti oriented "upwards" onto a variety of different substrates. For optimal visualization of the explant, glass coverslips can be coated with collagen, poly-lysine, or Cell-Tak. Alternatively, explants can be placed on similarly coated membranes. Explants are cultured in DMEM with 10% fetal bovine serum. Penicillin/fungizone antibiotics are added but aminoglycosides, such as gentamycin, should be avoided because they are toxic to the sensory hair cells. The amount of medium used is not crucial, but should be sufficient to cover the entire explant ($> 75 \, \mu$l). The medium should be changed every 3 to 4 days, but explants can be maintained in the same medium for up to 6 days if necessary. Explants should be maintained at $35–37°C$ in a humidified incubator with 5% CO_2.

Individual explants should be examined after approximately 12 hr *in vitro* to confirm that they have attached to the substrate and that they are relatively flat (Fig. 6). Curled or floating explants are usually viable, but will probably not allow good visualization of the sensory epithelium. Attached explants tend to flatten over the first 72–96 hr *in vitro*. The sensory epithelium can be visualized after fixation and staining with an antibody that detects a hair cell-specific protein, such as myosin VIIa.

Cochlear RNA Analysis

Standard methods of RNA analysis including Northern blot analysis and cDNA library construction can be applied to the cochlea despite its small size, if pools of cochleae are used to give adequate amounts of starting material. As a guide, 36 individual cochlear capsules from 18 pups at P13 yield approximately 300 μg of total RNA. Less RNA is obtained from adult cochlea because of the ossification: 24 individual cochleae from 12 adults may yield 70 μg of total RNA. These preparations are adequate to detect many genes in the cochlea and have the advantage that the yield is reasonable and the speed of dissection is rapid. Specific sub-regions

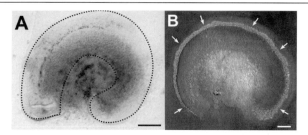

FIG. 6. Differential interference contrast (DIC) images of cochlear explant cultures. (A) Cochlear explant culture established from an E13 embryo after 12 hr *in vitro*. The base of the cochlear spiral is in the lower left and the developing organ of Corti spirals clockwise. (B) Cochlear explant culture established from an E13 embryo after 6 days *in vitro*. Hair cells have been fluorescently labeled with an antibody for myosin-VIIIa (expressed in hair cells) and superimposed onto the DIC image. Images kindly provided by Alain Dabdoub (A) and Jennifer Jones (B). Scale bar in A = 200 μm, scale bar in B = 100 μm.

within the cochlea may also be isolated, although this yields much less material and prolongs the time before tissue is frozen with an increased risk of some RNA degradation. To prepare poly(A)$^+$ selected mRNA for detection of low abundance mRNAs or for making cDNA libraries,[19] the numbers of pooled cochleae can be scaled up roughly 5–10-fold to generate enough material.

Acknowledgments

We thank Dr. L.C. Erway for valuable instruction on ABR analyses and Drs. A. Dabdoub and J. Jones for providing photographs in Figure 6. This work was supported in part by NIH (DC-03441), March of Dimes Birth Defects Foundation and a Hirschl Award (to DF) and by the Intramural Program at NIH (to MWK).

[25] Analysis of Estrogen Receptor Expression in Tissues

By MARGARET WARNER, LING WANG, ZHANG WEIHUA, GUOJUN CHENG, HIDEKI SAKAGUCHI, SHIGEHIRA SAJI, STEFAN NILSSON, THOMAS KIESSELBACH, and JAN-ÅKE GUSTAFSSON

Introduction

In 1995 with the discovery of the second estrogen receptor,[1] understanding of estrogen signaling acquired a new complexity. It is now clear that functions estrogen of are not confined to the female reproductive

[1] G. G. Kuiper, E. Enmark, M. Pelto-Huikko, S. Nilsson, and J.-Å. Gustafsson, *Proc. Natl. Acad. Sci. USA* **93**, 5925 (1996).

system, but that it also has important functions in nonreproductive tissues of both males and females.[2] The two estrogen receptors, ERα and ERβ, have distinct and sometimes opposing actions in multiple tissues.[3] In order to define the role of each receptor in estrogen-sensitive tissues, it is essential to have reliable methods to detect the expression of the two genes. Over the past several years, methods have been developed for the detection of ERα and its functions. The most common of these are: immunohistochemistry[4] and Western blotting[5] with specific antibodies; measurement of estrogen accumulation and retention in tissues;[6] measurement of the binding of estradiol in the 8S fraction after sedimentation in sucrose gradients of low salt tissue-extracts;[7] and the assessment of changes in mRNA levels of classic estrogen responsive genes.[8] There are several reasons why the presence of the second estrogen receptor has made it difficult to interpret results obtained with the use of these methods. The major difficulties include:

1. The low level of expression of ER in most tissues, which renders Western blotting problematic in tissue extracts.
2. The questionable affinity and specificity of available antibodies against ERα and ERβ and their use in Western blots of tissue extracts.
3. The secondary and tertiary structure of the receptors, which make several of the most interesting epitopes inaccessible to antibodies in immunohistochemistry, unless suitable retrieval techniques are employed.
4. Uncertainty regarding the subcellular location of the ER receptors, since in addition to the classic site in the nucleus, there is evidence that ERs also are located in the cell membrane and the cytosol.[9,10] This distribution means that nuclear staining cannot be used as a criterion for specificity of estrogen receptor antibodies.

[2] P. J. Shughrue and I. Merchenthaler, *Front Neuroendocrinol* **21**, 95 (2000).

[3] S. Nilsson, S. Makela, E. Treuter, M. Tujague, J. Thomsen, G. Andersson, E. Enmark, K. Pettersson, M. Warner, and J.-Å. Gustafsson, *Physiol. Rev.* **81**, 1535 (2001).

[4] E. V. Jensen, G. L. Greene, and E. R. DeSombre, *Prog. Clin. Biol. Res.* **249**, 283 (1987).

[5] J. D. Furlow, H. Ahrens, G. C. Mueller, and J. Gorski, *Endocrinology* **127**, 1028 (1990).

[6] E. V. Jensen, P. I. Brecher, S. Numata, S. Smith, and E. R. DeSombre, *Methods Enzymol.* **36**, 267 (1975).

[7] A. C. Tate, E. R. DeSombre, G. L. Greene, E. V. Jensen, and V. C. Jordan, *Breast Cancer Res. Treat.* **3**, 267 (1983).

[8] H. Rochefort, *Ciba Found Symp.* **191**, 254 (1995).

[9] K. L. Chambliss, I. S. Yuhanna, R. G. Anderson, M. E. Mendelsohn, and P. W. Shaul, *Mol. Endocrinol.* **16**, 938 (2002).

[10] M. Wehling, *Annu. Rev. Physiol.* **59**, 365 (1997).

5. The sedimentation of ERβ at 4S in sucrose gradients,[11] even in low salt tissue extracts, versus the sedimentation of ERα at 8S. Since the 4S region is close to nonreceptor binding peaks such as albumin, the ERβ-binding peak can be mistaken for nonspecific binding and therefore must be defined with antibodies.

6. The presence of nonestrogen binding or low affinity forms of ERβ,[12,13] which even if abundantly expressed, do not lead to tissue retention of estradiol.

7. The distinct profiles of genes regulated by ERα and ERβ, and the fact that the two receptors have opposite actions on some gene promoters.[14]

One of the most troublesome problems in studying estrogen action *in vivo*, particularly in complex tissues that contain multiple cell types, is the cellular localization of the two ERs and identification of cells that express both receptors. When ERα and ERβ are co-expressed in the same cell, the overall action of estradiol in that cell will depend on the relative ratio of the two receptors, and on the particular isoform of ERβ expressed in the cell. Also of importance are the expression patterns of coactivators and corepressors of ER as well as those of other ER interacting proteins (e.g., NFkB[15]) and genes whose promoters have AP-1[16] or Sp1[17] motifs. ERα and ERβ may oppose each other when the latter sites are present.

One issue that has not been completely resolved is the length of the N-terminal domain of ERβ in various tissues and the length of longest human ERβ. Several N-terminal variants of ERβ have been reported that are referred to by the number of amino acids in their sequence. In rodents, it is accepted that ERβ 549 and 530 are expressed in many tissues. The sequences of their N-termini have been established after purification from

[11] C. Palmieri, G. J. Cheng, S. Saji, M. Zelada-Hedman, A. Warri, Z. Weihua, S. Van Noorden, T. Wahlstrom, R. C. Coombes, M. Warner, and J.-Å. Gustafsson, *Endocr. Relat. Cancer* **1**, 1 (2002).

[12] S. Ogawa, S. Inoue, T. Watanabe, A. Orimo, T. Hosoi, Y. Ouchi, and M. Muramatsu, *Nucleic Acids Res.* **26**, 3505 (1998).

[13] K. Maruyama, H. Endoh, H. Sasaki-Iwaoka, K. Kanou, E. Shimaya, S. Hashimoto, S. Kato, and H. Kawashima, *Biochem. Biophys. Res. Commun.* **246**, 142 (1998).

[14] K. Paech, P. Webb, G. G. Kuiper, S. Nilsson, J.-Å. Gustafsson, P. J. Kushner, and T. S. Scanlan, *Science* **277**, 1508 (1997).

[15] E. Speir, Z. X. Yu, K. Takeda, V. J. Ferrans, and R. O. Cannon 3rd, *Circ. Res.* **87**, 1006 (2000).

[16] R. V. Weatherman and T. S. Scanlan, *J. Biol. Chem.* **276**, 3827 (2001).

[17] A. Zou, K. B. Marschke, K. E. Arnold, E. M. Berger, P. Fitzgerald, D. E. Mais, and E. A. Allegretto, *Mol. Endocrinol.* **13**, 418 (1999).

rat prostate extracts[18]. In addition, both proteins can also have an 18 amino acid (AA) insert described by Chu and Fuller[19] in their ligand-binding domains. These isoforms have been detected by Western blotting and by reverse transcriptase PCR (RT-PCR). No N-terminal sequences corresponding to ERβ 485 were found in the rat.[18] In human tissues, there is clear evidence for an ERβ 530 since antibodies raised against the first 18 AA of this sequence recognize a protein of the correct size in human tissue extracts.[20] There is one study showing the existence of human ERβ 548.[21] This protein results from an N-terminal extension of the 530 isoform found in rodents. In many labs the upstream ATG, which normally would be used for translation of this isoform in humans, is out of frame with the rest of the sequence and thus this isoform is not produced. Wilkinson et al.[21] isolated clones of ERβ 548 from a human genomic library and a specific antibody was raised against the first 18 AA of the sequence. Expression of the protein was found in human breast, kidney and lung. The reason for the differences between this laboratory and others regarding the existence of the ERβ 548 variant is not yet clear. It is possible that there is a polymorphism (e.g., a nucleotide deletion) in the 5' region of the human gene that changes the translational reading frame but no information on the frequency of this variant is yet available.

Although *in situ* hybridization has been used to identify cells that harbour ERα and ERβ mRNA, the strongest proof for the presence of functional receptors is detection of the proteins, and for such studies, good antibodies are essential. There are several antibodies available, and these have been used by many laboratories with varying degrees of success. In this chapter, we outline the problems encountered in the published literature on the quantification and localization of estrogen receptors and suggest guidelines that allow the comparison of results obtained in different laboratories. Tables I and II list the characteristics of the most widely used and commercially available antibodies, as well as the properties of more selective antisera that are available from KaroBio (Sweden) for noncommercial use. This list is not meant to be a complete survey of antisera from the literature, but it does illustrate the variety of antibodies available and the often times very different results that can be obtained with the same antibody.

[18] Z. Weihua, S. Makela, L. C. Andersson, S. Salmi, S. Saji, J. I. Webster, E. V. Jensen, S. Nilsson, M. Warner, and J.-Å. Gustafsson, *Proc. Natl. Acad. Sci. USA* **98**, 6330 (2001).

[19] S. Chu and P. J. Fuller, *Mol. Cell. Endocrinol.* **132**, 195 (1997).

[20] S. A. Fuqua, R. Schiff, I. Parra, W. E. Friedrichs, J. L. Su, D. D. McKee, K. Slentz-Kesler, L. B. Moore, T. M. Willson, and J. T. Moore, *Cancer Res.* **59**, 5425 (1999).

[21] H. A. Wilkinson, J. Dahllund, H. Liu, J. Yudkovitz, S. J. Cai, S. Nilsson, J. M. Schaeffer, and S. W. Mitra, *Endocrinology* **143**, 1558 (2002).

TABLE I
ERα ANTIBODIES

Antibody name and source	Immunohistochemistry	Western blotting
H222 Rat monoclonal, Abbott	Frozen and paraffin: nuclear staining[a], frozen rat brain[b]	Single protein at 67 kDa, rat brain[c], mouse uterus nuclear and cytoplasmic proteins (no size markers)[d]
ER21 Rabbit polyclonal against the first 21 AA of receptor, Geoffrey Greene	Rat uterus[e], retrieval with glycine buffer pH 3.1[e]	
HC-20 Rabbit polyclonal, Santa Cruz	Paraffin: human brain nuclear and cytosolic[f]; and lung, cytosolic and nuclear[g]	Proteins at 50 and 80 kDa in lung cancer and MCF-7 cells[g]
MC-20 C-terminal, Santa Cruz	Strong specific staining in mouse brain nuclear and cytoplasmic in human SON[h]	
Mouse monoclonal, DAKO	Paraffin: human breast nuclear staining[i]; prostate stromal cells[j]	Single 67 kDa protein from human breast cancers[i]
C-terminus, Upstate Biotechnologies	Frozen: rat brain[k] in parvalbumin and calbinding-negative cells of the cortex; human male and female tissues[l]	67 kDa nucleus, cytoplasm and plasma membrane in endothelial cells[m]
Mouse monoclonal AER320 AA495-595 Labvision Freemont Ca	Acrolein perfused: widespread staining in the cerebral cortex of rats[n]	
Polyclonal against N-terminal 25 AA of ERα, Zymed	Acrolein perfused: residual staining in the VMH of brain of ERα −/− mice[o] with nuclear staining in hippocampus[p] and forebrain[p]	67 kDa hippocampal cytosol
C-terminus, C1355, Dr. Margaret Shupnik	Spines of pyramidal neurons and synaptic vesicles[r]	
Polyclonal against ERα expressed in E. coli, Dr. Okamura		
6F11, mouse monoclonal, Novo Castra	Paraffin: human prostate[s] and breast[t] nuclear staining	

[a] A. N. Clancy, D. Zumpe, and R. P. Michael, Horm. Behav. **38**, 86 (2002).
[b] A. N. Clancy, D. Zumpe, and R. P. Michael, Neuroendocrinology **61**, 98 (1995).
[c] J. A. Butler, I. Kallo, M. Sjoberg, and C. W. Coen, J. Neuroendocrinol. 11, **325** (1999).

[d] S. Migliaccio, T. F. Washburn, S. Fillo, H. Rivera, A. Teti, K. S. Korach, and W. C. Wetsel, *Endocrinology* **139**, 4598 (1998).

[e] K. P. Nephew, X. Long, E. Osborne, K. A. Burke, A. Ahluwalia, and R. M. Bigsby, *Biol. Reprod.* **62**, 168 (2000).

[f] T. A. Ishunina and D. F. Swaab, *Neurobiol. Aging* **22**, 417 (2001).

[g] L. P. Stabile, A. L. Davis, C. T. Gubish, T. M. Hopkins, J. D. Luketich, N. Christie, S. Finkelstein, and J.M. Siegfried, *Cancer Res.* **622**, 141 (2002).

[h] T. A. Ishunina, F. P. Kruijver, R. Balesar, and D.F. Swaab, *J. Clin. Endocrinol. Metab.* **85**, 3283 (2000).

[i] P. T. Saunders, M. R. Millar, K. Williams, S. Macpherson, C. Bayne, C. O'Sullivan, T. J. Anderson, N. P. Groome, and W. R. Miller, *Br. J. Cancer* **86**, 250 (2002).

[j] L. G. Horvath, S. M. Henshall, C. S. Lee, D. R. Head, D. I. Quinn, S. Makela, W. Delprado, D. Golovsky, P. C. Brenner, G. O'Neill, R. Kooner, P. D. Stricker, J. J. Grygiel, J.-Å. Gustafsson, and R. L. Sutherland, *Cancer Res.* **61**, 5331 (2001).

[k] M. F. Kritzer, *Cereb. Cortex* **12**, 116 (2002).

[l] A. H. Taylor and F. J. Al-Azzaw, *Mol. Endocrinol.* **24**, 145 (2000).

[m] K. L. Chambliss, I. S. Yuhanna, R. G. Anderson, M. E. Mendelsohn, and P. W. Shaul, *Mol. Endocrinol.* **16**, 938 (2002).

[n] J. A. Butler, K. M. Sjoberg, and C. W. Coen, *J. Neuroendocrinol.* **11**, 325329 (1999).

[o] C. A. Moffatt, E. F. Rissman, M. A. Shupnik, and J. D. Blaustein, *J. Neurosci.* **18**, 9556 (1998).

[p] D. T. Solum and R. J. Handa, *Brain Res. Dev. Brain Res.* **128**, 165 (2001).

[q] B. Greco, E. A. Allegretto, M. J. Tetel, and J. D. Blaustein, *Endocrinology* **142**, 5172 (2001).

[r] M. M. Adams, S. E. Fink, R. A. Shah, W. G., Janssen S. Hayashi, T. A. Milner, B. S. McEwen, and J. H. Morrison, *J. Neurosci.* **22**, 3608 (2002).

[s] I. Leav, K. M. Lau, J. Y. Adams, J. E. McNeal, M. E. Taplin, J. Wang, H. Singh, and S. M. Ho, *Am. J. Pathol.* **159**, 79 (2001).

[t] E. V. Jensen, G. Cheng, C. Palmieri, S. Saji, S. Makela, S. Van Noorden, T. Wahlstrom, M. Warner, R. C. Coombes, and J.-Å. Gustafsson, *Proc. Natl. Acad. Sci. USA* **98**, 15197 (2001).

TABLE II
ERβ ANTIBODIES

Antibody name and source	Immunohistochemistry	Western blotting
Ligand binding domain rabbit, KaroBio AB Antibody 503, chicken KaroBio AB	Not recommended Frozen and paraffin: human rat and mouse tissues[d,e,f] Paraffin: Human prostate, nuclear staining[g]	Human, rat and mouse tissues[a,b,c] Not recommended
INS, chicken, anti-peptide, 18 AA insert, KaroBio AB	Nuclear staining in rat mammary gland[b], no signal in human tissue	Mammary gland[b] co-migrating with ERβ ins standard
Anti-peptide, AA 214-258, rabbit, Saunders et al.[h]	Paraffin: rat nuclear staining in multiple tissues[h]	
Anti-peptide, AA 99-116, rabbit, KaroBio AB	Paraffin: human nuclear and cytoplasmic, does not recognize rodent ERβ	Human breast cancer extracts 60 kDa[i]
N-terminus, AA 2-18, mouse monoclonal, Fuqua et al.[i]		MCF-7 cells 60 kDa protein[i]
Antipeptide, AA 130-143, rabbit, Fuqua et al.[j]	Frozen: human myometrium, nuclear staining[j]	Human myometrium, 60 kDa[j] and prostate, rat ovary 55 kDa[k]
N-terminus, goat, sequence around AA 118, Santa Cruz	Paraffin: human prostate nuclear staining[k], brain nuclear and cytoplasmic in SON, AVP neurons[l] and SCN[m]	
CX, sheep, KaroBio AB	Special retrieval conditions for use with human breast sections, unpublished[n]	60 kDa protein in human breast[a]
H-150, N-terminus, rabbit, Santa Cruz	Paraffin: hepatic stellate cells nuclear stain[o]	55 kDa protein in liver[o]
N-terminus, AA 1-14, Ligand Pharmaceuticals, San Diego	Frozen: rat brain neuronal staining in forebrain[p]	
AA 99-116, rabbit, Upstate	Paraffin: human nuclear staining in multiple issues but not in hippocampus[q] and breast cancer nuclear staining[r]	
C terminal, last 18 AA, rabbit, Zymed	Frozen: rat brain, in dopaminergic neurons[s] and in parvalbumin and calbindin-positive neurons of the cortex[t] Frozen[u] and paraffin: nuclear staining in LHRH neurons[v]	
Rabbit Panvera,		Many background proteins and intense protein at 50 kDa[w]
F domain, AA 449-465, rabbit	Paraffin: human prostate, nuclear staining of basal cells[x]	63 kDa protein[x]
1D5 C-terminus, AA 512-530, Saunders et al.[z]	Paraffin: human breast, nuclear staining[y]	60 and 52 kDa proteins in human breast cancer[y]
PAI-310, C-terminus, AA 467-485, Affinity BioReagents	Frozen: human breast, nuclear stain with weak cytoplasmic stain[z] Frozen: rat brain after acrolein perfusion nuclear staining in the PVN and SON[aa] or in the APVN not colocalized with vasopressin[ad]	54 kDa protein in nuclear, cytosolic and plasma membrane fractions of endothelial cells[ab] 80 and 55 kDa proteins in rat hypothalamus, ovary and GT1 neurons[ac]

[a]C. Palmieri, G. J. Cheng, S. Saji, M. Zelada-Hedman, A. Warri, Z. Weihua, S. Van Noorden, T. Wahlstrom, R. C. Coombes, M. Warner, and J.-Å Gustafsson, *Endocr. Relat. Cancer* **1**, 1 (2002).

[b]S. Saji, H. Sakaguchi, S. Andersson, M. Warner, and J.-Å. Gustafsson, *Endocrinology.* **142**, 3177 (2001).

[c]Z. Weihua, S. Makela, L. C. Andersson, S. Salmi, S. Saji, J. I. Webster, E. V. Jensen, S. Nilsson, M. Warner, and J.-Å. Gustafsson, *Proc. Natl. Acad. Sci. USA* **98**, 6330 (2001)

[d]S. Saji, E. V. Jensen, S. Nilsson, T. Rylander, M. Warner, and J.-Å. Gustafsson, *Proc. Natl. Acad. Sci. USA* **97**, 337 (2000).

[e]Y. Omoto, Y. Kobayashi, K. Nishida, E. Tsuchiya, H. Eguchi, K. Nakagawa, Y. Ishikawa, T. Yamori, H. Iwase, Y. Fujii, M. Warner, J.-Å. Gustafsson, and S. I. Hayashi, *Biochem. Biophys. Res. Commun.* **285**, 340 (2001).

[f]S. Salmi, R. Santti, J. A. Gustafsson, and S. Makela, *J. Urol.* **166**, 674 (2001).

[g]L. G. Horvath, S. M. Henshall, C. S. Lee, D. R. Head, D. I. Quinn, S. Makela, W. Delprado, D. Golovsky, P. C. Brenner, G. O'Neill, R. Kooner, P. D. Stricker, J. J. Grygiel, J. -Å. Gustafsson, and R. L. Sutherland, *Cancer Res.* **61**, 5331 (2001).

[h]P. T. Saunders, S. M. Maguire, J. Gaughan, and M. R. Millar. *J. Endocrinol.* **154**, R13 (1997).

[i]S. A. Fuqua, R. Schiff, I. Parra, W. E. Friedrichs, J. L. Su, D. D. McKee, K. Slentz-Kesler, L. B. Moore, T. M. Willson, J. T. Moore, *Cancer Res.* **59**, 5425 (1999).

[j]J. J. Wu, E. Geimonen, and J. Andersen, *Eur. J. Endocrinol.* **142**, 92 (2000).

[k]M. Royuela, M. de Miguel, F. R. Bethencourt, M. Sanchez-Chapado, B. Fraile, M. I. Arenas, and R. Paniagua, *J. Endocrinol.* **168**, 447 (2001).

[l]T. A. Ishunina and D. F. Swaab, *Neurobiol. Aging.* **22**, 417 (2001).

[m]F. P. Kruijver and D. F. Swaab. *Neuroendocrinology* **75**, 296 (2002).

[n]S. Saji, S. Horiguchi, N. Funada, M. Warner, J.-Å. Gustafsson, and M. Toi, *Cancer Res.* **62**, 4849 (2002).

[o]Y. Zhou, I. Shimizu, G. Lu, M. Itonaga, Y. Okamura, M. Shono, H. Honda, S. Inoue, M. Muramatsu, and S. Ito, *Biochem. Biophys. Res. Commun.* **286**, 1059 (2001).

[p]B. Greco, E. A. Allegretto, and J. D. Tetel, *Endocrinology* **142**, 5172 (2001).

[q]A. H. Taylor and F. J. Al-Azzaw, *Mol. Endocrinol.* **24**, 145 (2000).

[r]E. V. Jensen, M. Jordan, G. Cheng, C. Palmieri, S. Saji, S. Makela, S. Van Noorden, T. Wahlstrom, M. Warner, R. C. Coombes, and J.-Å. Gustafsson, *Proc. Natl. Acad. Sci. USA* **98**, 15197 (2001).

[s]L. M. Creutz and M. F. Kritzer, *J. Comp. Neurol.* **446**, 288 (2002).

[t]M. F. Kritzer, *Cereb Cortex.* **12**, 116 (2002).

[u]C. Orikasa, Y. Kondo, S. Hayashi, B. S. McEwen, and Y. Sakuma, *Proc. Natl. Acad. Sci. USA* **99**, 3306 (2002).

[v]E. Hrabovszky, A. Steinhauser, K. Barabas, P. J. Shughrue, S. L. Petersen, I. Merchenthaler, and Z. Liposits, *Endocrinology* **142**, 3261 (2001).

[w]L. P. Stabile, A. L. Davis, C. T. Gubish, T. M. Hopkins, J. D. Luketich, N. Christie, S. Finkelstein, and J. M. Siegfried, *Cancer Res.* **622**, 141 (2002).

[x]I. Leav, K. M. Lau, J. Y. Adams, J. E. McNeal, M. E. Taplin, J. Wang, H. Singh, and S. M. Ho. *Am. J. Pathol.* **159**, 79 (2001).

[y]P. T. Saunders, M. R. Millar, K. Williams, S. Macpherson, C. Bayne, C. O'Sullivan, T. J. Anderson, T. J. Groome, and W. R. Miller, *Br. J. Cancer* **86**, 250 (2002).

[z]T. A. Jarvinen, M. Pelto-Huikko, K. Holli, and J. Isola, *Am. J. Pathol.* **156**, 29 (2000).

[aa]X. Li, P. E. Schwartz, and E. F. Rissman, *Neuroendocrinology* **66**, 63 (1997).

[ab]K. L. Chambliss, I. S. Yuhanna, R. G. Anderson, M. E. Mendelsohn, and P. W. Shaul, *Mol. Endocrinol.* **16**, 938 (2002).

[ac]D. Roy, N. L. Angelini, and D. D. Belsham, *Endocrinology* **140**, 5045 (1999).

[ad]S. E. Alves, V. Lopez, B. S. McEwen, and N. G. Weiland, *Proc. Natl. Acad. Sci. USA* **95**, 3281 (1998).

Information Required for Inter-Laboratory Comparisons of Western Blots and Immunohistochemistry

One of the specificity checks that is essential both for immuno-histochemistry and Western blots is a pre-adsorption control. Sometimes this test is simply done by incubating the antibody with excess antigen prior to use. Other times, as in the case of steroid receptors, where the protein is large and many epitopes are not exposed on the surface of the folded protein, covalent linkage to a matrix is necessary to obtain efficient adsorption. Coupling to an insoluble matrix also can help when the antigen is a small peptide. In this case, in order to avoid covalent attachment to the support through amino groups in lysines, which may be important for antibody recognition, it is advisable to add a cysteine at the end of the peptide and to use the added SH group for coupling to the support.

To remove cross-reacting antibodies, an antiserum can be preincubated with tissue extracts from ER knockout mice.

Western Blotting

Except for the prostate and ovary in the case of ERβ, and the mature uterus in the case of ERα, the content of estrogen receptors in tissues is low. This is because most of the granulosa cells in the ovary and epithelial cells in the prostate express ERβ, and most cells in the adult uterus express ERα, but only a fraction of cells in other tissues harbour these receptors. From estrogen-binding studies, the content of ERα in the mature rat uterus and cytosol of MCF-7 cells is about 200 fmol/mg cytosolic protein.[22] This level means that analysis of 10–50 μg cytosolic protein on an SDS polyacrylamide gel is equivalent to 2–10 fmol of ERα, a range easily detectable with a good antibody. In most tissues, the content of ERα is less than one-tenth that in the uterus. This level means that more cytosolic protein must be analyzed on an SDS gel in order to reliably detect estrogen receptors and, unless the antibody is very selective, analysis of more protein invariably results in high backgrounds and nonspecific binding. In breast cancers the range of ERα concentrations is 10–2000 fmol/mg cytosolic protein.[22] More than 50% of samples have less than 200 fmol/mg cytosolic and thus are difficult to analyze by Western blotting unless an enrichment step is included in the work up of the sample.

[22] E. R. DeSombre, S. M. Thorpe, C. Rose, R. R. Blough, K. W. Andersen, B. B. Rasmussen, and W. J. King, *Cancer Res.* **46**, 4256s (1986).

Recombinant Proteins as Positive Controls

Recombinant ERα and ERβ1 (ERβ 530) proteins are available from Panvera (Madison, WI). These proteins have been quantified by ligand binding assays so it is possible to electrophorese known amounts of receptor on an SDS gel. All estrogen receptor proteins should be divided into small aliquots and thawed only once to avoid degradation. Loading of receptor standards is an essential control since a comparison of the intensity of the standard with that in the experimental sample is a good indication of whether the immunoreactive protein is likely to be an ER. If the estrogen receptor content of the tissue is 10 fmol/mg cytosolic protein and 100 μg protein is analyzed, then the intensity of the receptor protein signal cannot exceed that seen with 1 fmol of standard. More intense signals are most likely artefacts. This interpretation is also true for proteins of the incorrect size, which are sometimes designated as receptor breakdown products. If there is a suspicion of degradation during the preparation of cell extracts, then this notion should be confirmed by examining the stability of *in vitro* translated, [^{35}S]-labeled receptor, which can be added to the tissue during the homogenization step. The addition of two protease inhibitor cocktail tablets (Boehringer-Mannheim, Germany) per 50 ml to all buffers used in the isolation procedures is effective in preventing degradation of receptors.

Some common cross-reacting proteins can be eliminated easily or identified as such. For example, if tissue is not perfused before use, there can be a large protein corresponding to serum albumin with which many antibodies react. Since albumin has a mass of 67 kDa, it can be mistaken for ERα. Albumin can be removed from cell extracts by adsorption onto Sepharose blue gel. Another abundant protein that can be a problem when monoclonal antibodies are used particularly in mouse tissues and also in human tissues,[20] is IgG, which will interact with secondary antimouse antibodies and produce very intense signals. These proteins, which have a mass of ~ 50 kDa, are sometimes mistaken for receptor degradation products. In this case, a control blot in which the primary antibody is omitted will identify the signal as IgG.

Identification of ER by Protein Sequencing

Available estrogen receptor antibodies when tested with recombinant ERα or ERβ standards exhibit the appropriate specificity; however, this recognition does not indicate them to be high affinity antibodies. Thus, when extracts are used from tissues in which the receptor content is low, the antibodies may interact with both the estrogen receptor as well as background proteins. With the initial use of antibodies in tissues where estrogen receptors have not been characterized, it is advisable to obtain the N-terminal sequence of the protein that is recognized by the antibody on a Western blot. With today's ultra-sensitive machines, enough sequence can be obtained from sub-picomolar amounts of protein excised from a PVDF blot to permit an unambiguous identification of the protein. Identification can also be carried out using protein in the low femtomolar range by the acquisition of a peptide mass fingerprint. In this experiment, proteins are excised from SDS gels and digested in the gel with a protease such as trypsin to produce a set of sequence-specific peptides that are analyzed by MALDI-TOF mass spectrometry or by nanoflow electrospray mass spectrometry.[23] Protein identification can be achieved by an MS/MS analysis of single peptides.[24]

Tissue Specific Problems and Special Retrieval and Fixation Techniques

With most available antibodies, ERβ is easily detected in granulosa cells of the ovary and the epithelial cells of the prostate, two cell types where it is abundantly expressed. In tissues that contain less receptor and in which it is difficult to detect by immunohistochemistry, recognition can be improved with the use of harsher retrieval methods. In the adult uterus, ERβ is not detected by immunohistochemistry even though the mRNA is present. With the C-terminal ERβ antibody (Zymed), when citrate buffer is used for antigen retrieval, ERβ is easily detected in granulosa cells of wild type mice and rats, but no signal is detected in the uterus. When 0.8 M urea is used instead of citrate buffer as antigen retrieval solution, ERβ is detected in most of the epithelial and stromal cells of the uterus.[25] The urea retrieval procedure does not influence adversely the detection of ERβ in the ovarian granulosa cells. The specificity of the staining can be confirmed with sections from the uterus and granulosa cells of ERβ −/− mice in which no specific staining is found.

In immunohistochemical studies of human tissues, one problem that arises when C-terminal antibodies are used is the presence of two distinct C-termini, which result from the use of alternative exons.[12] For example, if

---

[23] A. Shevchenko, M. Wilm, O. Vorm, and M. Mann, *Mass Anal. Chem.* **68**, 850 (1996).

[24] M. Mann, R. C. Hendrickson, and A. Pandey, *Annu. Rev. Biochem.* **70**, 437 (2001).

[25] Z. Weihua, J. Ekman, Å. Almkvist, S. Saji, L. Wang, M. Warner, and J.-Å. Gustafsson, *Biol. of Reprod.* **67**, 616 (2002).

ERβ cx, the isoform with the alternative eighth exon, is the dominant isoform expressed in the tissue, the commercial C-terminal antibodies will give negative results while N-terminal antibodies will detect both ERβ and ERβ cx.

Another important consideration is optimization of the fixation procedure. Some epitopes are sensitive to high concentrations of aldehydes and others must be revealed with acrolein. Kritzer[26] has tested several protocols and found that for ERβ immunostaining, lower concentrations of paraformaldehyde (1–2%) as well as the inclusion of 15% picric acid are recommended. The brain is one organ in which it has been very difficult to obtain consistent intralaboratory immunohistochemical staining results with commercially available antibodies. Many of these reagents produce abundant nonspecific staining. The Affinity Bioreagents C-terminal antibody, produces very specific staining in frozen rat brain sections after perfusion with a solution of 2% paraformaldehyde, 2% acrolein.

Effect of Estrogen on Receptor Detection in the Brain

With the N-terminal ERβ antibody from Ligand Pharmaceuticals and the C-terminal ERα antibody from University of Virginia, the administration of estrogen to rats causes a marked reduction in the signals for both ERα and ERβ in certain brain regions.[27] In the BSTN and periventricular preoptic areas ERβ, but not ERα, is reduced, while in the medial amygdala and paraventricular nucleus ERα, but not ERβ, is reduced. Some of the intra-laboratory variations observed with ER localization in the brain may be due to the estrogen status of the animals.

Protocols

Immunohistochemistry with Paraffin-Imbedded Sections

The standard antigen retrieval technique is boiling in 10 mM citrate buffer (pH 6.0) in a microwave oven for 30 min. The cooled sections are then incubated in 1% (v/v) H_2O_2 for 30 min to quench endogenous peroxidase and then incubated with 1% (v/v) Triton X-100 in phosphate buffered saline (PBS) for 10 min. To block nonspecific binding of secondary antibodies, sections are incubated for 1 h at 4°C in normal serum prepared from the organismal source of the secondary antibodies.

[26] M. F. Kritzer, *Cereb. Cortex* **12**, 116 (2002).
[27] B. Greco, E. A. Allegretto, M. J. Tetel, and J. D. Blaustein, *Endocrinology* **142**, 5172 (2001).

Immunostaining with Chicken Anti-ERβ Antibody

Tissue sections are incubated for 10 min at room temperature with normal rabbit serum diluted at 1/20 in PBS containing 0.1% (w/v) bovine serum albumin and 0.1% (w/v) sodium azide (antibody diluent). Chicken anti-ERβ (1/1000) in antibody diluent is then applied and sections incubated overnight at 4°C. Negative controls should include substitution of the primary antibody with PBS, and primary antibody after adsorption with its antigen. Prior to addition of the secondary antibody, sections are rinsed in PBS containing 0.05% (v/v) Tween 20 (PBS-T). Rabbit anti-chicken IgG-HRP (1/1000 dilution) in PBS is applied to the sections. After 60 min, sections are washed with PBS-T. Peroxidase is developed with hydrogen peroxide and diaminobenzidine tetrahydrochloride; sections are counter-stained with hematoxylin. Sections are then dehydrated through graduated alcohol to xylene and mounted with Pertex.

Immunostaining with Anti-ERα Monoclonal Antibody (DAKO)

Tissue sections are incubated for 10 min in normal rabbit serum diluted at 1/20 in antibody diluent. This step is followed by overnight incubation at 4°C with ERα monoclonal antibody (Dako) at 1/100 dilution in antibody diluent. Negative controls should include substitution of the primary antibody with PBS. Prior to addition of secondary antibodies, sections are rinsed with PBS-T. Rabbit anti-mouse IgG at 1/100 dilution is applied to sections and after 20 min, slides are washed with PBS-T followed by a TBS wash. Mouse APAAP, 1/50 dilution, is applied for 30 min, and the alkaline phosphatase signal is developed with naphthol AS-MX and fast Blue BB.

Double Immunohistochemical Staining Using Fluorescence Conjugated Secondary Antibody

Frozen sections are fixed with ice-cold methanol, acetone and 2% paraformaldehyde. After treatment with 0.5% (v/v) Triton X-100 in PBS and 10% normal donkey serum, sections are incubated sequentially as follows: ERα 6F11 mouse antibody (1/100), FITC conjugated antimouse IgG (1/50), ERβins chicken antibody (1/500) and Cy3 conjugated antichicken IgY(1/200). Staining of DNA in the nucleus is done with 0.1 μg/ml of 4′,6-diamidino-2-phenylindole (DAPI). Sections are examined under the fluorescence microscope with suitable filters for FITC, Cy3 and DAPI, and images captured by CCD camera and analyzed with OpenLab 2.03 program (Improvision, Coventry, UK). Counting of stained cells should be done on at least six pictures produced from three different sections in each category. Controls must include preadsorption of the antibodies with their respective antigens. These preadsorbed antibodies must be used at the same dilutions as those before preadsorption.

Double Immunostaining for ERα and β

Tissue sections are incubated for 10 min at room temperature with normal rabbit serum diluted 1/20 in antibody diluent. This step is followed by an overnight incubation at 4°C with a mixture composed of ERα monoclonal antibody (1/100 dilution) and chicken ERβ IgY (1/1000 dilution). Prior to addition of secondary antibodies, sections are washed with PBS-T. A mixture of rabbit antimouse IgG (1/100 dilution) and rabbit antichicken IgG (1/1000 dilution) is added and after 30 min of incubation, slides are washed with TBS. A second mixture composed of APAAP (1/50) and rabbit antichicken IgG (1/1000) is added and incubation continued for 30 min. After washing, alkaline phosphatases signal is developed with naphthol AS-MX to produce a blue color, while the peroxidase is developed with hydrogen peroxide and 3-amino-9-ethyl carbazole to give a red colour.

Receptor Extraction for Western Blotting

With decavanadate-containing buffers, receptors can be extracted in low salt.[28] Tissue can be homogenized with a Polytron in four volumes of PBS containing 2 mM decavanadate, 1 mM EDTA, pH 7.4 and two protease inhibitor cocktail tablets (Boehringer-Mannheim, Germany) per 50 ml. The homogenates are then centrifuged for 1 hr at $105,000 \times g$ to obtain supernatants. Protein concentrations are measured using the Bio-Rad protein assay and bovine serum albumin as standard.

Western Blotting Analysis

SDS polyacrylamide gels, either 9% polyacrylamide or preformed gradient gels of 4–20% polyacrylamide may be used with a Tris–glycine buffer system. Transfer to PVDF membranes by electroblotting can be done either by semi-dry blotting or in a Tris–glycine buffer. Prestained molecular weight electrophoresis calibration standards are useful to monitor the efficiency of transfer.

---

[28] M. Fritsch, M. Aluker, and F. E. Murdoch, *Biochemistry* **38**, 6987 (1999).

After transfer is complete, membranes must be blocked by incubation for 1 hr with blocking buffer (10% (w/v) skim milk in PBS with 0.1% (v/v) NP-40) at room temperature (RT) prior to incubation at 4°C, overnight with primary antibodies, diluted in blocking buffer. This step is followed by 1 hr washing in blocking buffer, and then incubation with conjugated secondary antibodies, which are also diluted in blocking buffer, for 1 hr at RT. After sequential washing with blocking buffer, PBS with 0.1% NP-40 and PBS, signals can be developed with ECL. One of the advantages of using PVDF membranes is that they can be stained with Coomassie blue dye after blotting to ensure that roughly equal amounts of protein were examined in each lane of the gel.

Enrichment of Receptor

In tissues where ER level is very low, an enrichment step with heparin-sepharose can facilitate detection of the receptor on Western blots. Cytosol from such tissues can be prepared as described above and diluted 10-fold with 20 mM sodium phosphate buffer, pH 7.4, to reduce the ionic concentration. Heparin-sepharose (1 ml) is added and the mixture gently rotated for 1 hr at 5°C. Heparin-sepharose is recovered by centrifugation and washed 5 times with 20 mM sodium phosphate buffer. Proteins are eluted with 1 M NaCl, precipitated with 10% TCA, washed with methanol, and then used for immunoblotting.

Sucrose Density Gradient Centrifugation and ER-Binding Assay

Tissue frozen in liquid nitrogen and pulverized in a dismembrator (Braun Melsungen, Germany) for 45 sec at 1800 RPM, is added to a buffer containing 10 mM Tris chloride, pH 7.5, 1.5 mM EDTA and 5 mM sodium molybdate, using 1 ml buffer per 100 mg tissue. Cytosol is obtained by centrifugation of the homogenate at 204,000 × g for 1 hr at 4°C.

Tissue extracts are incubated for 3 hr at 0°C with 10 nM [^{3}H]estradiol, in the presence or absence of a 50-fold excess of radio inert estradiol, and the bound and unbound steroids are separated with dextran-coated charcoal. Sucrose density gradients [10–30% (w/v) sucrose] are prepared in a buffer containing 10 mM Tris–HCl, pH 7.5, 1.5 mM EDTA, 1 mM monothioglycerol, and 10 mM KCl. Samples of 200 μl are layered on 3.5 ml gradients and centrifuged at 40°C for 16 hr at 300,000 × g in a Beckman L-79K ultracentrifuge and an SW-60Ti rotor. Successive 100 μl fractions are collected from the bottom of the centrifuge tube by paraffin oil displacement, and assayed for radioactivity by liquid scintillation counting. For Western blotting, fractions are first precipitated with TCA, and the precipitate resuspended in methanol. Samples are placed on dry-ice for 30 min and the protein recovered by centrifugation. Protein pellets are dissolved in SDS sample buffer and resolved by SDS gel electrophoresis.

In rodent tissues the splice variant ERβins is expressed. This isoform has very little binding affinity for estradiol, but its affinity for 4-hydroxytamoxifen is similar to that of ERβ 530. ERβins along with ERβ 530 can be detected on sucrose gradients with 4-hydroxytamoxifen as ligand.

[26] *In Vivo* and *In Vitro* Reporter Systems for Studying Nuclear Receptor and Ligand Activities

By Alexander Mata de Urquiza and Thomas Perlmann

Introduction

Nuclear receptor (NR) ligands are generally small and lipophilic molecules making NRs highly attractive drug targets. These same properties render ligands challenging to researchers interested in NR biology. Novel ligands for orphan NRs are not easily identified by biochemical methods and often cannot be identified by expression cloning or related strategies. Moreover, while many signaling molecules, such as insulin and fibroblast-derived growth factors, are directly encoded by genes, NR ligands are usually derived from the enzymatic processing of vitamins or nutrients. As a consequence, *in situ* mRNA hybridization or immunological methods are not useful in analyzing the temporal and spatial distribution of ligands *in vivo*, unless the rate-limiting enzymatic steps in ligand synthesis are well defined. Thus, there has been a need for methods that facilitate *in vivo* characterization of known NR ligands in tissues. These methods should enable also the characterization of ligands for orphan NRs that have unidentified ligands.

One approach to better understand the biological roles played by NRs and their ligands is to localize the distribution of transcriptionally active receptors *in vivo*. An assay of this sort would allow the identification of target cells, provide important clues regarding the biological functions of NRs, and establish a basis for the isolation and identification of novel NR ligands. Several reports have described the use of transgenic mice to study the distribution of active retinoic acid receptor (RAR) during embryogenesis.[1–5] These animals carry a reporter transgene consisting of a β-galactosidase (*lacZ*) reporter gene under the transcriptional control of a retinoic acid (RA)-response element (RARE) from the *RARβ* gene. In cells where RA is synthesized and binds to the RAR, the activated receptor will induce transcription of the reporter gene via the RARE. The pattern of reporter gene expression in these animals suggests that active RAR is present in cells of the developing spinal cord, most prominently at the level of the developing limbs. This strategy has its limitations, as only NRs for which a strong natural response element has been characterized can be studied.[6] In addition, other NRs with related DNA-binding specificities may give rise to nonspecific activation of the reporter construct. Further, we describe alternative strategies that are based on transgenic mouse technology allowing NR activities to be assayed *in vivo*. We provide examples of complementary *in vitro* assays that are useful in verifying and extending results obtained in transgenic mouse experiments. Finally, we provide an example where further biochemical characterization allowed the identification of a novel and unexpected ligand for the retinoid X receptor (RXR).

Effector–Reporter Systems

In initial experiments, we developed a transgenic mouse assay in which the transcription of a *lacZ* reporter gene reflects the presence of active receptor.[7] In contrast to the RARE–*lacZ* system described in the "Introduction" section, the assay is not dependent on the presence of endogenous NRs but relies instead on the expression of a NR chimera. This so-called "effector protein" comprises a fusion between the DNA-binding domain of the yeast transcription factor GAL4, and the ligand-binding domain (LBD) of a NR. Ligand-activated effector protein is detected *in situ* by virtue of the ability to induce a reporter construct containing multiple GAL4 DNA-binding sites (UASs), preceding a minimal thymidine kinase

[1] W. Balkan, M. Colbert, C. Bock, and E. Linney, *Proc. Natl. Acad. Sci. USA* **89**, 3347 (1992).

[2] M. C. Colbert, E. Linney, and A. S. LaMantia, *Proc. Natl. Acad. Sci. USA* **90**, 6572 (1993).

[3] C. Mendelsohn, E. Ruberte, M. LeMeur, G. Morriss-Kay, and P. Chambon, *Development* **113**, 723 (1991).

[4] K. Reynolds, E. Mezey, and A. Zimmer, *Mech. Dev.* **36**, 15 (1991).

[5] J. Rossant, R. Zirngibl, D. Cado, M. Shago, and V. Giguere, *Genes Dev.* **5**, 1333 (1991).

[6] P. Ciana, G. Di Luccio, S. Belcredito, G. Pollio, E. Vegeto, L. Tatangelo, C. Tiveron, and A. Maggi, *Mol. Endocrinol.* **15**, 1104 (2001).

[7] L. Solomin, C. B. Johansson, R. H. Zetterström, R. P. Bissonnette, R. A. Heyman, L. Olson, U. Lendahl, J. Frisén, and T. Perlmann, *Nature* **395**, 398 (1998).

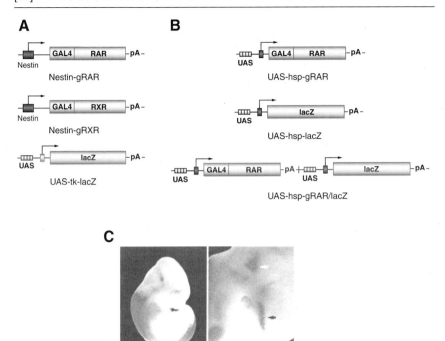

FIG. 1. Effector–reporter constructs and *in vivo* detection of RAR activity. (A) Effector plasmids expressing Gal4-RARα and Gal4-RXRα fusion proteins are driven by the nestin promoter. The reporter plasmid contains four Gal4-binding sites (UASs), followed by a minimal thymidine kinase (tk) promoter and a bacterial β-galactosidase (*lacZ*) gene. (B) FIND vectors expressing Gal4-RAR and *lacZ*, respectively, are driven by four UASs and a minimal heat shock protein (hsp) promoter. Effector and reporter constructs were combined in tandem on the same vector. (C) A representative transgenic embryo expressing FIND Gal4-RARα effector and *lacZ* reporter collected after 11.5 (E11.5) days of gestation. Left panel: X-gal staining reveals strong reporter activation at limb levels of the developing spinal cord. Right panel: At higher magnification, blue-stained cells are detected in the developing forebrain (black arrow) and at the midbrain/hindbrain boundary (white arrow). Reproduced with permission from A. Mata de Urquiza, L. Solomin, and T. Perlmann, *Proc. Natl. Acad. Sci. USA* **96**, 13270 (1999). Copyright [1999] National Academy of Sciences, USA. (See Color Insert.)

promoter and a bacterial *lacZ* reporter gene (Fig. 1A). Thus, the presence of ligand will activate the effector fusion protein, subsequently leading to transcription of the reporter. Effector protein expression was controlled in our early experiments by a nestin promoter, thereby limiting expression to neuronal precursor cells of the central nervous system (CNS) (Fig. 1A). The assay was used to study RAR and RXR activities during mouse embryogenesis.[7] The resulting pattern of reporter gene expression is in agreement with previous results.[1-5] Moreover, our data defined RXR as a ligand-induced receptor *in vivo* in the developing spinal cord, presumably resulting from activation by 9-*cis* retinoic acid.[7]

Although the initial system was useful for detecting nuclear receptor activation in tissues, we wished to refine and expand the applicability of the effector–reporter assay. In the first version of the system, the effector transgene was expressed under a tissue-specific promoter, limiting the sites and stages where active receptors could be detected. A ubiquitous transgenic promoter driving GAL4 effector protein expression would allow reporter induction in a broader range of tissues; however, to minimize potential phenotypic side effects of effector overexpression, it would be beneficial to limit the expression to tissues in which the receptor is normally active. It would also be advantageous to combine effector and reporter genes in a single mouse line to circumvent the necessity of intercrossing separate transgenic effector and reporter mouse strains.

A refined version of the effector–reporter system [referred to as the feedback-inducible nuclear-receptor-driven (FIND) expression system],[8] relies on the same type of GAL4 effector fusion protein and a *lacZ* reporter gene that is preceded by a minimal promoter from the heat shock protein (hsp) 68 gene and four UASs (Fig. 1B). Since the hsp promoter is expressed ubiquitously, ligand detection will not become restricted to certain tissues, as was the case when the nestin promoter was used. Instead, autoregulated gene expression will occur in any tissue where ligands are present to activate the effector protein.

The FIND system was tested using a GAL4-RAR chimera as effector gene. Transgenic E11.5 and E12.5 embryos were collected and retinoid-induced reporter gene expression was analyzed (Fig. 1C). Prominent activation was detected in the developing spinal cord, especially at the level of the developing limbs. This finding is consistent with previous reports showing that the brachial and lumbar regions of the spinal cord are "hot spots" for retinoid synthesis and action.[1-5,9-11] Reporter gene expression additionally was detected in the developing proximal forelimb bud, and in the optic cup, also in accordance with previous results.[1,3,12] Interestingly, staining was observed in the midbrain/hindbrain boundary, a structure not previously reported to contain retinoids at this stage (Fig. 1C, white arrow). One of the retinoic acid synthesizing enzymes, Raldh3, was recently shown

[8] A. Mata de Urquiza, L. Solomin, and T. Perlmann, *Proc. Natl. Acad. Sci. USA* **96**, 13270 (1999).

[9] P. McCaffery and U. C. Drager, *Proc. Natl. Acad. Sci. USA* **91**, 7194 (1994).

[10] S. Sockanathan and T. M. Jessell, *Cell* **94**, 503 (1998).

[11] M. Wagner, B. Han, and T. M. Jessell, *Development* **116**, 55 (1992).

[12] A. S. LaMantia, M. C. Colbert, and E. Linney, *Neuron* **10**, 1035 (1993).

to be expressed in this region, thus identifying a relevant rate-limiting enzyme for retinoid synthesis at this location.[13,14] An additional novel site of retinoid signaling was identified in the ventral forebrain (Fig. 1C, black arrow). The consistencies between the results obtained in this and previous studies suggest that the FIND version of the effector–reporter system functions as expected. Moreover, similar results were obtained with a green fluorescent protein reporter, allowing analyses in intact living embryos or cells cultured *in vitro*.[8]

Generally, the *in vivo* results obtained using either of the two described transgenic mouse assays should be interpreted with caution. It is well established that several nuclear receptors are activated by ligand-independent mechanisms, which in transgenic experiments could lead to reporter induction in absence of ligand. In addition, it should be emphasized that the FIND system in its present form might not be applicable to all nuclear receptors. The current FIND system includes only the LBD of the NR, while some receptors may require additional or full-length sequences in order to become activated. It is also evident that the complexity of natural promoters can influence substantially the ligand-induced responses of NRs in ways that would not be detected in the FIND assay. Finally, an important and general problem with analyses of transgenic mice is the effect of silencing or nonspecific reporter gene activation due to influences from surrounding genomic DNA. Thus, several independent integration events (transgenic lines) need to be analyzed to confirm an observed pattern of reporter gene induction. A way of minimizing such effects could be to flank the transgenes with so-called insulator or boundary elements. Insulators are naturally occurring genomic DNA sequences, first discovered in *Drosophila* but later also in vertebrates, that minimize the effects of surrounding genomic DNA. Such sequences have been used successfully, both *in vivo* and *in vitro*, to limit negative integration site effects.[6,15–18]

With the above considerations in mind, the assay described in the next section provides a new tool for studying the activation by ligands, or by ligand-independent mechanisms, of potentially any nuclear receptor LBD. Importantly, by defining tissues where ligands are generated, the system may

[13] R. J. Haselbeck, I. Hoffmann, and G. Duester, *Dev. Genet.* **25**, 353 (1999).
[14] F. A. Mic, A. Molotkov, X. Fan, A. E. Cuenca, and G. Duester, *Mech. Dev.* **97**, 227 (2000).
[15] A. C. Bell, A. G. West, and G. Felsenfeld, *Science* **291**, 447 (2001).
[16] H. C. Zhan, D. P. Liu, and C. C. Liang, *Hum. Genet.* **109**, 471 (2001).
[17] Q. Li, M. Zhang, H. Han, A. Rohde, and G. Stamatoyannopoulos, *Nucleic Acids Res.* **30**, 2484 (2002).
[18] F. Recillas-Targa, M. J. Pikaart, B. Burgess-Beusse, A. C. Bell, M. D. Litt, A. G. West, M. Gaszner, and G. Felsenfeld, *Proc. Natl. Acad. Sci. USA* **99**, 6883 (2002).

be useful in the isolation of new ligands for nuclear receptors, which is an important advantage since ligands have not been identified for many NRs. We also envision that the system can be applied as an *in vivo* model for the characterization of drug candidates by defining the target tissue(s) and pharmacokinetic properties of such compounds.

Protocol for an *In Vitro* Coculture Assay

As explained earlier, the results from the FIND system must be interpreted with caution, and assays allowing validation of transgenic data are highly desirable. We have found that GAL4-LBD chimeras can also be used in less sophisticated, but useful *in vitro* assays designed to complement and extend the findings obtained in transgenic mice.[8,19–21] In brief, cells in culture are transiently cotransfected with plasmids expressing a GAL4-NR effector and a UAS luciferase reporter gene, and subsequently cocultured with small pieces of tissue dissected from either embryonal or adult tissues (Fig. 2). Ligands present in the tissue pieces are released and taken up by the cocultured transfected cells, leading to NR activation and reporter gene expression. The protocol used is as follows:

1. Cultured cells are grown and transfected using an established transfection protocol.
2. Target tissues from embryonal or adult animals are dissected and cut into pieces, roughly 3 mm^3, and kept in serum-free growth medium on ice until use, but no longer than 6 hr.
3. Tissue pieces are added to cultured cells, and incubated for up to 24 hr. Coculturing of tissue explants has, in certain cases, been toxic to transfected cells. This problem can be addressed by reducing the exposure time to tissue pieces. The cell growth medium that has been in contact with the tissue during dissection is preferably added to the transfected cells, thereby ensuring that ligands released from the tissue will be available.
4. After incubation, tissue and medium are removed, and the cells are harvested and lysed. Cell extracts are assayed for reporter gene activity as described.[22]

[19] A. Mata de Urquiza, S. Liu, M. Sjoberg, R. H. Zetterstrom, W. Griffiths, J. Sjovall, and T. Perlmann, *Science* **290**, 2140 (2000).
[20] H. Toresson, A. Mata de Urquiza, C. Fagerstrom, T. Perlmann, and K. Campbell, *Development* **126**, 1317 (1999).
[21] R. H. Zetterström, E. Linquist, A. Mata de Urquiza, A. Tomac, U. Eriksson, T. Perlmann, and L. Olson, *Eur. J. Neurosci.* **11**, 407 (1999).
[22] T. Perlmann and L. Jansson, *Genes Dev.* **9**, 769 (1995).

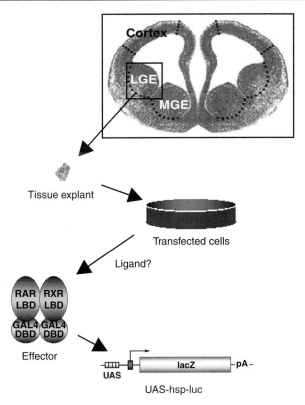

FIG. 2. An *in vitro* coculture assay for detection of ligands in tissue explants. Coronal section through the developing mouse forebrain (E12.5) showing the relative positions of the medial and lateral ganglionic eminences (MGE and LGE, respectively), and the developing cortex. Defined tissue pieces are dissected out and placed on cells transfected with Gal4-RARα effector and luciferase (luc) reporter plasmids *in vitro*. Potential ligand(s) present in the tissue and secreted into the culture medium are subsequently taken up by the cells. Binding of ligand to RAR will activate the effector chimera, leading to reporter expression.

Both embryonic and adult tissues have been used in these experiments. When adult tissue is used, results are improved by perfusion with saline prior to dissection, presumably because this treatment prevented toxicity from contaminating blood. Perfusion is not necessary when embryonic tissues are used. We tested a wide range of different tissues, including brain, heart, lung, liver, kidney, muscle, testis, and fat, and in general, most tissues are compatible with commonly used cultured cells. The human embryonic kidney 293T cell line is generally more robust than, for example, JEG-3 cells. Certain tissues, for example spleen, pancreas, and intestine, are toxic to cells in culture,[23] making these tissues difficult to analyze in this type of assay. Transfections and coculturing are performed under serum-free conditions, to lower background reporter activity resulting from low levels of activating NR ligands present in serum.

Conditioned Medium from Tissue or Primary Cell Cultures

We have found that medium preconditioned with tissue can be used as a way of assaying NR ligand activities. The use of preconditioned medium is probably effective, only when relatively large amounts of NR ligand activities are present in a given tissue. The main advantage of this approach is that large quantities of conditioned medium can be produced allowing more careful and reproducible analyses of particular NR ligand activity, e.g., by treating many wells of cells transfected with different NR expression vectors, and culturing in the presence and absence of NR antagonists and other compounds. The experiments are performed as follows:

1. Tissue from adult mice is dissected and cut into small pieces, and then incubated 24 hr with serum-free medium at 37°C in a large Petri dish. In a typical experiment, tissue from three mouse brains is incubated in 50 ml of medium.
2. Conditioned medium is collected from the dish, and fresh medium is added to the tissue pieces for another overnight incubation.
3. The decanted medium is centrifuged at 3000 rpm for 5–10 min to remove cell debris. The supernatant is collected and frozen for future use.

NR ligands may be unstable and/or sensitive to light. Therefore, all preparations should be kept on ice, and care should be taken to avoid unnecessary exposure to light. The time between dissection and cell coculture should be as short as possible to minimize loss of NR ligand activities. If necessary, the pH of the medium can be adjusted prior to use on transfected cells.

Another possible source of NR ligand activity is from tissue culture cells or primary cells cultured *in vitro*. For example, we have shown that primary glial cell cultures from the developing mouse forebrain are a source of RAR-ligand activity.[20] Interestingly, this activity was observed in cells cultured from a region of the developing forebrain (the lateral ganglionic eminence) that stained positive in FIND-RAR transgenic embryos (Fig. 1C, black arrow). Thus, the combined experiments identified glial cells as the retinoid-synthesizing cell type in the developing forebrain. Clearly, this

---

[23] A. Mata de Urquiza and Å. Wallén, unpublished results (1999).

strategy may prove useful in the identification of the relevant cell types synthesizing NR ligands in a given tissue *in vivo.*

Isolation of DHA, a Novel RXR Ligand

Although the characterization of NR ligand activities in tissues, utilizing both the transgenic and *in vitro* assays described earlier, has the potential of elucidating important signaling events, the goal in some cases may be to identify biochemically a novel NR ligand activity. In the case described below, characterization comprised the biochemical isolation and identification of a previously undescribed ligand that bound and activated RXR.

The function of RXR as a ligand-inducible signaling receptor remains elusive. Its proposed natural ligand, 9-*cis* RA, has proven difficult to identify in mammalian tissue. Nonetheless, the significance of RXR signaling *in vivo* has been suggested from several studies.[7,24–29] We therefore wished to analyze the importance of retinoid receptors in the adult CNS by searching for endogenous ligands that activate RAR and RXR. Using the previously described effector–reporter assay in transfected cells, we compared the activation patterns of GAL4-RAR and GAL4-RXR in cells cocultured with serum-free media that had been incubated overnight with various tissues from adult brain (Fig. 3A). Little or no RAR activation was detected when reporter cells were cocultured with adult brain tissue. Instead, a substantial RXR activation was revealed.[21] Additional experiments, with an RXR-selective antagonist, indicated that the adult brain-conditioned medium directly activated RXR. To characterize the putative RXR ligand(s), biochemical isolation, purification, and mass spectrometry of brain-conditioned medium was performed as follows:

1. Conditioned medium was prepared with adult mouse brain tissue as described in the previous section.
2. About 80 ml of conditioned medium was acidified with 1 M hydrochloric acid (HCl) (0.1 M HCl final concentration), and mixed with an equal volume of hexane.

[24] J. Botling, D. S. Castro, F. Oberg, K. Nilsson, and T. Perlmann, *J. Biol. Chem.* **272**, 9443 (1997).

[25] H. C. Lu, G. Eichele, and C. Thaller, *Development* **124**, 195 (1997).

[26] B. Mascrez, M. Mark, A. Dierich, N. B. Ghyselinck, P. Kastner, and P. Chambon, *Development* **125**, 4691 (1998).

[27] R. Mukherjee, P. J. Davies, D. L. Crombie, E. D. Bischoff, R. M. Cesario, L. Jow, L. G. Hamann, M. F. Boehm, C. E. Mondon, A. M. Nadzan, J. R. J. Pateniti, and R. A. Heyman, *Nature* **386**, 407 (1997).

[28] J. J. Repa, S. D. Turley, J. A. Lobaccaro, J. Medina, L. Li, K. Lustig, B. Shan, R. A. Heyman, J. M. Dietschy, and D. J. Mangelsdorf, *Science* **289**, 1524 (2000).

[29] B. Roy, R. Taneja, and P. Chambon, *Mol. Cell. Biol.* **15**, 6481 (1995).

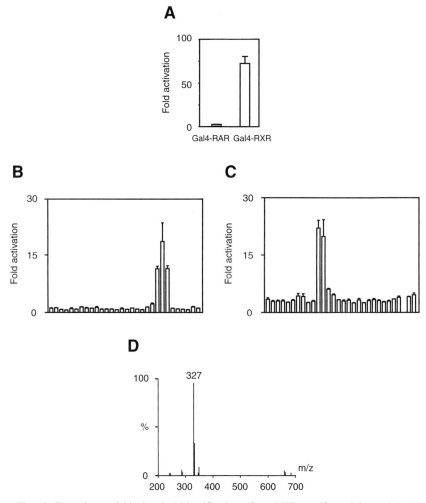

FIG. 3. Detection and biochemical identification of an RXR-specific activity enriched in adult mouse brain. (A) JEG-3 cells were transfected with a luciferase reporter plasmid and either Gal4-RARα or Gal4-RXRα expression vectors, and treated with brain-conditioned medium. (B) Brain-conditioned medium was extracted as described and separated by normal-phase HPLC, and individual fractions were tested on cells transfected with Gal4-RXRα and UAS-luc vectors. (C) Active fractions from (B) were pooled and fractionated by reverse-phase HPLC. Values in (A)–(C) are shown as fold induction after normalization to β-gal. (D) Peak fractions from (C) were analyzed by mass spectrometry, recording negative-ion nano-ES spectra. The active fraction is dominated by a very intense peak at m/z 327.2. Minor peaks are observed at m/z 283.2, 339.2, and 655.5, the latter representing dimers of the m/z 327.2 ion. Reproduced with permission from A. Mata de Urquiza, S. Liu, M. Sjöberg, R. H. Zetterström, W. Griffiths, J. Sjövall, and T. Perlmann, *Science* **290**, 2140 (2000). Copyright [2000] American Association for the Advancement of Science.

3. The mixture was vortexed for 5 min, then centrifuged at 13,000 rpm for 10 min at 15°C.

4. The upper organic phase was collected into a new tube, and the lower water phase reextracted with the same volume of hexane as in step 2. Shaking and centrifugation were repeated as before.

5. The two organic phases were pooled and evaporated under a slow stream of nitrogen gas. The residue was dissolved in 1–2 ml of hexane, and then concentrated under nitrogen to a final volume of about 200 μl. If the preparation appeared turbid, the solution was centrifuged through a prewetted eppendorf tube filter prior to injection into the HPLC column.

6. One hundred and fifty microliters were injected into a normal-phase HPLC column (Genesis Silica 4 μm, 4.6 mm × 25 cm, Jones Chromatography Ltd., UK). Elution was performed using a linear gradient from hexane to hexane–dichloromethane–isopropanol (85:10:5, v/v), both containing 1% acetic acid, in 30 min at a flow rate of 0.5 ml/min. Starting 17 min after injection, 30 fractions of 0.25 ml were collected using a Gilson fraction collector.

7. Activity assays were performed with cells transfected with effector and reporter constructs (see Fig. 3B legend), using an aliquot of each column fraction. Each aliquot was evaporated and redissolved in a small volume of 100% EtOH prior to addition to the transfected cells (Fig. 3B).

8. Fractions containing activity were pooled and evaporated to dryness. The residue was dissolved in 50 μl of 80% methanol, and 30 μl of this solution was injected into a reverse-phase HPLC column (Genesis C18 4 μm, 3.0 mm × 25 cm, Jones Chromatography Ltd., UK). A mobile phase of methanol–isopropanol–water (80:10:10, v/v) containing 1% acetic acid, at a flow rate of 0.3 ml/min was used for separation.

9. Aliquots from each fraction were again analyzed in the transfected cell assay (Fig. 3C). Active fractions, along with the preceding and following fractions, were taken for analysis by nano-electrospray (nano-ES) mass spectrometry.

10. Negative-ion nano-ES spectra were recorded on an AutoSpec-OATOFFPD (Micromass, Manchester, England) hybrid double focusing magnetic sector–orthogonal acceleration (OA) time-of-flight (TOF) tandem mass spectrometer. Fractions were diluted 40-fold in methanol and 1–5 μl loaded into gold-coated borosilicate capillaries (Protona A/S, Odense, Denmark). Mass spectra were recorded as magnet scans over a m/z range of 70–1000 at an instrument resolution of 3000 (10% valley definition) (Fig. 3D).

High-resolution (10,000, 10% valley definition) accurate mass measurements were made using voltage scans over a minimum m/z range, just great enough to cover the peaks of interest and calibrant ions. Collision-induced dissociation (CID) spectra were recorded by selecting the precursor $[M-H]^-$ ions with the double focusing sectors of the instrument, and focusing them into the fourth field-free region gas cell containing xenon collision gas. The collision energy was 400 eV, and the collision gas pressure was sufficient to attenuate the precursor ion beam by 75%. Undissociated precursor ions and fragment ions were pulsed into an OATOF mass analyzer.

The activity was thus identified as docosahexaenoic acid (DHA), a long-chain polyunsaturated fatty acid (PUFA) that accumulates in the postnatal and adult central nervous system.[30,31] Interestingly, we could show that other PUFAs also activate RXR, although DHA is the most efficient. DHA is specific for RXR, as it does not activate RAR or the receptors for thyroid hormone or vitamin D, and behaves as a *bona fide* NR ligand that promotes the interaction between RXR and the coactivator SRC-1 *in vitro*.[19] A recent study using X-ray crystallography demonstrated that DHA and 9-*cis* RA can adopt similar conformations in the ligand-binding pocket of RXR.[32] Thus, by producing conditioned medium with tissue from adult mouse brain, we were able to detect, isolate, and identify a natural and novel RXR ligand. The success of these efforts was greatly influenced by the great abundance of DHA in the adult brain. A high-affinity ligand present locally in small quantities would provide a much greater challenge in a similar purification effort. Nonetheless, the results underscore the potential of using conditioned medium for more advanced biochemical characterization of NR ligand activities.

Conclusions

The transgenic system described here, combined with *in vitro* assays and biochemical characterization, have provided new tools for the characterization of signaling events *in vivo*. During the last few years, intense efforts have been made in the field of NR research to better understand the roles played

---

[30] M. Neuringer, G. J. Anderson, and W. E. Connor, *Ann. Rev. Nutr.* **8**, 517 (1988).

[31] N. Salem, Jr., H.-Y. Kim, and J. A. Yergey, *in* "Health Effects of Polyunsaturated Fatty Acids in Seafoods" (A. P. Simopoulos, R. R. Kifer, and R. E. Martin, eds.), Chapter 15, p. 263. Academic Press, Boston, 1986.

[32] P. F. Egea, A. Mitschler, and D. Moras, *Mol. Endocrinol.* **16**, 987 (2002).

by NRs in the development and physiology of higher eukaryotic organisms. For example, NRs such as the peroxisome proliferator-activated receptors (PPARs), liver X receptors (LXRs), and farnesoid X receptor (FXR) have been implicated in processes regulating cholesterol, fatty acid, and sugar metabolism, dysregulation of which lead to diseases such as diabetes, obesity, and atherosclerosis.[33–35] Most known NRs are still lacking an identified ligand, and although expression patterns are readily characterized, the *in vivo* roles of NRs are, in many cases, far from understood. As outlined here, by establishing an assay in which cells containing active NRs are identified, new and valuable insights are provided and the identification of entirely novel signaling pathways is accomplished. Further improvements in established methods, as well as the development of techniques to search for natural NR ligands *in vivo*, will be an essential complement to the screening efforts currently in progress, in particular within pharmaceutical and biotech companies. Although synthetic ligands are invaluable as tools, the identification of natural ligands will undoubtedly prove vital for achieving a complete understanding of the roles played by NRs *in vivo*.

[33] A. Chawla, J. J. Repa, R. M. Evans, and D. J. Mangelsdorf, *Science* **294**, 1866 (2001).
[34] T. T. Lu, J. J. Repa, and D. J. Mangelsdorf, *J. Biol. Chem.* **276**, 37735 (2001).
[35] E. D. Rosen and B. M. Spiegelman, *J. Biol. Chem.* **276**, 37731 (2001).

[27] Methods to Characterize *Drosophila* Nuclear Receptor Activation and Function *In Vivo*

By TATIANA KOZLOVA and CARL S. THUMMEL

Introduction

The completion of the *Drosophila* genome sequence has revealed the presence of 21 members of the nuclear receptor superfamily, many of which have orthologs in *C. elegans* and mammals.[1] Of these receptors, only the EcR/USP heterodimer has a known ligand, the major molting hormone in *Drosophila*, 20-hydroxyecdysone (referred to here as 20E).[2] As with vertebrate nuclear receptors, a major focus of current research is the

[1] M. D. Adams *et al.*, *Science* **287**, 2185 (2000).
[2] L. M. Riddiford, *in* "The Development of *Drosophila melanogaster*" (M. Bate and A. Martinez Arias, eds.), p. 899. Cold Spring Harbor Laboratory Press, Cold Spring Harbor, 1993.

identification of hormones for the orphan members of the *Drosophila* nuclear receptor family. In addition, new tools are required to facilitate functional studies of nuclear receptor action during the *Drosophila* life cycle.

In this chapter, we describe the use of transgenic animals that express a fusion of the yeast GAL4 DNA-binding domain (DBD) and the ligand-binding domain (LBD) of a nuclear receptor, as a means of determining when and where the corresponding receptor is activated during development. This approach has provided new insights into ecdysteroid activation of the EcR/USP heterodimer and should, as well, facilitate the identification of ligands for orphan nuclear receptors. Finally, we describe a new approach for functional genomics in *Drosophila* that has been used to characterize the stage-specific functions of nuclear receptors. This method utilizes regulated expression of double-stranded RNA (dsRNA) corresponding to the gene of interest, directing targeted degradation of the corresponding mRNA by RNAi.

The GAL4-LBD System in *Drosophila*

Understanding the molecular mechanisms by which a systemic hormonal signal is refined into different stage- and tissue-specific biological responses requires characterization of the temporal and spatial patterns of nuclear receptor activation during development. A method to achieve this goal was described by Solomin et al. as part of their effort to characterize the patterns of retinoic acid signaling in the mouse central nervous system (CNS).[3] This method involves the establishment of transgenic animals that express the yeast GAL4 DBD fused to a nuclear receptor LBD, combined with a second transgenic construct that carries a GAL4-dependent promoter driving a reporter gene. The temporal and spatial patterns of reporter gene expression in these transgenic animals indicate where and when the LBD has been activated by its ligand, providing a direct means of following the patterns of hormone signaling in the context of a developing organism. We have adapted this system for use in *Drosophila*, allowing us to follow the patterns of *Drosophila* nuclear receptor activation during development.[4]

[3] L. Solomin, C. B. Johansson, R. H. Zetterstrom, R. P. Bissonnette, R. A. Heyman, L. Olson, U. Lendahl, J. Frisen, and T. Perlmann, *Nature* **395**, 398 (1998).
[4] T. Kozlova and C. S. Thummel, *Development* **129**, 1739 (2002).

Criteria for Ligand-Dependent Activation of GAL4-LBD Fusion Proteins

Several criteria must be met, if the GAL4-LBD system is to function as expected in *Drosophila*:

1. If GAL4-LBD transcriptional activity is controlled by a hormonal ligand, then no activation (or minimal background activation) is expected when the hormone titer is low, and activation should increase dramatically coincident with hormone pulses. For orphan nuclear receptors, changes in activation over time could suggest that their activity is regulated by one or more ligands.
2. Activation of GAL4-LBD fusion proteins should be achieved in organ culture by physiological concentrations of the corresponding ligand(s). In the case of orphan nuclear receptors, either candidate chemicals or lipophilic extracts from the stage when activation is detected *in vivo* can be applied in culture to induce ectopic activation.
3. Activation of the GAL4-LBD fusion protein should be reduced or eliminated if the ligand is removed either by a mutation in the biosynthetic pathway *in vivo* or after antagonist treatment in organ culture.

Characterization of the activation patterns of GAL4-EcR and GAL4-USP by ecdysteroids, both *in vivo* and in organ culture, fulfills these criteria, indicating that the system is working as desired.[4,5] These criteria can also be used as guidelines for assaying activation of GAL4-LBD fusions derived from orphan nuclear receptors.

The Constructs

A fragment encoding the DBD from the yeast GAL4 transcription factor (amino acids 1–147) was amplified by PCR and cloned into the pCaSpeR-hs-act P element transformation vector,[6] resulting in the formation of pCaSpeR-hs-GAL4act. This vector contains the *hsp70* heat-inducible promoter upstream from the GAL4 sequences, followed by 3'-UTR sequences from the *Drosophila actin 5C* gene. Addition of these 3' sequences should result in increased stability of the mRNA and subsequent increased production of the encoded protein.[6] Two unique restriction sites, for EcoRI and BamHI, are present in pCaSpeR-hs-GAL4act for the insertion of sequences encoding nuclear receptor LBDs.

[5] T. Kozlova, unpublished results (2002).
[6] C. S. Thummel and V. Pirrotta, *Dros. Inf. Serv.* **71**, 150 (1992).

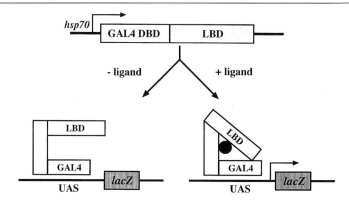

FIG. 1. Schematic representation of the GAL4-LBD system. The upper diagram depicts a fusion gene comprising the coding region for the yeast GAL4-DBD fused in-frame with the LBD of EcR or USP. This construct is introduced into the *Drosophila* genome under the control of the *hsp70* promoter. Heat-induced GAL4-LBD protein can bind to GAL4 upstream activating sequences (UASs) that are present in a second transgenic construct. In the absence of ligand, the GAL4-LBD protein does not activate *lacZ* reporter gene transcription; however, in the presence of its ligand (black circle) it can switch into an active conformation and induce *lacZ* expression. (Reprinted with permission from Ref. 4.)

In order to monitor the patterns of EcR/USP activation during *Drosophila* development, the coding region for either the EcR LBD, including the hinge region and F domain (amino acids 330–878),[7] or the USP LBD, including the hinge region (amino acids 170–508),[8] were subcloned into pCaSpeR-hs-GAL4act. These P element constructs were separately introduced into the fly genome by standard transformation procedures (Fig. 1).[9] The *hs-GAL4-EcR* and *hs-GAL4-USP* transformants were crossed with flies that carry a GAL4-dependent promoter driving a *lacZ* reporter gene, which encodes a nuclear β-galactosidase (*UAS-nlacZ*). The choice of the reporter is described in more detail in the following section. The *hsp70* promoter was selected in order to provide precise temporal control, reducing potential lethality that might be caused by overexpression of the GAL4-LBD fusion proteins. Indeed, both GAL4-EcR and GAL4-USP behave as stage-specific dominant negatives in our assays, likely interfering with ecdysteroid signaling *in vivo*.[4] The *hsp70* promoter also directs widespread expression of the GAL4-LBD proteins upon heat induction, allowing one to potentially monitor activation throughout the animal. Expression of the

---

[7] M. R. Koelle, W. S. Talbot, W. A. Segraves, M. T. Bender, P. Cherbas, and D. S. Hogness, *Cell* **67**, 59 (1991).
[8] A. E. Oro, M. McKeown, and R. M. Evans, *Nature* **347**, 298 (1990).

GAL4-LBD fusion proteins was determined by heat-treating *hs-GAL4-EcR* and *hs-GAL4-USP* late third instar larvae for 30 min at 37°C. After a 4-hr recovery at 25°C, these animals were processed for immunostainings with anti-GAL4 polyclonal antibodies (Santa Cruz Biotechnology, Inc.) at 1:100 dilution using standard protocols.[10] GAL4-EcR and GAL4-USP fusion proteins were detected in both the nucleus and cytoplasm, and were at comparable levels in all tissues examined. Transcriptional activation by these fusion proteins, however, should only occur at times and in places where the appropriate hormonal ligand is present (Fig. 1).

Histochemical Detection of β-Galactosidase and Choice of the Reporter

Histochemical detection of β-galactosidase was performed by staining dissected larval tissues with X-gal as described.[11] X-gal stains of larval tissues fixed with 4% formaldehyde (Polysciences) were allowed to develop overnight, while tissues fixed with 1% glutaraldehyde (Sigma) were allowed to develop for several hours. Glutaraldehyde fixation results in increased sensitivity, but the background from the reporter should be monitored carefully using this protocol. Because of the relatively long staining time, even a low level of background β-galactosidase expression would complicate the interpretation of the results. Consequently, several independent *UAS-nlacZ* reporter lines were assayed in the third larval instar for their basal level of β-galactosidase expression. We found that the *UAS-nlacZ* reporter P{w^{+mC} = UAS-NZ}J312 (stock 3956 Bloomington Stock Center) has the lowest level of background expression at this stage in development, with detectable β-galactosidase in the larval salivary glands, cells surrounding the larval mouthhooks, and a few cells in the CNS. Another reporter line, 7.4, a homozygous viable *UAS-nlacZ* insertion on the third chromosome (kindly provided by Y. N. Jan), gives high background staining in the larval salivary glands and the epidermis of third instar larvae. This reporter, however, has the lowest background expression during embryogenesis, and is our reporter of choice for this stage.[5] Therefore, one reporter line might not be suitable for analysis of different developmental stages and the best ones should be identified experimentally. We have also attempted to use cytoplasmic or nuclear *UAS-GFP* reporters to monitor GAL4-EcR and GAL4-USP activation in living animals. While activation can be detected in certain tissues without dissection, the reduced sensitivity of these reporters precluded their routine use in our experiments. The use of transformant

[9] G. M. Rubin and A. C. Spradling, *Science* **218**, 348 (1982).
[10] N. Patel, *Methods Cell Biol.* **44**, 445 (1994).
[11] M. Mlodzik and Y. Hiromi, *Methods Neurosci.* **9**, 397 (1992).

lines that carry enhanced GFP reporters should facilitate the analysis of GAL4-LBD activation in living animals.

Detection of GAL4-LBD Activation In Vivo

In order to establish appropriate parameters for induction of *UAS-nlacZ in vivo*, we used a hs-GAL4 driver, expressing the full-length GAL4 transcriptional activator under the control of the *hsp70* promoter (P{w^{+mC} = GAL4-Hsp70.PB}89-2-1, stock 1799 obtained from the Bloomington Stock Center). *hs-GAL4*; *UAS-nlacZ* late third instar larvae were heat treated for 30 min at 37°C, allowed to recover for 4 and 6 hr at 25°C, and processed for histochemical staining with X-gal. β-Galactosidase was first detected 4 hr after heat treatment in late third instar larvae of this genotype. This expression was strong in all analyzed tissues by 6 hr after heat induction, defining this as an optimal time to assay reporter activity. Because there is 4–5 hr delay between GAL4-LBD induction and reporter activation (as assayed by histochemical staining), the absolute timing of GAL4-LBD activation remains unclear. We assume, for example, that if third instar larvae are heat treated at 3–4 hr before pupariation, then the reporter gene expression that we detect in ~2 hr prepupae represents the response in animals ~2 hr before puparium formation. It might be possible to perform activation assays in a shorter time frame, if one uses *in situ* hybridization to detect reporter gene expression rather than histochemical staining.

The ability of the GAL4-EcR and GAL4-USP constructs to faithfully reflect ecdysteroid signaling *in vivo* was tested during the third instar larval stage. *hs-GAL4-EcR*; *UAS-nlacZ* and *hs-GAL4-USP*; *UAS-nlacZ* third instar larvae were staged on food containing 0.5% bromophenol blue as described.[12] Heat treatments were performed by incubating plastic culture vials with food in a 37°C waterbath for 30 min. For inducing GAL4-LBD fusions in mid-third instar larvae, animals were maintained on blue food, heat treated, allowed to recover for 6–7 hr at 25°C, and selected as wandering dark blue gut animals. For inducing GAL4-LBD fusion proteins at the onset of metamorphosis, partial blue gut and white gut third instar larvae were transferred to vials with regular food, heat treated, and allowed to recover at 25°C. Animals that formed white prepupae between 3 and 4 hr after heat treatment were selected from this population,

---

[12] A. J. Andres and C. S. Thummel, in "*Drosophila melanogaster*: Practical Uses in Cell and Molecular Biology" (L. S. B. Goldstein and E. A. Fyrberg, eds.), p. 565. Academic Press, New York, 1994.

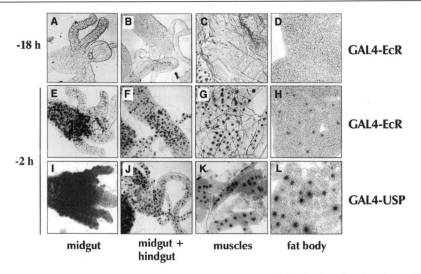

Fig. 2. GAL4-EcR and GAL4-USP are widely activated by the late larval ecdysteroid pulse. *hs-GAL4-EcR; UAS-nlacZ* (GAL4-EcR) or *hs-GAL4-USP; UAS-nlacZ* (GAL4-USP) third instar larvae were heat treated, allowed to recover for several hours, staged at either 12–18 hr before puparium formation (−18 hr; A–D) or ∼2 hr before puparium formation (−2 hr; E–L), and processed for histochemical staining. Blue precipitate indicates expression of nuclear β-galactosidase. GAL4-EcR (A–D) and GAL4-USP (data not shown) are not activated in third instar larvae 12–18 hr before puparium formation. In contrast, both GAL4-EcR (E–H) and GAL4-USP (I–L) are widely activated in late third instar larvae, as depicted in the midgut, hindgut, muscles, and fat body. (Reprinted with permission from Ref. 4.) (See Color Insert.)

allowed to age for an additional 2–3 hr, and processed for histochemical staining.

As expected, no GAL4-EcR or GAL4-USP activation was detected in tissues from mid-third instar larvae (Figs. 2A–D and data not shown) consistent with the low ecdysteroid titer at this stage in development. In contrast, activation was readily detected in most larval tissues at ∼2 hr before puparium formation (Fig. 2E–H for GAL4-EcR and Figs. 2I–L for GAL4-USP). Therefore, as expected, both GAL4-EcR and GAL4-USP are widely activated by the late larval ecdysteroid pulse that triggers the onset of metamorphosis.

It is possible to restrict ecdysteroid signaling to the anterior part of *Drosophila* larvae by using ligature to separate an endocrine organ, the ring gland, from posterior tissues. Following heat treatment, late third instar larvae (∼5 hr before pupariation) were ligated with a hair separating the anterior 1/3–1/2 of the animal from the posterior portion. In some of the ligated larvae, the anterior portion undergoes pupariation within

4–5 hr while the posterior portion remains larval. When activation of GAL4-EcR and GAL4-USP was assayed in these animals, strong widespread β-galactosidase expression was observed only in the pupariated anterior portion. No activation was detected posterior to the ligature, consistent with direct endocrine activation of the fusion proteins (data not shown).

Detection of GAL-LBD Activation in Organ Culture

hs-GAL4-EcR; *UAS-nlacZ* and *hs-GAL4-USP*; *UAS-nlacZ* third instar larvae reared on blue food were heat treated at 37°C for 30 min, allowed to recover for 2–3 hr at 25°C, and staged animals were dissected in oxygenated Grace's Insect Medium (BRL). We typically use partial blue gut third instar larvae in these experiments,[12] because GAL4-EcR and GAL4-USP only show low background activation at this stage *in vivo*, although animals from other stages can also be used.[4] Dissected larval organs from 4 to 5 animals per sample were cultured in 300 μl of oxygenated Grace's medium using various concentrations of 20E (Sigma), essentially as described.[12] Samples were processed for histochemical staining after 12 hr in culture. We used 5×10^{-8}, 5×10^{-7}, 5×10^{-6}, and 5×10^{-5} M 20E in culture experiments, and found that activation was easily detectable at all of these hormone concentrations (data not shown). The activation observed in culture, even with the highest 20E concentrations, however, is less robust than that detected in response to the endogenous late larval pulse of ecdysteroids. It is likely that this effect is due to the stress inherent in maintaining larval organs in culture, although it may also reflect the state of competence of individual tissues. Activation of GAL4-EcR induced by 10^{-7} M 20E is efficiently reduced in the presence of 10^{-5} M ecdysteroid receptor antagonist cucurbitacin B (kindly provided by L. Dinan).[5,13]

Strengths and Weaknesses of the Method

There are several potential weaknesses that must be kept in mind when working with the GAL4-LBD system. Despite the fact that this system has been used widely to monitor ligand-dependent activation of nuclear receptors in transfected tissue culture cells, and appears to function as expected *in vivo*, the GAL4-LBD constructs direct the synthesis of artificial recombinant proteins. Thus, negative results remain difficult to interpret.

[13] L. Dinan, P. Whiting, J. P. Girault, R. Lafont, T. S. Dhadialla, D. E. Cress, B. Mugat, C. Antoniewski, and J. A. Lepesant, *Biochem. J.* **327**, 643 (1997).

For example, it is possible that essential cofactors may not be able to interact with the GAL4-LBD fusion protein in the same way that they can interact with the natural endogenous receptor. In addition, the fusion proteins are missing the N-terminal A/B domain found in most nuclear receptors—a domain that may be required for hormone-dependent activation *in vivo*. This deficiency may, however, be compensated for in receptors that function as homodimers, because the GAL4-LBD fusion protein could dimerize with an endogenous full-length receptor that provides an A/B domain. An effort to compare the activation patterns directed by GAL4-LBD and A/B-GAL4-LBD constructs would allow one to determine how this might impact the assay system. It also should be taken into account that GAL4-LBD proteins are often expressed ectopically (depending on the choice of promoter), and it is important to ensure that the endogenous receptor is present in those tissues where activation of the corresponding GAL4-LBD is detected. Finally, efforts could be made to improve the sensitivity of detecting reporter gene expression directed by the GAL4-LBD system in *Drosophila*. For example, the sensitivity of GAL4-LBD activation could be increased by introducing a UAS-GAL4 construct to create a positive feedback loop, increasing the level of reporter gene expression.[14]

The GAL4-LBD system provides a powerful and versatile tool to study the mechanisms of hormone action during development, allowing the researcher to analyze the spatial and temporal activation patterns of any nuclear receptor LBD *in vivo*. *Drosophila* also offers additional advantages for use of this system. Genetically manipulating the background in which GAL4-LBD activation is assayed (either in the whole animal or in mosaic clones) should facilitate the identification of additional components of hormone-dependent pathways, such as transcriptional cofactors or nuclear receptor partners. For example, we demonstrated that, as expected for a heterodimeric EcR partner, GAL4-USP activation is significantly reduced in an *EcR* mutant background.[4] Ligation experiments similar to the one described earlier could be used to determine if the activation of an orphan nuclear receptor LBD is dependent on a diffusible hormone. In addition, a combination of *in vivo* analysis, organ culture experiments, and biochemical approaches offers a relatively straightforward way to identify possible ligands for orphan nuclear receptors. Finally, GAL4-LBD fusions can act as specific dominant negatives, disrupting endogenous-signaling pathways (as was shown for GAL4-EcR and GAL4-USP).[4] These transgenes can thus be used to selectively inactivate nuclear receptor signaling during development.

[14] B. A. Hassan, N. A. Bermingham, Y. He, Y. Sun, Y. N. Jan, H. Y. Zoghbi, and H. J. Bellen, *Neuron* **25**, 549 (2000).

Inhibition of Nuclear Receptor Function by Inducible
Expression of dsRNA

With the completion of the *Drosophila* genome sequence, new methods
are required that facilitate the transition from gene sequence to gene
function. An ideal tool for achieving this goal is RNA interference, or
RNAi. As shown originally by Fire *et al.*, injection of dsRNA corresponding
to the transcribed region of a particular gene will effectively and specifically
interfere with the expression of that gene in *C. elegans*.[15] Subsequent studies
revealed that this method is applicable in a wide range of organisms,
including *Drosophila*.[16,17] More recently, this technique was expanded by
Tavernarakis *et al.*, who showed that inducible expression of snapback
dsRNA from a transgenic construct could effectively induce RNAi.[18]
Importantly, this method worked in a broader range of tissues than those
affected by injected dsRNA. In addition, it provided an opportunity to
control where and when RNAi would occur in the context of an intact
organism. Fortier and Belote were the first to test this approach in
Drosophila, showing that they could reproduce known mutant phenotypes
associated with a sex differentiation gene.[19] Subsequent studies by a number
of labs, including our own, have provided a strong foundation for designing
optimal transgenic constructs for RNAi in *Drosophila*. In particular, we
have used this method to inactivate nuclear receptor gene expression at later
stages in the *Drosophila* life cycle, allowing us to assign stage-specific
functions that could not be addressed with more conventional loss-of-
function genetic tools.[20]

The Construct

The establishment of transgenic *Drosophila* lines depends on P
transposable element vectors that can be used to create single stable
insertions in the fly genome.[9] These vectors carry a marker gene to select for
transformant lines, often the *white* gene that directs the synthesis of red
pigment in the adult eye. For inducible RNAi, an inverted repeat from the
gene of interest is inserted into a P element vector downstream from a

[15] A. Fire, S. Xu, M. K. Montgomery, S. A. Kostas, S. E. Driver, and C. C. Mello, *Nature* **391**,
806 (1998).
[16] J. R. Kennerdell and R. W. Carthew, *Cell* **95**, 1017 (1998).
[17] L. Misquitta and B. M. Paterson, *Proc. Natl. Acad. Sci. USA* **96**, 1451 (1999).
[18] N. Tavernarakis, S. L. Wang, M. Dorovkov, A. Ryazanov, and M. Driscoll, *Nat. Genet.* **24**,
180 (2000).
[19] E. Fortier and J. M. Belote, *Genesis* **26**, 240 (2000).
[20] G. Lam and C. S. Thummel, *Curr. Biol.* **10**, 957 (2000).

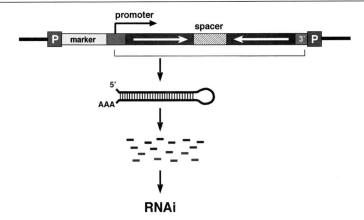

FIG. 3. Schematic representation of a transgenic insertion designed to express snapback dsRNA. A P element vector is depicted, inserted in *Drosophila* genomic DNA (heavy black lines). P element ends (blue boxes, P), the *white* marker gene (yellow box), the inducible promoter sequence (green box), the inverted repeats (arrows), spacer sequence (hatched box), and 3′ trailer (red box) are all depicted. The transcribed region is marked by a bracket. This forms a snapback dsRNA that is rapidly degraded into ~21 nucleotide RNAs that direct RNAi. (See Color Insert.)

regulated promoter (Fig. 3). As described originally for injected dsRNA, and later for transgenic expression of dsRNA, a ~1 kb region of dsRNA is an ideal length.[15,18] Regions of 700 bp, however, have been used effectively.[19,21] Selection of the sequences that comprise the inverted repeat depends upon the target gene of interest, but can derive from anywhere within an exon. It is a good idea to select a region that is unique to the gene of interest, testing this by doing a BLAST search against the *Drosophila* genome sequence. Common repeats, such as CAX repeats, are to be avoided. The inverted repeat can be in either orientation. Fortier and Belote report that switching from one orientation to the other can facilitate the cloning of inverted repeat sequences in bacteria.[19] All reported constructs include a poly(A) addition site at the 3′ end of the inverted repeat sequence, although this may not be necessary for effective RNAi (Fig. 4).

The Promoter

Two different promoters have been used to direct expression of dsRNA in *Drosophila*. One of these derives from the GAL4 system, which

[21] J. R. Kennerdell and R. W. Carthew, *Nat. Biotechnol.* **18**, 896 (2000).

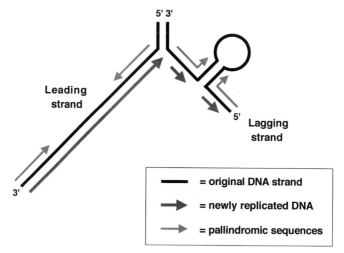

Fig. 4. Inverted repeat sequences lead to defects at the replication fork. A DNA replication fork is depicted with inverted repeat sequences represented by red arrows. The leading strand is replicated continuously, preventing the formation of a snapback. The lagging strand, in contrast, snaps back on itself with a small loop caused by a short spacer region. See text for details. (See Color Insert.)

allows precise spatial and temporal regulation of expression.[22] The yeast GAL4 transcription factor can activate expression of any target gene by binding to a GAL4-UAS positioned near a minimal promoter. In this system, two different P element transformant lines are established, one of which expresses GAL4 in a particular spatial and temporal pattern, and the other of which carries a target gene under the control of multiple UAS elements. Expression of GAL4 has no detectable effects on *Drosophila* development and the target gene is not expressed in the absence of GAL4. Only upon crossing these two lines will the target gene be activated, in a temporal and spatial pattern that reflects that of GAL4. Many stocks have been isolated that express GAL4 in a defined pattern during development. These stocks are available to the community through either the stock center or individual labs, making this a valuable method for targeting gene expression in *Drosophila*.[23]

[22] A. H. Brand and N. Perrimon, *Development* **118**, 401 (1993).

[23] A. H. Brand, A. S. Manoukian, and N. Perrimon, *in* "*Drosophila melanogaster*: Practical Uses in Cell and Molecular Biology" (L. S. B. Goldstein and E. A. Fyrberg, eds.), p. 635. Academic Press, New York, 1994.

A number of labs have successfully used the GAL4 system to drive the expression of snapback dsRNA in *Drosophila*.[19,21,24–27] In cases where the targeted gene was already associated with a known mutant phenotype, that phenotype was reproduced with dsRNA expressed under GAL4 control. In addition, these studies show a clear dose effect, in which higher levels of dsRNA directs more efficient RNAi. One of the most careful studies was performed by Fortier and Belote, who showed that increasing the copy number of the inverted repeat transgene, or using stronger GAL4 drivers, resulted in a stronger loss-of-function phenotype.[19] They could also achieve this effect by shifting the animals to 29°C, which allows GAL4 to act as a more effective activator. Thus, by using the appropriate conditions, the investigator can achieve a range in the efficiency of gene inactivation. Including an intron in the spacer region of the transgenic construct has also been reported to increase the efficiency of RNAi, as described in the next section.

Temporal control can be achieved by using the *hsp70* promoter to express dsRNA.[20] This promoter drives high levels of widespread gene expression throughout the animal. It does not, therefore, provide the tissue-specific patterns of expression that can be achieved by use of the GAL4-dependent promoter. However, the high levels of dsRNA driven by the *hsp70* promoter tend to generate strong loss-of-function mutant phenotypes. In addition, in cases where the targeted gene is widely expressed, or when the investigator is interested in inactivating gene expression at a certain stage in development, the *hsp70* promoter is ideal.

The Spacer

It is difficult to clone inverted repeat sequences in bacteria without having a sizable spacer region between the repeats. This is because of problems that arise at the replication fork during DNA synthesis.[28] In the absence of a spacer, or with a short spacer as shown in Fig. 4, the leading strand can be replicated efficiently. DNA synthesis is processive on this strand and, thus, the template strand of DNA is replicated before it can snapback on itself. In contrast, on the lagging strand, the inverted repeat can snapback with zero-order kinetics, blocking the progression of discontinuous DNA synthesis. This results in a strong selection for plasmids that have eliminated part or all of the inverted repeat. Including a sizable

[24] E. Giordano, R. Rendina, I. Peluso, and M. Furia, *Genetics* **160**, 637 (2002).
[25] S. Martinek and M. W. Young, *Genetics* **156**, 1717 (2000).
[26] S. Kalidas and D. P. Smith, *Neuron* **33**, 177 (2002).
[27] M. D. Adams and J. J. Sekelsky, *Nat. Rev. Genet.* **3**, 189 (2002).
[28] D. R. Leach, *Bioessays* **16**, 893 (1994).

spacer region, of at least several hundred base pairs, prevents this problem and allows clones to be readily established in recombination-deficient strains of bacteria. Giordano *et al.*, also provide evidence that inverted repeats without a spacer region may be unstable in the fly genome.[24] A recent paper describes the use of an intron as a spacer region and reports strong loss-of-function phenotypes for three different targeted genes.[26] They propose that splicing may improve the efficiency of RNAi; however, no control constructs were made that were missing intronic sequences. This observation is, nonetheless, worthy of further investigation. It should be noted that including introns should not enhance RNAi driven by the *hsp70* promoter because heat treatment blocks RNA splicing.[29]

Timing of Heat-Induced dsRNA Expression from the hsp70 Promoter

A single 30 min heat treatment at 37°C results in a high degree of lethality in late third instar larvae carrying a heat-inducible inverted repeat derived from the EcR gene.[20] An additional heat treatment 6 hr later increased the severity of this phenotype, consistent with observations using the GAL4 system that more dsRNA is more efficient at directing RNAi. RNA corresponding to the length of the snapback could be detected immediately after heat treatment by Northern-blot hybridization.[20] This RNA was, however, significantly reduced by 2 hr after heat treatment and undetectable by 4 hr. It is likely that the dsRNA is rapidly processed to the ~21 nucleotide RNAs that are the active components of RNAi.[30] Two sequential heat treatments are thus used to direct the most severe loss-of-function phenotypes, and these phenotypes can be observed ~12 hr after the final heat treatment.[20] Further studies are required to determine how long heat-induced RNAi will persist. However, it is clear that phenotypes can, in some cases, be seen within hours after heat treatment.

Interpreting Phenotypic Effects

An understanding of where and when the target gene is expressed allows the investigator to choose the right experimental design for RNAi—which promoter to use (*hsp70* or GAL4 dependent), and where and when to look for mutant phenotypes. Just like all P element insertions, inducible expression of snapback dsRNA is subject to position effects, giving different

[29] H. J. Yost and S. Lindquist, *Cell* **45**, 185 (1986).
[30] B. L. Bass, *Cell* **101**, 235 (2000).
[31] M. A. Bhat, S. Izaddoost, Y. Lu, K. O. Cho, K. W. Choi, and H. J. Bellen, *Cell* **96**, 833 (1999).

levels of background and inducible expression depending on the site of insertion in the genome. It is thus critical to select a transformant line that directs the most effective RNAi. This can be most easily achieved by selecting a time when the target mRNA is expressed, and examining the effect of inducible RNAi on the levels of endogenous target mRNA by Northern-blot hybridization. Western-blot analysis or antibody stains of tissues can also be performed, if an antibody is available for the encoded protein. These studies allow one to determine how effectively gene activity has been disrupted. As mentioned earlier, more effective RNAi can be achieved by: (1) increasing the strength and/or copy number of the GAL4 driver, (2) increasing the copy number of the snapback transgene, (3) using higher temperatures and multiple heat treatments with the *hsp70* promoter, and (4) including introns in constructs expressed using the GAL4 system.

Strengths and Weaknesses of the Method

Several shortcomings must be kept in mind when using RNAi to disrupt *Drosophila* gene function. First, as with any reverse genetic approach, the investigator may not know what mutant phenotype(s) to expect or, indeed, whether a visible phenotype is associated with disruption of the gene of interest. It is thus critical to first establish the proper parameters for directing efficient RNAi (as described earlier), and then design a series of studies that are most likely to reveal gene function. Second, mutant phenotypes that arise from inducible RNAi represent only a partial loss-of-function. Stable preexisting protein may partially rescue RNAi-induced mutant phenotypes. In addition, it is rare that investigators have reported a complete absence of target gene expression by RNAi. Phenotypes that arise from this method thus need to be interpreted with this caveat in mind.

A number of unique advantages associated with regulated expression of dsRNA for RNAi have established this method in the *Drosophila* community.[27] Inducible expression of dsRNA allows the researcher to induce RNAi at any stage in the life cycle, or restrict the effects of RNAi to one cell type. It also provides what appear to be uniform effects throughout the tissue or animal, unlike the mosaic response often seen in animals injected with dsRNA. RNAi will also target both maternal and zygotic mRNA for degradation, circumventing the need for analyzing germline mutant clones for some genes.[31] Finally, this method provides a new way to study the later functions of genes which, when mutated, result in early lethality. In effect, this provides a means of generating dominant temperature-sensitive alleles for any coding region in the *Drosophila* genome. In the past, these mutations have only been obtained as fortuitous alleles in open-ended genetic screens. The simplicity of establishing P element transformants makes this approach a valuable new tool for functional genomic studies in *Drosophila*.

Conclusions

Taken together, the use of GAL4-LBD fusions to characterize the temporal and spatial patterns of nuclear receptor activation, and the selective inactivation of nuclear receptor gene function by inducible RNAi, provide a powerful combination to dissect nuclear receptor-signaling pathways in *Drosophila*. It is likely that the patterns of nuclear receptor activation will provide new insights into ecdysteroid signaling through the EcR/USP heterodimer, as well as new directions for understanding orphan receptor function. The hypotheses that arise from these studies can, in turn, be tested by selectively inactivating the corresponding gene by inducible RNAi. These methods should greatly facilitate our ability to characterize nuclear receptor signaling pathways in this insect model system.

Acknowledgments

We thank S. Lewis (University of Toronto) for explaining the effects of inverted repeat sequences on DNA replication. We are indebted to L. Dinan (University of Exeter) for providing cucurbitacin B and Y. N. Jan (UCSF) for the *UAS-nlacZ* insertion on the third chromosome. T. K. is a Research Associate and C. S. T. is an Investigator with the Howard Hughes Medical Institute.

Author Index

Numbers in parentheses are footnote reference numbers and indicate that an author's work is referred to although the name is not cited in the text.

A

Aarsonson, D., 175, 179(18), 188(18)
Aasland, R., 247
Abagyan, R., 115
Abeil, J. F., 95
Abel, M. G., 300
Acena-Nagel, M., 350
Adam, J., 441
Adams, J. Y., 453, 455
Adams, M. D., 95, 475, 487, 488(27)
Adams, M. M., 453
Adams, S. L., 315
Addis, P. B., 31
Aden, Y., 26, 33(10)
Adoutte, A., 104, 105(24)
Afshari, C. A., 300
Agami, R., 348
Agard, D. A., 82, 87, 119, 225
Agbayani, A., 95
Ahima, R. S., 300
Ahlborg, G., 25, 26, 26(4), 32
Ahluwalia, A., 453
Ahrens, H., 449
Akimenko, M. A., 411
Akiyama, T. E., 407
Akiyoshi-Shibata, M., 368
Akmaev, V., 305
Alami, M., 87
Al-Azzaw, F. J., 453, 455
Albericio, F., 236
Albers, M., 117
Alberti, S., 407
Aldaz, C. M., 306, 307(36)
Alfarano, C., 315
Ali, S., 176, 176(37), 177, 181
Allan, G. F., 119
Alland, L., 248
Allegretto, E. A., 450, 453, 455, 459
Allen, W. V., 412

Allgood, V. E., 174, 176(9)
Allis, C. D., 211, 247, 287, 295
Almkvist, Å., 458
Altmann, M., 181
Altschul, S. F., 97, 107(13), 306
Aluker, M., 461
Alvelius, G., 31
Alves, S. E., 455
Amacher, S. L., 408
Amanatides, P. G., 95
Amatruda, J. F., 409, 410, 419(9)
Amling, M., 379, 404(5,6)
Amma, L., 426, 436, 448(19)
Amsterdam, A., 408
An, H. J., 95
Andersen, J., 455
Andersen, J. R., 315
Andersen, K. W., 456
Anderson, C. M., 224, 244(4)
Anderson, G. J., 474
Anderson, J. P., 108
Anderson, R. G., 449, 453, 455
Anderson, T. J., 453, 455
Andersson, G., 449
Andersson, L. C., 451, 455
Andersson, O., 25, 26, 32
Andersson, S., 455
Andersson, U., 31
Andre, M., 411
Andres, A. J., 480, 482(12)
Andrews-Pfannkoch, C., 95
Angelin, B., 407
Angelini, N. L., 455
Angelov, D., 284
Angrand, P. O., 383(64), 404
Anlag, K., 407
Anselmo, D., 173
Antoniewski, C., 482
Aparicio, O., 315
Apolito, C. J., 69

491

Subject Index

A

B

C

ISBN: 0-12-182267-2

90000

9 780121 822675

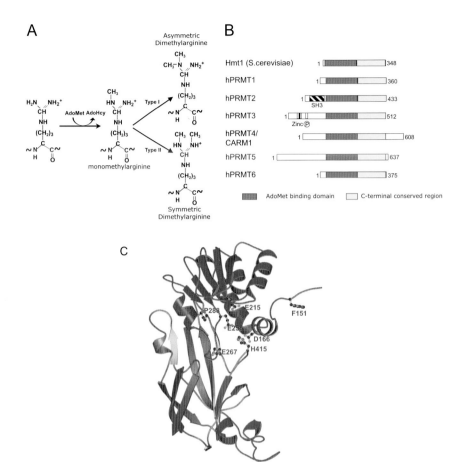

WEI XU ET AL., CHAPTER 12, FIG. 3. Structural features of PRMTs. (A) Methylation of arginine residues by PRMTs converts S-adenosyl methionine (AdoMet) into S-adenosyl homocysteine (AdoHcy). (B) Schematic representation of seven PRMT homologs. The yeast hmt1 and six known human arginine methyltransferases (PRMT1 → 6) share a conserved "core" region of ∼310 amino acids, including AdoMet-binding domain (purple) and a little less-conserved domain (green). (C) The structural model of CARM1 (amino acid 148–458) generated with SWISS-MODEL and the Swiss-Pdb Viewer[32] based on the 3D structures of Hmt1 and PRMT3 (Accession code 1G6Q and 1F3L). CARM1 is folded into two domains connected at proline 288. The N-terminal residues E215, E258, E267 constitute arginine amino methylation catalytic active sites, which are conserved between PRMTs. The two terminal amino groups of the substrate arginine interact with the side chains of E258 and E267. The carboxylate oxygen of D166 forms a hydrogen bond to H415, which is involved in elimination of proton after methyl transfer. The hydrophobic phenyl ring of F151 in N-terminal helix is essential to lock the AdoHcy in position and almost buries the AdoHcy. The movement of N-terminal helix would allow AdoMet into and AdoHcy out of the binding site.

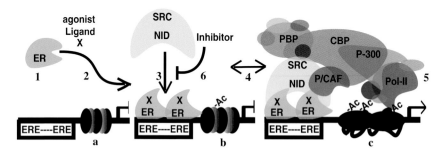

TIMOTHY R. GEISTLINGER AND R. KIPLIN GUY, CHAPTER, 13, FIG. 1. Schematic model of the dynamic assembly of the transcription–activation complex by agonist-bound ER and function of SRC-binding inhibitors. (1) In the absence of ligand, chromatin is unmodified (a) and transcription at the DNA estrogen response element (ERE) is unaltered. (2) Binding of agonist ligand to the LBD of ER induces a conformational change in ER leading to translocation and homodimerization on the ERE. (3) Liganded ER on the ERE recruits SRCs using the NR box ($L_1XXL_2L_3$) of the NID, resulting in chromatin modification (b). (4) Subsequently, the ER–SRC complex recruits other proteins, such as PBP/DRIP/TRAP, CBP/p300, and RNA Pol-II, to form the activation complex where chromatin is fully modified (c), and (5) transcription of the ERE gene commences. (6) Direct competitive inhibition of SRC binding to NR will block the initial step (3) of activation complex formation and thus prevent transcription.

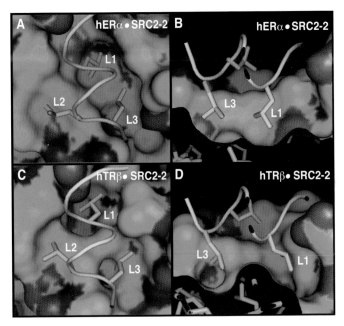

TIMOTHY R. GEISTLINGER AND R. KIPLIN GUY, CHAPTER 13, FIG. 2. Panel A: Cocrystal structure of the SRC-binding pocket in E_2–hERα–SRC2-2[12] oriented to show a view from above the pocket. The three leucine side chains of the SRC2-2 ($L_1XXL_2L_3$) motif (yellow wire frame) and labeled L_1 L_2 L_3. The E_2–hERα receptor van der Waals radii surface was generated and color coded according to atom type: red–oxygen, blue–nitrogen, green–carbon and hydrogen, orange–sulfur ($PyMol^{TM}$). Panel B: Cross section of the same binding site as in A, approximately through the middle of the L_1 and L_3 subpockets. Panel C: Cocrystal structure of the SRC-binding pocket in T_3–hTRβ–SRC2-2[2] oriented in the same view as that of panel A. Panel D: Cross section of the same binding site in B, oriented as in C. These views demonstrate clear significant differences in steric structure between the hTRβ and hERα SRC-binding pockets.

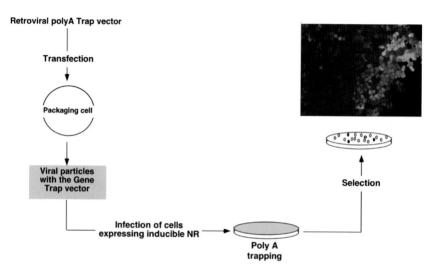

HANNA KOUTNIKOVA *ET AL.*, CHAPTER 17, FIG. 3. The construction of the trapping library. Packaging cells are transfected with the retroviral gene trap vector for the production of the corresponding viral particles. The cells chosen for the gene trapping are infected with these retroviruses carrying the gene trap vector and are further selected for the expression of the selectable marker, which only will be produced if spliced to an endogenous polyA signal. Therefore, each cell clone selected will have integrated the gene trap vector into a gene that contains a polyA signal. As the promoterless reporter gene is now under the control of the promoter of the gene whose polyA has been trapped, each cell clone will show different degrees of reporter expression depending on the inherent activity of the promoter of the gene trapped by the polyA trap vector.

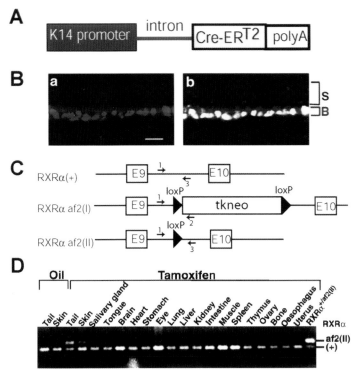

DANIEL METZGER ET AL., CHAPTER 22, FIG. 1. Characterization of the K14-Cre-ER[T2] transgenic line. (A) Structure of the K14-Cre-ER[T2] transgene. The human K14 promoter, the Cre-ER[T2] coding sequence and the simian virus 40 polyadenylation signal (polyA) are represented by black, grey and open boxes, respectively. The rabbit β-globin intron is depicted by a line. (B) Immunohistochemical pattern of Cre-ER[T2] expression in the tail epidermis of K14-Cre-ER[T2] transgenic mice. The red color (a and b) corresponds to the staining of Cre-ER[T2], and the blue color (b) corresponds to the DAPI staining of the nuclei. The white color of the basal keratinocyte nuclei (b) results from the superimposition of the red color of the anti-Cre signal and the blue color of the DAPI staining. B and S, basal and suprabasal layers, respectively. Scale bar, 25 μm. (C) Genomic structure of the RXRα WT, the RXRα af2(I) target allele and the recombined RXRα af2(II) allele, and PCR strategy to identify the RXRα af2(I) allele and to analyze Cre-mediated excision of the floxed marker. (D) PCR detection of Cre-mediated DNA excision in mice. PCR was performed on DNA isolated from the indicated organs one day after oil- and Tam-treated K14-Cre-ER[T2 (tg/0)]/RXRα[+/af2(I)] mice, as well as from tail of RXRα[+/af2(II)] mice, as indicated. The position of the PCR products amplified from the WT RXRα allele (+) and the RXRα af2(II) allele are shown.

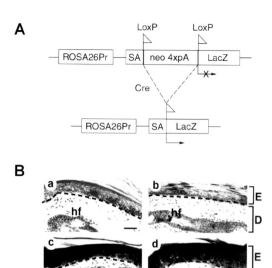

DANIEL METZGER *ET AL.*, CHAPTER 22, FIG. 2. Characterization of Cre recombinase activity in skin of K14-Cre-ER^{T2} transgenic mice. (A) Structure of the ROSA fl allele before and after Cre-mediated recombination. SA, splice acceptor site; neo 4xpA, neomycin resistance cassette with 4 polyadenylation sites. (B) Kinetics of β-galactosidase expression in tail epidermis of K14-Cre-ER$^{T2(tg/0)}$/ROSA$^{fl/+}$ bigenic mice injected daily with Tam from day 0 to 4. X-Gal-stained tail sections collected just before the first Tam injection (a) and 5 (b), 30 (c) and 60 (d) days after Tam treatment, are presented. D and E, dermis and epidermis, respectively; hf, hair follicle. Scale bar, 16 μm.

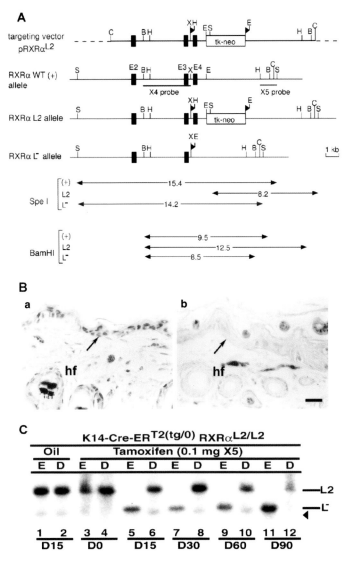

DANIEL METZGER *ET AL.*, CHAPTER 22, FIG. 3. Temporally controlled RXRα ablation in epidermal keratinocytes. (A) Schematic diagram of the pRXRα^{L2} targeting vector, the RXRα WT (+) genomic locus, the floxed RXRα L2 allele, and the RXRα L⁻ allele obtained after Cre-mediated excision of exon 4. Restriction enzyme sites and the location of X4 and X5 probes are indicated. The numbers in the lower part of the diagram are in kilobases (kb). Abbreviations: B, *Bam*HI; C, *Cla*I; E, *Eco*RI; H, *Hin*dIII; S, *Spe*I; X, *Xba*I. The dashed line corresponds to backbone vector sequences. Arrowheads represent LoxP sites. (B) Immunohistochemical detection of RXRα in skin sections from Tam treated RXRα$^{L2/L2}$ (a) and K14-Cre-ER$^{T2(tg/0)}$/RXRα$^{L2/L2}$ (b) mice. Arrows point to the dermal–epidermal junction. hf, hair follicle. Scale bar, 16 μm. (C) Efficiency of K14-Cre-ER^{T2}-mediated RXRα recombination in adult skin. L2 and L⁻ RXRα alleles were identified by Southern blot on DNA extracted from epidermis "E" or dermis "D" isolated from tail of K14-Cre-ER$^{T2(tg/0)}$/RXRα$^{L2/L2}$ mice before (D0) and 15, 30, 60 and 90 days after the first Tam administration, and 15 days after oil (vehicle) administration, as indicated. The position of the DNA segments corresponding to the RXRα L2 and L⁻ alleles are indicated. The arrowhead points to a nonspecific signal.

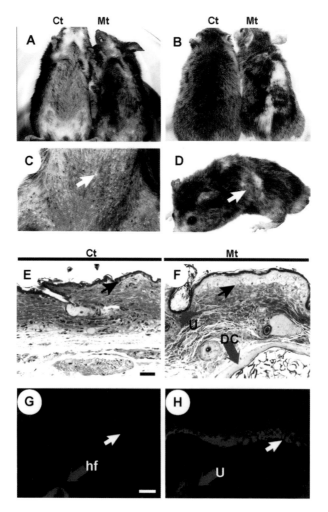

DANIEL METZGER *ET AL.*, CHAPTER 22, FIG 4. Abnormalities generated by Tam-induced K14-Cre-ER^{T2}-mediated disruption of RXRα in skin of adult mouse. (A) Ventral view of a female K14-Cre-ER$^{T2(tg/0)}$/RXRα$^{L2/+}$ "control" (ct) mouse (left), and a female K14-Cre-ER$^{T2(tg/0)}$/RXRα$^{L2/L2}$ "mutant" (mt) mouse (right), 16 weeks after the first Tam treatment. (B) Dorsal views of the same animals. (C) Higher magnification of the ventral region of the K14-Cre-ER$^{T2(tg/0)}$/RXRα$^{L2/L2}$ "mutant" mouse, with arrow pointing to one of the cysts. (D) Dorsal view of a female K14-Cre-ER$^{T2(tg/0)}$/RXRα$^{L2/L2}$ "mutant" mouse, 28 weeks after the first Tam treatment; arrow points to a skin lesion. (E and F) Histological analysis of sections of ventral skin 16 weeks after the first Tam treatment, taken from "control" (E) and "mutant" (F) mice. Scale bar (in E): E and F, 60 μm; (G and H) Keratin 6 (K6) immunohistochemistry on "control" (G) and "mutant" (H) skin sections (16 weeks after the first Tam treatment). The red color corresponds to the staining of the K6 antibody, and the cyan color corresponds to DAPI staining. Scale bar (in G): G and H, 25 μm. Arrows in E–H point to the dermal–epidermal junction. hf, hair follicles; u, utriculi; DC, dermal cysts.

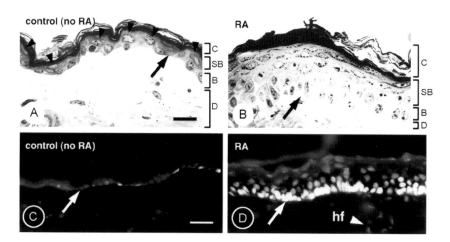

DANIEL METZGER *ET AL.*, CHAPTER 22, FIG. 6. Histology and proliferative response of wild-type skin upon topical retinoid treatment. Dorsal skin was topically treated for four consecutive days with RA (see text) (B and D) or acetone vehicle (A and C, control) and analyzed 24 hr after the last RA application. (A and B) Histology of control and RA-treated epidermis. Semi-thin sections (2 µm-thick) were stained with toluidine blue. Arrowheads point to the spinous and granular keratinocytes in control epidermis. (C and D) Immunohistochemical detection of the proliferation marker Ki67 (white color) on skin sections counterstained with DAPI (blue color). Arrows point to the dermal–epidermal junction. RA, retinoic acid. B, basal layer; C, cornified layer; D, dermis; hf, hair follicle; SB, suprabasal layers (spinous and granular keratinocytes). Scale bar (in A): 15 µm in A and B; (in C) 25 µm in C and D.

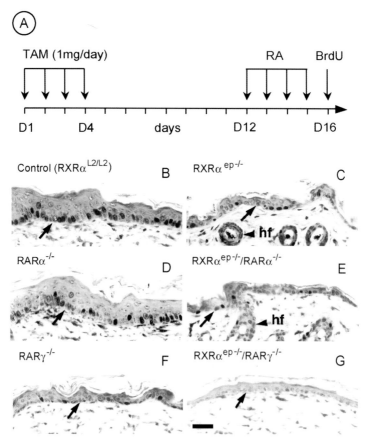

DANIEL METZGER *ET AL.*, CHAPTER 22, FIG. 7. RA-induced proliferative response in the skin of retinoid receptor mutants. (A) Scheme of the experimental protocol. I.P. injection of Tam (1 mg) was done from day 1 (D1) to day (D4). 0.4 ml of 40 nmoles RA in acetone was topically applied on ~9 cm^2 shaved back skin once a day from D12 to D15. BrdU (50 mg/kg) was injected on D16, 2 hr before skin sampling. (B to G) Representative skin sections from mice of the indicated genotype, labelled with BrdU (brown color). Arrows point to the dermal–epidermal junction. hf, hair follicles. Scale bar (in G): 50 μm in B to G.

DANIEL METZGER *ET AL.*, CHAPTER 22, FIG. 8. Conditional mutagenesis of RARγ and RA-induced proliferative response in mice lacking RARγ in suprabasal layers (RAR$\gamma^{sb-/-}$ mice). (A) Schematic drawing of RARγ wild-type (+), L3, partially excised L2Neo and fully excised L$^-$ alleles. Sizes of *Nsi*I fragments obtained for each allele are in kilobases (kb). Black boxes, exons 7 to 13 (E7–13); neo, neomycin resistance gene; N, *Nsi*I. Arrowhead flags represent loxP sites. (B) Experimental protocol. (C) Efficiency of RARγ gene disruption in mice bearing the CMV-Cre-ER^T transgene. RARγ wild-type (+), L3, L2Neo and L$^-$ alleles were analyzed on tail epidermis and dermis genomic DNA before and after Tam administration. Note that the minute amount of RARγ excised L$^-$ allele present in dermis most probably originates from contaminating epidermal keratinocytes.[32] M: DNA ladder. (D to G) Immunohistological

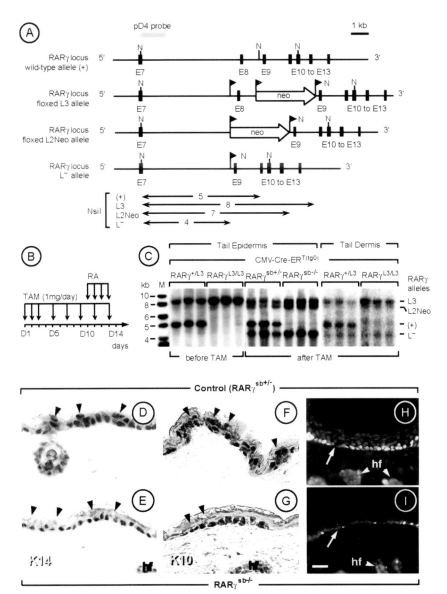

detection of RARγ (brown color), K14 (D and E; yellow false color) and K10 (F and G; yellow false color) on back skin sections taken at D9 from control (D and F; RARγ$^{sb+/−}$) and mutant (E and G; RARγ$^{sb−/−}$) mice. Arrowheads in D and F point to suprabasal keratinocytes that express RARγ, whereas arrowheads in E and G point to suprabasal keratinocytes that do not express RARγ. (H and I) Immunohistological detection of the Ki67 proliferation marker on skin sections taken at D14 from control (H) and mutant (I) animals (white color, Ki67 signal; blue color, DAPI nuclear staining). Arrows point to the dermal–epidermal junction. hf, hair follicles. Scale bars (in I): 50 μm.

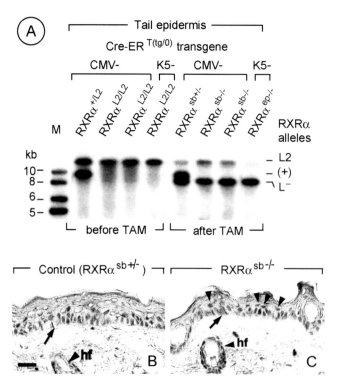

DANIEL METZGER ET AL., CHAPTER 22, FIG. 9. Conditional mutagenesis of RXRα in epidermis suprabasal layers, and RA-response in RXRα$^{sb-/-}$ mice. (A) Efficiency of RXRα gene disruption in mice bearing the CMV-Cre-ER^T transgene as compared to the K5-Cre-ER^T transgene (as indicated). The experimental protocol is the same as in Fig. 8B. RXRα wild-type (+), L2 and L$^-$ alleles were analyzed on tail epidermis genomic DNA before and after Tam administration. M: DNA ladder. (B and C) Immunohistochemical detection of RXRα on skin sections from control (B, RXRα$^{sb+/-}$) and experimental (C, RXRα$^{sb-/-}$) mice, after RA treatment. Arrowheads indicate nuclei of suprabasal keratinocytes expressing RXRα. Arrows point to the dermal–epidermal junction. hf, hair follicles. Scale bar (in B): 50 μm.

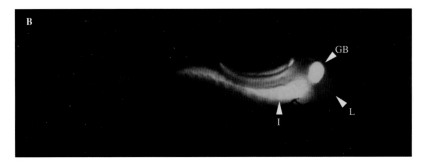

A

EX (505 nm)

$CH_3(CH_2)_{14}$-C-O-CH_2

$(CH_2)_4$-C-O-CH

CH_2O-P-OCH_2CH_2NHC$(CH_2)_5$NH

PLA2

EM (515 nm)

H_3C H_3C

NO_2

NO_2

B

GB

I

L

Shiu Ying Ho *et al.*, Chapter 23, Fig. 1. PED6 as a biosensor to visualize lipid metabolism in live zebrafish larva. (A) The structure of PED6. The emission of BODIPY-labeled acyl chain at *sn*-2 position is quenched by dinitrophenyl group at *sn*-3 position when this molecule is intact. Upon PLA$_2$ cleavage at *sn*-2 position, BODIPY-labeled acyl chain emits green fluorescence. (B) 5 dpf zebrafish larva labeled with PED6 (0.3 μg/ml) for 6 hr. Arrow shows liver (L), gall bladder (GB) and intestine (I).

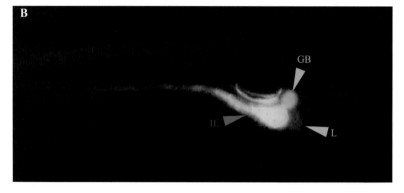

SHIU YING HO *ET AL.*, CHAPTER 23, FIG. 3. BODIPY FR-PC label in zebrafish larvae. (A) The structure of BODIPY FR-PC. When the molecule is intact in the cell, excitation at 505 nm results in orange (568 nm) emission due to fluorescence resonance energy transfer (FRET) between the two BODIPY labeled moieties. Upon PLA$_2$ cleavage at *sn*-2 position, BODIPY moiety at *sn*-1 position showed green (515 nm) emission when excited at 505 nm. (B) BODIPY FR-PC (5 μg/ml) labeled 5 dpf zebrafish larva. The liver (L) and gall bladder (GB) showed green fluorescence (green arrow), indicating the accumulation of cleaved products. Uncleaved orange BODIPY FR-PC (orange arrow) is observed only in the intestinal epithelium (IE).

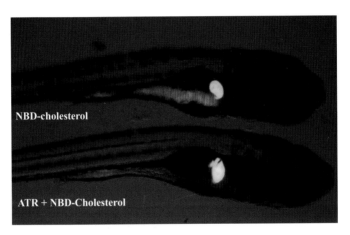

SHIU YING HO *ET AL.*, CHAPTER 23, FIG. 5. Atorvastatin (ATR) interferes with NBD-cholesterol labeling. Wild-type larvae had reduced fluorescence in the intestinal lumen, while gallbladder fluorescence was preserved after 1.5 hr labelling of NBD-cholesterol (2 μg/ml) solubilized with tilapia's bile.

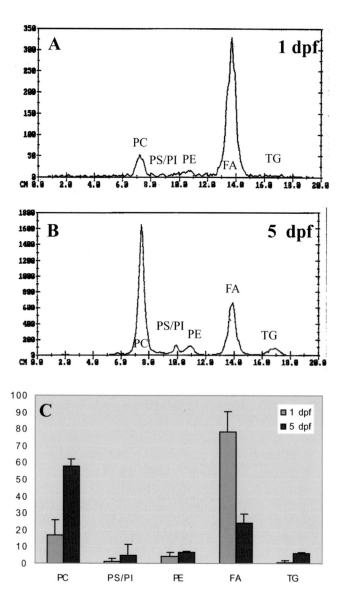

SHIU YING HO *ET AL.*, CHAPTER 23, FIG. 6. Lipomics during development. Larvae (1 and 5 dpf) were incubated with radioactive oleic acid for 20 hr, followed by lipid extraction and TLC. The solvent chloroform:ethanol:triethylamine:water (30:34:30:8) was used to develop the TLC plate. The radioactivities were then scanned. The major metabolites derived from oleic acid (FA) are phosphatidylcholine (PC), phosphatidylserine (PS), phosphatidylinositol (PI), phosphatidylethanolamine (PE), and triacylglycerol (TG). (A) One representative lipomics of 1 dpf larva. (B) One representative lipomics of 5 dpf larva. (C) Comparison lipomics between 1 and 5 dpf larvae ($n = 9$, mean $\pm$ SD).

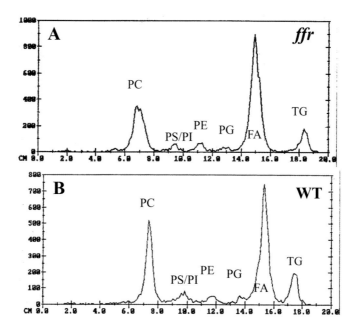

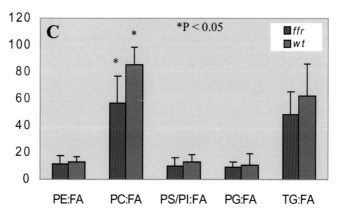

SHIU YING HO *ET AL.*, CHAPTER 23, FIG. 8. Lipomics of *fat-free* and wild-type. Both *fat-free* and wild-type larvae (4 dpf) were incubated with radioactive oleic acid for 20 hr, followed by lipid extraction and TLC. The solvent chloroform:ethanol:triethylamine:water (30:34:30:8 v/v) was used to develop the TLC plate. The radioactivities were then scanned. The major metabolites derived from oleic acid (FA) are phosphatidylcholine (PC), phosphatidylserine (PS), phosphatidylinositol (PI), phosphatidylethanolamine (PE), phosphatidylglycerol (PG), and triacylglycerol (TG). (A) One representative lipomics of *fat-free* larva. (B) One representative lipomics of wild-type larva. (C) Comparison lipomics between *fat-free* and wild-type larva. *Fat-free* has significantly decreased PC production when expressed as metabolite/FA ($n = 9$, mean ± SD).

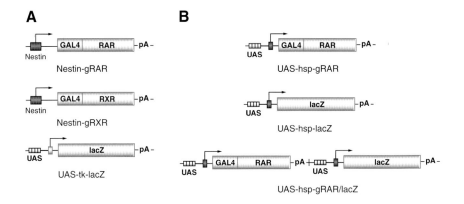

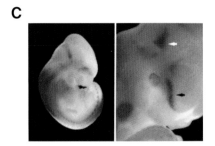

ALEXANDER MATA DE URQUIZA AND THOMAS PERLMANN, CHAPTER 26, FIG. 1. Effector–reporter constructs and *in vivo* detection of RAR activity. (A) Effector plasmids expressing Gal4-RARα and Gal4-RXRα fusion proteins are driven by the nestin promoter. The reporter plasmid contains four Gal4-binding sites (UASs), followed by a minimal thymidine kinase (tk) promoter and a bacterial β-galactosidase (*lacZ*) gene. (B) FIND vectors expressing Gal4-RAR and *lacZ*, respectively, are driven by four UASs and a minimal heat shock protein (hsp) promoter. Effector and reporter constructs were combined in tandem on the same vector. (C) A representative transgenic embryo expressing FIND Gal4-RARα effector and *lacZ* reporter collected after 11.5 (E11.5) days of gestation. Left panel: X-gal staining reveals strong reporter activation at limb levels of the developing spinal cord. Right panel: At higher magnification, blue-stained cells are detected in the developing forebrain (black arrow) and at the midbrain/hindbrain boundary (white arrow). Reproduced with permission from A. Mata de Urquiza, L. Solomin, and T. Perlmann, *Proc. Natl. Acad. Sci. USA* **96**, 13270 (1999). Copyright [1999] National Academy of Sciences, USA.

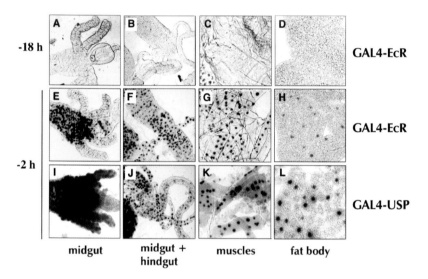

| midgut | midgut + hindgut | muscles | fat body |

TATIANA KOZLOVA AND KARL S. THUMMEL, CHAPTER 27, FIG. 2. GAL4-EcR and GAL4-USP are widely activated by the late larval ecdysteroid pulse. *hs-GAL4-EcR*; *UAS-nlacZ* (GAL4-EcR) or *hs-GAL4-USP*; *UAS-nlacZ* (GAL4-USP) third instar larvae were heat treated, allowed to recover for several hours, staged at either 12–18 hr before puparium formation (−18 hr; A–D) or ~2 hr before puparium formation (−2 hr; E–L), and processed for histochemical staining. Blue precipitate indicates expression of nuclear β-galactosidase. GAL4-EcR (A–D) and GAL4-USP (data not shown) are not activated in third instar larvae 12–18 hr before puparium formation. In contrast, both GAL4-EcR (E–H) and GAL4-USP (I–L) are widely activated in late third instar larvae, as depicted in the midgut, hindgut, muscles, and fat body. (Reprinted with permission from Ref. 4.)

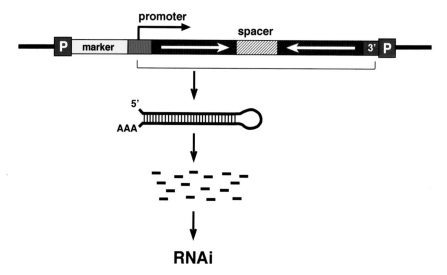

RNAi

TATIANA KOZLOVA AND CARL S. THUMMEL, CHAPTER 27, FIG. 3. Schematic representation of a transgenic insertion designed to express snapback dsRNA. A P element vector is depicted, inserted in *Drosophila* genomic DNA (heavy black lines). P element ends (blue boxes, P), the *white* marker gene (yellow box), the inducible promoter sequence (green box), the inverted repeats (arrows), spacer sequence (hatched box), and 3′ trailer (red box) are all depicted. The transcribed region is marked by a bracket. This forms a snapback dsRNA that is rapidly degraded into ~21 nucleotide RNAs that direct RNAi.

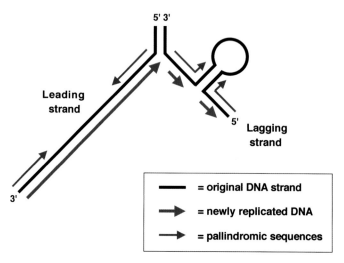

TATIANA KOZLOVA AND CARL S. THUMMEL, CHAPTER 27, FIG. 4. Inverted repeat sequences lead to defects at the replication fork. A DNA replication fork is depicted with inverted repeat sequences represented by red arrows. The leading strand is replicated continuously, preventing the formation of a snapback. The lagging strand, in contrast, snaps back on itself with a small loop caused by a short spacer region. See text for details.

M